Livestock Diseases and Management

This book series discusses the various infectious diseases affecting the livestock, principle of the disease control, and specific disease management. It discusses the existing strategies to control infectious disease includes animal management programs, vaccination, targeted antimicrobial use, and food hygiene.

Despite public health and veterinary public health improvement within the last century, animal populations remain vulnerable to health threats caused by infectious diseases.

It reviews the current understanding of the zoonotic, emerging, and transboundary animal infections in relation to their transmission, epidemiology, clinical and pathological effects, diagnosis and treatment.

It also examines the recent advancements in the veterinary diagnostics including the existing capabilities, constraints, opportunities, and future potentials. In addition, it elaborates on the conventional and recombinant vaccines that are used in the veterinary medicines and the molecular approaches that have led to the development of new vaccines in recent years. A volume focusing on the various water- and food-borne diseases and its impact on the domestic animals is also a part of this series.

The book series examines the emergence of antimicrobial resistance in livestock, ongoing global surveillance, and monitoring program, its impact on the animal-human interface and strategies for combating resistance.

Naveen Kumar • Yashpal Singh Malik
Shailly Tomar • Sayeh Ezzikouri
Editors

Advances in Antiviral Research

Springer

Editors
Naveen Kumar
WOAH Reference Laboratory for Avian Influenza
National Institute of High Security Animal Diseases (NIHSAD)
Bhopal, Madhya Pradesh, India

Yashpal Singh Malik
College of Animal Biotechnology
Guru Angad Dev Veterinary and Animal Sciences University (GADVASU)
Ludhiana, Punjab, India

Shailly Tomar
Dept. of Biosciences and Bioengineering
Indian Institute of Technology (IIT)
Roorkee, Uttarakhand, India

Sayeh Ezzikouri
Viral Hepatitis Laboratory
Institut Pasteur du Maroc
Casablanca, Morocco

ISSN 2662-4346 ISSN 2662-4354 (electronic)
Livestock Diseases and Management
ISBN 978-981-99-9194-5 ISBN 978-981-99-9195-2 (eBook)
https://doi.org/10.1007/978-981-99-9195-2

This Springer imprint is published by the registered company Springer Nature Singapore Pte Ltd.
The registered company address is: 152 Beach Road, #21-01/04 Gateway East, Singapore 189721, Singapore

Foreword

In an ongoing battle against viral infections, where the forces of science and nature collide, a stalwart group of individuals stand as beacons of hope. They are the antiviral researchers, dedicated to unraveling the intricate interactions between pathogens and hosts. Their pursuit is uncompromising, driven by an unwavering commitment to outpace the ever-evolving nature of viruses that threaten our health and well-being. The search for novel solutions is never-ending in this ever-changing environment. The next generation of antiviral researchers steps onto this challenging terrain as torchbearers of progress incorporating many of the data and computational advances in other fields and bringing them to bear on these elusive targets.

The book "Advances in Antiviral Research" emerges as a guide, providing a comprehensive platform meticulously designed to illuminate the path for the next generation of researchers and students at every stage of their journey. This book covers a wide range of creative approaches to accelerating the discovery of new pharmaceuticals using computer-aided drug screening, as well as the drug repurposing strategy for reducing the time it takes to reach the market for public health applications. Furthermore, emerging methods including cryo-electron microscopy, metabolomics, proteomics, and artificial intelligence have piqued the interest of antiviral researchers, offering promising avenues for the identification of new drug targets. To reduce the reliance on animal experiments in antiviral research, this book explores the growing exploration of organoids, often known as mini-organs, as this approach offers enormous potential to advance drug discovery and personalized treatment.

This book goes beyond traditional methods by offering an insight into the development of broad-spectrum antiviral strategies. These strategies strive to combat multiple viruses concurrently, potentially overcoming the limitations of conventional antiviral therapies and the urgent challenge posed by the emergence of antiviral resistance. Unlike direct-acting antivirals that exclusively target viral proteins, this book emphasizes the importance of host-directed epigenetic and epitranscriptomic modifications. By manipulating the cellular processes exploited by viruses, this innovative approach reduces the emergence of drug resistance and extends their efficacy to a wide range of viral families. Additionally, this book provides an in-depth exploration of drug application policies, regulations, and procedures as well as a careful analysis of clinical trials regulations in different countries. This global perspective provides readers with an in-depth understanding of the complex landscape encompassing drug development, regulatory protocols, and public acceptance of new therapies.

This collection is more than just a book; it represents the goal of nourishing the scientific community and igniting the fires of curiosity in the brains of aspiring researchers. Undergraduate and graduate students, the architects of the future, will find this book treasure trove of knowledge. If they use this knowledge as a compass, they will be able to overcome the complex obstacles that lie ahead. The common goal of this book is to foster collaboration, stimulate innovation, and catalyze progress in the field of antiviral research.

I extend my congratulations to the editors of "Advances in Antiviral Research" for their steadfast dedication to pushing the boundaries of knowledge. I express my gratitude to these luminaries for providing the clarity and direction for this publication to serve as a roadmap. May this book stand as a guiding light for the ensuing generation of antiviral researchers, illuminating their path, nurturing their fervor, and propelling them toward the development of more efficacious therapies. In unity, we strive for a world where viruses cease to evoke fear, but instead provide challenges that must be tackled with tenacity.

Department of Biological Sciences
Lilly Hall of Life Sciences, Purdue University,
West Lafayette, IN, USA

Richard J. Kuhn

Preface

Emerging and re-emerging viral infections continue to pose a global threat to both public health and animal welfare. Antiviral research is a light of hope, constantly striving to keep pace with the ever-changing nature of viruses. As the next generation of researchers embarks on this vital field of study, they encounter a multitude of challenges. New tools and methodologies are rapidly emerging, but their limited descriptions in research articles frequently leave prospective scientists unsure of the best processes to use for their intended aims. This dilemma not only impedes progress but also leads to the frustrating phenomenon of failure to reproduce an experiment. In acknowledgement of these challenges, we present "Advances in Antiviral Research." Our mission is to serve as a comprehensive platform, offering researchers at various phases of their careers a thorough and enlightening review of the latest protocols and methodologies in antiviral research. This book has been meticulously crafted to address the needs of the next generation of antiviral researchers by showcasing a diverse array of methods and tools.

The authors of this book are highly experienced researchers, each with a proven background in antiviral research spanning a wide range of viruses, including the emerging and re-emerging viruses. Their collective expertise has allowed us to curate a selection of cutting-edge advancements, making this a valuable resource for researchers seeking to make significant strides in their antiviral pursuits. Among the unique features of this book are the updated fundamental to sophisticated tools and techniques. These invaluable resources are intended to expedite the development of new antiviral drugs, enabling the next generation of researchers to navigate their research endeavors with confidence and ease. Not only do we want to help seasoned investigators, but we also aim to nurture the growth of the scientific community by providing undergraduate and graduate students with first-hand insights into fundamental to advanced methods for drug design approaches. Moreover, these methods' evaluation in in vitro and in vivo models will provide students with the expertise they need to confront the most complicated challenges in antiviral research.

In essence, "Advances in Antiviral Research" strives to be a compendium of novel methods and tools compiled by experienced researchers. Our aspiration is to foster collaboration, inspire innovation, and accelerate progress in the field of antiviral research. We hope that this book will be a guiding beacon, leading researchers toward new discoveries and, ultimately, contributing to the development of more effective antiviral therapies.

As editors, we are humbled by the opportunity to contribute to the growth and advancement of antiviral research. We express our deep gratitude to all the esteemed authors whose expertise has shaped this compilation. We hope that this book will serve as a lighthouse, illuminating the path for the next generation of antiviral researchers and allowing them to make a lasting contribution to this vital field.

Bhopal, Madhya Pradesh, India — Naveen Kumar
Ludhiana, Punjab, India — Yashpal S. Malik
Roorkee, Uttarakhand, India — Shailly Tomar
Casablanca, Morocco — Sayeh Ezzikouri

Contents

Editors and Contributors

About the Editors

Naveen Kumar is Scientist at National Institute of High Security Animal Diseases, Bhopal, India. He is an evolutionary biologist specialized in virus evolution and with a background in synthetic biology, molecular and clinical virology. His primary research interests focus on viral genomics, nanoparticle-based diagnostics, therapeutics, and vaccines for emerging and re-emerging zoonotic viruses, primarily influenza and SARS-CoV-2 viruses. He also employs selective genetic and evolutionary tools to comprehend the dynamics of virus evolution. He has been conferred with many prestigious honors, including the Indian Virological Society's Young Scientist Award 2016, the SERB-International Travel Grant for attending ASV-2018 at the University of Maryland in the USA, and IWBBIO-2023 at Gran Canaria, Spain, as well as the International Federation of Biosafety's Biorisk Management Certification 2019. He has developed a number of diagnostic technologies, including one for which a patent has been granted on synthetic-biology inspired diagnostics. He holds editorial positions in many international prestigious journals, such as Associate Editor and Review Editor. He has served as a referee for a variety of international journals. He has also published more than 50 research articles in the peer-reviewed international journals and authored or co-authored numerous book chapters. He is a member of many national and international scientific societies and organizations importantly, American Society of Virology (ASV), American Society of Microbiolgy (ASM), Indian Virological Society (IVS), and Society of Biosafety, India.

Yashpal Singh Malik is currently the Dean of the College of Animal Biotechnology, Guru Angad Dev Veterinary and Animal Sciences University, Ludhiana, India. He is a recipient of the prestigious position "ICAR National Fellow" at the ICAR-Indian Veterinary Research Institute. His areas of expertise are rotaviral diseases, viral disease epidemiology, microbial biodiversity, host–virus interactions, and pathogen diagnostics. He has pursued advanced studies in molecular virology at the University of Minnesota, USA; University of Ottawa, Ontario, Canada; and Wuhan Institute of Virology, Wuhan, China. He is the recipient of several prestigious national, state, and academy awards and honors, including the ICAR-Jawaharlal Nehru Award. He has authored 7 books, 62 book chapters, and over 255 research and review articles. He has been associated with societies of international repute, like the Secretary General of the World Society for Virology (USA) and at national level serving as Secretary General of Indian Virological Society. Being a member in "One Health Group in Federation of Asian Veterinary Association" (FAVA) for 2021–2025, he is the Indian flag bearer on the international forum. He is also the Editor in Chief for various journals. He has been awarded a prestigious Fellowship by the Academy of Microbiological Sciences (AMSc), National Academy of Veterinary Sciences, and National Academy of Dairy Sciences.

Shailly Tomar presently holds the position of a Professor within the Department of Biosciences and Bioengineering at the Indian Institute of Technology, Roorkee, situated in India. Her current research endeavors are focused on ailments induced by positive-sense single-stranded RNA viruses, encompassing diseases like chikungunya, dengue, influenza, and SARS-CoV-2. Her research group is dedicated to the pursuit of antiviral therapeutics against infectious (+) ssRNA viruses, as well as the development of vaccines based on virus-like particles (VLPs). To achieve these goals, they employ an array of scientific techniques including structural biology, cell biology, molecular biology, biochemistry, computational biology, protein–protein interactions, and molecular biophysics. At her laboratory at IIT Roorkee, known as the Molecular Virology Lab, she specializes in various aspects of virology including virus propagation, quantification, detection, antiviral testing, and the structure-based identification of antiviral compounds. The lab performs virus isolation from clinical samples in Biosafety Level 2 (BSL2) and Biosafety Level 3 (BSL3) containment facilities. With more than 15 years of teaching experience, she imparts knowledge in subjects such as Fundamentals of Biotechnology, Virology, Advanced Virology, Genetics and Molecular Biology, Biophysical Techniques, and Separation and Analysis of Biomolecules. She has contributed to the academic community through over 70 research articles published in internationally recognized peer-reviewed journals, as well as the authorship or co-authorship of numerous books and book chapters. Furthermore, she has taken an active role in organizing various conferences and workshops, including the "Indo-US International Conference on Recent Advances in Structural Biology and Drug Discovery," "Recent Advancements in Biophysical Techniques and Virology," and the "48th National Seminar on Crystallography" at IIT Roorkee. She holds membership in several prominent international scientific societies and organizations, notably including the American Society for Virology (ASV), the American Society for Microbiology (ASM), the Indian Crystallography Society (ICA), the Indian Biophysical Society (IBS), and the Indian Virology Society (IVS). Her outstanding contributions to the field of virology have been acknowledged with prestigious awards, including the DST Young Scientist Award (2006), an Institute Research Fellowship for Outstanding Young Faculty (2018), a Research Council (SRC) Grant from the Karolinska Institute, Sweden (2019), and the distinction of being a Visiting Scientist at Purdue University, USA, as part of the OVDF DST SERB program (2019).

Sayeh Ezzikouri is a Research Director and Head of Viral Hepatitis Laboratory, Pasteur Institute of Morocco. He has earlier served as Researcher in the Virology Unit at Pasteur Institute of Morocco (2007–2016), Visiting Professor at Joint Faculty of Veterinary Medicine, Kagoshima University, Japan (2013–2015), and Visiting Scientist at Joint Faculty of Veterinary Medicine, Kagoshima University, Japan (2020). His research interest included but not limited to the public health, virus–host interactions, drug design, and HLA-based peptide vaccine design with specialization in viral hepatitis A, B, C, D, and E, non-primates hepacivirus, HIV, HPV, Zika virus, SARS-CoV-2, and cancer. He has served as referee for a number of international journals. He has more than 12 years of teaching experience in virology and molecular biology. He has authored/co-authored over 100 research articles in the peer-reviewed international journals. He is recipient of many national and international awards. He is a member of numerous national and international scientific societies and organizations.

Contributors

Islam Abbadi Virology Unit, Viral Hepatitis Laboratory, Institut Pasteur du Maroc, Casablanca, Morocco

Laboratory of Virology, Microbiology, Quality and Biotechnology/Eco-toxicology and Biodiversity, University Hassan II Casablanca Faculty des Sciences Techniques, Mohammedia, Morocco

Karima Abounouh Virology Unit, Viral Hepatitis Laboratory, Institut Pasteur du Maroc, Casablanca, Morocco

Laboratory of Genetics and Molecular Pathology, Medical School, University Hassan II, Casablanca, Morocco

Heba Alshater Forensic Medicine and Clinical Toxicology Department, Menoufia University Hospital, Shbien El-Kom, Egypt

Haya Altawalah Department of Microbiology, Faculty of Medicine, Kuwait University, Kuwait City, Kuwait

Virology Unit, Yacoub Behbehani Center, Sabah Hospital, Ministry of Health, Kuwait City, Kuwait

Sakshi Arjun Department of Pharmaceutical Technology (Formulations), National Institute of Pharmaceutical Education and Research, (NIPER) Guwahati, Guwahati, India

S. Arutkumaran ICAR-Indian Veterinary Research Institute, Bareilly, Uttar Pradesh, India

Manidipa Banerjee Kusuma School of Biological Sciences, Indian Institute of Technology Delhi, New Delhi, India

Sandeep Bhatia National Institute of High Security Animal Diseases, Bhopal, Madhya Pradesh, India

Naveen Chella Department of Pharmaceutical Technology (Formulations), National Institute of Pharmaceutical Education and Research, (NIPER) Guwahati, Guwahati, India

Shweta Choudhary Indian Institute of Technology, Roorkee, India

Ujjwal Kumar De Division of Veterinary Medicine, ICAR-Indian Veterinary Research Institute, Izatnagar, Bareilly, Uttar Pradesh, India

Division of Medicine, ICAR-Indian Veterinary Research Institute, Izatnagar, Bareilly, Uttar Pradesh, India

Rajib Deb ICAR-National Research Centre on Pig, Guwahati, Assam, India

Hind Dehbi Laboratory of Genetics and Molecular Pathology, Medical School, University Hassan II, Casablanca, Morocco

Dhruv Desai University of Pennsylvania, Philadelphia, PA, USA

Debajit Dey School of Medicine, University of Maryland, Baltimore, MD, USA

Preeti Dhaka Department of Biosciences and Bioengineering, Indian Institute of Technology Roorkee, Roorkee, Uttarakhand, India

Kuldeep Dhama Division of Pathology, ICAR-Indian Veterinary Research Institute, Izatnagar, Bareilly, Uttar Pradesh, India

Phelipe Magalhães Duarte Postgraduate Program in Animal Bioscience, Federal Rural University of Pernambuco (UFRPE), Recife, Pernambuco, Brazil

Sarah El-Nakeep Hepatogastroentrology Unit, Internal Medicine Department, Faculty of Medicine, Ain Shams University, Cairo, Egypt

Wahiba Ezzemani Virology Unit, Viral Hepatitis Laboratory, Institut Pasteur du Maroc, Casablanca, Morocco

Laboratoire de Biologie et Santé (URAC34), Départment de Biologie, Faculté des Sciences Ben Msik, Hassan II University of Casablanca, Casablanca, Morocco

Sayeh Ezzikouri Virology Unit, Viral Hepatitis Laboratory, Institut Pasteur du Maroc, Casablanca, Morocco

Mohamed Fawzy Department of Virology, Faculty of Veterinary Medicine, Suez Canal University, Ismailia, Egypt

Sudesh Gaidhani Pharmacology, Central Council for Research in Ayurvedic Sciences, Ministry of AYUSH, Government of India, New Delhi, India

Balasubramanian Ganesh Laboratory Division, Indian Council of Medical Research-National Institute of Epidemiology, Ministry of Health & Family Welfare, Chennai, Tamil Nadu, India

Suresh Samadhan Ghoke Department of Veterinary Epidemiology and Preventive Medicine, COVAS, Udgir, Latur, Maharashtra, India

National Institute of High Security Animal Diseases, Bhopal, Madhya Pradesh, India

Souvik Ghosh One Health Center for Zoonoses and Tropical Veterinary Medicine, Basseterre, St. Kitts, West Indies

Fadila Guessous Mohammed VI University of Health Sciences (UM6SS), Casablanca, Morocco

Department of Microbiology, Immunology and Cancer Biology, School of Medicine, University of Virginia, Charlottesville, VA, USA

Vivek Kumar Gupta ICAR-National Research Centre on Pig, Guwahati, Assam, India

Pouneh Hajipour Faculty of Veterinary Medicine, Shahid Bahonar University of Kerman, Kerman, Iran

Ahmed Hesham Department of Chemistry, Suez Canal University, Ismailia, Egypt

Middle East for Vaccines (MEVAC), El-Salihya El-Gededa, El-Sharkia, Egypt

Inderpal Kaur College of Animal Biotechnology, Guru Angad Dev Veterinary and Animal Sciences University, Ludhiana, Punjab, India

Rahul Kaushik Biotechnology Research Center, Technology Innovation Institute, Masdar City, Abu Dhabi, UAE

Laboratory for Structural Bioinformatics, Biosystems Dynamics Research Center, Riken, Japan

Anass Kettani Laboratoire de Biologie et Santé (URAC34), Départment de Biologie, Faculté des Sciences Ben Msik, Hassan II University of Casablanca, Casablanca, Morocco

Manoj Kumar National Institute of High Security Animal Diseases, Bhopal, India

Naveen Kumar ICAR-National Institute of High Security Animal Diseases, Bhopal, India

Prashant Kumar Amity Institute of Virology and Immunology, Amity University, Noida, Uttar Pradesh, India

Pravindra Kumar Department of Biosciences and Bioengineering, Indian Institute of Technology Roorkee, Roorkee, Uttarakhand, India

Thomas Launey Biotechnology Research Center, Technology Innovation Institute, Masdar City, Abu Dhabi, UAE

Mustapha Lkhider Laboratory of Virology, Microbiology, Quality and Biotechnology/Eco-toxicology and Biodiversity, University Hassan II Casablanca Faculty des Sciences Techniques, Mohammedia, Morocco

Milan Kumar Lokshman Kusuma School of Biological Sciences, Indian Institute of Technology Delhi, New Delhi, India

Sonalika Mahajan ICAR-Indian Veterinary Research Institute Izatnagar Bareilly, Bareilly, India

Yashpal Singh Malik College of Animal Biotechnology, Guru Angad Dev Veterinary and Animal Sciences University, Ludhiana, Punjab, India

M. Manu College of Animal Biotechnology, Guru Angad Dev Veterinary and Animal Sciences University, Ludhiana, Punjab, India

Adarsh Mishra College of Animal Biotechnology, Guru Angad Dev Veterinary and Animal Sciences University, Ludhiana, Punjab, India

Pankhuri Narula Kusuma School of Biological Sciences, Indian Institute of Technology Delhi, New Delhi, India

Manikandan Natesan Laboratory Division, Indian Council of Medical Research-National Institute of Epidemiology, Ministry of Health & Family Welfare, Chennai, Tamil Nadu, India

Senthilkumar Natesan Indian Institute of Public Health Gandhinagar, Lekawada, Gandhinagar, Gujarat, India

Sanketkumar Nehul Indian Institute of Technology, Roorkee, India

Ahd Ouladlahsen Faculté de Médecine et de Pharmacie, Université Hassan II, Casablanca, Morocco

Service des maladies infectieuses, CHU Ibn Rochd, Casablanca, Morocco

Amruta Prabhakar Padakanti Department of Pharmaceutical Technology (Formulations), National Institute of Pharmaceutical Education and Research, (NIPER) Guwahati, Guwahati, India

Seema Rani Pegu ICAR-National Research Centre on Pig, Guwahati, Assam, India

Rushika Perera Center for Vector-Borne Infectious Diseases, Department of Microbiology, Immunology, and Pathology, Colorado State University, Fort Collins, CO, USA

Ali A. Rabaan Molecular Diagnostic Laboratory, Johns Hopkins Aramco Healthcare, Dhahran, Saudi Arabia

Swaraj Rajkhowa ICAR-National Research Centre on Pig, Guwahati, Assam, India

Upasana Sahu National Institute of High Security Animal Diseases, Bhopal, India

Rachid Saile Laboratoire de Biologie et Santé (URAC34), Département de Biologie, Faculté des Sciences Ben Msik, Hassan II University of Casablanca, Casablanca, Morocco

Sina Salajegheh Tazerji Department of Clinical Sciences, Faculty of Veterinary Medicine, Science and Research Branch, Islamic Azad University, Tehran, Iran

Young Researchers and Elites Club Science and Research Branch, Islamic Azad University, Tehran, Iran

Gyanendra Singh Sanger ICAR-National Research Centre on Pig, Guwahati, Assam, India

Aniket Sanyal National Institute of High Security Animal Diseases, Bhopal, India

M'hammed Sarih Service de Parasitologie et des Maladies Vectorielles, Institut Pasteur du Maroc, Casablanca, Morocco

Anuradha Sharma College of Animal Biotechnology, Guru Angad Dev Veterinary and Animal Sciences University, Ludhiana, Punjab, India

Department of Molecular Biology and Genetic Engineering, Lovely Professional University, Phagwara, Punjab, India

Gaurav Kumar Sharma ICAR-Indian Veterinary Research Institute Izatnagar Bareilly, Bareilly, India

Ankur Singh Department of Biosciences and Bioengineering, Indian Institute of Technology Roorkee, Roorkee, Uttarakhand, India

Ashutosh Singh National Institute of High Security Animal Diseases, Bhopal, India

Jitin Singla Department of Biosciences and Bioengineering, Indian Institute of Technology Roorkee, Roorkee, Uttarakhand, India

Department of Computer Science and Engineering, Indian Institute of Technology Roorkee, Roorkee, Uttarakhand, India

Paul S. Soma Center for Vector-Borne Infectious Diseases, Department of Microbiology, Immunology, and Pathology, Colorado State University, Fort Collins, CO, USA

Joyshikh Sonowal ICAR-National Research Centre on Pig, Guwahati, Assam, India

Richa Sood National Institute of High Security Animal Diseases, Bhopal, Madhya Pradesh, India

Rajkumar Thangarasu Laboratory Division, Indian Council of Medical Research-National Institute of Epidemiology, Ministry of Health & Family Welfare, Chennai, Tamil Nadu, India

Shailly Tomar Department of Biosciences and Bioengineering, Indian Institute of Technology Roorkee, Roorkee, Uttarakhand, India

Bhupendra Nath Tripathi Indian Council of Agricultural Research, New Delhi, India

Chapter 1
Antivirals: Approaches and the Way Forward

Adarsh Mishra, Inderpal Kaur, Anuradha Sharma, M. Manu, Ujjwal Kumar De, Naveen Kumar, and Yashpal Singh Malik

Abstract Viral diseases pose significant challenges in human and veterinary medicine, as effective treatment options are limited. The development of antiviral drugs becomes increasingly challenging in the face of emerging and reemerging viral diseases. The genetic changes and rapid evolution of certain viruses further complicate the development of universal pathogen-specific drug therapies. Hence, there is a pressing need for research and development in the field of antiviral drugs. The novel approaches, such as computer-aided drug designing (CADD), have expedited the drug development process, notably reducing the preclinical phase. Phyto-medicines offer promising alternatives as antivirals, while organic polymers like extracellular polymeric substances (EPS) have demonstrated effectiveness in combating viruses. Additionally, exosomes have shown potential as antivirals and in antiviral delivery, particularly when loaded with drugs and specifically targeted against viral diseases. The repurposing of drugs for antiviral therapeutics has garnered significant attention in both research institutions and pharmaceutical industries due to its numerous advantages. Combination antiviral therapies not only reduce toxicity but also address issues of drug resistance. Lastly, extensive clinical trials will be crucial for the successful repurposing of drugs as antiviral therapeutics.

A. Mishra · I. Kaur · M. Manu · Y. S. Malik (✉)
College of Animal Biotechnology, Guru Angad Dev Veterinary and Animal Sciences University, Ludhiana, Punjab, India
e-mail: malikyps@gadvasu.in

A. Sharma
College of Animal Biotechnology, Guru Angad Dev Veterinary and Animal Sciences University, Ludhiana, Punjab, India

Present address: Department of Molecular Biology and Genetic Engineering, Lovely Professional University, Phagwara, Punjab, India

U. K. De
Division of Veterinary Medicine, ICAR-Indian Veterinary Research Institute, Izatnagar, Bareilly, Uttar Pradesh, India

N. Kumar
National Institute of High Security Animal Diseases, Bhopal, India

N. Kumar et al. (eds.), *Advances in Antiviral Research*, Livestock Diseases and Management, https://doi.org/10.1007/978-981-99-9195-2_1

Keywords Antivirals · Novel approaches · Computer-aided drug designing (CADD) · Phyto-medicine · Extracellular polymeric substance (EPS) · Exosomes · Drug repurposing · Therapeutics

1.1 Introduction

Viruses, being acellular entities, exhibit host specificity. Some exclusively affect humans while others target animals. Certain viral pathogens are zoonotic in nature affecting both human and animal health (Namusisi et al. 2021; Kumar et al. 2022a). Many of these viral pathogens are continuously emerging and reemerging in nature. They undergo natural selection, leading to a wide range of host susceptibility, tissue tropism, pathogenicity, and virulence (Kumar et al. 2021). Frequently, viruses undergo a shift in their phenotype, transitioning from an existing form to an altered one (Sood et al. 2021). This may involve infecting new hosts and new age groups within previously susceptible hosts or adopting a modified pathogenic form.

In recent times, the world has encountered numerous deadly viral pathogens. Some have already undergone evolutionary changes, while others are currently in the process of evolution (Kumar et al. 2022b). Developing effective therapeutic and preventive strategies against these pathogens is always a challenge, particularly when a disease appears for the first time or reemerges in an altered form within an uncommon host.

With recent advancements in science, the process of discovering antiviral drug has taken a new direction, incorporating innovative approaches such as computational biology tools for drug design and repurposing existing drugs. The availability of informatics tools has significantly reduced the time required for preclinical drug development. Computer-aided drug designing (CADD) encompasses both structure-based and ligand-based drug designing strategies to discover new drugs. The concept of drug repurposing has provided an opportunity to utilize drugs that are already available, existing, preexisting or extensively studied drugs for a different purpose or target. This approach has not only reduced the cost of drug development but also yielded significant results (Kumar et al. 2023). According to an estimate by Rudrapal et al. 2020, approximately 30% of US Food and Drug Administration (FDA)-approved drugs are repurposed drugs (Rudrapal et al. 2020).

Moreover, alternative medicines like phyto-medicines have shown promise as antiviral therapeutics. Synergistic combination of phyto-medicines or herbal medicines with other antivirals is a potential avenue worth exploring. Additionally, extracellular polymeric substances have emerged as a novel concept in antiviral therapeutics. The use of extracellular vesicles or exosomes also holds promise for delivering antiviral drugs.

In this chapter we converse on recent approaches in the development of antivirals for important viral pathogens.

1.2 Historical Perspectives

The discovery of interferon by scientists Alick Isaacs and Jean Lindenmann in 1957 played a significant role in advancing antiviral drug research (Isaacs and Lindenmann 1957). According to information from the American Chemical Society presented in periodic graphics (Brunning 2022), the first FDA-approved antiviral drug against herpes simplex virus (HSV) infection was the nucleoside analogue 5-iodo-2′-deoxyuridine (idoxuridine). Its approval for topical application came in 1963 after demonstrating clinical importance as shown by Dr. Herbert Kaufman in 1962 (Kaufman et al. 1962; Kaufman 1962). Another nucleoside analogue, adenine arabinoside (vidarabin), was described for the first time by M. P. de Garilhe and J. De Rudder in 1964 and later confirmed to be clinically effective by Dr. Richard Whitley in 1976 (Field and De Clercq 2004). In 1977, vidarabin became the first systemically used antiviral drug against herpesvirus infections. Acyclovir, approved in 1981, was the first selective antiviral drug for herpesvirus infections.

The first anti-influenza drug, amantadine, was reported by C.E. Hoffmann and others and received the FDA approval in 1966. Later, its derivative, rimantadine, was approved as another anti-influenza drug in 1993. The sialic acid analogue, zanamivir, was developed by M. von Itzstein and colleagues in 1993 (1993) and marketed as "Relenza" in 1999. Another anti-influenza drug, oseltamivir (trade name tamiflu), was reported by C.U. Kim and co-workers (1997). Subsequently, both oseltamivir and zanamivir were approved for treating influenza A. Azidothymidine, approved in 1987, was the first antihuman immunodeficiency virus (HIV) drug. Saquinavir, the first protease inhibitor, was approved for HIV treatment. In response to the recent severe acute respiratory syndrome coronavirus 2 (SARS-CoV-2) pandemic in 2019, remdesivir received approval in 2020 as an antiviral therapy against the virus.

1.3 Mechanism of Action

While the development of effective antiviral drugs to combat viral infections poses significant challenges that persist over time, it is crucial to comprehend the fundamental mechanism of action underlying these drugs. Rather than simply inactivating a virus, antiviral agents aim to target specific stages of viral replication such as inhibition of virus attachment and entry into host cells, disrupting virus uncoating, impeding virus replication, suppressing viral protein synthesis, and hindering virus assembly and release.

1.3.1 Antiviral Agents Targeting Virus Attachment and Entry

The initial and crucial step in the replication process and establishment of viral infections is the attachment and entry of the virus. This attachment to the cell surface involves complex interactions between the virus and the cell. Many antiviral agents target the attachment of the virus to the cell, including compounds like pyrrolopyridine (BMS-488043 and BMS-663068), which can hinder virus-cell interactions during entry and effectively reduce HIV-1 virus level (Da et al. 2011; Yang et al. 2010; Chen et al. 2014; Ballana and Este 2015). Maraviroc (MVC), an approved HIV treatment drug, acts as an antagonist to the CCR5 receptor (Dorr et al. 2005). Antiviral agents such as fusion inhibitors or entry inhibitors can prevent virus entry by obstructing the fusion of viral proteins with the cell membrane. These inhibitors impede the conformational changes necessary for the fusion process. Specifically, designed fusion inhibition peptides have been shown to efficiently hinder the entry of SARS-CoV-2 into the host cells (Kandeel et al. 2021; Zhu et al. 2022). Furthermore, enfuvirtide, a peptide-based fusion inhibitor derived from GP-41 protein sequence, successfully prevents the HIV-1 fusion with the host cells and is an approved drug for HIV-1 treatment (Joly et al. 2010; Monteiro et al. 2021).

Hydroxychloroquine/chloroquine inhibits the entry of SARS-CoV-2 virus into cells through endosome-mediated mechanisms facilitated by endosomal protease-cathepsin L (Agarwal et al. 2020; Satarker et al. 2020). Several fusion inhibitor peptides have been discovered for different virus families, including *Arenaviridae*, *Coronaviridae*, *Filoviridae*, *Orthomyxoviridae*, and *Paramyxoviridae* (Duzgunes et al. 2021).

1.3.2 Antiviral Agents Targeting Virus Uncoating

Uncoating plays a crucial role in the virus replication, making it a promising target for the development of antiviral drugs. It involves the breakdown of the virus capsid and the release of its genetic material. Antiviral agents, such as amantadine and rimantadine, act by inhibiting uncoating, specifically by blocking the M2 ion channel that prevents the release of the influenza genome (Fleming 2001; Saito et al. 2006; Hubsher et al. 2012; Sood et al. 2018).

1.3.3 Inhibitors of Virus Replication

In general, the majority of antiviral drugs focus on either targeting viral proteins or enzymes crucial for virus replication cycle, such as nucleic acid polymerases, integrases, proteases, and neuraminidases. Alternatively, they may target molecules essential for cellular functions that are necessary for viral replication and survival.

1.3.3.1 Polymerase Inhibitors

Polymerases serve as the essential enzymes for the replication of viruses. There are two types of virus polymerases: viral DNA polymerases and RNA polymerases. These enzymes are important for copying the genetic information of viruses, enabling them to replicate. Polymerase inhibitors work by blocking the activity or function of these enzymes, thereby impeding the virus replication process. Generally, two classes of antiviral agents are utilized to target polymerases: nucleoside inhibitors and non-nucleoside inhibitors. Nucleoside inhibitors and nucleotide analogues bind to the active site of the polymerase enzyme, effectively obstructing its activity. Non-nucleoside inhibitors, on the other hand, bind to allosteric binding sites within the enzyme, preventing its function. Polymerase inhibitors are further categorized as DNA polymerase inhibitors and RNA polymerase inhibitors.

Most of the approved antiviral drugs are nucleoside analogues that inhibit DNA polymerase. Examples include acyclovir, penciclovir, abacavir, brivudine, and zalcitabine whereas cidofovir, tenofovir, and adefovir (Oberg 2006). Nucleoside analogues act like natural nucleotides and induce chain termination of nascent DNA (Abdullah Al Awadh 2022). Nucleotide analogues, also known as nucleoside monophosphates, are preferred for rational drug design over nucleoside analogues due to their synthesis omitting the step of pyrophosphorylation necessary for nucleoside analogue activation (Goraz and De Jonghe 2021). Pyrophosphate analogues are preferred as polymerase inhibitors since they can act without the need of metabolic transformations.

Foscarnet is an antiviral agent that selectively inhibits the viral DNA polymerases by binding to the pyrophosphate binding site (Crumpacker 1992). Ribavirin, an example of an RNA polymerases inhibitor, hinders the formation of guanosine triphosphate, thereby blocking viral mRNA capping. Allosteric inhibitors of polymerases bind to sites other than the active sites, altering the spatial conformation of the enzyme and inhibited polymerase activity. For instance, pimodivir, an inhibitor of RNA polymerases, binds to allosteric sites of the RdRp protein, interfering with RNA polymerase activity and impeding viral (Tian et al. 2021).

Nucleoside and nucleotide reverse transcriptase inhibitors are primarily used in the treatment of HIV. Azidothymidine (zidovudine), the first antiretroviral drug, is a nucleoside analogue and reverse transcriptase inhibitor that competitively inhibits the enzyme and acts as a DNA synthesis chain terminator (Cihlar and Ray 2010). Examples of nucleoside reverse transcriptase inhibitors include azidothymidine, zalcitabine, stavudine, lamivudine, emtricitabine, didanosine, and abacavir. Nucleotide reverse transcriptase inhibitors encompass cidofovir, adefovir, tenofovir disoproxil fumarate, and tenofovir alafenamide fumarate (Wonganan et al. 2020). Non-nucleoside reverse transcriptase inhibitors (NNRTs) are another important class of antiviral agents for retroviral therapy. These agents bind to different active sites of the reverse transcriptase enzyme and inactivate it by allosteric mechanisms. Approved NNRTs include delavirdine, efavirenz, etravirine, nevirapine, and rilpivirine (Sluis-Cremer 2018).

1.3.3.2 Integrase Enzyme Inhibitors

Integrase enzyme enables the integration of retrovirus-proviral DNA into the genome of the host cell. However, this integration process can be hindered by the integrase inhibitors, which effectively block the enzyme and prevent the integration of the provirus into the host genome, consequently impeding the virus replication. Inhibitors targeting integrase play a crucial role and can be categorized into two categories: those that inhibit the strand transfer activity of integrases and those that hinder the binding of integrase. In the case of HIV, integrase strand transfer inhibitors specifically target the integrase enzyme, blocking the step where viral DNA is transferred and integrated into the host cell genome (Scarsi et al. 2020). Several integrase inhibitors, such as bictegravir, dolutegravir, elvitegravir, cabotegravir, and raltegravir have received approval from the FDA (Trivedi et al. 2020a).

1.3.4 Inhibitors of Viral Assembly

Viral proteases are enzymes that play essential roles in the proteolytic processing of polyprotein precursors necessary for the expression of the viral genome. They are also involved in the processing of various proteins crucial for replication, structural morphogenesis, and the assembly of viral particles (Patick and Potts 1998). The validation of viral proteases as drug targets came after laboratory experiments involving mutated proteases resulted in noninfectious virus progeny (Kaplan et al. 1993). Protease inhibitors (PRI) are antiviral agents that effectively hinder virus replication and assembly by binding to the catalytic site of the protease enzyme. Currently, the protease inhibitors being used include nelfinavir, amprenavir, fosamprenavir, darunavir, saquinavir, atazanavir, indinavir, lopinavir ritonavir, tipranavir, simeprevir, atazanavir, and darunavir (Majerova and Konvalinka 2022). Among the most recent PRI inhibitors used in the treatment of HIV are darunavir, lopinavir, and atazanavir (Weber et al. 2021). Apart from HIV, proteases from other viruses, such as herpesvirus, hepatitis C virus, picornaviruses, SARS-COV-2, etc., are also targeted for the design of PRI antiviral drugs. These drugs can be effective when used alone or in combination with other antiviral agents (Patick and Potts 1998; Zhao et al. 2021; Liu et al. 2020).

1.3.5 Inhibitors of Virus Release

The release of virus can occur through either lysis of the infected cells or the process of budding. In the case of the influenza virus, the neuraminidase glycoprotein plays an important role in cleaving sialic acid residues of cell receptors, facilitating the release of the virus to infect new cells (Colman et al. 1983). Several neuraminidase inhibitors have demonstrated effectiveness and safety against most influenza strains.

These include oseltamivir (Tamiflu), zanamivir, laninamivir, and peramivir (Ison 2011, 2013; Jefferson et al. 2014; McKimm-Breschkin 2013; Laborda et al. 2016; Yang et al. 2016; Sood et al. 2018). Ongoing studies are exploring various approaches to develop new neuraminidase inhibitors, utilizing recent advancements in chemical synthesis pathways (Mahal et al. 2021).

1.3.6 Immunomodulation

In addition to antiviral agents, various immune-boosting agents like nitazoxanide are administered alongside other drugs to enhance the immune system and improve the body's ability to combat viral infections (Kausar et al. 2021).

1.4 Antiviral Drug Developments

Recently, studies have been focused on developing antiviral strategies against DNA and RNA viruses. The following section discusses antiviral therapeutics for important viral infections in humans and animals.

1.4.1 DNA Viruses

1.4.1.1 Herpesviruses

Herpesviridae, a family of DNA viruses, includes several pathogenic viruses that affect humans, avian species, and animals. Herpes simplex virus 1, herpes simplex virus 2, cytomegalovirus (CMV), Epstein-Barr virus (EBV), varicella-zoster virus, human herpesviruses 6, 7, and 8 are the major human pathogens within the *Herpesviridae* family. These DNA viruses attach to host cell, undergo DNA replication, and interact with the host's DNA through cellular recombination and repair mechanisms (Poole and James, 2008). Herpes simplex viruses 1 and 2 cause oral ulcers, keratitis, genital infections, encephalitis, and severe illness in newborns. CMV infection is primarily a congenital infection that leads to hearing loss and severe complications in individuals with weakened immune systems. Human herpesviruses 6 and 7 cause febrile exanthem, posttransplant complications such as interstitial pneumonitis and encephalitis. EBV and human herpesvirus 8 are associated with malignancies like Burkitt's lymphoma, Hodgkin's lymphoma, pharyngeal carcinoma, Kaposi sarcoma, etc. (Poole and James 2018).

The antiviral drugs developed to treat herpesvirus infections work by inhibiting DNA polymerase activity through various mechanisms, such as targeting the terminase complex, helicase primase complex, and the protein kinase enzyme (Poole and

James 2018). Acyclovir, valacyclovir, and famciclovir are highly effective nucleoside analogues that inhibit the DNA polymerase activity and are commonly used to treat herpes infections (Klysik et al. 2020; Sadowski et al. 2021). Acyclovir, the first drug in this class, was discovered in 1974 and became available to physicians in 1982 after successful clinical trials and preclinical investigations, stemming from an antiviral drug screening program initiated in the 1960s at Burroughs Wellcome. While acyclovir is highly effective against herpes simplex virus, its effectiveness is comparatively lower against varicella-zoster, Epstein-Barr virus, and cytomegalovirus (King 1988).

Other nucleoside analogue inhibitors of DNA polymerase for herpesvirus include brincidofovir, valomaciclovir, valganciclovir, ganciclovir, and cidofovir. Valganciclovir and ganciclovir are specifically recommended for cytomegalovirus treatment but possess activity against all herpes viruses. Foscarnet is another antiviral drug used to treat herpesvirus by directly interfering with DNA polymerase activity. Letermovir is an inhibitor of terminase complex and effectively targets cytomegalovirus.

These antiherpetic drugs have limitations and side effects. For example, the oral uptake of acyclovir results in reduced bioavailability. The most common side effect of acyclovir and related drugs is kidney toxicity, which is generally manageable even with long-term administration. Ganciclovir has been reported to interfere with cellular DNA polymerases in addition to virus and infected cell DNA polymerases, leading to toxicity and myelosuppression as major side effects. Foscarnet exhibits selectivity in its action as its activity, with higher activity against viral DNA polymerases compared to cellular polymerases, but its dosage requires continuous monitoring, and high dosages can cause nephrotoxicity (Wagstaff and Bryson 1994). Additionally, these drugs may develop resistant, especially in immunosuppressed individuals (Poole and James 2018).

1.4.1.2 Poxviruses

The members of the *Poxviridae* family affect both human and animals. Smallpox virus, also known as variola, is one of the most studied viruses among this family and caused severe disease and death. While it has been globally eradicated since 1980, it played a crucial role in the structural characterization and development of antiviral drugs against poxviruses (https://www.cusabio.com/infectious-diseases/poxvirus.html; Accessed on March 16, 2023). Monkeypox, cowpox, camelpox, ectromelia, myxoma, rabbit fibroma, vaccinia, contagious ecthyma, sheeppox, goatpox, swinepox, bovine papular stomatitis, pseudocowpox, lumpy skin disease, and fowlpox are various diseases caused by pox viruses (De Clercq 2002; Haller et al. 2014; Saghazadeh and Rezaei 2022). Since the 1950s, vaccinia virus has been used as a model to developing anti-poxviral drugs. Thiosemicarbazones were subsequently identified as the first anti-poxviral compounds, followed by the discovery of various compounds such as interferons, viral dehydrogenase, hydrolase, polymerase enzyme inhibitors, cellular kinase inhibitors, and inhibitors of viral egress (De Clercq 2010). In 2018, the US FDA-approved two drugs, tecovirimat and brincidofovir, for the treatment of smallpox virus (Delaune and Iseni 2020).

1.4.1.3 Other DNA Viruses Affecting Human Health

Hepatitis B virus (HBV), a virus with partially double stranded DNA causes severe acute and chronic liver infections, affecting approximately 290 million individuals worldwide (Chien & Liaw 2022). The first approved drug for treating chronic hepatitis B (CHB) was lamivudine. Adefovir dipivoxil, the first nucleotide analogue approved by the FDA in 2002, was later replaced by tenofovir for CHB treatment. Currently, the preferred first-line drugs for CHB includes entecavir (ETV), tenofovir disoproxil fumarate (TDF), and tenofovir alafenamide (TAF) (Chien and Liaw 2022). Several new antiviral immunomodulatory therapies are being developed to achieve functional cure and fulfill the World Health Organization's (WHO) goal of HBV elimination by 2030 (Jeng et al. 2023).

Parvovirus, a DNA virus belonging to the *Parvoviridae* family, can be fatal in immunocompromised humans and cause congenital complications. Currently, there is no approved specific antiviral drug for parvovirus. However, several compounds have exhibited antiviral efficacy against human parvovirus infection, such as the nucleoside analogue cidofovir (Bonvicini et al. 2017), hydroxyurea (Bonvicini et al. 2017), and brincidofovir (Bua et al. 2019). Recently, a purine derivative compound has also demonstrated effectiveness against this infection (Ning et al. 2022). Additionally, immunoglobulin therapy has shown efficacy in treating parvovirus infection in humans (Mogensen et al. 2023).

Papillomaviruses, another group of DNA virus, are responsible for various infections, including oncogenicity in humans. In recent studies, in silico approaches have been used to investigate the efficacy of antiviral peptides against human papillomavirus infection (Saravanan and Monisha 2023).

1.4.1.4 Other DNA Viruses Affecting Animal Health

Poxviruses infect not only humans but also animals. Lumpy skin disease (LSD), caused by LSD virus (LSDV), affects large ruminants like cattle and buffaloes. It leads to skin nodular lesions, lameness, and often mortality in affected herds. At present, there is no approved antiviral drug for treating LSD in bovines. However, a recent study has revealed the effectiveness of a natural compound propolis encapsulated within the alginate nanoparticles against LSD (Farag et al. 2020). In vitro study has also shown that ivermectin can inhibit LSD replication (Toker et al. 2022). Furthermore, poxvirus infection in wild birds has recently been treated effectively with ribavirin (Marcello et al. 2023).

Canine parvoviral enteritis, caused by canine parvovirus 2 of the *Parvoviridae* family, is a highly prevalent viral infection in dogs. Immunoglobulin therapy, especially canine interferon lambda 3, has been found to be effective against canine parvovirus infection (Kim et al. 2022).

1.4.2 RNA Viruses

These viruses form the most diverse group characterized by possessing either segmented or non-segmented RNA as their genetic material. Many of these RNA viruses cause severe life-threatening diseases that can affect both humans and animals. Although there are some antiviral therapeutics available against certain RNA viruses, research and development efforts are still ongoing to discover potent antiviral drugs against this diverse group of RNA viruses. The following subsections discuss some of important RNA viruses and potential treatment options.

1.4.2.1 Influenza Viruses

Antiviral drugs approved for influenza virus infection treatment can be categorized into four groups: M2 ion channel blockers (amantadine and rimantadine), neuraminidase inhibitors (zanamivir, oseltamivir, peramivir, and laninamivir), RNA-dependent RNA polymerase (RdRp) inhibitor (favipiravir), and acidic endonucleases inhibitor (baloxavir marboxil).

The M2 ion channel blockers, such as amantadine and rimantadine, were developed several decades ago but are no longer approved for use due to their limited effectiveness against influenza B virus. The neuraminidase inhibitors, including zanamivir and oseltamivir, are most widely used as antiviral drugs against influenza viruses and importantly, they can inhibit both influenza A and B viruses. Notably, oseltamivir is effective not only against human influenza but also against avian influenza. Other antiviral drugs in this category are peramivir and laninamivir. Peramivir is the only anti-influenza drug which has been approved for parenteral administration by the FDA.

Similarly, favipiravir, an RdRp inhibitor, is the only anti-influenza drug that hinders the synthesis of viral RNA. It is worth noting that while laninamivir, a neuraminidase inhibitor, and favipiravir, an RdRp inhibitor, have received approval for use in Japan, they have yet to be approved by the FDA (Batool et al. 2023). In comparison to the aforementioned three categories, a novel generation of anti-influenza drug has emerged: baloxavir marboxil, an inhibitor of polymerase acidic endonucleases. This drug was recently approved by the FDA in 2018. It inhibits the acidic endonucleases in the viral RdRp complex, thereby effectively blocking viral mRNA and protein synthesis (Baker 2019).

1.4.2.2 Coronaviruses

Among coronaviruses, severe acute respiratory syndrome coronavirus (SARS-CoV), Middle East respiratory syndrome coronavirus (MERS-CoV), and severe acute respiratory syndrome coronavirus-2 (SARS-CoV-2) continue to pose a global health threat (Malik et al. 2020). Since the first detection, SARS-CoV-2 has evolved over a short period of time generating multiple variants capable of jeopardizing the efficacy of vaccines as well as antibody therapeutics (Kumar et al. 2022b, 2023). However, considerable global efforts have been put together for the development of potent antivirals. The utilization of favipiravir, a repurposed drug for COVID-19, remains as a topic of debate (Hassanipour et al. 2021; Joshi et al. 2021). Interestingly, a combination approach involving the repurposed drug remdesivir, along with other immunotherapies such as dexamethasone, tocilizumab, and baricitinib, were found effective against COVID-19 (Chakraborty et al. 2021). Additionally, the broad-spectrum antiviral activity of the molnupiravir cytidine analogue against SARS-CoV-2 has been established (Murakami et al. 2023). Recent clinical studies have shown that the combination of nirmatrelvir (SARS-CoV2 main protease inhibitor) and ritonavir (an HIV-1 protease inhibitor) can reduce hospitalization and mortality in patients infected with SARS-CoV-2 (Saravolatz et al. 2023). Furthermore, remdesivir, an RNA-dependent RNA polymerase inhibitor, has been granted emergency use authorization to treat COVID-19 and has been found to be effective against subvariants of the omicron variant (Imai et al. 2023). Natural products, which have been tested and found effective against the SARS-CoV-2, have been discussed in detail elsewhere (Chakravarti et al. 2021).

1.4.2.3 Zika Virus

Zika virus (ZIKV), a flavivirus, is responsible for causing Zika virus infection, a neurological arboviral disease known as Guillain-Barre syndrome in humans. This virus was initially detected in rhesus macaque in Uganda in 1947, and its first appearance in human hosts occurred in the Republic of Tanzania and Uganda in 1952. Subsequently, it spread to various countries, especially in Central America and Latin America, as well as in Southeast Asia. Recognizing its significance, the World Health Organization (WHO) announced Zika virus as a “Public Health Emergency of International Concern” on February 1, 2016. To date, no universally effective clinically approved therapeutic or prophylactic strategy against the virus exists. Extensive research is underway to identify suitable antiviral approaches for controlling the virus. Bernatchez et al. (2020) outlined various potential targets for drug development against ZIKV, including inhibitors of virus receptors, RNA-dependent RNA polymerase (RdRp), proteases, and virus assembly.

Similarly, lycorine, an alkaloid, has shown promise as a potential inhibitor of viral RdRp, RNA replication, and protein synthesis in laboratory mice (Chen et al. 2020b). Other drugs, such as niclosamide (Xu et al. 2016) and its derivatives (Li et al. 2020) have been tested for their antiviral potential against ZIKV using a drug repurposing approach (described in subsequent section of the chapter). Additionally, compounds like finasteride, brequinar (Adcock et al. 2017), gemcitabine, saliphenylhalamide, obatoclax (Kuivanen et al. 2017), adenosine triphosphate analogs (Hercik et al. 2017), a non-nucleoside RNA polymerase inhibitor (TPB) compound (Pattnaik et al. 2018), candesartan cilexetil (Loe et al. 2019), pinocembrin (Lee et al. 2019), aurintricarboxylic acid (Park et al. 2019), and remdesivir (Konkolova et al. 2020) have been evaluated for their ZIKV antiviral potential.

Recently, cinnamic acid has demonstrated in vitro inhibition of ZIKV (Chen et al. 2021). Similarly, through an informatics approach, derivatives of dasabuvir, efavirenz, and tipranavir have been identified as potential inhibitors of viral RdRp and methyltransferase proteins (Ezeh et al. 2022). However, the effectiveness of these compounds can only be confirmed with standard drug trials. One clinically approved antiviral drug against this virus is sofosbuvir (Sacramento et al. 2017).

1.4.2.4 Ebola Virus

Ebola virus (EBOV), a member of the *Filoviridae* family, has been causing a highly lethal disease in humans since its first outbreak in 1976 in the Democratic Republic of the Congo (Africa). Nonhuman primates are also susceptible to this virus. Currently, five species of EBOV have been identified. Among them, four species, namely, Ebola virus (*Zaire ebolavirus*), Sudan virus (*Sudan ebolavirus*), Tai Forest virus (*Tai Forest ebolavirus*, formerly *Cote d'Ivoire ebolavirus*), and Bundibugyo virus (*Bundibugyo ebolavirus*) have been reported to cause Ebola disease in humans. The fifth species, Reston virus (*Reston ebolavirus*), infects only nonhuman primates but not humans (https://www.ncdhhs.gov/divisions/public-health/ebola-information).

Zaire ebolavirus, commonly known as Ebola virus, is the most well-known species that infects humans. As per the information available from Centers for Disease Control and Prevention (CDC) (https://www.cdc.gov/vhf/ebola/history/chronology.html), recent outbreaks of Ebola caused by *Zaire ebolavirus* have been reported in 2021 in the Democratic Republic of the Congo and Guinea, as well as in 2022 in the Democratic Republic of the Congo. The first documented case of human-to-human transmission of Ebola outside of African continent was reported in Spain in 2014. More recently, an outbreak of Ebola caused by *Sudan ebolavirus* was reported in Uganda in September 2022.

Currently, there are two FDA-approved treatment strategies available for Ebola disease. The first is the use of Inmazeb, a combination therapy consisting of three monoclonal antibodies namely, atoltivimab, maftivimab, and odesivimab-ebgn. Inmazeb blocks the virus attachment and entry into host cells by targeting the virus's surface glycoprotein. It was approved in October 2020 for the treatment of *Zaire ebolavirus* infection (https://www.fda.gov/news-events/press-announcements/fda-approves-first-treatment-ebola-virus). The second treatment option is Ebanga, a monoclonal antibody used against ebolavirus infection. Ebanga received FDA approval in December 2020 (https://www.fda.gov/drugs/news-events-human-drugs/fda-approves-treatment-ebola-virus). However, it is important to note that according to the FDA, the efficacy of both the drugs has not been evaluated for other species of EBOV apart from *Zaire ebolavirus*. Various studies have focused on the development of antiviral drugs targeting EBOV proteins such as VP35, VP40, GP, VP30, VP24, and viral polymerase (Liu et al. 2022).

1.4.2.5 Rabies Virus

Rabies, a fatal viral disease, is caused by the rabies virus, which belongs to the *Lyssavirus* in the *Rhabdoviridae* family. In addition to post-exposure prophylaxis, immunoglobulin therapy is most commonly administered at the site of bite for post-exposure treatment. Recent studies have demonstrated the effectiveness of the antiviral drug favipiravir as post-exposure therapy for rabies, both in vitro and in vivo mice model, either alone (Yamada et al. 2016, 2019) or in combination with post-exposure vaccination (Kimitsuki et al. 2023). Monoclonal antibodies have also been shown efficacious in Syrian hamster challenge models (Chao et al. 2020). Intriguingly, the use of phototherapy with a photothermal agent, in conjunction with post-exposure prophylaxis, has demonstrated promise for treating rabies in mouse model (Bai et al. 2023). Furthermore, an experimental study on mice exposed to category III-like rabies has shown that a combination of the antiviral drug favipiravir and rabies vaccine provides effective and complete protection against the disease (Kimitsuki et al. 2023).

1.4.2.6 Other Medically Important RNA Viruses

Japanese encephalitis (JE) is a severe neurological disease that affects both humans and animals such as pigs and horses. It is caused by a member of the *Flaviviridae* family, with water birds acting as reservoirs of the virus. It is primarily a vector-borne disease, transmitted by the *Culex* species. Human infections with Japanese encephalitis virus (JEV) often occur without symptoms. Currently, there is no approved treatment for JE. However, researchers have reported the potential of alternative medicines such as belladonna compounds (Saxena et al. 2023), as well as andrographolide, berberine, curcumin, resveratrol, and retinoic acid (Maurya et al. 2023). Repurposed drugs like manidipine, cilnidipine, benidipine hydrochloride, pimecrolimus, and nelfinavir mesylate (Wang et al. 2017), along with tubacin (Lu et al. 2017), have also been investigated for the treatment of JE. Furoquinoline derivatives (Huang et al. 2016) and pyrazole analogues have been identified as potential anti-JE therapeutics (Gupta et al. 2022).

Dengue fever is one of the vector-borne viral diseases transmitted by the *Aedes aegypti* mosquito and poses a significant threat to human health. The dengue virus has five serotypes, and currently, there are no approved antiviral therapeutics specifically targeting dengue. However, ongoing research focuses on new drug development as well as repurposing existing drugs. The nonstructural proteins NS3 and NS5 are considered potential drug targets against the virus (Obi et al. 2021). Among the repurposed drugs, doxycycline has shown effectiveness against dengue virus by inhibiting serine proteases, thereby interfering with viral entry and replication (Rothan et al. 2014). More recently, riamilovir, a nucleoside analogue, was reported from the Republic of Guinea as a potential treatment for dengue (Maltsev et al. 2023).

1.4.2.7 Other RNA Viruses Affecting Animal Health

Newcastle disease virus (NDV), a major poultry pathogen, belongs to the *Paramyxoviridae* family and causes substantial mortality in affected flocks. Despite regular vaccination, outbreaks continue to be reported under field conditions. There is an urgent need for effective therapeutic measures to control poultry mortality. Studies have shown the efficacy of natural compounds such as *Chlorella vulgaris* (CV) microalgae extract against NDV (Fathy et al. 2023). Similarly, β-carboline alkaloid compounds including 1-formyl-β-carboline derivatives (Wang et al. 2021) and 9-butyl-harmol have been demonstrated effectiveness against NDV (Wang et al. 2023). Additionally, sialic acid derivatives have been shown as potential anti-NDV agents (Rota et al. 2023).

Avian influenza virus, a member of the *Orthomyxoviridae* family, is one of the most serious pathogens affecting poultry and is also of zoonotic importance. Recent research highlights that a few of the phytochemical compounds, such as harmane, harmalol, harmaline, strychnine sulfate (Hegazy et al. 2023), and extract of santonica flower (*Artemisia cina*) (Hegazy et al. 2022) acts as potent antiviral drugs against influenza A virus.

1.5 Novel Approaches in Antiviral Drug Development

Currently, with advancements in virology research, there has been an accelerated focus on the development of drugs and effective delivery strategies. This involves employing rational drug design to create innovative antiviral formulations, exploring alternative medicinal options such as phyto-medicines as alternatives to conventional antiviral strategies, utilizing organic polymers like extracellular polymeric substances (EPS) for their antiviral properties, and investigating the potential of exosomes for antiviral therapeutics purposes. Below, we delve into some of these novel approaches in antiviral drug development.

1.5.1 Computer-Aided Drug Designing (CADD)

Computer-aided drug designing (CADD) is a promising tool in the realm of drug discovery. Currently, the field of rational drug designing, in conjunction with computational approaches, has ushered in a new era, expediting the drug discovery process. Within CADD, structure-based and ligand-based approaches are extensively employed. Moreover, this process allows for the exploration of the absorption, distribution, metabolism, excretion, and toxicity (ADMET) properties of an ideal therapeutic drug.

Recent studies have utilized CADD to analyze drug targets against papain-like cysteine proteases and main proteases, which are essential in the replication of SARS-CoV-2 (Gao et al. 2023; Yu et al. 2022). For instance, a sulfur-containing compound, specifically, thiophene exhibits potential as an antiviral drug against papain-like cysteine proteases of MERS-CoV, SARS-CoV, and SARS-CoV-2 (Jeong et al. 2022). Additionally, various phytochemicals have been tested against papain-like cysteine proteases and main proteases of SARS-CoV-2 (Cayona and Creencia 2022a, b; Mitra et al. 2022; Shree et al. 2022). Computational tools have also been employed to develop a novel consensus-based framework for examining antibody therapeutics against SARS-CoV-2 including its multiple variants (Kumar et al. 2023).

1.5.2 Phyto-medicines as Antivirals

Phytochemicals possess antiviral properties that hinder various stages of viral infection within host cells. They are capable of impeding virus attachment, entry, viral genome replication, transcription, protein expression, virus assembly, and release from host cells. Several phytochemicals, such as cimicifugin (Wang et al. 2012), guggulsterone (Bouslama et al. 2019), curcumin (Chen et al. 2013), and caffeic acid (Weng et al. 2019), act on viruses through diverse mechanisms, including binding to the virus envelope, virus penetration, and promoting virus attachment. Other phytocompounds like dicaffeoylquinic acid (Ooi et al. 2006), isocorilagin (Chen et al. 2020a), c-methylated flavonoid (Dao et al. 2010), and geranin (Choi et al. 2019) have been identified as inhibitors of virus by targeting multiple enzymes that are crucial in virus replication and propagation. The synthesis of viral mRNA can be imperfect at the transcriptional level, leading to a hindered expression of viral coat proteins. Studies have identified carnosic acid (Shin et al. 2013) and phenanthrenes (Shi et al. 2017) as effective agents at controlling virus propagation by inhibiting virus transcription. In some in vitro studies, compounds such as palmitoyl EGCG (de Oliveira et al. 2013), brevilin A (Zhang et al. 2019), and 7-methoxy-1-methyl 4,9 dihydro-3H-pyrido(3,4b) indole (Bag et al. 2013, 2014) have demonstrated the ability to block RNA and inhibit viral protein translation.

During viral infection, newly synthesized viral components, including the genome and proteins, contribute to the generation of viruses. The final stage of the viral infection cycle involves the assembly and release of the viruses from the host cell. Preventing virus assembly and release can significantly reduce the spread of viruses between cells. Several compounds, such as proanthocyanidin A2 (Zhang et al. 2018b), dipotassium glycyrrhizate (Wang et al. 2013), platycodin D (Zhang et al. 2018a), and pyrithione (Guo et al. 2017) have been discovered to inhibit the release from viruses from host cells (Fig. 1.1).

Phytochemicals can be categorized into various types, including phenolics, alkaloids, carotenoids, organosulfur compounds, and terpenoids. Among these, phenolics, alkaloids, and terpenoids exhibit antiviral activities. Phenolic compounds can be further subdivided into phenolic acids, flavonoids, coumarins, stilbenes, and tannins. Flavonoids are classified based on chemical structure, i.e., anthocyanidins, flavonones, flavonols, flavones, and isoflavonoids.

In a study conducted by Cho et al. (2015), the extract of *Epimedium koreanum* Nakai was found to possess antiviral properties. Quercetin, a flavonoid and the active compound in the herb, demonstrated potent antioxidant activity and protected BALB/c mice against influenza A subtypes, including H1N1, H5N2, H7N3, and H9N2. Another compound, kaempferol glucoside derived from *Impatiens balsamina*, was found to target coronavirus ion channel proteins and hinder virus release from infected host cell, making it a potential antiviral drug candidate (Dhama et al. 2018).

Polyphenols, aromatic compounds with strong antioxidant activity due to their aromatic rings, have been studied for their antiviral properties. Curcumin, abundant in turmeric, and EGCG from green tea have shown antiviral activity against the Zika virus (Mounce et al. 2017; Carneiro et al. 2016). Curcumin has also been demonstrated to possess antiviral effects against bovine herpes virus (Zhu et al. 2015a, b). These phytochemicals are believed to bind to the lipid envelope of viruses, inhibiting virus entry and replication. Additionally, polyphenol extracted from *Acacia*

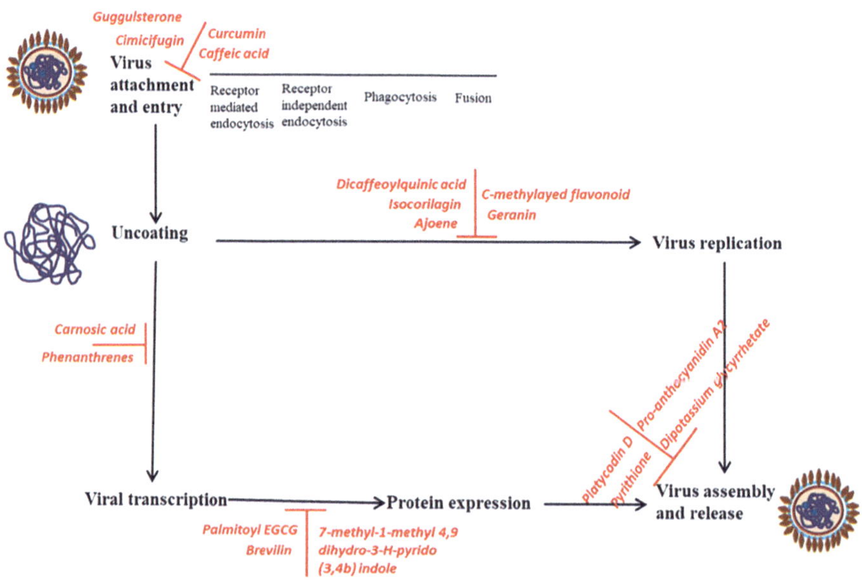

Fig. 1.1 Therapeutic potential of bioactive compounds on regulatory steps of virus infection

arabica was found to possess antiviral properties against influenza virus, H9N2 (Ghoke et al. 2018).

Rabies, a lethal disease for both animals and humans, was addressed in in vivo studies using an Ayurvedic formulation of *Datura metel*, which exhibited potential as an antirabies agent (Roy et al. 2018). *Datura metel* is rich in the alkaloid atropine, known for its anticholinergic properties.

Foot and mouth disease (FMD) is one of the contagious viral diseases that spread through aerosols and affects all cloven-hoofed animals, namely, bovines, sheep, goats, pigs, camelids, deer, etc. Outbreaks of FMD have been observed in livestock regions worldwide. To control the virus's spread, control measures for FMD and drugs targeting specific viral proteins have been developed. Apigenin, a dietary flavonoid naturally found in plants like celery, basil, and others, possesses anti-inflammatory, antioxidant, and anticancer properties. In vitro testing against FMD virus revealed that apigenin inhibits virus replication and disrupts translation process (Qian et al. 2015).

Ocimum basilicum, commonly known as basil and widely cultivated in India, holds significant importance in daily life and the pharmaceutical industry. Numerous scientific studies have reported various health benefits associated with it. Chiang et al. (2005) discovered that ursolic acid, a terpenoid compound derived from *Ocimum basilicum*, inhibits the infection and replication of enteroviruses, thereby controlling the spread of the virus.

Digitoxin, a steroidal glucoside present in the *Digitalis purpurea* plant, is a toxic compound used for the treatment of cardiac failure. It affects the concentration of cations within cells by interacting with transmembrane proteins, such as sodium potassium ATPase. Previous research has shown that alterations in sodium concentration impact the replication cycle of the herpes simplex virus (HSV). Consequently, Su et al. (2008) reported that digitoxin and structurally similar analogues, such as ouabain octahydrate and digoxin affect HSV-1 replication and release.

A phenolic compound found in cereals, fruits, and vegetables, p-coumaric acid, possesses free radical scavenging, anticancer, antidiabetic, and anti-inflammatory properties. Kang and Lee (2015) discovered that treatment of porcine reproductive and respiratory syndrome virus (PRRSV) infection with *Sasa quelpaertensis* Nakai extract (SQE) suppressed the virus in cultured porcine alveolar macrophages (PAMs). SQE, which contains a high concentration of p-coumaric acid, effectively inhibits virus replication, transcription, protein expression, and release, thereby significantly reducing virus propagation. The experiment also resulted in increased synthesis of pro-inflammatory cytokine.

In a study conducted by Bagla et al. (2012), it was demonstrated that a crude extract derived from *Podocarpus henkelii* exhibited potent inhibition of canine distemper virus and lumpy skin disease virus replication, reducing it by more than 75% at a concentration of 3 mg/mL. The crude extract contains numerous bioactive compounds that can act independently or in combination with others, targeting the virus replication cycle. A similar outcome was observed with an extract from *Achyrocline bogotensis*, which demonstrated anti-rotavirus activity in vitro (Tellez et al. 2015).

The results indicated that the components of the extract adhere to the virus's outer structure, thereby preventing its entry. Comparable findings were also observed with hydrolyzable tannins present in *Terminalia chebula* Retz, such as chebulagic acid and punicalagin, which bind to viral coat proteins and prevent virus binding and infection, as described by Lin et al. (2011). Additionally, Wang et al. (2016) discovered that oblongifolin M, a bioactive compound derived from *Garcinia oblongifolia*, exhibits antiviral activity against enteroviruses. It functions by inhibiting the activity of the viral internal ribosome entry site (IRES), which has an impact on ERp57. A list of natural compounds used as antivirals is provided in Table 1.1, and some of these antivirals are currently undergoing clinical trials, as shown in Table 1.2.

1.5.3 Extracellular Polymeric Substances as Antivirals

Extracellular polymeric substances (EPS) are natural organic polymers released by specific microbes. These substances are primarily composed of carbohydrates and proteins, although lipids and nucleic acid also play significant roles. Extensive research over the years has demonstrated the antiviral properties of EPS. For instance, sulfated polysaccharides extracted from various algae and cyanobacteria have been found to inhibit mumps (Gerber et al. 1958), influenza B viruses (Ahmadi et al. 2015), HSV-1, HSV-2, and coxsackievirus B5 (Deig et al. 1974). Nakashima et al. (1987) reported an in vitro antiviral effect of EPS extracted from the red alga *Schizymenia pacifica* against HIV, which was attributed to its ability to affect the reverse transcriptase activity of the virus and thus prevented virus replication.

EPS exerts both direct and indirect effects on viruses. In the direct pathway, EPS interacts with the virus surface through electrostatic interactions, effectively covering it and inhibiting virus entry, acting as a virucide. Furthermore, in certain cases, EPS prevents virus replication within the host cell, providing protection to the host cell. In the indirect pathway, EPS elicits an immune response and contributes to the elimination of virus-infected host cells (Bello-Morales et al. 2022).

1.5.4 Exosomes as Antiviral Agents

Exosomes, small endosomes ranging in size from 40 to 100 nm, are secreted by various cell types. These bilayer membranes contain proteins such as tetraspanins, transport proteins, heat shock proteins, lipids including sphingolipids, cholesterols, ceramides, saturated fatty acids, and nucleic acids, as well as nucleic acids like mRNA and miRNA (Simons and Raposo 2009; Qin and Xu 2014). Previously, exosomes were primarily thought to eliminate unwanted substances from cells (Pan et al. 1985). However, recent research has shed light on their role in intracellular

Table 1.1 List of natural compounds, source, and mechanism against various viruses

Name of virus	Type of compound	Active compound	Plant	Bioactivity	References
Influenza virus	Flavonoid	Quercetin	*Epimedium koreanum Nakai*	Immunomodulatory effect, secretion of type 1 IFN and pro-inflammatory cytokines	Cho et al. (2015)
Coronavirus	Flavonoid	Kaempferol glucoside	*Impatiens balsamina*	Inhibition of ion channel proteins	Dhama et al. (2018)
Zika virus	Polyphenol	Curcumin, epigallocatechin (EGCG)	*Curcuma longa* Green tea	Inhibit virus entry and replication	Mounce et al. (2017), Carneiro et al. (2016)
Rabies virus	Alkaloid	Atropine	*Datura metel*	Anticholinergic agent	Roy et al. (2018)
Foot and mouth disease virus	Flavonoid	Apigenin	–	Suppress FMDV IRES-mediated translational activity	Qian et al. (2015)
Enterovirus	Terpenoids	Ursolic acid	*Ocimum basilicum*	Affects replication	Chiang et al. (2005)
Bovine herpes virus	Polyphenol	Curcumin	*Curcuma longa*	Impair virus post-binding entry process and inhibit replication	Zhu et al. (2015a, b)
Herpes simplex virus	Steroidal glycosides	Digitoxin	*Digitalis purpurea*	Inhibit viral release	Su et al. (2008)
PRRSV	Phenolic compound	p-coumaric acid	*Sasa quelpaertensis*	Suppression of replication, pro-inflammatory cytokine production	Kang and Lee (2015)
Canine distemper virus	–	Crude extract	*Podocarpus henkelii*	Inhibit virus replication	Bagla et al. (2012)
Lumpy skin disease virus	–	Crude extract	*Podocarpus henkelii*	Inhibit virus replication	Bagla et al. (2012)
Rotavirus	–	Crude extract	*Achyrocline bogotensis*	Prevent virus binding and penetration	Lin et al. (2011), Tellez et al. (2015)
Enterovirus	Polycyclic polyprenylated acylphloroglucinols	Oblongifolin M	*Garcinia oblongifolia*	Downregulation of ERp57	Wang et al. 2016

Table 1.2 List of natural antivirals under clinical trials, registered in www.clinicaltrials.gov for treatment of various viral infections

S. no.	Condition/ disease	Treatment under study	Clinical Trial. gov identifier	Responsible party	Sponsor	Start date	Completion date
1.	Coronavirus	Chinese herbal medicine	NCT04380870	Katherine Taromina, Center for Integrated Care	Center for Integrated Care	May 11, 2020	Continue
2.	HIV	*Euphorbia kansui* extract powder	NCT02531295	Sulggi A. Lee, University of California, San Francisco	University of California, San Francisco	June 1, 2019	Continue
3.	Coronavirus	Herbal medicine: Shen Cao Gan Jiang Tang (Gan Cao Gan Jiang Tang with Ginseng)	NCT05055427	Nguyen Van Dan, University of Medicine and Pharmacy at Ho Chi Minh City	University of Medicine and Pharmacy at Ho Chi Minh City	August 20, 2021	Continue
4.	Plantar wart	Banaba leaf extract	NCT05592041	Deraya University	Deraya University	October 20, 2022	January 10, 2023
5.	Coronavirus	Chinese herbal medicine	NCT05360004	Zhaoxiang Bian, Hong Kong Baptist University	Hong Kong Baptist University	May 2022	June 2022
6.	Coronavirus	Vitamin D3 50,000 IU, vitamin C/zinc dietary supplement: vitamin K2/D, microcrystalline cellulose capsule, medium chain triglyceride oil	NCT04780061	The Canadian College of Naturopathic Medicine	The Canadian College of Naturopathic Medicine	July 12, 2021	July 7, 2022
7.	Coronavirus	Kan Jang capsules	NCT04847518	Swedish Herbal Institute AB	Swedish Herbal Institute AB	May 26, 2021	March 2022
8.	Coronavirus	Eucalyptus Oil	NCT05398965	Bumi Herman, Hasanuddin University	Hasanuddin University	November 18, 2020	March 1, 2022

S. no.	Condition/ disease	Treatment under study	Clinical Trial. gov identifier	Responsible party	Sponsor	Start date	Completion date
9.	SARS-CoV2	Herbal oil, vitamin D (cholecalciferol)	NCT05092698	Federal Research Clinical Center of Federal Medical and Biological Agency, Russia	Federal Research Clinical Center of Federal Medical and Biological Agency, Russia	May 1, 2020	January 31, 2022
10.	Coronavirus	Xagrotin	NCT05017493	Biomad AS	Biomad AS	July 1, 2020	February 8, 2021
11.	Coronavirus	Jing-Guan-Fang (JGF): Chinese herbal decoction	NCT05236309	Chung-Hua Hsu, Taipei City Hospital	Taipei City Hospital	January 1, 2020	November 18, 2021
12.	Coronavirus	Dietary supplement: artemisinin	NCT05004753	Mateon Therapeutics	Mateon Therapeutics	September 30, 2020	April 30, 2021
13.	Coronavirus	Jing-Si-Herbal-Tea	NCT04967755	Yao-Kuang Wu, Taipei Tzu Chi Hospital, Buddhist Tzu Chi Medical Foundation	Yao-Kuang Wu	June 1, 2021	September 30, 2021
14.	Viral infection	Reduxium	NCT04525456	Longevity Therapeutics, Inc.	Longevity Therapeutics, Inc.	October 7, 2020	January 27, 2021
15.	SARS-CoV-2	Quercetin	NCT04851821	Riadh Boukef, Hôpital Universitaire Sahloul	Hôpital Universitaire Sahloul	January 4, 2021	June 30, 2021
16.	SARS	Quercetin	NCT04853199	Riadh Boukef, Hôpital Universitaire Sahloul	Hôpital Universitaire Sahloul	June 1, 2021	August 30, 2021
17.	HIV	Dietary supplement: *Artemisia annua*, *Moringa oleifera*	NCT03366922	Mbarara University of Science and Technology	Mbarara University of Science and Technology	December 10, 2017	August 10, 2020
18.	Coronavirus	Quercetin, bromelain, zinc, vitamin C	NCT04468139	Amr Kamel Khalil Ahmed, Ministry of Health, Saudi Arabia	Ministry of Health, Saudi Arabia	June 20, 2020	July 30, 2020

(continued)

Table 1.2 (continued)

S. no.	Condition/ disease	Treatment under study	Clinical Trial. gov identifier	Responsible party	Sponsor	Start date	Completion date
19.	Coronavirus	Brazilian green propolis extract (EPP-AF)	NCT04480593	Marcelo Silveira, D'Or Institute for Research and Education	D'Or Institute for Research and Education	June 2, 2020	August 30, 2020
20.	Coronavirus SARS-CoV-2	Dietary supplement: *Nigella sativa* oil	NCT04401202	Dr Abdulrahman E. Koshak, King Abdulaziz University	King Abdulaziz University	May 21, 2020	December 31, 2020
21.	Coronavirus SARS-CoV2	Honeydrug: *Nigella sativa*/ black cumin drug: placebos	NCT04347382	Sohaib Ashraf, Sheikh Zayed Federal Postgraduate Medical Institute	Sohaib Ashraf	April 30, 2020	August 30, 2020
22.	Coronavirus	Oral administration of colchicine plus herbal phenolic monoterpene fractions	NCT04392141	Dr. Ali Mostafaie, PhD, Kermanshah University of Medical Sciences	Kermanshah University of Medical Sciences	April 1, 2020	November 1, 2020
23.	Human papillomavirus	Dietary supplement: AHCC 3 grams once a day	NCT02405533	Judith A. Smith, The University of Texas Health Science Center, Houston	The University of Texas Health Science Center, Houston	April 2015	November 2019
24.	Influenza-like illness	Biobran/MGN-3	NCT04646980	Ahmed Farouk Elsaid, Zagazig University	Zagazig University	November 28, 2018	February 28, 2019
25.	Influenza	Polyherbal formulation	NCT03365336	Composite Interceptive Med Science	Composite Interceptive Med Science	February 1, 2018	May 15, 2018
26.	Hepatitis chronic	Traditional Chinese medicine (TCM)	NCT03018821	Xi'an Traditional Medicine Hospital	Xi'an Traditional Medicine Hospital	January 2017	December 2017
27.	Hepatitis C	Selenium-containing food supplement herbal tablets	NCT02215525	Egyptian Military Medical Services	Egyptian Military Medical Services	April 2014	May 2015

S. no.	Condition/ disease	Treatment under study	Clinical Trial. gov identifier	Responsible party	Sponsor	Start date	Completion date
28.	Vulvar intraepithelial neoplasiagenital warts	SR-T100 gel with 2.3% of SM in *Solanum undatum* plant extract	NCT01676792	National Cheng-Kung University Hospital	National Cheng-Kung University Hospital	April 2011	October 2013
29.	Hepatitis C, chronic	Oral iodine compound {RENESSANS}	NCT01463592	MTI Medical Private Limited, Pakistan	MTI Medical Private Limited, Pakistan	June 2010	December 2013
30.	Chronic hepatitis C infection	Fuzheng Huayu	NCT00854087	Tarek Hassanein, SCTI Research Foundation	Tarek Hassanein	August 2010	August 2013
31.	Hepatitis C	St John's Wort	NCT01663922	St Stephens Aids Trust	St Stephens Aids Trust	August 2012	December 2012
32.	Hepatitis C	SHO-SAIKO-TO	NCT00590564	Memorial Sloan Kettering Cancer Center	Memorial Sloan Kettering Cancer Center	October 2002	February 2011
33.	Hand-foot-mouth disease	Chinese herbal medicines	NCT01145664	Beijing YouAn Hospital	Beijing YouAn Hospital	May 2010	August 2010
34.	Hepatitis C	Herbal remedies	NCT00010816	National Center for Complementary and Integrative Health (NCCIH)	National Center for Complementary and Integrative Health (NCCIH)	September 1998	June 2003

communication and their involvement in various diseases, including cancer and infectious diseases. During virus infections, exosomes and viruses share similarities in their characteristics and functions within a host. Both entities carry biologically active compounds, and their envelopes help protect and deliver materials to either support normal cell functions or initiate infection (Mcnamara and Dittmer 2020). Nonetheless, they have evolutionary differences and possess distinct membranes with unique or overlapping protein sites. In some instances, viruses exploit extracellular vesicles, including exosomes, to conceal their genetic material from host recognition.

Recent studies have provided evidence supporting the effectiveness of specific exosomes as antiviral agents. Yao et al. (2018) reported that macrophage exosomes deliver INF-α to liver cells, conferring resistance against hepatitis B virus (HBV) and displaying anti-HBV activity. Similarly, Wu et al. (2021) confirmed the anti-HBV activity of INF-α, highlighting the impact of exosome-mediated transfer of INF-α on virus replication and expression. Similarly, Zhu et al. (2015a, b) identified exosome-mediated interferon-inducible transmembrane proteins 3 as a potential candidate against the dengue virus infection. Chettimada et al. (2018) shown the immunomodulatory potential of exosomes in anti-retroviral drug therapy.

During the current COVID-19 pandemic, researchers are extensively exploring the use of exosomes to develop effective therapies. MSC-derived exosomes have shown promising results in combating SARS-CoV-2-induced pneumonia, exhibiting immunomodulatory and anti-inflammatory activities (Akbari and Rezaie 2020). In addition to this, clinical trials have investigated the aerosol and intravenous administration of MSC-derived exosomes for the treatment of SARS-CoV-2 pneumonia. Significant outcomes were achieved with a single dose of MSC-derived exosomes, leading to reduced cytokine storms and improved oxygen levels (Sengupta et al. 2020).

1.5.5 Exosomes for Drug Delivery

Recent studies have exploited exosomes as vehicles for drug delivery to infected cells. This is due to their specific biological markers, inherent ability to cross biological barriers, and capacity to evade host immune system (Sancho-Albero et al. 2020). Moreover, exosomes offer stability, lower toxicity, and reduced immunogenicity compared to synthetic drug delivery system. Recently, Rezakhani et al. (2021) utilized bone marrow mesenchymal stem cells (MSC)-derived exosomes loaded with specific mRNA and miRNA as potential drug delivery carriers. Figure 1.2 illustrates the methods employed to prepare drug-loaded exosomes for efficient drug delivery.

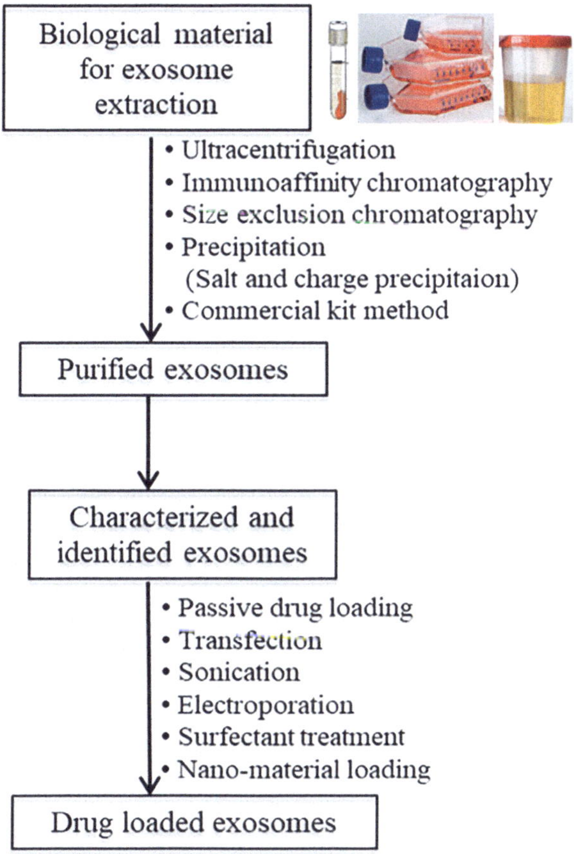

Fig. 1.2 Flow chart represents the general method to synthesize drug-loaded exosomes for efficient drug delivery

1.6 Repurposing of Drugs as Antivirals

Due to the high rate of failure in the development of new antiviral drugs and the substantial investments needed to create novel drugs and their combinations, pharmaceutical industries have increasingly adopted the strategy of drug repurposing (DR). Previous studies estimating the cost of drug development and clinical trials have found it to range from $350 million to $2 billion per drug over a span of 15–17 years (Paul et al. 2010; Trivedi et al. 2020b). Another contributing factor to the declining approval rates is the growing number of drug failures at different stages of clinical trials (Bedair and Mansour 2021). Consequently, repurposing or reprofiling existing drugs has generated attention as a viable approach for both common and uncommon infectious diseases.

Drug repositioning or drug reprofiling involves utilizing the mechanistic information of currently approved medications to address new disease conditions. Repurposing existing drugs as antiviral therapeutics offers significant advantages, including low toxicity and a reduced risk of failure due to extensive studies on structure-activity relationship, preclinical evaluation, and clinical trials on target hosts, thereby mitigating the associated toxicity concerns. This approach has gained substantial momentum in situations where time is limited, and there is an urgent need to develop treatment strategies for virus-associated pandemics. The global community has witnessed the essential role of drug repurposing in creating more effective treatments and enabling the development of numerous medications to combat the SARS-CoV-2 epidemic.

In order to be prepared for future pandemics with viral infections, the drug repurposing approach is being reconsidered for expedited development of improved and efficient antiviral therapeutic strategies. However, before embarking on future studies, challenges related to intellectual property rights should be taken into account when exploring the drug repurposing approach. Despite the considerable advantages of drug repurposing, there are key obstacles to overcome, such as determining the appropriate route of administration and effective concentrations required for clinically achievable antiviral activity after successful repurposing, in order to address the toxicity issues. One feasible solution to this problem could be the development of combination therapies that leverage the synergistic effects of two or more from different classes, each with distinct mechanisms of action.

1.6.1 Strategies

There are various strategies of repurposing drugs as antivirals, including the same target-new virus approach, same target-new indication approach, and new target-new indication approach.

Same target-new virus approach: Drugs specific to a particular viral or cellular functional pathway are explored to treat different viruses. The drugs exhibit antiviral activity by acting on a viral target that shares structural homology and common enzymatic pathways required for virus replication. For instance, favipiravir and sofosbuvir, which are approved for treating influenza virus and hepatitis C virus (HCV) infections, respectively, have demonstrated potential for repurposing against EBOV and ZIKV utilizing their viral RNA polymerase inhibiting activities (Mercorelli et al. 2018).

Same target-new indication approach: In this approach, the focus is on the pharmacological property of drugs. An approved drug is used to modulate a protein or pathway that plays a crucial role associated with a viral disease. In this approach, this drug can be repurposed for a new indication as an antiviral agent. For instance, imatinib, a cellular Abelson (ABL) kinase inhibitory anticancer drug has been found effective against coronaviruses (Coleman et al. 2016).

New target-new indication approach: In this approach, a drug with promising bioactivity is found to have a new molecular target essential for replication of virus. Various examples include teicoplanin, ivermectin, itraconazole, and nitazoxanide. These drugs have been found to have targets in virus-infected cells, and inhibiting these targets has detrimental effects on viral replication (Strating et al. 2015; Colson and Raoult 2016).

1.6.2 Computational Tools for Drug Repurposing for Antivirals

The use of computational tools in drug repurposing studies offers both time efficiency and cost-effectiveness (Peyvandipour et al. 2018). These tools provide insights into the three-dimensional structures of compounds and their binding affinities with target proteins in a three-dimensional space (Sahoo et al. 2021). For instance, the structures of various targets of the coronavirus were identified by X-ray crystallography and NMR spectroscopy, those that include spike (s), envelop (E), membrane (M), nucleocapsid (N) proteins, as well as protease, hemagglutinin esterase, and helicase (Prajapat et al. 2020). In the field of drug repurposing, numerous databases and computational tools are available, such as DrugBank, DrugPredict, e-Drug3D, Promiscuous, Mantra2.0, PharmDB, DRAR-CPI, repoDB, and repurpose, among others.

1.6.3 Drug Repurposing for RNA Virus Infections

Several drugs have been repurposed based on their availability to inhibit RNA-dependent-RNA polymerase (RdRp) and protease enzymes for the treatment of infections caused by RNA viruses. A number of FDA-approved RdRp inhibitors have shown effectiveness against a broad spectrum of RNA viruses such as hydroxychloroquine, remdesivir, ribavirin, favipiravir, galidesivir, kaletra, sofosbuvir, tenofovir, and ritonavir (Gangadharan et al. 2022). Moreover, Piplani et al. (2021) have predicted the potential repurposing of certain natural products, such as silibinin (an antioxidant and hepatoprotectant), digoxin (a cardiac drug of arrhythmia), and ivermectin (an antiparasitic), based on their binding affinity to RdRp. These natural products hold promise for the treatment of RNA viruses, including SARS-CoV-2 (Piplani et al. 2021). Additionally, Zhang et al. (2022) have identified other examples of repurposing drugs for the treatment of Zika virus, while amiodarone, bepridil, teicoplanin, amiodarone, and favipiravir have been repurposed for Ebola and GS-5734 for MERS-CoV (Zhang et al. 2022).

1.6.4 *Drug Repurposing for DNA Virus Infections*

In the drug repurposing study, various approved or investigational drugs have been found as potential antivirals against viruses. For instance, statins, cardiac glycosides, antiparasitic drugs (emetine and nitazoxanide), kinase inhibitors, and antihypertensive drugs (manidipine) have been found to have antiviral activity against human cytomegalovirus. Similarly, dopamine receptor antagonist, the antifungal terbinafine and calcium-channel blockers show promise as antivirals against hepatitis B virus. Furthermore, lopinavir/ritonavir, which targets the HIV protease, is being considered as a potential antiviral therapeutics option for human papillomavirus-related preinvasive cervical malignancies (Mercorelli et al. 2018).

1.7 Conclusion

In the context of today's globalized world, various human activities such as urbanization, deforestation, and global migration have led to environmental disturbances and territorial encroachment. These factors have disrupted the ecological balance in natural habitats. Moreover, the rise in global temperatures has imposed environmental stressors on the natural microflora and fauna, posing risks to human and animal health. This has resulted in the emergence of new infections or existing infections in altered form, leading to global pandemics.

Viruses, being one of the infectious agents, constantly face selection pressures from environmental stressors, including the drugs used against them. Consequently, viruses often develop alternative or bypass strategies to counteract these drugs. Therefore, it is crucial to continuously invest efforts in research and development of new antiviral drugs, including drug repurposing. Another approach is the use of combination drug therapy to combat these organisms. Rational drug design is opening up new possibilities in the field of antiviral drug development. Many of these drugs are currently in various stages of research, ranging from preclinical to clinical phases. Furthermore, the success of antiviral therapeutics depends on appropriate formulation and delivery strategies. Therefore, the use of suitable antiviral drugs, along with proper administration strategies, is essential for effective therapeutic interventions against viral pathogens.

References

Abdullah Al Awadh A (2022) Nucleotide and nucleoside-based drugs: past, present, and future. Saudi J Biol Sci 29(12):103481. https://doi.org/10.1016/j.sjbs.2022.103481

Adcock RS, Chu YK, Golden JE, Chung DH (2017) Evaluation of anti-Zika virus activities of broad-spectrum antivirals and NIH clinical collection compounds using a cell-based, high-throughput screen assay. Antivir Res 138:47–56. https://doi.org/10.1016/j.antiviral.2016.11.018

Agarwal M, Ranjan P, Baitha U, Mittal A (2020) Hydroxychloroquine as a chemoprophylactic agent for COVID-19: a clinico-pharmacological review. Front Pharmacol 11:593099. https://doi.org/10.3389/fphar.2020.593099

Ahmadi A, Zorofchian Moghadamtousi S, Abubakar S, Zandi K (2015) Antiviral potential of algae polysaccharides isolated from marine sources: a review. Biomed Res Int 2015:825203. https://doi.org/10.1155/2015/825203

Akbari A, Rezaie J (2020) Potential therapeutic application of mesenchymal stem cell-derived exosomes in SARS-CoV-2 pneumonia. Stem cell Res Ther 11(1):356. https://doi.org/10.1186/s13287-020-01866-6

Bag P, Ojha D, Mukherjee H, Halder UC, Mondal S, Chandra NS, Nandi S, Sharon A, Sarkar MC, Chakrabarti S, Chattopadhyay D (2013) An indole alkaloid from a tribal folklore inhibits immediate early event in HSV-2 infected cells with therapeutic efficacy in vaginally infected mice. PLoS One 8(10):e77937. https://doi.org/10.1371/journal.pone.0077937

Bag P, Ojha D, Mukherjee H, Halder UC, Mondal S, Biswas A, Sharon A, Van Kaer L, Chakrabarty S, Das G, Mitra D, Chattopadhyay D (2014) A dihydro-pyrido-indole potently inhibits HSV-1 infection by interfering the viral immediate early transcriptional events. Antivir Res 105:126–134. https://doi.org/10.1016/j.antiviral.2014.02.007

Bagla VP, McGaw LJ, Eloff JN (2012) The antiviral activity of six South African plants traditionally used against infections in ethnoveterinary medicine. Vet Microbiol 155(2–4):198–206. https://doi.org/10.1016/j.vetmic.2011.09.015

Bai Y, Huang P, Feng N, Li Y, Huang J, Jin H, Zhang M, Sun J, Li N, Zhang H, Xia X, Tang BZ, Wang H (2023) Treat the "untreatable" by a photothermal agent: triggering heat and immunological responses for rabies virus inactivation. Adv Sci 10(2):e2205461. https://doi.org/10.1002/advs.202205461

Baker DE (2019) Baloxavir Marboxil. Hosp Pharm 54(3):165–169. https://doi.org/10.1177/0018578719841044

Ballana E, Este JA (2015) BMS-663068, a safe and effective HIV-1 attachment inhibitor. Lancet HIV 2(10):e404–e405. https://doi.org/10.1016/S2352-3018(15)00160-5

Batool S, Chokkakula S, Song MS (2023) Influenza treatment: limitations of antiviral therapy and advantages of drug combination therapy. Microorganisms 11(1):183. https://doi.org/10.3390/microorganisms11010183

Bedair A, Mansour FR (2021) Insights into the FDA 2018 new drug approvals. Curr Drug Discov Technol 18(2):293–306. https://doi.org/10.2174/1570163816666191202104315

Bello-Morales R, Andreu S, Ruiz-Carpio V, Ripa I, López-Guerrero JA (2022) Extracellular polymeric substances: still promising antivirals. Viruses 14(6):1337. https://doi.org/10.3390/v14061337

Bernatchez JA, Tran LT, Li J, Luan Y, Siqueira-Neto JL, Li R (2020) Drugs for the treatment of Zika virus infection. J Med Chem 63(2):470–489. https://doi.org/10.1021/acs.jmedchem.9b00775

Bonvicini F, Bua G, Conti I, Manaresi E, Gallinella G (2017) Hydroxyurea inhibits parvovirus B19 replication in erythroid progenitor cells. Biochem Pharmacol 136:32–39. https://doi.org/10.1016/j.bcp.2017.03.022

Bouslama L, Kouidhi B, Alqurashi YM, Chaieb K, Papetti A (2019) Virucidal effect of guggulsterone isolated from Commiphora Gileadensis. Planta Med 85(16):1225–1232. https://doi.org/10.1055/a-1014-3303

Brunning A (2022) Periodic graphics: a short history of antiviral drugs. Chem Eng News 100(4) Published on: January 30, 2022. https://cen.acs.org/biological-chemistry/infectious-disease/Periodic-Graphics-short-history-antiviral/100/i4

Bua G, Conti I, Manaresi E, Sethna P, Foster S, Bonvicini F, Gallinella G (2019) Antiviral activity of brincidofovir on parvovirus B19. Antivir Res 162:22–29. https://doi.org/10.1016/j.antiviral.2018.12.003

Carneiro BM, Batista MN, Braga ACS, Nogueira ML, Rahal P (2016) The green tea molecule EGCG inhibits Zika virus entry. Virology 496:215–218. https://doi.org/10.1016/j.virol.2016.06.012

Cayona R, Creencia E (2022a) Discovery of a "cocktail" of potential SARS-COV-2 main protease inhibitors through virtual screening of known chemical components of Vitex negundo L. ("Lagundi"). Med Chem 18(3):364–381. https://doi.org/10.2174/1573406417666210618132003
Cayona R, Creencia E (2022b) Phytochemicals of *Euphorbia hirta* L. and their inhibitory potential against SARS-CoV-2 main protease. Front Mol Biosci 8:801401. https://doi.org/10.3389/fmolb.2021.801401
Chakraborty C, Sharma AR, Bhattacharya M, Agoramoorthy G, Lee SS (2021) The drug repurposing for COVID-19 clinical trials provide very effective therapeutic combinations: lessons learned from major clinical studies. Front Pharmacol 12:704205. https://doi.org/10.3389/fphar.2021.704205
Chakravarti R, Singh R, Ghosh A, Dey D, Sharma P, Velayutham R, Roy S, Ghosh D (2021) A review on potential of natural products in the management of COVID-19. RSC Adv 11(27):16711–16735. https://doi.org/10.1039/d1ra00644d
Chao TY, Zhang SF, Chen L, Tsao E, Rupprecht CE (2020) *In vivo* efficacy of SYN023, an anti-rabies monoclonal antibody cocktail, in post-exposure prophylaxis animal models. Trop Med Infect Dis 5(1):31. https://doi.org/10.3390/tropicalmed5010031
Chen TY, Chen DY, Wen HW, Ou JL, Chiou SS, Chen JM, Wong ML, Hsu WL (2013) Inhibition of enveloped viruses infectivity by curcumin. PLoS One 8(5):e62482. https://doi.org/10.1371/journal.pone.0062482
Chen K, Risatti C, Bultman M, Soumeillant M, Simpson J, Zheng B, Fanfair D, Mahoney M, Mudryk B, Fox RJ, Hsaio Y, Murugesan S, Conlon DA, Buono FG, Eastgate MD (2014) Synthesis of the 6-azaindole containing HIV-1 attachment inhibitor pro-drug, BMS-663068. J Org Chem 79(18):8757–8767. https://doi.org/10.1021/jo5016008
Chen H, Lao Z, Xu J, Li Z, Long H, Li D, Lin L, Liu X, Yu L, Liu W, Li G, Wu J (2020a) Antiviral activity of lycorine against Zika virus *in vivo* and *in vitro*. Virology 546:88–97. https://doi.org/10.1016/j.virol.2020.04.009
Chen F, Yang L, Huang Y, Chen Y, Sang H, Duan W, Yang J (2020b) Isocorilagin, isolated from Canarium album (lour.) Raeusch, as a potent neuraminidase inhibitor against influenza A virus. Biochem Biophys Res Commun 523(1):183–189. https://doi.org/10.1016/j.bbrc.2019.12.043
Chen Y, Li Z, Pan P, Lao Z, Xu J, Li Z, Zhan S, Liu X, Wu Y, Wang W, Li G (2021) Cinnamic acid inhibits Zika virus by inhibiting RdRp activity. Antivir Res 192:105117. https://doi.org/10.1016/j.antiviral.2021.105117
Chettimada S, Lorenz DR, Misra V, Dillon ST, Reeves RK, Manickam C, Morgello S, Kirk GD, Mehta SH, Gabuzda D (2018) Exosome markers associated with immune activation and oxidative stress in HIV patients on antiretroviral therapy. Sci Rep 8(1):7227. https://doi.org/10.1038/s41598-018-25515-4
Chiang LC, Ng LT, Cheng PW, Chiang W, Lin CC (2005) Antiviral activities of extracts and selected pure constituents of Ocimum basilicum. Clin Exp Pharmacol Physiol 32(10):811–816. https://doi.org/10.1111/j.1440-1681.2005.04270.x
Chien RN, Liaw YF (2022) Current trend in antiviral therapy for chronic hepatitis B. Viruses 14(2):434. https://doi.org/10.3390/v14020434
Cho WK, Weeratunga P, Lee BH, Park JS, Kim CJ, Ma JY, Lee JS (2015) Epimedium koreanum Nakai displays broad spectrum of antiviral activity in vitro and in vivo by inducing cellular antiviral state. Viruses 7(1):352–377. https://doi.org/10.3390/v7010352
Choi JG, Kim YS, Kim JH, Chung HS (2019) Antiviral activity of ethanol extract of GeraniiHerba and its components against influenza viruses via neuraminidase inhibition. Sci Rep 9(1):12132. https://doi.org/10.1038/s41598-019-48430-8
Cihlar T, Ray AS (2010) Nucleoside and nucleotide HIV reverse transcriptase inhibitors: 25 years after zidovudine. Antivir Res 85(1):39–58. https://doi.org/10.1016/j.antiviral.2009.09.014
Coleman CM, Sisk JM, Mingo RM, Nelson EA, White JM, Frieman MB (2016) Abelson kinase inhibitors are potent inhibitors of severe acute respiratory syndrome coronavirus and

Middle East respiratory syndrome coronavirus fusion. J Virol 90(19):8924–8933. https://doi.org/10.1128/JVI.01429-16

Colman PM, Varghese JN, Laver WG (1983) Structure of the catalytic and antigenic sites in influenza virus neuraminidase. Nature 303(5912):41–44. https://doi.org/10.1038/303041a0

Colson P, Raoult D (2016) Fighting viruses with antibiotics: an overlooked path. Int J Antimicrob Agents 48(4):349–352. https://doi.org/10.1016/j.ijantimicag.2016.07.004

Crumpacker CS (1992) Mechanism of action of foscarnet against viral polymerases. Am J Med 92(2A):3S–7S. https://doi.org/10.1016/0002-9343(92)90329-a

Da LT, Quan JM, Wu YD (2011) Understanding the binding mode and function of BMS-488043 against HIV-1 viral entry. Proteins 79(6):1810–1819. https://doi.org/10.1002/prot.23005

Dao TT, Tung BT, Nguyen PH, Thuong PT, Yoo SS, Kim EH, Kim SK, Oh WK (2010) C-methylated flavonoids from Cleistocalyx operculatus and their inhibitory effects on novel influenza A (H1N1) neuraminidase. J Nat Prod 73(10):1636–1642. https://doi.org/10.1021/np1002753

De Clercq E (2002) Cidofovir in the treatment of poxvirus infections. Antivir Res 55(1):1–13. https://doi.org/10.1016/s0166-3542(02)00008-6

De Clercq E (2010) Historical perspectives in the development of antiviral agents against poxviruses. Viruses 2(6):1322–1339. https://doi.org/10.3390/v2061322

de Oliveira A, Adams SD, Lee LH, Murray SR, Hsu SD, Hammond JR, Dickinson D, Chen P, Chu TC (2013) Inhibition of herpes simplex virus type 1 with the modified green tea polyphenol palmitoyl-epigallocatechin gallate. Food Chem Toxicol 52:207–215. https://doi.org/10.1016/j.fct.2012.11.006

Deig EF, Ehresmann DW, Hatch MT, Riedlinger DJ (1974) Inhibition of herpesvirus replication by marine algae extracts. Antimicrob Agents Chemother 6(4):524–525. https://doi.org/10.1128/AAC.6.4.524

Delaune D, Iseni F (2020) Drug development against smallpox: present and future. Antimicrob Agents Chemother 64(4):e01683–e01619. https://doi.org/10.1128/AAC.01683-19

Dhama K, Karthik K, Khandia R, Munjal A, Tiwari R, Rana R, Khurana SK, Ullah S, Khan RU, Alagawany M, Farag MR, Dadar M, Joshi SK (2018) Medicinal and therapeutic potential of herbs and plant metabolites/extracts countering viral pathogens – current knowledge and future prospects. Curr Drug Metab 19(3):236–263. https://doi.org/10.2174/138920021966618012914525 2

Dorr P, Westby M, Dobbs S, Griffin P, Irvine B, Macartney M, Mori J, Rickett G, Smith-Burchnell C, Napier C, Webster R, Armour D, Price D, Stammen B, Wood A, Perros M (2005) Maraviroc (UK-427,857), a potent, orally bioavailable, and selective small-molecule inhibitor of chemokine receptor CCR5 with broad-spectrum anti-human immunodeficiency virus type 1 activity. Antimicrob Agents Chemother 49(11):4721–4732. https://doi.org/10.1128/AAC.49.11.4721-4732.2005

Duzgunes N, Fernandez-Fuentes N, Konopka K (2021) Inhibition of viral membrane fusion by peptides and approaches to peptide design. Pathogens (Basel, Switzerland) 10(12):1599. https://doi.org/10.3390/pathogens10121599

Ezeh MI, Okonkwo OE, Okpoli IN, Orji CE, Modozie BU, Onyema AC, Ezebuo FC (2022) Chemoinformatic design and profiling of derivatives of Dasabuvir, Efavirenz, and Tipranavir as potential inhibitors of Zika virus RNA-dependent RNA polymerase and methyltransferase. ACS Omega 7(37):33330–33348. https://doi.org/10.1021/acsomega.2c03945

Farag TK, El-Houssiny AS, Abdel-Rahman EH, Hegazi AG (2020) A new approach to the treatment of lumpy skin disease infection in cattle by using propolis encapsulated within alg nps. Adv Ani Vet Sci 8(12):1346–1355. https://nexusacademicpublishers.com/uploads/files/AAVS_8_12_1346-1355.pdf

Fathy RR, Soliman RA, El-Hadary A, Osman M, Magdy Y, El-basuni SS (2023) Assessment of antiviral activity for ethanolic chlorella vulgaris extract against Newcastle Disease Virus (NDV) infection in Sasso chicken. J Appl Vet Sci 8(2):67–79. https://doi.org/10.21608/javs.2023.184929.1206

Field HJ, De Clercq E (2004) Antiviral drugs–a short history of their discovery and development. Microbiol Today 31:58–61. Published on: May 1 2004. https://socgenmicrobiol.org.uk/pubs/micro_today/pdf/050402.pdf

Fleming DM (2001) Managing influenza: amantadine, rimantadine and beyond. Int J Clin Pract 55(3):189–195

Gangadharan S, Ambrose JM, Rajajagadeesan A, Kullappan M, Patil S, Gandhamaneni SH, Veeraraghavan VP, Nakkella AK, Agarwal A, Jayaraman S, Surapaneni KM (2022) Repurposing of potential antiviral drugs against RNA-dependent RNA polymerase of SARS-CoV-2 by computational approach. J Infect Public Health 15(11):1180–1191. Advance online publication. https://doi.org/10.1016/j.jiph.2022.09.007

Gao H, Dai R, Su R (2023) Computer-aided drug design for the pain-like protease (PLpro) inhibitors against SARS-CoV-2. Biomed Pharmacother 159:114247. https://doi.org/10.1016/j.biopha.2023.114247

Gerber P, Dutcher JD, Adams EV, Sherman JH (1958) Protective effect of seaweed extracts for chicken embryos infected with influenza B or mumps virus. Proc Soc Exp Biol Med 99(3):590–593. https://doi.org/10.3181/00379727-99-24429

Ghoke SS, Sood R, Kumar N, Pateriya AK, Bhatia S, Mishra A, Dixit R, Singh VK, Desai DN, Kulkarni DD, Dimri U, Singh VP (2018) Evaluation of antiviral activity of Ocimum sanctum and Acacia arabica leaves extracts against H9N2 virus using embryonated chicken egg model. BMC Complement Altern Med 18(1):174. https://doi.org/10.1186/s12906-018-2238-1

Groaz E, De Jonghe S (2021) Overview of biologically active nucleoside phosphonates. Front Chem 8:616863. https://doi.org/10.3389/fchem.2020.616863

Guo C, Zhu Z, Wang X, Chen Y, Liu XJVM (2017) Pyrithione inhibits porcine reproductive and respiratory syndrome virus replication through interfering with NF-κB and heparanase. Vet Microbiol 201:231–239. https://doi.org/10.1016/j.vetmic.2017.01.033

Gupta A, Gawandi S, Vandna, Yadav I, Mohan H, Desai VG, Kumar S (2022) Analysis of fluoro based pyrazole analogues as a potential therapeutics candidate against Japanese encephalitis virus infection. Virus Res 323:198955. Advance online publication. https://doi.org/10.1016/j.virusres.2022.198955

Haller SL, Peng C, McFadden G, Rothenburg S (2014) Poxviruses and the evolution of host range and virulence. Infect Genet Evol 21:15–40. https://doi.org/10.1016/j.meegid.2013.10.014

Hassanipour S, Arab-Zozani M, Amani B, Heidarzad F, Fathalipour M, Martinez-de-Hoyo R (2021) The efficacy and safety of Favipiravir in treatment of COVID-19: a systematic review and meta-analysis of clinical trials. Sci Rep 11(1):11022. https://doi.org/10.1038/s41598-021-90551-6

Hegazy A, Mostafa I, Elshaier YAMM, Mahmoud SH, Abo Shama NM, Shehata M, Yahya G, Nasr NF, El-Halawany AM, Ali MA, Ali MA, Mraheil MA, El-Shazly AM, Mostafa A (2022) Robust antiviral activity of Santonica flower extract (*Artemisia cina*) against avian and human influenza A viruses: *in vitro* and Chemoinformatic studies. ACS Omega 7(45):41212–41223. https://doi.org/10.1021/acsomega.2c04867

Hegazy A, Mahmoud SH, Elshaier YAMM, Shama NMA, Nasr NF, Ali MA, El-Shazly AM, Mostafa I, Mostafa A (2023) Antiviral activities of plant-derived indole and β-carboline alkaloids against human and avian influenza viruses. Sci Rep 13(1):1612. https://doi.org/10.1038/s41598-023-27954-0

Hercik K, Kozak J, Sala M, Dejmek M, Hrebabecky H, Zbornikova E, Smola M, Ruzek D, Nencka R, Boura E (2017) Adenosine triphosphate analogs can efficiently inhibit the Zika virus RNA-dependent RNA polymerase. Antivir Res 137:131–133. https://doi.org/10.1016/j.antiviral.2016.11.020

Huang SH, Lien JC, Chen CJ, Liu YC, Wang CY, Ping CF, Lin YF, Huang AC, Lin CW (2016) Antiviral activity of a novel compound CW-33 against Japanese encephalitis virus through inhibiting intracellular calcium overload. Int J Mol Sci 17(9):1386. https://doi.org/10.3390/ijms17091386

Hubsher G, Haider M, Okun MS (2012) Amantadine: the journey from fighting flu to treating Parkinson disease. Neurology 78(14):1096–1099. https://doi.org/10.1212/WNL.0b013e31824e8f0d

Imai M, Ito M, Kiso M, Yamayoshi S, Uraki R, Fukushi S, Watanabe S, Suzuki T, Maeda K, Sakai-Tagawa Y, Iwatsuki-Horimoto K, Halfmann PJ, Kawaoka Y (2023) Efficacy of antiviral agents against omicron subvariants BQ.1.1 and XBB. N Engl J Med 388(1):89–91. https://doi.org/10.1056/NEJMc2214302

Isaacs A, Lindenmann J (1957) Virus interference. I. The interferon. Proc R Soc London Ser B 147(927):258–267

Ison MG (2011) Antivirals and resistance: influenza virus. Curr Opin Virol 1(6):563–573. https://doi.org/10.1016/j.coviro.2011.09.002

Ison MG (2013) Clinical use of approved influenza antivirals: therapy and prophylaxis. Influenza Other Respir Viruses 7(Suppl 1):7–13. https://doi.org/10.1111/irv.12046

Jefferson T, Jones MA, Doshi P, Del Mar CB, Hama R, Thompson MJ, Spencer EA, Onakpoya I, Mahtani KR, Nunan D, Howick J, Heneghan CJ (2014) Neuraminidase inhibitors for preventing and treating influenza in adults and children. Cochrane Database Syst Rev 2014(4):CD008965. https://doi.org/10.1002/14651858.CD008965.pub4

Jeng WJ, Papatheodoridis GV, Lok ASF (2023) Hepatitis B. Lancet (London, England) 401(10381):1039–1052. https://doi.org/10.1016/S0140-6736(22)01468-4

Jeong K, Kim J, Chang J, Hong S, Kim I, Oh S, Jeon S, Lee JC, Park HJ, Kim S, Lee W (2022) Chemical screen uncovers novel structural classes of inhibitors of the papain-like protease of coronaviruses. iScience 25(10):105254. https://doi.org/10.1016/j.isci.2022.105254

Joly V, Jidar K, Tatay M, Yeni P (2010) Enfuvirtide: from basic investigations to current clinical use. Expert Opin Pharmacother 11(16):2701–2713. https://doi.org/10.1517/14656566.2010.522178

Joshi S, Parkar J, Ansari A, Vora A, Talwar D, Tiwaskar M, Patil S, Barkate H (2021) Role of favipiravir in the treatment of COVID-19. Int J Infect Dis 102:501–508. https://doi.org/10.1016/j.ijid.2020.10.069

Kandeel M, Yamamoto M, Tani H, Kobayashi A, Gohda J, Kawaguchi Y, Park BK, Kwon HJ, Inoue JI, Alkattan A (2021) Discovery of new fusion inhibitor peptides against SARS-CoV-2 by targeting the spike S2 subunit. Biomol Ther 29(3):282–289. https://doi.org/10.4062/biomolther.2020.201

Kang H, Lee C (2015) Sasa quelpaertensis Nakai extract suppresses porcine reproductive and respiratory syndrome virus replication and modulates virus-induced cytokine production. Arch Virol 160(8):1977–1988. https://doi.org/10.1007/s00705-015-2469-0

Kaplan AH, Zack JA, Knigge M, Paul DA, Kempf DJ, Norbeck DW, Swanstrom R (1993) Partial inhibition of the human immunodeficiency virus type 1 protease results in aberrant virus assembly and the formation of noninfectious particles. J Virol 67(7):4050–4055. https://doi.org/10.1128/JVI.67.7.4050-4055.1993

Kaufman HE (1962) Clinical cure of herpes simplex keratitis by 5-iodo-2-deoxyuridine. Proc Soc Exp Biol Med 109:251–252. https://doi.org/10.3181/00379727-109-27169

Kaufman H, Martola EL, Dohlman C (1962) Use of 5-iodo-2′-deoxyuridine (IDU) in treatment of herpes simplex keratitis. Arch Ophthalmol 68:235–239. https://doi.org/10.1001/archopht.1962.00960030239015

Kausar S, Said Khan F, Ishaq Mujeeb Ur Rehman M, Akram M, Riaz M, Rasool G, Hamid Khan A, Saleem I, Shamim S, Malik A (2021) A review: mechanism of action of antiviral drugs. Int J Immunopathol Pharmacol 35:20587384211002621. https://doi.org/10.1177/20587384211002621

Kim CU, Lew W, Williams MA, Liu H, Zhang L, Swaminathan S, Bischofberger N, Chen MS, Mendel DB, Tai CY, Laver WG, Stevens RC (1997) Influenza neuraminidase inhibitors possessing a novel hydrophobic interaction in the enzyme active site: design, synthesis, and structural analysis of carbocyclic sialic acid analogues with potent anti-influenza activity. J Am Chem Soc 119(4):681–690. https://doi.org/10.1021/ja963036t

Kim DH, Han SH, Go HJ, Kim DY, Kim JH, Lee JB, Park SY, Song CS, Lee SW, Choi IS (2022) Antiviral activity of canine interferon lambda 3 expressed using a recombinant adenovirus against canine coronavirus, canine parvovirus, and canine distemper virus. Vet Res Commun 46(4):1363–1368. https://doi.org/10.1007/s11259-022-10000-1

Kimitsuki K, Khan S, Kaimori R, Yahiro T, Saito N, Yamada K, Nakajima N, Komeno T, Furuta Y, Quiambao BP, Virojanapirom P, Hemachudha T, Nishizono A (2023) Implications of the antiviral drug favipiravir on rabies immunoglobulin for post-exposure prophylaxis of rabies in mice model with category III-like exposures. Antivir Res 209:105489. https://doi.org/10.1016/j.antiviral.2022.105489

King DH (1988) History, pharmacokinetics, and pharmacology of acyclovir. J Am Acad Dermatol 18(1 Pt 2):176–179. https://doi.org/10.1016/s0190-9622(88)70022-5

Klysik K, Pietraszek A, Karewicz A, Nowakowska M (2020) Acyclovir in the treatment of herpes viruses – a review. Curr Med Chem 27(24):4118–4137. https://doi.org/10.2174/0929867325666180309105519

Konkolova E, Dejmek M, Hřebabecký H, Šála M, Böserle J, Nencka R, Boura E (2020) Remdesivir triphosphate can efficiently inhibit the RNA-dependent RNA polymerase from various flaviviruses. Antivir Res 182:104899. https://doi.org/10.1016/j.antiviral.2020.104899

Kuivanen S, Bespalov MM, Nandania J, Ianevski A, Velagapudi V, De Brabander JK, Kainov DE, Vapalahti O (2017) Obatoclax, saliphenylhalamide and gemcitabine inhibit Zika virus infection in vitro and differentially affect cellular signaling, transcription and metabolism. Antivir Res 139:117–128. https://doi.org/10.1016/j.antiviral.2016.12.022

Kumar N, Kaushik R, Tennakoon C, Uversky VN, Mishra A, Sood R, Srivastava P, Tripathi M, Zhang KYJ, Bhatia S (2021) Evolutionary signatures governing the codon usage bias in coronaviruses and their implications for viruses infecting various bat species. Viruses 13(9):1847. https://doi.org/10.3390/v13091847

Kumar N, Uversky VN, Tomar S, Li KSM, Chappell K, Lau SKP (2022a) Editorial: emerging and re-emerging viral zoonoses. Front Microbiol 13:978259. https://doi.org/10.3389/fmicb.2022.978259

Kumar N, Kaushik R, Singh A, Uversky VN, Zhang KYJ, Sahu U, Bhatia S, Sanyal A (2022b) Bayesian molecular dating analyses combined with mutational profiling suggest an independent origin and evolution of SARS-CoV-2 omicron BA.1 and BA.2 sub-lineages. Viruses 14(12):2764. https://doi.org/10.3390/v14122764

Kumar N, Kaushik R, Zhang KYJ, Uversky VN, Sahu U, Sood R, Bhatia S (2023) A novel consensus-based computational pipeline for screening of antibody therapeutics for efficacy against SARS-CoV-2 variants of concern including omicron variant. Proteins 91(6):798–806. https://doi.org/10.1002/prot.26467

Laborda P, Wang SY, Voglmeir J (2016) Influenza neuraminidase inhibitors: synthetic approaches, derivatives and biological activity. Molecules (Basel, Switzerland) 21(11):1513. https://doi.org/10.3390/molecules21111513

Lee JL, Loe MWC, Lee RCH, Chu JJH (2019) Antiviral activity of pinocembrin against Zika virus replication. Antivir Res 167:13–24. https://doi.org/10.1016/j.antiviral.2019.04.003

Li Z, Xu J, Lang Y, Fan X, Kuo L, D'Brant L, Hu S, Samrat SK, Trudeau N, Tharappel AM, Rugenstein N, Koetzner CA, Zhang J, Chen H, Kramer LD, Butler D, Zhang QY, Zhou J, Li H (2020) JMX0207, a niclosamide derivative with improved pharmacokinetics, suppresses Zika virus infection both *in vitro* and *in vivo*. ACS Infect Dis 6(10):2616–2628. https://doi.org/10.1021/acsinfecdis.0c00217

Lin LT, Chen TY, Chung CY, Noyce RS, Grindley TB, McCormick C, Lin TC, Wang GH, Lin CC, Richardson CD (2011) Hydrolyzable tannins (chebulagic acid and punicalagin) target viral glycoprotein-glycosaminoglycan interactions to inhibit herpes simplex virus 1 entry and cell-to-cell spread. J Virol 85(9):4386–4398. https://doi.org/10.1128/JVI.01492-10

Liu Y, Liang C, Xin L, Ren X, Tian L, Ju X, Li H, Wang Y, Zhao Q, Liu H, Cao W, Xie X, Zhang D, Wang Y, Jian Y (2020) The development of coronavirus 3C-like protease (3CLpro) inhibitors from 2010 to 2020. Eur J Med Chem 206:112711. https://doi.org/10.1016/j.ejmech.2020.112711

Liu CH, Hu YT, Wong SH, Lin LT (2022) Therapeutic strategies against Ebola virus infection. Viruses 14(3):579. https://doi.org/10.3390/v14030579

Loe MWC, Lee RCH, Chu JJH (2019) Antiviral activity of the FDA-approved drug candesartan cilexetil against Zika virus infection. Antivir Res 172:104637. https://doi.org/10.1016/j.antiviral.2019.104637

Lu CY, Chang YC, Hua CH, Chuang C, Huang SH, Kung SH, Hour MJ, Lin CW (2017) Tubacin, an HDAC6 selective inhibitor, reduces the replication of the Japanese encephalitis virus via the decrease of viral RNA synthesis. Int J Mol Sci 18(5):954. https://doi.org/10.3390/ijms18050954

Mahal A, Duan M, Zinad DS, Mohapatra RK, Obaidullah AJ, Wei X, Pradhan MK, Das D, Kandi V, Zinad HS, Zhu Q (2021) Recent progress in chemical approaches for the development of novel neuraminidase inhibitors. RSC Adv 11(3):1804–1840. https://doi.org/10.1039/d0ra07283d

Majerova T, Konvalinka J (2022) Viral proteases as therapeutic targets. Mol Asp Med 88:101159. https://doi.org/10.1016/j.mam.2022.101159

Malik YS, Kumar N, Sircar S, Kaushik R, Bhat S, Dhama K, Gupta P, Goyal K, Singh MP, Ghoshal U, El Zowalaty ME, O R V, Yatoo MI, Tiwari R, Pathak M, Patel SK, Sah R, Rodriguez-Morales AJ, Ganesh B, Kumar P et al (2020) Coronavirus disease pandemic (COVID-19): challenges and a global perspective. Pathogens (Basel, Switzerland) 9(7):519. https://doi.org/10.3390/pathogens9070519

Maltsev OV, Kasyanenko KV, Zhdanov KV, Malyshev NA, Kolomoets EV, Konomou VK (2023) The experience in treatment of dengue fever using antiviral drug riamilovir in the Republic of Guinea (case report). Ter Arkh 95(1):85–89. https://doi.org/10.26442/00403660.2023.01.202054

Marcello GM, Holder KA, Hallager S, Norton BB, Backues KA, Tyler AE, Zeitlin A, Murray-Hiteshew E, Murray S (2023) Intralesional application of ribavirin in two American flamingos (Phoenicopterus ruber) with poxvirus infection. J Comp Pathol 201:49–52. https://doi.org/10.1016/j.jcpa.2022.11.004

Maurya VK, Kumar S, Ansari S, Sachan AK, Singh U, Paweska JT, Abdel-Moneim AS, Saxena SK (2023) Antiviral and anti-inflammatory activity of natural compounds against Japanese encephalitis virus via inhibition of NS5 protein and regulation of key immune and inflammatory signaling pathways. J Med Virol 95(3):e28675. https://doi.org/10.1002/jmv.28675

McKimm-Breschkin JL (2013) Influenza neuraminidase inhibitors: antiviral action and mechanisms of resistance. Influenza Other Respir Viruses 7(Suppl 1):25–36. https://doi.org/10.1111/irv.12047

McNamara RP, Dittmer DP (2020) Modern techniques for the isolation of extracellular vesicles and viruses. J Neuroimmune Pharmacol 15(3):459–472. https://doi.org/10.1007/s11481-019-09874-x

Mercorelli B, Palù G, Loregian A (2018) Drug repurposing for viral infectious diseases: how far are we? Trends Microbiol 26(10):865–876. https://doi.org/10.1016/j.tim.2018.04.004

Mitra D, Verma D, Mahakur B, Kamboj A, Srivastava R, Gupta S, Pandey A, Arora B, Pant K, Panneerselvam P, Ghosh A, Barik DP, Mohapatra PKD (2022) Molecular docking and simulation studies of natural compounds of *Vitex negundo* L. against papain-like protease (PLpro) of SARS CoV-2 (coronavirus) to conquer the pandemic situation in the world. J Biomol Struct Dyn 40(12):5665–5686. https://doi.org/10.1080/07391102.2021.1873185

Mogensen TH, Skouboe MK, Mølle I (2023) Successful use of interferon alfa-2a for persistent parvovirus B19 infection. Lancet Infect Dis 23(4):e160–e165. https://doi.org/10.1016/S1473-3099(22)00685-5

Monteiro A, Yu KOA, Hicar MD (2021) Peptide-based fusion inhibitors for preventing the six-helix bundle formation of class I fusion proteins: HIV and beyond. Curr HIV Res 19(6):465–475. https://doi.org/10.2174/1570162X19666210908115231

Mounce BC, Cesaro T, Carrau L, Vallet T, Vignuzzi M (2017) Curcumin inhibits Zika and chikungunya virus infection by inhibiting cell binding. Antivir Res 142:148–157. https://doi.org/10.1016/j.antiviral.2017.03.014

Murakami N, Hayden R, Hills T, Al-Samkari H, Casey J, Del Sorbo L, Lawler PR, Sise ME, Leaf DE (2023) Therapeutic advances in COVID-19. Nat Rev Nephrol 19(1):38–52. https://doi.org/10.1038/s41581-022-00642-4

Nakashima H, Kido Y, Kobayashi N, Motoki Y, Neushul M, Yamamoto N (1987) Purification and characterization of an avian myeloblastosis and human immunodeficiency virus reverse transcriptase inhibitor, sulfated polysaccharides extracted from sea algae. Antimicrob Agents Chemother 31(10):1524–1528. https://doi.org/10.1128/AAC.31.10.1524

Namusisi S, Mahero M, Travis D, Pelican K, Robertson C, Mugisha L (2021) A descriptive study of zoonotic disease risk at the human-wildlife interface in a biodiversity hot spot in South Western Uganda. PLoS Negl Trop Dis 15(1):e0008633. https://doi.org/10.1371/journal.pntd.0008633

Ning K, Roy A, Cheng F, Xu P, Kleiboeker S, Escalante CR, Wang J, Qiu J (2022) High-throughput screening identifies inhibitors for parvovirus B19 infection of human erythroid progenitors. J Virol 96(2):e0132621. https://doi.org/10.1128/JVI.01326-21

Oberg B (2006) Rational design of polymerase inhibitors as antiviral drugs. Antivir Res 71(2–3):90–95. https://doi.org/10.1016/j.antiviral.2006.05.012

Obi JO, Gutiérrez-Barbosa H, Chua JV, Deredge DJ (2021) Current trends and limitations in dengue antiviral research. Trop Med Infect Dis 6(4):180. https://doi.org/10.3390/tropicalmed6040180

Ooi LSM, Wang H, He Z, Ooi VEC (2006) Antiviral activities of purified compounds from Youngia japonica (L.) DC (Asteraceae, Compositae). J Ethnopharmacol 106(2):187–191. https://doi.org/10.1016/j.jep.2005.12.028

Pan BT, Teng K, Wu C, Adam M, Johnstone RM (1985) Electron microscopic evidence for externalization of the transferrin receptor in vesicular form in sheep reticulocytes. J Cell Biol 101(3):942–948. https://doi.org/10.1083/jcb.101.3.942

Park JG, Ávila-Pérez G, Madere F, Hilimire TA, Nogales A, Almazán F, Martínez-Sobrido L (2019) Potent inhibition of Zika virus replication by Aurintricarboxylic acid. Front Microbiol 10:718. https://doi.org/10.3389/fmicb.2019.00718

Patick AK, Potts KE (1998) Protease inhibitors as antiviral agents. Clin Microbiol Rev 11(4):614–627. https://doi.org/10.1128/CMR.11.4.614

Pattnaik A, Palermo N, Sahoo BR, Yuan Z, Hu D, Annamalai AS, Vu HLX, Correas I, Prathipati PK, Destache CJ, Li Q, Osorio FA, Pattnaik AK, Xiang SH (2018) Discovery of a non-nucleoside RNA polymerase inhibitor for blocking Zika virus replication through in silico screening. Antivir Res 151:78–86. https://doi.org/10.1016/j.antiviral.2017.12.016

Paul SM, Mytelka DS, Dunwiddie CT, Persinger CC, Munos BH, Lindborg SR, Schacht AL (2010) How to improve R&D productivity: the pharmaceutical industry's grand challenge. Nat Rev Drug Discov 9(3):203–214. https://doi.org/10.1038/nrd3078

Peyvandipour A, Saberian N, Shafi A, Donato M, Draghici S (2018) A novel computational approach for drug repurposing using systems biology. Bioinformatics (Oxford, England) 34(16):2817–2825. https://doi.org/10.1093/bioinformatics/bty133

Piplani S, Singh PK, Winkler DA, Petrovsky N (2021) Computationally repurposed drugs and natural products against RNA dependent RNA polymerase as potential COVID-19 therapies. Mol Biomed 2(1):28. https://doi.org/10.1186/s43556-021-00050-3

Poole CL, James SH (2018) Antiviral therapies for herpesviruses: current agents and new directions. Clin Ther 40(8):1282–1298. https://doi.org/10.1016/j.clinthera.2018.07.006

Prajapat M, Sarma P, Shekhar N, Avti P, Sinha S, Kaur H, Kumar S, Bhattacharyya A, Kumar H, Bansal S, Medhi B (2020) Drug targets for corona virus: A systematic review. Indian J Pharmacol 52(1):56–65. https://doi.org/10.4103/ijp.IJP_115_20

Qian S, Fan W, Qian P, Zhang D, Wei Y, Chen H, Li X (2015) Apigenin restricts FMDV infection and inhibits viral IRES driven translational activity. Viruses 7(4):1613–1626. https://doi.org/10.3390/v7041613

Qin J, Xu Q (2014) Functions and application of exosomes. Acta Pol Pharm 71(4):537–543. https://pubmed.ncbi.nlm.nih.gov/25272880/

Rezakhani L, Kelishadrokhi AF, Soleimanizadeh A, Rahmati S (2021) Mesenchymal stem cell (MSC)-derived exosomes as a cell-free therapy for patients infected with COVID-19: real opportunities and range of promises. Chem Phys Lipids 234:105009. https://doi.org/10.1016/j.chemphyslip.2020.105009

Rota P, La Rocca P, Bonfante F, Pagliari M, Piccoli M, Cirillo F, Ghiroldi A, Franco V, Pappone C, Allevi P, Anastasia L (2023) Design, synthesis, and antiviral evaluation of sialic acid derivatives as inhibitors of Newcastle disease virus hemagglutinin-neuraminidase: a translational study on human parainfluenza viruses. ACS Infect Dis 9(3):617–630. https://doi.org/10.1021/acsinfecdis.2c00576

Rothan HA, Mohamed Z, Paydar M, Rahman NA, Yusof R (2014) Inhibitory effect of doxycycline against dengue virus replication in vitro. Arch Virol 159(4):711–718. https://doi.org/10.1007/s00705-013-1880-7

Roy S, Samant L, Ganjhu R, Mukherjee S, Chowdhary A (2018) Assessment of *in vivo* antiviral potential of Datura metel Linn. extracts against rabies virus. Pharm Res 10(1):109–112. https://doi.org/10.4103/pr.pr_121_17

Rudrapal M, Khairnar SJ, Jadhav AG (2020) Drug repurposing (DR): an emerging approach in drug discovery. In: Drug repurposing hypothesis, molecular aspects and therapeutic applications. IntechOpen, London. https://doi.org/10.5772/intechopen.93193

Sacramento CQ, de Melo GR, de Freitas CS, Rocha N, Hoelz LV, Miranda M, Fintelman-Rodrigues N, Marttorelli A, Ferreira AC, Barbosa-Lima G, Abrantes JL, Vieira YR, Bastos MM, de Mello Volotão E, Nunes EP, Tschoeke DA, Leomil L, Loiola EC, Trindade P, Rehen SK et al (2017) The clinically approved antiviral drug sofosbuvir inhibits Zika virus replication. Sci Rep 7:40920. https://doi.org/10.1038/srep40920

Sadowski LA, Upadhyay R, Greeley ZW, Margulies BJ (2021) Current drugs to treat infections with herpes simplex Viruses-1 and -2. Viruses 13(7):1228. https://doi.org/10.3390/v13071228

Saghazadeh A, Rezaei N (2022) Poxviruses and the immune system: implications for monkeypox virus. Int Immunopharmacol 113(Pt A):109364. https://doi.org/10.1016/j.intimp.2022.109364

Sahoo BM, Ravi Kumar BVV, Sruti J, Mahapatra MK, Banik BK, Borah P (2021) Drug Repurposing Strategy (DRS): emerging approach to identify potential therapeutics for treatment of novel coronavirus infection. Front Mol Biosci 8:628144. https://doi.org/10.3389/fmolb.2021.628144

Saito R, Li D, Sato M, Suzuki H (2006) Amantadine. Virus Rep 3(1):40–47

Sancho-Albero M, Sebastián V, Sesé J, Pazo-Cid R, Mendoza G, Arruebo M, Martín-Duque P, Santamaría J (2020) Isolation of exosomes from whole blood by a new microfluidic device: proof of concept application in the diagnosis and monitoring of pancreatic cancer. J Nanobiotechnol 18(1):150. https://doi.org/10.1186/s12951-020-00701-7

Saravanan D, Monisha M (2023) Computational study for identification of antiviral peptides targeting oncogenic human papillomavirus (HPV) infections. J Nanomed 6(1):1056. https://meddocsonline.org/journal-of-nanomedicine/computational-study-for-identification-of-antiviral-peptides-targeting-oncogenic-human-papillomavirus-infections.pdf

Saravolatz LD, Depcinski S, Sharma M (2023) Molnupiravir and Nirmatrelvir-ritonavir: Oral coronavirus disease 2019 antiviral drugs. Clin Infect Dis 76(1):165–171. https://doi.org/10.1093/cid/ciac180

Satarker S, Ahuja T, Banerjee M, E VB, Dogra S, Agarwal T, Nampoothiri M (2020) Hydroxychloroquine in COVID-19: potential mechanism of action against SARS-CoV-2. Curr Pharmacol Rep 6(5):203–211. https://doi.org/10.1007/s40495-020-00231-8

Saxena SK, Kumar S, Maurya VK, Nayak D, Kaushik S, Manchanda RK, Gadugu S (2023) Antiviral and anti-inflammatory activity of novel belladonna formulation against Japanese encephalitis virus via inhibition of p65 nuclear translocation and TNF-α mediated NF-kB signaling. Biotechnol Genet Eng Rev:1–23. Advance online publication. https://doi.org/10.1080/02648725.2023.2166258

Scarsi KK, Havens JP, Podany AT, Avedissian SN, Fletcher CV (2020) HIV-1 integrase inhibitors: a comparative review of efficacy and safety. Drugs 80(16):1649–1676. https://doi.org/10.1007/s40265-020-01379-9

SenGupta V, SenGupta S, Lazo A, Woods P, Nolan A, Bremer N (2020) Exosomes derived from bone marrow mesenchymal stem cells as treatment for severe COVID-19. Stem Cells Dev 29(12):747–754. https://doi.org/10.1089/scd.2020.0080

Shi T, McAllister DA, O'Brien KL, Simoes EAF, Madhi SA, Gessner BD, Polack FP, Balsells E, Acacio S, Aguayo C, Alassani I, Ali A, Antonio M, Awasthi S, Awori JO, Azziz-Baumgartner E, Baggett HC, Baillie VL, Balmaseda A, Barahona A et al (2017) Global, regional, and national disease burden estimates of acute lower respiratory infections due to respiratory syncytial virus in young children in 2015: a systematic review and modelling study. Lancet (London, England) 390(10098):946–958. https://doi.org/10.1016/S0140-6736(17)30938-8

Shin HB, Choi MS, Ryu B, Lee NR, Kim HI, Choi HE, Chang J, Lee KT, Jang DS, Inn KS (2013) Antiviral activity of carnosic acid against respiratory syncytial virus. Virol J 10:303. https://doi.org/10.1186/1743-422X-10-303

Shree P, Mishra P, Selvaraj C, Singh SK, Chaube R, Garg N, Tripathi YB (2022) Targeting COVID-19 (SARS-CoV-2) main protease through active phytochemicals of ayurvedic medicinal plants – *Withania somnifera* (Ashwagandha), *Tinospora cordifolia* (Giloy) and *Ocimum sanctum* (Tulsi) – a molecular docking study. J Biomol Struct Dyn 40(1):190–203. https://doi.org/10.1080/07391102.2020.1810778

Simons M, Raposo G (2009) Exosomes–vesicular carriers for intercellular communication. Curr Opin Cell Biol 21(4):575–581. https://doi.org/10.1016/j.ceb

Sluis-Cremer N (2018) Future of nonnucleoside reverse transcriptase inhibitors. Proc Natl Acad Sci U S A 115(4):637–638. https://doi.org/10.1073/pnas.1720975115

Sood R, Kumar N, Bhatia S, Chanu KV, Gupta CL, Pateriya AK, Mishra A, Khandia R, Mawale N, Singh VP (2018) Neuraminidase inhibitors susceptibility profiles of highly pathogenic influenza A (H5N1) viruses isolated from avian species in India (2006-2015). Antivir Res 158:143–146. https://doi.org/10.1016/j.antiviral.2018.08.007

Sood R, Kumar N, Gokhe SS, Pateriya AK, Bhat S, Bhatia S, Panickan S, Mishra A, Murugkar HV, Dixit R, Shrivastava D, Singh P, Tripathi M, Singh VP (2021) Identification and molecular characterization of H9N2 viruses carrying multiple mammalian adaptation markers in resident birds in central-western wetlands in India. Infect Genet Evol 94:105005. https://doi.org/10.1016/j.meegid.2021.105005

Strating JR, van der Linden L, Albulescu L, Bigay J, Arita M, Delang L, Leyssen P, van der Schaar HM, Lanke KH, Thibaut HJ, Ulferts R, Drin G, Schlinck N, Wubbolts RW, Sever N, Head SA, Liu JO, Beachy PA, De Matteis MA, Shair MD et al (2015) Itraconazole inhibits enterovirus replication by targeting the oxysterol-binding protein. Cell Rep 10(4):600–615. https://doi.org/10.1016/j.celrep.2014.12.054

Su CT, Hsu JTA, Hsieh HP, Lin PH, Chen TC, Kao CL, Lee CN, Chang SY (2008) Anti-HSV activity of digitoxin and its possible mechanisms. Antivir Res 79(1):62–70. https://doi.org/10.1016/j.antiviral.2008.01.156

Tellez MA, Téllez AN, Vélez F, Ulloa JC (2015) In vitro antiviral activity against rotavirus and astrovirus infection exerted by substances obtained from Achyrocline bogotensis (Kunth) DC. (Compositae). BMC Complement Altern Med 15(1):428. https://doi.org/10.1186/s12906-015-0949-0

Tian L, Qiang T, Liang C, Ren X, Jia M, Zhang J, Li J, Wan M, YuWen X, Li H, Cao W, Liu H (2021) RNA-dependent RNA polymerase (RdRp) inhibitors: the current landscape and repurposing for the COVID-19 pandemic. Eur J Med Chem 213:113201. https://doi.org/10.1016/j.ejmech.2021.113201

Toker EB, Ates O, Yeşilbağ K (2022) Inhibition of bovine and ovine capripoxviruses (lumpy skin disease virus and Sheeppox virus) by ivermectin occurs at different stages of propagation in vitro. Virus Res 310:198671. https://doi.org/10.1016/j.virusres.2021.198671

Trivedi J, Mahajan D, Jaffe RJ, Acharya A, Mitra D, Byrareddy SN (2020a) Recent advances in the development of integrase inhibitors for HIV treatment. Curr HIV/AIDS Rep 17(1):63–75. https://doi.org/10.1007/s11904-019-00480-3

Trivedi J, Mohan M, Byrareddy SN (2020b) Drug repurposing approaches to combating viral infections. J Clin Med 9(11):3777. https://doi.org/10.3390/jcm9113777

von Itzstein M, Wu WY, Kok GB, Pegg MS, Dyason JC, Jin B, Van Phan T, Smythe ML, White HF, Oliver SW (1993) Rational design of potent sialidase-based inhibitors of influenza virus replication. Nature 363(6428):418–423. https://doi.org/10.1038/363418a0

Wagstaff AJ, Bryson HM (1994) Foscarnet. A reappraisal of its antiviral activity, pharmacokinetic properties and therapeutic use in immunocompromised patients with viral infections. Drugs 48(2):199–226. https://doi.org/10.2165/00003495-199448020-00007

Wang KC, Chang JS, Lin LT, Chiang LC, Lin CC (2012) Antiviral effect of Cimicifugin from Cimicifuga foetida against human respiratory syncytial virus. Am J Chin Med 40(5):1033–1045. https://doi.org/10.1142/S0192415X12500760

Wang ZW, Sun N, Wu CH, Jiang JB, Bai YS, Li HQ (2013) In vitro antiviral activity and underlying molecular mechanisms of dipotassiumglycyrrhetate against porcine reproductive and respiratory syndrome virus. Antivir Ther 18(8):997–1004. https://doi.org/10.3851/IMP2662

Wang M, Dong Q, Wang H, He Y, Chen Y, Zhang H, Wu R, Chen X, Zhou B, He J, Kung HF, Huang C, Wei Y, Huang JD, Xu H, He ML, Oblongifolin M (2016) Oblongifolin M, an active compound isolated from a Chinese medical herb Garcinia oblongifolia, potently inhibits enterovirus 71 reproduction through downregulation of ERp57. Oncotarget 7(8):8797–8808.10.18632/oncotarget.7122

Wang S, Liu Y, Guo J, Wang P, Zhang L, Xiao G, Wang W (2017) Screening of FDA-approved drugs for inhibitors of Japanese encephalitis virus infection. J Virol 91(21):e01055–e01017. https://doi.org/10.1128/JVI.01055-17

Wang C, Wang T, Dai J, An Z, Hu R, Duan L, Chen H, Wang X, Chu Z, Liu H, Wang J, Li N, Yang Z, Wang J (2021) 1-formyl-β-carboline derivatives block Newcastle disease virus proliferation through suppressing viral adsorption and entry processes. Biomol Ther 11(11):1687. https://doi.org/10.3390/biom11111687

Wang C, Wang T, Hu R, Duan L, Hou Q, Han Y, Dai J, Wang W, Ren S, Liu H, Wang X, Xiao S, Li N, Wang J, Yang Z (2023) 9-Butyl-Harmol exerts antiviral activity against Newcastle disease virus through targeting GSK-3β and HSP90β. J Virol 97(3):e0198422. https://doi.org/10.1128/jvi.01984-22

Weber IT, Wang YF, Harrison RW (2021) HIV protease: historical perspective and current research. Viruses 13(5):839. https://doi.org/10.3390/v13050839

Weng JR, Lin CS, Lai HC, Lin YP, Wang CY, Tsai YC, Wu KC, Huang SH, Lin CW (2019) Antiviral activity of sambucusFormosanaNakaiethanol extract and related phenolic acid constituents against human coronavirus NL63. Virus Res 273:197767. https://doi.org/10.1016/j.virusres.2019.197767

Wonganan P, Limpanasithikul W, Jianmongkol S, Kerr SJ, Ruxrungtham K (2020) Pharmacokinetics of nucleoside/nucleotide reverse transcriptase inhibitors for the treatment and prevention of HIV infection. Expert Opin Drug Metab Toxicol 16(7):551–564. https://doi.org/10.1080/17425255.2020.1772755

Wu W, Wu D, Yan W, Wang Y, You J, Wan X, Xi D, Luo X, Han M, Ning Q (2021) Interferon-induced macrophage-derived exosomes mediate antiviral activity against hepatitis B virus through miR-574-5p. J Infect Dis 223(4):686–698. https://doi.org/10.1093/infdis/jiaa399

Xu M, Lee EM, Wen Z, Cheng Y, Huang WK, Qian X, Tcw J, Kouznetsova J, Ogden SC, Hammack C, Jacob F, Nguyen HN, Itkin M, Hanna C, Shinn P, Allen C, Michael SG, Simeonov A, Huang W, Christian KM, Goate A, Brennand KJ, Huang R, Xia M, Ming GL, Zheng W, Song H, Tang H (2016) Identification of small-molecule inhibitors of Zika virus infection and induced neural cell death via a drug repurposing screen. Nat Med 22(10):1101–1107. https://doi.org/10.1038/nm.4184

Yamada K, Noguchi K, Komeno T, Furuta Y, Nishizono A (2016) Efficacy of Favipiravir (T-705) in rabies Postexposure prophylaxis. J Infect Dis 213(8):1253–1261. https://doi.org/10.1093/infdis/jiv586

Yamada K, Noguchi K, Kimitsuki K, Kaimori R, Saito N, Komeno T, Nakajima N, Furuta Y, Nishizono A (2019) Reevaluation of the efficacy of favipiravir against rabies virus using in vivo imaging analysis. Antivir Res 172:104641. https://doi.org/10.1016/j.antiviral.2019.104641

Yang Z, Zadjura LM, Marino AM, D'Arienzo CJ, Malinowski J, Gesenberg C, Lin PF, Colonno RJ, Wang T, Kadow JF, Meanwell NA, Hansel SB (2010) Utilization of *in vitro* Caco-2 permeability and liver microsomal half-life screens in discovering BMS-488043, a novel HIV-1 attachment inhibitor with improved pharmacokinetic properties. J Pharm Sci 99(4):2135–2152. https://doi.org/10.1002/jps.21948

Yang J, Liu S, Du L, Jiang S (2016) A new role of neuraminidase (NA) in the influenza virus life cycle: implication for developing NA inhibitors with novel mechanisms of action. Rev Med Virol 26(4):242–250. https://doi.org/10.1002/rmv.1879

Yao Z, Qiao Y, Li X, Chen J, Ding J, Bai L, Shen F, Shi B, Liu J, Peng L, Li J, Yuan Z (2018) Exosomes exploit the virus entry machinery and pathway to transmit alpha interferon-induced antiviral activity. J Virol 92(24):e01578–e01518. https://doi.org/10.1128/JVI.01578-18

Yu W, Zhao Y, Ye H, Wu N, Liao Y, Chen N, Li Z, Wan N, Hao H, Yan H, Xiao Y, Lai M (2022) Structure-based design of a dual-targeted covalent inhibitor against papain-like and main proteases of SARS-CoV-2. J Med Chem 65(24):16252–16267. https://doi.org/10.1021/acs.jmedchem.2c00954

Zhang M, Du T, Long F, Yang X, Sun Y, Duan M, Zhang G, Liu Y, Zhou EM, Chen W, Chen J (2018a) Platycodin D suppresses type 2 porcine reproductive and respiratory syndrome virus. In primary and established cell lines. Viruses 10(11):e010657. https://doi.org/10.3390/v10110657

Zhang M, Wu Q, Chen Y, Duan M, Tian G, Deng X, Sun Y, Zhou T, Zhang G, Chen W, Chen J (2018b) Inhibition of proanthocyanidin A2 on porcine reproductive and respiratory syndrome virus replication in vitro. PLoS One 13(2):e0193309. https://doi.org/10.1371/journal.pone.0193309

Zhang X, Xia Y, Yang L, He J, Li Y, Xia C (2019) Brevilin A, a sesquiterpene lactone, inhibits the replication of influenza A virus *in vitro* and *in vivo*. Viruses 11(9):835. https://doi.org/10.3390/v11090835

Zhang R, Oerlemans R, Wang C, Zhang L, Groves MR (2022) Drug repurposing techniques in viral diseases. IntechOpen. https://doi.org/10.5772/intechopen.101443

Zhao B, Hu L, Song Y, Patil K, Ramani S, Atmar RL, Estes MK, Prasad BVV (2021) Norovirus protease structure and antivirals development. Viruses 13(10):2069. https://doi.org/10.3390/v13102069

Zhu L, Ding X, Zhang D, Yuan C, Wang J, Ndegwa E, Zhu G (2015a) Curcumin inhibits bovine herpesvirustype 1 entry into MDBK cells. Acta Virol 59(3):221–227. https://doi.org/10.4149/av_2015_03_221

Zhu X, He Z, Yuan J, Wen W, Huang X, Hu Y, Lin C, Pan J, Li R, Deng H, Liao S, Zhou R, Wu J, Li J, Li M (2015b) IFITM3-containing exosome as a novel mediator for anti-viral response in dengue virus infection. Cell Microbiol 17(1):105–118. https://doi.org/10.1111/cmi.12339

Zhu Y, Dong X, Liu N, Wu T, Chong H, Lei X, Ren L, Wang J, He Y (2022) SARS-CoV-2 fusion-inhibitory lipopeptides maintain high potency against divergent variants of concern including omicron. Emerging Microbes Infect 11(1):1819–1827. https://doi.org/10.1080/22221751.2022.2098060

Chapter 2
Bioinformatics Databases and Tools Available for the Development of Antiviral Drugs

Ankur Singh, Preeti Dhaka, Pravindra Kumar, Shailly Tomar, and Jitin Singla

Abstract Viruses are responsible for spreading diseases and severe threats to humankind. Emerging viruses and their enigmatic genomic mutations are a serious concern. Lack of precise elucidation and availability of insufficient data is the main barrier to overcoming virus pandemics and epidemics. To tackle the viruses and their emerging variants, the combination of virology and knowledge of bioinformatics leads to light for healthy beings. Viroinformatics is the umbrella term that covers both virology and bioinformatics-based applications. It comprises web servers, tools, and databases flooded with information related to viruses. Resources reported in the chapter include information about coronaviruses, flaviviruses, alphaviruses, influenza viruses, and human immunodeficiency virus (HIV). The chapter also includes the applications of databases and web servers specific to viruses to contribute towards effective vaccines and drug development.

Keywords Viruses · Bioinformatics · Databases · Webservers · Drug discovery

---

A. Singh · P. Dhaka · P. Kumar · S. Tomar (✉)
Department of Biosciences and Bioengineering, Indian Institute of Technology Roorkee, Roorkee, Uttarakhand, India
e-mail: shailly.tomar@bt.iitr.ac.in

J. Singla (✉)
Department of Biosciences and Bioengineering, Indian Institute of Technology Roorkee, Roorkee, Uttarakhand, India

Department of Computer Science and Engineering, Indian Institute of Technology Roorkee, Roorkee, Uttarakhand, India
e-mail: jsingla@bt.iitr.ac.in

N. Kumar et al. (eds.), *Advances in Antiviral Research*, Livestock Diseases and Management, https://doi.org/10.1007/978-981-99-9195-2_2

2.1 Introduction

Viruses are the smallest biological entities infamous for being infectious agents for all living organisms, from unicellular bacteria to multicellular chordates (Cossart and Helenius 2014). Viruses cause most contagious diseases in animals and humans as they replicate inside host organisms by hijacking their machinery (Lin and Guttman 2010). They usually have a protein coat and contain genetic information in DNA or RNA. Viruses generally follow host tropism, but can sometimes cross species barriers and infect other species (Geoghegan et al. 2017). Viruses have led to several disease epidemics over the centuries, including the Spanish flu, AIDS, influenza, diarrhoea, encephalitis, chikungunya fever, dengue fever, and, in recent times, the outbreaks of Ebola, Zika, and severe acute respiratory syndrome (SARS) are some examples of viral diseases that created dreadfulness in living beings (Mushegian 2020). The viruses are divided into two categories: DNA and RNA. The mutation rate in the viruses is very high, ranging from 10^{-6} to 10^{-4} switches per nucleotide per cell for RNA viruses, whereas 10^{-8} to 10^{-6} switches per nucleotide per cell for DNA viruses. Therefore, it is very tough to analyse the biodiversity and infectivity range of the viruses. Vaccinations are an effective method to induce active immunity against viral disease prevention. However, vaccines are unavailable for many infectious viruses, and some viruses evolve to generate new variants that escape developed immunity. Therefore, developing antiviral chemotherapeutic approaches to combat and control emerging infectious viral diseases is equally imperative.

Antivirals are drugs or remedies that help to cure viral infections. They are usually small molecules or peptide sequences, which can be virus-specific, broad-spectrum, or targeting host proteins. A significant challenge for antiviral-based therapeutics is high mutation rates during the virus replication cycles. Conserved regions also mutate frequently, resulting in high variable sequences that make viruses more infectious and drug-resistant. The viral replication process also depends on the host pathways and their responses.

Moreover, using antiviral drugs over the years for a few viral diseases has led to the rapid development of antiviral drug resistance. Developing antivirals for current and emerging viral infections involves numerous sequence and structure-based computational techniques. It consists of specific databases and biophysical methods for computational-assisted antiviral drug discovery. This chapter highlights the databases and tools most important for carrying out various tasks, such as structure-based molecular modelling, virtual screening and docking analysis, binding site prediction, protein-protein interaction studies, antibody-based epitope prediction, rational drug discovery via machine learning, and functional genomics-related studies. Hence, researchers are constantly developing new strategies to design antivirals, as shown in Fig. 2.1. The functional mechanism of antivirals is to allosterically impede the function of viral proteins such as the replication enzymes like polymerases or proteases and the structural proteins (glycoproteins) responsible for preventing the entry step of the virus (Kausar et al. 2021). Novel antiviral strategies include

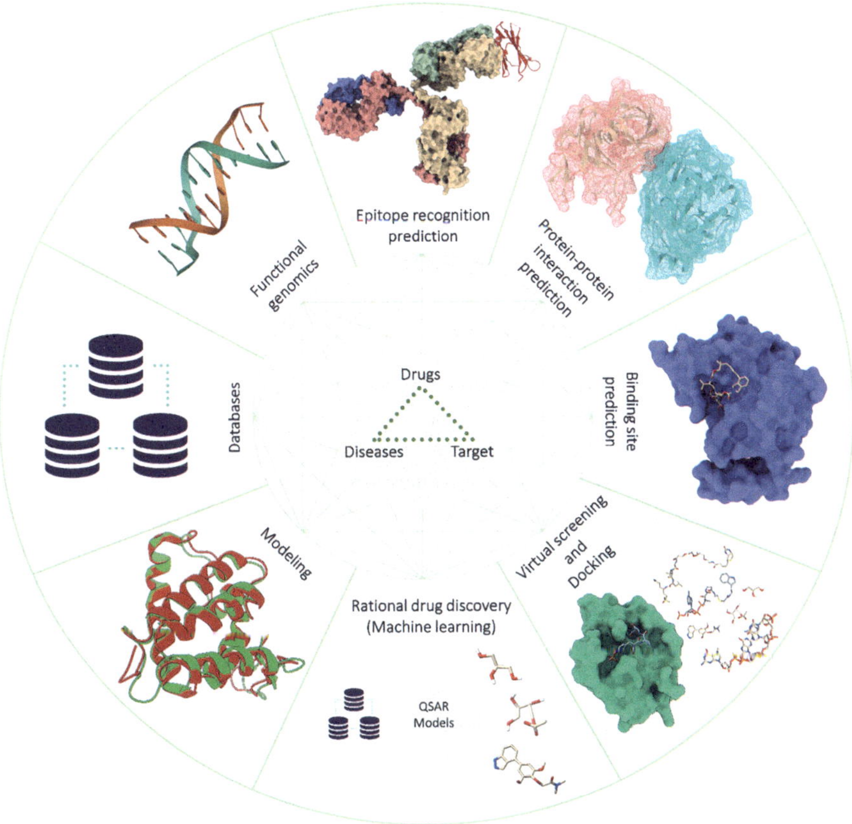

Fig. 2.1 Schematic representation of experimental and computational strategies for developing antivirals. The light green lines represent the integrative nature of data, models, and techniques involved in drug discovery. 1IGT, 2XNI, 1G03, and 3S81 PDB IDs were used to generate the schematic

gene silencing to interfere with viral gene expression that enhances the immune system response and targets host factors responsible for the replication of viruses. The efficient use of computational tools is the only way to analyse the infectivity of the viruses by identifying viral sequences, their encoded functional elements, functional annotation and prediction, their structure-function relationships, host-virus interactions, and many more. Many bioinformatics tools and databases are available for comprehending valuable information in developing and identifying potential antivirals. In addition, a popular method of drug discovery involves drug repurposing of previously approved drug molecules. Researchers can use computational biology to perform a high-throughput screening of such pre-approved small molecules against known viral proteins. This chapter discusses strategies, outlines the various bioinformatics tools/databases, and highlights the virus-specific bioinformatics tools for antiviral design.

2.2 Databases and Resources to Advance Viral Studies and Research

Drug discovery is a multifaceted and laborious process involving multiple phases, including target identification, lead generation/optimization, and preclinical and clinical trials. One of the key challenges in drug discovery is identifying the probable drug targets and the development of safe and effective therapeutics to treat various diseases, including viral infections. Researchers use multiple databases and tools to address these challenges to access enormous amounts of biological and chemical data. These resources can help accelerate the drug discovery process by providing improved knowledge about the molecular mechanisms of primary diseases and the potential targets for drug development.

Antiviral drug resistance is a critical challenge in developing effective therapeutics against viral infections. The overuse of antiviral drugs has contributed to the development of drug-resistant viruses. Therefore, designing and developing new antiviral drugs require computational techniques to gear up the drug discovery process and reduce the threat of resistance. In silico, structure-based techniques are computational methods used to discover new antiviral drugs. Some of the most critical databases and tools used in structure-based drug discovery are discussed below.

2.2.1 Viral Bioinformatics Resource Center (VBRC)

VBRC (Hsu et al. 2007) is an NIH-sponsored bioinformatics resource center that delivers resources and informational and investigative tools to the researchers. VBRC is an online resource that provides access to a curated viral genome database and various tools for bioinformatics studies. The main goal of this resource centre is to promote research against various emerging and re-emerging viruses, especially against pathogens that can be potential bioterrorism threats. It included numerous virus families such as *Flaviviridae*, *Paramyxoviridae*, *Poxviridae*, *Arenaviridae*, *Bunyaviridae*, *Togaviridae*, and *Filoviridae*.

2.2.2 Virus Variation Resource (NCBI-VVR)

NCBI-VVR (Resch et al. 2009) is an online database for viruses, such as Ebola, West Nile virus, rotavirus, Zika, influenza, dengue, MERS, and SARS coronaviruses. It permits the researchers to submit the viral sequences accompanied by significant metadata, for example, time of collection, source of isolation, host, geographical location, disease severity, and many more. This dataset also permits the integration and analysis of viral sequences by constructing a phylogenetic tree and multiple sequence alignment.

2.2.3 Virus Pathogen Database and Analysis Resource (ViPR)

ViPR (Pasznik et al. 2019) is a freely accessible database supported by the National Institute of Allergy and Infectious Diseases (NIAID) and Bioinformatics Resource Center's (Pickett et al. 2012b) (BRC) programme. This database contains information for wide-ranging human pathogenic virus families, viz., *Poxviridae*, *Reoviridae*, *Rhabdoviridae*, *Togaviridae*, *Flaviviridae*, *Filoviridae*, *Hepeviridae*, *Herpesviridae*, *Paramyxoviridae*, *Picornaviridae*, *Arenaviridae*, *Bunyaviridae*, *Caliciviridae*, *Coronaviridae*, and many more. This database comprises numerous types of information such as viral sequence archives, annotation of genes and proteins (Rost et al. 2004), BLAST (Altschul et al. 1990) comparison, multiple sequence alignment, determination of sequence variation, phylogenetic tree construction, immune epitope locations, 3D protein structures, surveillance metadata, and some fresh data resulting from relative genomics study. This database is beneficial and assists researchers in designing diagnostics and therapeutics against pathogenic viruses. Some other databases, including those above for virus studies, are illustrated in Table 2.1.

2.2.4 Virus-Host Interaction Databases

In viruses, virus-host interactions are the key factors in determining its pathogenicity as well as the evasion mechanism of the immune system of the host (Bhutkar et al. 2022). There are various tools and databases available to understand these interactions. PhEVER (Palmeira et al. 2011) is one of the databases that facilitate the analysis of lateral gene transfers between virus-virus and virus-host through evolutionary and phylogenetic information. VIGOR (Wang et al. 2010) is another online tool for gene prediction developed by J. Craig Venter Institute in 2010. At first, its capacity lay in predicting genes of small viral genomes, for example, rotavirus, influenza, rhinovirus, and coronaviruses. However, an update in 2012 expanded the capabilities of VIGOR and presently, VIGOR can be efficiently used for the gene prediction of twelve more viruses; respiratory syncytial virus, mumps virus, rubella virus, Venezuelan equine encephalitis virus, measles virus, yellow fever virus, norovirus, parainfluenza virus, Japanese encephalitis virus, metapneumovirus, Sendai virus, and alphaviruses. With the help of VIGOR, researchers can predict the start and stop codons, stop codon leakage, protein-coding regions, RNA editing, ribosomal shunting, frameshifts, embedded genes, and overlapping genes in the viral genomes.

Virus-PLoc is an easily accessible, user-friendly highly accurate web server that can be used to predict the subcellular localization of viral proteins within the host cells. Similarly, iLoc-Virus (Xiao et al. 2011) and pLoc-mVirus (Cheng et al. 2017) are also web servers through which subcellular localization of viral proteins at one or multiple sites can be predicted. AVPpred (Sharma et al. 2022b) is a web server

Table 2.1 Virus-specific online repositories and databases

Databases/Repositories	URL	Information available
VirOligo	https://doi.org/10.1093/nar/30.1.203	Virus-specific oligonucleotides
VIDA	https://doi.org/10.1093/nar/29.1.133	Animal viruses with *Poxviridae*, *Coronaviridae*, *Herpesviridae*, *Papillomaviridae*, and *Arteriviridae*
VBRC	https://4virology.net/	Viruses from *Arenaviridae*, *Filoviridae*, *Bunyaviridae*, *Paramyxoviridae*, *Flaviviridae*, *Poxviridae*, *Togaviridae*, virus families
PBRC	https://doi.org/10.1093/nar/gki110	Viruses of the *Poxviridae* family
IVDB, IRD, FluGenome, NCBI-IVR, ISED, OpenFluDB, ATIVS, GISAID	https://www.ncbi.nlm.nih.gov/genomes/FLU/Database/nph-select.cgi?go=database	Viruses of influenza group
NCBI-VVR	https://doi.org/10.1186/1471-2180-9-65	Ebola virus, Zika virus, influenza virus, dengue virus, rotavirus, West Nile virus, and coronavirus

that predicts exceedingly operative antiviral peptides. The algorithm of this web server uses the physicochemical properties of peptides that include the composition of amino acids, sequence alignment, peptide motifs, etc. The web server was designed by collecting 1245 experimentally validated antiviral peptides against some crucial human viruses like HCV, HIV, influenza, SARS, and many more. VIPS (Hong et al. 2013) is also a web server that is designed to predict internal ribosomal entry site (IRES), RNA secondary structure prediction, comparison, and pseudoknot prediction. STRING (Cook et al. 2018) is a database well-known for predicting protein-protein interaction between virus-virus and virus-host proteins. The prediction of these interactions is mainly resulting from five key sources high-throughput lab experiments, genomic context predictions, automated text mining, (conserved) coexpression, and preceding knowledge stored in databases. Thus, these tools can provide breakthrough information in developing vaccines, diagnostics, and antivirals.

2.3 Databases for Enveloped Positive Sense RNA Virus

2.3.1 Databases for Coronaviruses

The COVID-19 pandemic and future pandemics are of utmost concern. Databases for coronaviruses provide freedom for drug discovery and help open research that ultimately supports the battle against COVID-19. Databases related to coronaviruses include Global Initiative on Sharing All Influenza Data (GISAID) database (Shu and McCauley 2017), Nextstrain (Hadfield et al. 2018), Nextclade (Aksamentov et al. 2021), CoV-Hipathia (COVID-19 pathway interpretation and analysis), and Artic (Table 2.2). The GISAID is the largest sequence database repository for SARS-CoV-2 globally. Nextstrain provides available information to acquire public

Table 2.2 Databases for coronaviruses

Database	URL	Information available
GISAID	https://doi.org/10.2807/1560-7917.ES.2017.22.13.30494	Largest repository for SARS-CoV-2 sequences worldwide
Nextstrain	https://doi.org/10.1093/bioinformatics/bty407	Information of pathogen genome data
Nextclade	https://doi.org/10.21105/joss.03773	After comparing differences between query sequences, it reports potential sequence quality issues in your data
CoV-Hipathia	https://doi.org/10.1093/bib/bbaa261	It helps to interpret human signalling via gene mutations and observe changes in the gene expression level
Artic	https://artic.network/	A material-based network is available to provide information about datasets of bioinformatics along with laboratory protocols

health impending pathogen genome data. It is a powerful analytical and visualization tool with freely accessible updated data. It is helpful for epidemiological understanding and improving outbreak response. Nextclade reports potential sequence quality issues in your data after comparing differences between query sequences and a reference sequence and utilizes these differences to assign query sequences to clades. CoV-Hipathia is an automated web tool that helps interpret human signalling via observing gene mutations and related changes in the gene expression level that lead to detailed information about human signalling pathways. Artic is a set of material-based networks available to provide information about bioinformatics datasets and laboratory protocols. Other databases like Virus Pathogen and Analysis Resource (Pickett et al. 2012a) (ViPR), VirHostNet, ViralZone (Hulo et al. 2011), Virus Taxonomy (Lefkowitz et al. 2018), and NCBI Viral Genomes (Palmeira et al. 2011) are also devoted resources to COVID-19. ViPR includes data and information about the sequence of viral proteins, host factors, antiviral drugs, and the 3D structure of immune epitopes. Additionally, it contains information about report analysis of epidemiology based on bioinformatics resources, news, and publications. VirHostNet has host-virus interaction-related networks, including information on protein-protein interactions (PPIs) and inhibiting drugs. ViralZone is a general database created by the Expert Protein Analysis System (ExPASy) and the Swiss Institute of Bioinformatics. It hosts bioinformatics resources and tools, including the Virosaurus (a portmanteau of viruses and thesaurus). Virus Taxonomy describes the nomenclature of possible mutants of viruses. NCBI Viral Genomes delivers links related to the coronavirus blast tool and articles in PubMed on coronaviruses. Other than these mentioned databases, there is some specific COVID-19 Pathway Collection, including Breaking-Cas (guide RNAs for CRISPR-Cas experiments with the SARS-CoV-2 genomes). WikiPathways is another disease-related pathways database. Antibody-specific databases comprise more than 1400 published and patented nanobodies and antibodies responsible for binding with at least one coronavirus. CoV-AbDab is one such database that primarily consolidates antibodies for SARS-CoV-2 along with other viruses of *Coronaviridae* family.

2.3.2 Databases for Flaviviruses

Flavivirus is a genus of positive-strand RNA viruses, comprising the dengue virus, West Nile virus, tick-borne encephalitis virus, Zika virus, yellow fever virus, and quite a few other viruses responsible for causing encephalitis. Yellow fever virus belongs to flaviviruses that cause yellow jaundice. Databases available for flaviviruses include Flavitrack (Misra and Schein 2007) that is the depository of properly annotated sequences of viruses responsible for causing diseases. It is helpful towards antiviral drug designing and vaccine production. Flavitrack can also track evolutionary phenotypes, conserved motif identification, and interpretation of mutational and structural data. It contains complete protein and genome sequences of more than 590 flaviviruses with their mutations and all the literature references. Some internal tools of Flavitrack help in the comparative study of query and depository sequences along with their similarity scores.

DGraph Clusters (Braun et al. 2021) is a programme that calculates property distance (PD) scores and displays it graphically based on physicochemical property (PCP) having similarities from another unaligned set of FASTA files. Similarly, "PD graphs" demonstrate the relationship between the sequences. These graphs also propose the evolutionary relationships of different virus sequences from their ancestral sequences. FLAVIdB (Olsen et al. 2011) is an assembled database with antigenic data of flaviviruses and their automated application-based analysis of vaccine and immunological studies, which is pulled together by assortment, integration, and annotation of various data from GenBank (Benson et al. 2007), GenPept (Agarwala et al. 2016), UniProt (Bateman et al. 2015), The Immune Epitope Database (Vita et al. 2019) (IEDB), and Protein Data Bank (Berman et al. 2000) (PDB). This programme is one of the most accurate data mining systems that can be used for drug discovery against flaviviruses. It harbours 12,858 entries of an antigenic flavivirus sequence, near about 200 verified B-cell epitopes and T-cell type epitopes. This programme also contains four illustrative molecular structures of the envelope protein of the dengue virus. The data mining system comprises tools available for analysing conservation and variability and B-cell and T-cell epitope prediction. FLAVIdB characterizes a different group of databases where tools and data are collectively used to build the infrastructures for data mining and precisely designed to support vaccine targets discovery for rational vaccine design.

The Database-Integrated Genome Screening (DIGS) tool, as proposed by Zhu et al. in 2018, utilizes the well-established essential local alignment search tool (BLAST) programme suite (Altschul et al. 1990) for conducting comprehensive genome screening. This integrated framework aims to provide detailed and accurate information about various genomic elements. The DIGS system employs My Structured Query Language (MySQL) as its underlying relational database management system. The output generated by the BLAST programme suite is meticulously recorded and stored within the MySQL database, explicitly using the Community Server version 8.0.26. By combining the power of BLAST and the efficiency of MySQL, the DIGS system offers a sophisticated and scientifically robust approach to genome screening and data management, enabling researchers to access valuable genomic insights with high precision and reliability. DengueNet (Lawrence 2002) provides information validated by the World Health Organization (WHO) globally on dengue haemorrhagic and dengue fever. A set of databases is responsible for delivering web retrieval interfaces, visualization, and analysis tools (presently for West Nile and dengue viruses) and is virus-specific, called NCBI Virus Variation Resources (NCBI-VVR). West Nile virus NCBI-VVR (Resch et al. 2009) is a database part of NCBI-VVR with detailed analysed sequences. Likewise, Dengue NCBI-VVR is a comprehensive sequence database with analysis tools specifically designed for the dengue virus.

2.3.3 Databases for Alphaviruses

Alphaviruses include arthropod-borne RNA viruses like chikungunya, Sindbis, Venezuelan equine encephalitis, Una, Ross River, etc. Insects like mosquitoes transmit these viruses mainly during seasonal changes. As the climate shifts, the distribution of these viruses can expand, increasing their medical importance and the risk of infections in affected regions. Understanding these connections can help us better prepare and respond to potential outbreaks. In past decades, the expansion of mosquitoes of the *Aedes* genus has been mostly associated with disease outbreaks. Changes in the climate favoured the expansion of vectors and led to virus emergence in tropical and temperate regions. The Viral Genome Database (VGDB) (Hiscock and Upton 2000) and VirGen (Kulkarni-Kale et al. 2004), along with ViPR, harbour comprehensive information about the genome of alphaviruses (Table 2.3). Analysis tools like BLAST, phylogenetic tree construction, and multiple sequence alignment (MSA) are used in the comparative analysis of the data repository ViPR. ViPR harbours sequences, protein structures, gene/protein annotations, clinical and surveillance, and epitope metadata specific to the viruses. It comprises the statistical information for many human pathogenic virus families, including *Togaviridae*.

Table 2.3 Databases for flavivirus and alphavirus

Database	References/URL	Information available
Flavitrack	https://doi.org/10.1093/bioinformatics/btm383	Depository of properly annotated sequences
DGraph Clusters	https://doi.org/10.1177/1177932211020316	It checks property distance (PD) scores based on a physicochemical property (PCP)
FLAVIdB	http://cvc.dfci.harvard.edu/flavi/	An assembled database with antigenic data of flaviviruses
DengueNet	https://ghdx.healthdata.org/record/who-denguenet	It provides information validated the World Health Organization (WHO) globally on dengue haemorrhagic and dengue fever
NCBI-VVR	https://doi.org/10.1186/1471-2180-9-65	It provides web retrieval interfaces, analysis and visualization tools
ViPR	https://doi.org/10.1093/nar/gkr859	It harbours sequences, protein structures, gene/protein annotations, clinical and surveillance, and epitope metadata specific to the viruses

2.4 Databases for Enveloped Negative-Sense RNA Viruses

2.4.1 Databases for Influenza Virus

Influenza virus is among the most widespread viruses that severely infect humans. The effects of this virus were observed in 2009 globally, after the first pandemic of influenza A virus (swine flu, H1N1). Influenza research has been supported by a comprehensive array of 11 specialized tools and web servers, notably encompassing the Influenza Virus Database (IVDB), Influenza Research Database (IRD), and NCBI Influenza Virus Resource (NCBI-IVR) (Squires et al. 2012). These mentioned databases can be used for various purposes, such as multiple sequence alignment, their BLAST, sequence database, sequence distribution tools, and phylogenetic tree construction. Others include a web server named ATVIS (Liao et al. 2009) (analysis tools for influenza virus surveillance) that helps in the data analysis of different serological samples stored details, while EpiFlu (Shu and McCauley 2017) helps with influenza sequences accompanying clinical and epidemiological metadata. Another model is FluTE (Holtzapple et al. 2020), which spreads awareness and preparedness about influenza-related pandemics or epidemics. Efforts towards the cure of virus spread include Influenza Primer Design Resource (IPDR) (Bose et al. 2008), responsible for influenza diagnostics primer designing, influenza sequences, and epitope database (ISED) (Yang et al. 2009) harbour information on viruses from whole Asia, GiRaF (Nagarajan and Kingsford 2011) explains about the identification of influenza genotype, and FluGenome (Lu et al. 2007) identifies the influenza A virus genotype and helps in data assortment and analysis. IRD is a comprehensive repository of information on how the influenza virus interacts with its host and other pathogens, along with their sequence BLAST and alignment and phylogenetic tools. OpenFluDB (Liechti et al. 2010) is mainly an influenza virus isolate-centric database with details related to host type, associated virus, and collection records.

2.4.2 Databases for Nipah Virus

Nipah virus belongs to the *Paramyxoviridae* family and is an emerging pathogen because of its high fatality rate. Nipah is a threat to humankind and livestock as it causes severe infections in the respiratory tract, like respiratory ailments, and causes encephalitis. Various research groups are focused on tackling the disease spread, but until now, no approved antiviral drug has been available. The databases and tools available that help towards the discovery of antivirals against Nipah are anti-Nipah (Rajput et al. 2018), NipahVR (Gupta et al. 2020), Global Alert and Response Information Resources (GAR) (Mackenzie et al. 2014). The World Organisation for Animal Health (OIE) and the Food and Agriculture Organization of the United Nations (FAO) report Nipah affecting animal-related studies. NipahVR contains a bulk of information for the development of multi-targeted putative therapeutics. Additionally, it gives a detailed description of phylogenomics, primers for diagnostics purposes, therapeutics (based on siRNAs, sgRNAs, miRNAs), genomes, codon usage, and epitope description for vaccine development. Anti-Nipah web server is a freely available software that can forecast the antiviral activity of drug compounds against the Nipah virus.

Using these databases and tools, researchers can accelerate drug discovery, reduce costs, and create more potent treatments for viral infections. For example, these technologies made the recent development of COVID-19 vaccines and antiviral drugs possible.

2.5 Bioinformatics Tools for Antiviral Development

Certain tools and databases are available to analyse the specific characteristics of the viruses, such as gene identification, functional annotation, phylogeny, host-virus interactions, and structure-function relationships. However, it is quite challenging to analyse the accurate characteristics of the viruses because of the higher rate of mutations and recombination and the use of translational frameshifts or overlapping reading frames to encode multiple proteins from their small genome size. For specific studies of a target virus, it is now crucial to categorize and store data on viral sequences in databases and repositories appropriate to a given virus family.

2.5.1 Molecular Epidemiology and Phylogenetic Studies

Phylogenetic and epidemiological analysis of viruses are the most basic and essential studies. Phylogenetic analysis helps to compare and analyse the evolutionary relationships of virus isolates based on genome/amino acid sequences to describe their specific family. General purpose phylogeny programmes such as PHYLogeny Inference Package (Lim and Zhang 1999) (PHYLIP), Protein Analysis Through Evolutionary Relationships (Mi et al. 2021) (PANTHER), Princeton Protein Orthology Database (P-POD) (Heinicke et al. 2007), PFam (Finn et al. 2008), TreeFam (Schreiber et al. 2014), and PhyloFacts (Krishnamurthy et al. 2006) are primarily used for molecular epidemiology and phylogeny analysis of the viruses.

2.5.2 Epitope Recognition

Identifying epitopes has become an effective way to design new vaccines against viruses. Epitopes are part of antigens and are known as antigenic determinants that are recognized by the immune system of host cells and targeted via antibodies (T and B cells). Based on their conformation and paratope interactions, epitopes are divided into conformational and linear types. Linear are recognized by T cells, whereas conformational epitopes are recognized by B cells. Many software are available specifically for designing and prediction of epitopes. EPIMHC (Reche et al. 2005), FIMM (Schönbach et al. 2005), IEDB (Vita et al. 2019), MHCPEP (Brusic et al. 1998), SYFPEITHI (Rammensee et al. 1999), and MHCBN (Bhasin et al. 2003) are most commonly used software for the prediction of T-cell epitopes (Linear), whereas Bcipep, Epitome, and AntiJen are some of the databases used to predict B-cell epitopes (conformational) (Table 2.4). The peptide-specific database (MHCPEP) bonds only with the major histocompatibility complex (MHC class I and MHC II) molecules and specifically binds at the T-cell peptide recognition site. This database is a composite of around 13,000 peptide sequences. SYFPEITHI is a very widely used database. It contains a library of motifs related to individual peptides and MHC-based ligands from mice, primates, cattle, and other species besides humans. The FIMM database focuses on cellular immunology and contains all the data pertinent to molecular immunology. It also includes data on peptides, MHC-associated peptides, protein antigens, and the associated diseases. It is a popular database, providing output through lists and tables. The MHCBN database houses the recent information on T-cell epitopes for immunological research. The immune epitope database (IEBD) is an online tool focused on investigating and predicting T- and B-cell epitopes. Analysis and peptide EPIMHC aid MHC binding along with their interaction predictions. Conversely, the identification epitope for B cells with MHC binders (T-cell epitopes), evaluated via a server, is named Bcipep (Saha et al. 2005). Another tool is Epitome for predicting epigenetic events with multicell deep ensemble learning for the novel cell types. AntiJen (Toseland et al. 2005) software helps in vector classification and identification of amino acid propensity along with the prediction of B-cell epitope prediction.

2.5.3 Designing siRNA Antibody

RNA interference (RNAi) has evolved as an influential modality to obstruct virus infections. Many target-specific repositories with experimentally validated siRNA/shRNA results are available. It includes HIVsirDB (Tyagi et al. 2011), which contains information about the human immunodeficiency virus (HIV-1), and VIRsiRNAdb (Thakur et al. 2012) is full of critical experimental validations of other important human viruses. Besides these, siVirus is a web-based software mainly used to design siRNA antiviral against HIV-1, human coronavirus (HCV), influenza virus, and coronavirus (SARS). Another support vector machine (SVM) tool for the prediction of viral efficacy (siRNA) is VIRsiRNApred (Qureshi et al. 2013), which is easy to use as well. ViRome and Vistor are software created mainly for studying viral siRNA sequencing datasets (Watson et al. 2013). A Perl script tool named Paparazzi (Vodovar et al. 2011) was found beneficial for rebuilding the entire viral genome of a virus using viral-derived siRNAs.

Table 2.4 Epitope prediction tools

Tools	URL	Methods	MHC class
EpiDOCK	https://doi.org/10.1093/protein/gzt018	Structure-based	II
MotifScan	https://www.hiv.lanl.gov/content/immunology/motif_scan/motif_scan	Sequence motif	I, II
Rankpep	https://doi.org/10.1016/s0198-8859(02)00432-9	Motif matrix (MM)	I, II
MAPPP	https://bio.tools/mapp	MM	I
PREDIVAC	https://doi.org/10.1186/1471-2105-14-52	MM	II
SYFPEITHI	https://doi.org/10.1007/s002510050595	MM	I, II
PEPVAC	https://doi.org/10.1093/nar/gki357	MM	I
MHCPred 2.0	https://doi.org/10.2165/00822942-200605010-00008	Quantitative structure-activity relationship (QSAR)-based	I, II
EpiTOP	https://doi.org/10.1093/bioinformatics/btq324	QSAR-based	II
BIMAS	https://www-bimas.cit.nih.gov/molbio/hla_bind/	Quantitative affinity matrix (QAM)	I
TEPITOPE	https://doi.org/10.1371/journal.pone.0030483	QAM	II
ProPred	https://doi.org/10.1093/bioinformatics/17.12.1236	QAM	II
IL4pred	http://webs.iiitd.edu.in/raghava/il4pred/index.php	Support vector machine (SVM)	II
SVMHC	https://doi.org/10.1093/nar/gkl284	SVM	I, II
MULTIPRED2	https://doi.org/10.1016/j.jim.2010.11.009	Artificial neural network (ANN)	I, II
NetMHC-3.0	https://doi.org/10.1093/nar/gkn202	ANN	I
NetMHCIIpan-4.1	https://doi.org/10.1093/nar/gkaa379	ANN	II
NetCTL	https://doi.org/10.1186/1471-2105-8-424	ANN	I

2.5.4 Active/Binding Site Prediction

An active site is an area within the protein/enzyme where substrate/ligands bind through temporary bonds and undergo a chemical reaction. Active site prediction is one of the critical aspects of therapeutics against any viruses. Protein Data Bank (PDB) has a significant quantity of available protein structures, though it does not provide structural information of all the proteins. Any protein must undergo an expensive and time-consuming experimental, structural, and functional characterization process. Therefore, it is effective to anticipate the active site of proteins by using computational techniques. The prediction of any protein's active site residues is possible using various programmable tools and online resources. An online tool, PAR-3D (Goyal et al. 2007), estimates critical interacting residues (conserved/variable) present in the active site of a protein and compares the target protein functionally and structurally with similar protein families. COACH (Yang et al. 2013) is a meta server for predicting the protein-ligand binding site. COACH predicts complementary ligand binding sites using two related techniques, TM-SITE and S-SITE, concerning the provided protein structure. The TM-SITE helps to frame structural comparisons of binding pocket residues of template and query proteins, while the S-SITE gives results of binding-specific sequence profile alignment. Further, it also identifies the templates/reference for the ligand binding sites using BioLiP (Yang et al. n.d.) and helps to prepare protein databases with their functions via Iterative Threading ASSEmbly Refinement (I-TASSER) (Roy et al. 2010) protein structure, and their structure-based function annotations are also predicted. Using COFACTOR (Roy et al. 2012) web server, protein function annotations are automatically performed based on their structural comparison. To predict the active site on the surface of the protein, a meta server like FINDSITE (Brylinski and Skolnick 2011) and ConCavity (Capra et al. 2009) helps to allocate the binding site for the ligand after interaction analysis on the surface of the protein. There are many other web servers available for active site predictions, for example, ConSurf (Ashkenazy et al. 2016) (protein profile prediction based on evolutionarily conserved sequences) and CRESCENDO (Wang et al. 2015) (discovers both functional and structural constraints in evolution to forecast protein active sites). The computed atlas of surface topography of proteins (CASTp) (Tian et al. 2018) is the online available server to allocate both the topological and ordered protein structures and predict the internal cavities, cross channels (tunnels), as well as surface pockets of proteins. PocketFinder (Laurie and Jackson 2005), PocketPicker (Weisel et al. 2007), 3DLigandSite (Wass et al. 2010), FINDSITE, LIGSITE (Huang and Schroeder 2006), and metaPocket (Huang 2009) servers are also helpful for the active site prediction of a protein.

2.5.5 Protein-Protein Interaction (PPI) Site Prediction

Interactions between the proteins play vital roles in several biological and developmental processes, including analysis of the protein function for target protein and the binding ability of drug molecules, cell-to-cell communications, regulation of metabolic rates, and the developmental process control system. The classical and experiments-based investigations to determine PPIs are expensive and seek more time for validations. Therefore, many computer-based techniques, tools, and web servers have been designed to predict PPI sites. It includes cons-PPISP (Chen and Zhou 2005), HOMCOS (Kawabata 2016), HotPOINT (Tuncbag et al. 2010), ISIS (Ofran and Rost 2007), KFC server (Darnell et al. 2008), ODA (Fernandez-Recio et al. 2005), PINUP (Liang et al. 2006), and meta-PPISP (Qin and Zhou 2007) as shown in Table 2.5.

2.5.6 High-Throughput Virtual Screening and Molecular Docking

Drug discovery is a process to identify novel potential therapeutic compounds to cure the disease. This process combines bioinformatics and experiments-based translational and clinical approaches in which screening, identification of top hit compounds, and their synthesis, characterization, and development of therapeutic assays are involved. In silico high-throughput virtual screening, molecular docking, and different simulations studies are mainly engaged in drug discovery to identify potential hits from large libraries of compounds that can interact with the active site

Table 2.5 Protein-protein interaction site prediction tools

Tools	URL	Information available
cons-PPISP	https://doi.org/10.1002/prot.20514	Consensus neural network method
HOMCOS	https://doi.org/10.1093/nar/gkn218	Homology modelling of complex structures
HotPOINT	https://doi.org/10.1093/nar/gkq323	Prediction of PPIs using an empirical model
ISIS	https://doi.org/10.1093/bioinformatics/btl303	Predict PPIs from sequence
KFC server	https://doi.org/10.1093/nar/gkn346	Automated decision-tree approach
meta-PPISP	https://doi.org/10.1093/bioinformatics/btm434	A meta web server that is the combination of three separate web servers (Pronate, PINUP, and cons-PPISP)
ODA	https://doi.org/10.1002/prot.20285	Identification of optimal surface patches having the lowest docking de-solvation energy
PINUP	https://doi.org/10.1093/nar/gkl454	Empirical scoring function

pocket of proteins/enzymes. The two main approaches used for the virtual screening process are ligand-based and structure-based approaches (also known as docking). The ligand-based virtual screening comprises the well-known description of compounds. These descriptions help identify similar molecules using similar screening techniques from the pharmacophore libraries. The structure-based virtual screening is a promising and extensively exploited technique in drug discovery. This technique comprises a computational programme explored to screen the top hit molecules from different compound libraries and calculate the interaction of potential compounds with the active site residues of proteins. This technique uses one or more scoring functions to compute the binding interactions between the molecules and the active/binding site of the target protein.

Virtual screening is an indispensable tool for executing drug discovery as it is a high-speed and cost-effective technique. There are various tools and software available for ligand/structure-based virtual screening. The ligand-based virtual screening includes various tools and web servers such as ChemDes (Dong et al. 2015) (an online way to compute various molecular fingerprints and descriptors) and ChemMine tool (Backman et al. 2011), clustering small molecules based on their conformational resemblances and physicochemical properties. Other tools like macromolecule extended atom-pair fingerprint (MXFP) (Capecchi et al. 2019) are personalized to analyse the compounds based on their molecular size, shape, volume, and pharmacophores. The pepMMsMIMIC (Floris et al. 2011) is an online peptide-mimetic compounds virtual screening tool that defines the three-dimensional (3D)-similarity based on multi-conformers. The 3D-similarity search approach helps mimic the chemical structures of PPI inhibitory compounds using shape-similarity techniques and pharmacophore. PharmaGist (Schneidman-Duhovny et al. 2008) and Pharmit (Sunseri and Koes 2016) help in virtual screening and pharmacophore studies based detection of small molecules. Rchempp (Klambauer et al. 2015) helps design cyberspace drug candidates and chemical probes. SwissSimilarity (Zoete et al. 2016) tool includes the virtual screening of drug libraries based on 2D molecular fingerprints and 3D superimposition and non-superimposition methodology. Another ultrafast web server, Ultrafast Shape Recognition-Virtual Screening (USR-VS) (Li et al. 2016), uses shape recognition approaches for large-scale virtual screening. Library-specific database software is used for screening the available subclass of the ZINC database (Irwin and Shoichet 2005) named ZINCPharmer (Koes and Camacho 2012). DeepScreen (Liu et al. 2019) and EEPScreen employ freely accessible 2D structural depictions of compounds as input rather than conservative descriptors of restricted performance. DEEPScreen fundamentally has 2D depictions and, based on that, produces more accurate predictions. Ligand similarity using Clique Algorithm (LiSiCA) compares 2D and 3D relationships of target compounds with the reference compounds in ligand-based virtual screening. Similarly, the structure-based virtual screening includes various web servers and tools such as CaverWeb (Stourac et al. 2019), DOCK blaster (Irwin et al. 2009), DOCKovalent (London et al. 2014), DockThor (Guedes et al. 2021), EasyVS (Pires et al. 2020), e-LEA3D (Douguet 2010), ezCADD (Tao et al. 2019), iScreen (Tsai et al. 2011), MTiOpenScreen (Labbé et al.

2015), Pharmit (Sunseri and Koes 2016), Sanjeevini (Jayaram et al. 2012), Automatic Molecular Mechanics Optimization for in silico screening (AMMOS2) (Labbe et al. 2017), CSM–Lig (Pires and Ascher 2016), dockNmine (Gheyouche et al. 2019), MedusaDock (Wang and Dokholyan 2019), MoMA-LigPath (Devaurs et al. 2013), PRODIGY-LIG (Vangone et al. 2019), SwissDock (Grosdidier et al. 2011), PyRx (Dallakyan and Olson 2015), and many more. CaverWeb is a graphical, easy-to-use server used for in silico analysis of protein channels and tunnels. It also predicts the ligand transport pathways. A freely accessible facility for structure-based ligand screening includes the database DOCK blaster, while virtual screening of electrophilic small-molecule libraries against the target protein structure can be performed using DOCKovalent. This server helps to allocate the covalent bindings between the small molecules and the macromolecules. DockThor delivers ready-for-docking 3D configurations for wild-type proteins and carefully chosen mutations of the target proteins. It also helps in screening as well as docking of drug libraries. EasyVS is a web server built to verify and screening of drug libraries. It also provides analysis and visualization of docked structures. e-LEA3D is a de novo approach towards drug discovery and designing, which is performed to discover innovative ligands. It comprises both ligand- and structure-based assessments. The de novo method is a better substitute for the blind virtual screening of bigger compound libraries. MTiOpenScreen application comprises virtual screening with various chemical libraries on very exciting protein targets along with their PPI-based studies. Pharmit is a web server used to screen compounds from large compound databases, including their molecular shape, pharmacophore, and energy minimization studies. Sanjeevini provides accessible services for the discovery of target-directed lead molecules. Automatic Molecular Mechanics Optimization for in silico Screening (AMMOS2) executes the accuracy of protein-ligand connections with high ranks of elasticity of the protein receptor. It utilizes a physics-derived force field to predict the interactions. CSM-Lig personalized to calculate the binding efficacy of small-molecule protein complexes on the basis of complementarity in conformation and their chemistry using graph-based structural signs. dockNmine integrates multiple molecular docking energy scores and improves the interactions of protein-ligand complexes. MedusaDock is the web server that delivers a graphical edge for accurate and efficient protein-ligand molecular docking. Accompanied by the binding interactions, MoMA-LigPath provides ligand unbinding pathways to collect useful information about protein-ligand interactions. PROtein binDIng enerGY (PRODIGY-LIG) is the best server focused on determining binding affinities to predict the crystallographic complex structures. SwissDock performs docking studies and designs inhibitors for the target site as required. The PyRx tool is most commonly used for structure-based virtual screening. It collectively follows the algorithm of Open Babel, Autodock Vina, and Autodock Tools for drug library screening against the target site of the protein molecules. There are numerous examples of PyRx for virtual screening of drug libraries to discover probable antivirals against many viruses, such as alphaviruses (Aggarwal et al. 2015, 2017; Sharma et al. 2016, 2018; Puranik et al. 2019; Kumar et al. 2021b; Pareek et al. 2022; Mahajan et al. 2022), Flaviviruses (Bhutkar et al. 2023), coronaviruses (Ruchi et al. 2020; Choudhary et al. 2020; Kumar et al. 2021a; Sharma et al. 2022a; Dhaka et al. 2023; Choudhary et al. 2022; Rani et al. 2022), and many more.

2.5.7 *Quantitative Structure-Activity Relationship (QSAR) for Drug Discovery*

QSAR guides lead optimization and screens and enriches compounds based on their drug-likeness properties, removing non-viable candidates and predicting the toxic molecules. It utilizes mathematical modelling methods and identifies the correlations between structural characteristics and biological activities of chemical compounds. It is an essential computational method for rational drug-designing process. The fundamental principle of the structure-function relationship states that a slight change in structural conformation might alter the physicochemical properties and pharmacokinetics of the ligand. This method aims to improve the biological activities and physicochemical properties of existing principal molecules. The two datasets are required for QSAR-based studies, such as test and training datasets. The training dataset trains the models, whereas the test dataset helps assess the trained models. Other than models, five significant steps are also involved in this study. The primary crucial step includes generating 3D models based on the molecular structure of ligands. These 3D models are required for the calculation of geometric descriptors. The second important step leads to constructing the geometric descriptors. The third step helps select the most significant descriptors using feature selection approaches. The fourth essential step utilizes the best descriptor sets for designing QSAR-based models. The last and most crucial step approves the authentication of the model by calculating the activity of ligands with the training dataset, which helps to select the best-suited (best-fitted) model. The fitness level and performance of the model mainly depend on the two statistical measures such as R^2 and Q^2. R^2 is known as the coefficient of determination for multiple regression. R^2 should be high for a good-fit model. Q^2 is called a squared correlation coefficient and is used to measure the analytical capability and robustness of the model. The value of $R^2 - Q^2$ should be ≤ 0.3 for the best probability of the model. However, the robustness and predictability of the QSAR model depend not only on R^2 and Q^2 values but on other parameters also, such as chi-square (χ^2), root-mean-squared error (RMSE), slopes k and k' of the regression lines through the origin, correlation coefficient R between predicted and observed activities, and the chemical space of training and test sets. Thus, it can be stated that the QSAR technology is mainly used to build models to calculate the various biological events of chemical compounds.

There are various examples of the application of the QSAR technique for designing potential antivirals, such as compound derivatives against HIV-1 protease enzyme, nucleoside analogues with anti-HIV activity, nucleoside analogues to inhibit filo-, pneumo-, and paramyxoviruses and many more. Specific tools used to predict QSAR include multivariate image analysis-quantitative structure-activity relationship (MIA-QSAR) (Pereira et al. 2022), DPubChem (Soufan et al. 2018), Random Forest QSAR (RFQSAR) (Lee et al. 2017), QSAR-Co (Ambure et al. 2019), and Cloud 3D-QSAR (Wang et al. 2021). MIA-QSAR is a freely accessible method to predict interactions via 2D image analysis. For QSAR-based modelling and, after that, virtual screening of potential drug targets, DPubChem fulfils the demand. RFQSAR determines structure-based function targets that involve structure-activity relationships. QSAR-Co is an open software for developing classification-based QSAR multiple models to predict structure-function relationships. Cloud 3D-QSAR is a web-based principal investigator of structure-based activity relationship prediction models and helps in drug discovery quantitatively. It is helpful to search, design, and predict new molecule structures. Some detailed information about QSAR/machine learning-based software for drug discovery has been shown in Table 2.6.

2.5.8 *Other Popular Tools for Viruses*

Viruses have been lethal to humans for decades, and more than 220 virus species are known to infect humans. However, clinically approved antiviral drugs are commercially available against ten viruses only. Designing and developing an antiviral requires considering new and effective approaches that will help improve various parameters such as affinity, toxicity, pharmacology, and drug resistance captivity. Considering these points, in silico computational tools are widely used to predict the aforementioned parameters. Some of the most common databases and tools that are frequently used for therapeutics against many viruses are illustrated in Table 2.7.

2.6 Conclusion

The area of virology concerns the study of viruses, including their structures, replication mechanisms, and interactions with their hosts. However, due to the virus's complexity and the enormous quantity of data generated through research, it can be challenging to understand the underlying mechanisms of viral infections and develop effective treatments. Bioinformatics is a field that utilizes computational techniques and tools to evaluate large and complex biological datasets. By using bioinformatics approaches, researchers can analyse genomic data, protein structures, and other biological data to expand understanding into viruses' behaviour and identify potential drug targets. Databases and web servers are critical resources for virologists and bioinformaticians. These resources provide information on viral genomes, protein structures, and other aspects of viral biology. Some of the most widely used databases, web servers in virology, include NCBI's GenBank, UniProt, and the Protein Data Bank. The combined approach of virology and bioinformatics

Table 2.6 Tools for antiviral discovery using QSAR/machine learning approaches

Tools	URL	Information available
ChemSAR	https://doi.org/10.1186/s13321-017-0215-1	Machine learning used for QSAR model building
DPubChem	https://doi.org/10.1038/s41598-018-27495-x	Machine learning/QSAR
DeepScreening	https://doi.org/10.1093/database/baz104	Deep learning model construction
MLViS	https://doi.org/10.1371/journal.pone.0124600	Machine learning/QSAR
OCHEM	https://doi.org/10.1186/1758-2946-2-S1-P5	Machine learning/QSAR
ChemProt-3	https://doi.org/10.1093/database/bav123	QSAR models
DIA-DB	https://doi.org/10.1021/acs.jcim.0c00107	Structure similarity search against antidiabetic drugs
HitPickV2	https://doi.org/10.1093/bioinformatics/bty759	Specific target prediction using 2D similarity search
MuSSel	https://doi.org/10.1021/acs.jcim.8b00698	Specific target prediction using 2D similarity search
PPB2	https://doi.org/10.1021/acs.jcim.8b00524	Polypharmacology browser
RFQSAR	https://doi.org/10.6084/m9.figshare.c.3964425.v1	Ligand-based screening
SEA	https://doi.org/10.1038/nbt1284	Specific target prediction using 2D similarity search and statistical models used for score correction
SuperPred	https://doi.org/10.1093/nar/gku477	ATC code prediction and specific target prediction using 2D similarity search
SwissTargetPrediction	https://doi.org/10.1093/nar/gku293	Specific target prediction using both 2D and 3D similarity search
TargetNet	https://doi.org/10.1007/s10822-016-9915-2	Specific target prediction using QSAR model
WDL-RF	https://doi.org/10.1093/bioinformatics/bty070	Prediction of bioactivities against G protein-coupled receptors using machine learning approach

has enabled researchers to ascertain new drug targets for treating viral infections. Researchers can design drugs that target these specific molecules by using computational tools to analyse large datasets and identify key protein residues and pathways involved in viral infections. This method can cut the time and budget of drug discovery, allowing researchers to identify potential drug candidates more quickly and efficiently.

In conclusion, utilizing databases, web servers, and bioinformatics tools is a potent strategy that can aid in developing new antiviral therapeutics. These tools facilitate virologists and bioinformaticians in comprehending viral biology and

Table 2.7 Common tools for antiviral drug discovery

Tools	URL	Information available
ACFIS	https://doi.org/10.1093/nar/gkw393	Fragment-based drug discovery
CaverWeb	https://doi.org/10.1093/nar/gkz378	Docking in protein channels and tunnels using Vina
DOCK Blaster	https://doi.org/10.1021/jm9006966	DOCK 3.6 for docking
DOCKovalent	https://doi.org/10.1038/nchembio.1666	Covalent docking
e-LEA3D	https://doi.org/10.1093/nar/gkq322	Docking with PLANTS, de novo design
ezCADD	https://doi.org/10.1021/acs.jcim.8b00633	Docking with Smina or Vina, virtual screening. Protein-ligand interactions visualization using ezLigPlot
iScreen	https://doi.org/10.1007/s10822-011-9438-9	Docking with PLANTs
MTiOpenScree	https://doi.org/10.1093/nar/gkv306	Virtual screening with Autodock or Autodock Vina
Pharmit	https://doi.org/10.1093/nar/gkw287	Receptor-based pharmacophore search utility database
Sanjeevini	https://doi.org/10.1186/1471-2105-13-S17-S7	Virtual screening
AMMOS2	https://doi.org/10.1093/nar/gkx397	Rescoring and post-docking energy optimization
CompScore	https://doi.org/10.1021/acs.jcim.9b00343	Rescoring
CSM-Lig	https://doi.org/10.1093/nar/gkw390	Binding affinity prediction of a protein-molecule complex on the basis of structural signatures
dockNmine	https://doi.org/10.3390/ijms20205062	Assembly and analysis of virtual and experimental interface data
MedusaDock	https://doi.org/10.1021/ci8001167	Flexible docking with constraints
MoMA-LigPath	https://doi.org/10.1093/nar/gkt380	Ligand unbinding
nAPOLI	https://doi.org/10.1109/tcbb.2019.2892099	Analysis of protein-ligand interactions
PlayMolecule	https://doi.org/10.1021/acs.jcim.7b00190	Prepare proteins, machine learning, rescoring, pocket search, and interaction prediction of a ligand with a pathway
PRODIGY-LIG	https://doi.org/10.1093/bioinformatics/btw514	Binding affinity
Screening Explorer	https://doi.org/10.1021/acs.jcim.6b00283	Analysis of screening results, consensus scoring
SwissDock	https://doi.org/10.1093/nar/gkr366	Protein-ligand docking as well as prediction of advanced binding affinity

(continued)

Table 2.7 (continued)

Tools	URL	Information available
AnchorQuery	https://doi.org/10.1002/pro.3303	Receptor-based pharmacophore screening
farPPI	https://doi.org/10.1093/bioinformatics/bty879	Rescoring PPI inhibitors with MM/PB(GB) SA
ACID	https://doi.org/10.1186/s13321-019-0394-z	Strategy for consensus inverse docking. LEDOCK, docking with AutoDock Vina, PSOVina, and PLANTS
CRDS	https://doi.org/10.1093/bioinformatics/btz656	Consensus inverse docking
DIA-DB	https://doi.org/10.1021/acs.jcim.0c00107	Inverse docking with Vina
EDMON v3	http://atome.cbs.cnrs.fr/ATOME_V3/SERVER/EDMon_v3.html	Combine machine learning on the docking outputs of PLANTS and @TOME. Inverse screening
GOMoDo	https://doi.org/10.1371/journal.pone.0074092	Docking with AutoDock Vina or HADDOCK
GUT-DOCK	http://gut-dock.miningmembrane.com/	Docking with AutoDock Vina.
idTarget	https://doi.org/10.1093/nar/gks496	Inverse docking with MEDock

identifying novel therapeutic targets, thereby advancing the development of cures for viral diseases.

Acknowledgments The authors duly acknowledge the financial support from the Science and Engineering Research Board (SERB), Department of Science and Technology (DST), and Government of India under the scheme Intensification of Research in High Priority Area (IRHPA) (Project no. IPA/2020/000054). The authors also thank the DST-sponsored Translation and Structural Bioinformatics Center at Department of Biosciences and Bioengineering (BSBE), IIT Roorkee (reference number BT/PR40141/BTIS/137/16/2021).

References

Agarwala R, Barrett T, Beck J, Benson DA, Bollin C, Bolton E, Bourexis D, Brister JR, Bryant SH, Canese K, Charowhas C, Clark K, Dicuccio M, Dondoshansky I, Federhen S, Feolo M, Funk K, Geer LY, Gorelenkov V, Hoeppner M, Holmes B, Johnson M, Khotomlianski V, Kimchi A, Kimelman M, Kitts P, Klimke W, Krasnov S, Kuznetsov A, Landrum MJ, Landsman D, Lee JM, Lipman DJ, Lu Z, Madden TL, Madej T, Marchler-Bauer A, Karsch-Mizrachi I, Murphy T, Orris R, Ostell J, O'sullivan C, Panchenko A, Phan L, Preuss D, Pruitt KD, Rodarmer K, Rubinstein W, Sayers E, Schneider V, Schuler GD, Sherry ST, Sirotkin K, Siyan K, Slotta D, Soboleva A, Soussov V, Starchenko G, Tatusova TA, Todorov K, Trawick BW, Vakatov D, Wang Y, Ward M, Wilbur WJ, Yaschenko E, Zbicz K (2016) Database resources of the National Center for Biotechnology Information. Nucleic Acids Res 44:D7–D19. https://doi.org/10.1093/NAR/GKV1290

Aggarwal M, Sharma R, Kumar P, Parida M, Tomar S (2015) Kinetic characterization of trans-proteolytic activity of Chikungunya virus capsid protease and development of a FRET-based HTS assay. Sci Rep 5:14753. https://doi.org/10.1038/srep14753

Aggarwal M, Kaur R, Saha A, Mudgal R, Yadav R, Dash PK, Parida M, Kumar P, Tomar S (2017) Evaluation of antiviral activity of piperazine against Chikungunya virus targeting hydrophobic pocket of alphavirus capsid protein. Antiviral Res 146:102–111. https://doi.org/10.1016/J.ANTIVIRAL.2017.08.015

Aksamentov I, Roemer C, Hodcroft EB, Neher RA (2021) Nextclade: clade assignment, mutation calling and quality control for viral genomes. J Open Source Softw 6:3773. https://doi.org/10.21105/JOSS.03773

Altschul SF, Gish W, Miller W, Myers EW, Lipman DJ (1990) Basic local alignment search tool. J Mol Biol 215:403–410. https://doi.org/10.1016/S0022-2836(05)80360-2

Ambure P, Halder AK, González Díaz H, Cordeiro MNDS (2019) QSAR-Co: an open source software for developing robust multitasking or multitarget classification-based QSAR models. J Chem Inf Model 59(6):2538–2544. https://doi.org/10.1021/acs.jcim.9b00295

Ashkenazy H, Abadi S, Martz E, Chay O, Mayrose I, Pupko T, Ben-Tal N (2016) ConSurf 2016: an improved methodology to estimate and visualize evolutionary conservation in macromolecules. Nucleic Acids Res 44:W344. https://doi.org/10.1093/NAR/GKW408

Backman TWH, Cao Y, Girke T (2011) ChemMine tools: an online service for analyzing and clustering small molecules. Nucleic Acids Res 39:W486–W491. https://doi.org/10.1093/NAR/GKR320

Bateman A, Martin MJ, O'Donovan C, Magrane M, Apweiler R, Alpi E, Antunes R, Arganiska J, Bely B, Bingley M, Bonilla C, Britto R, Bursteinas B, Chavali G, Cibrian-Uhalte E, Da Silva A, De Giorgi M, Dogan T, Fazzini F, Gane P, Castro LG, Garmiri P, Hatton-Ellis E, Hieta R, Huntley R, Legge D, Liu W, Luo J, Macdougall A, Mutowo P, Nightingale A, Orchard S, Pichler K, Poggioli D, Pundir S, Pureza L, Qi G, Rosanoff S, Saidi R, Sawford T, Shypitsyna A, Turner E, Volynkin V, Wardell T, Watkins X, Zellner H, Cowley A, Figueira L, Li W, McWilliam H, Lopez R, Xenarios I, Bougueleret L, Bridge A, Poux S, Redaschi N, Aimo L, Argoud-Puy G, Auchincloss A, Axelsen K, Bansal P, Baratin D, Blatter MC, Boeckmann B, Bolleman J, Boutet E, Breuza L, Casal-Casas C, De Castro E, Coudert E, Cuche B, Doche M, Dornevil D, Duvaud S, Estreicher A, Famiglietti L, Feuermann M, Gasteiger E, Gehant S, Gerritsen V, Gos A, Gruaz-Gumowski N, Hinz U, Hulo C, Jungo F, Keller G, Lara V, Lemercier P, Lieberherr D, Lombardot T, Martin X, Masson P, Morgat A, Neto T, Nouspikel N, Paesano S, Pedruzzi I, Pilbout S, Pozzato M, Pruess M, Rivoire C, Roechert B, Schneider M, Sigrist C, Sonesson K, Staehli S, Stutz A, Sundaram S, Tognolli M, Verbregue L, Veuthey AL, Wu CH, Arighi CN, Arminski L, Chen C, Chen Y, Garavelli JS, Huang H, Laiho K, McGarvey P, Natale DA, Suzek BE, Vinayaka CR, Wang Q, Wang Y, Yeh LS, Yerramalla MS, Zhang J (2015) UniProt: a hub for protein information. Nucleic Acids Res 43:D204–D212. https://doi.org/10.1093/NAR/GKU989

Benson DA, Karsch-Mizrachi I, Lipman DJ, Ostell J, Wheeler DL (2007) GenBank. Nucleic Acids Res 35:D21–D25. https://doi.org/10.1093/NAR/GKL986

Berman HM, Westbrook J, Feng Z, Gilliland G, Bhat TN, Weissig H, Shindyalov IN, Bourne PE (2000) The Protein Data Bank. Nucleic Acids Res 28:235–242. https://doi.org/10.1093/NAR/28.1.235

Bhasin M, Singh H, Raghava GPS (2003) MHCBN: a comprehensive database of MHC binding and non-binding peptides. Bioinformatics 19:665–666. https://doi.org/10.1093/BIOINFORMATICS/BTG055

Bhutkar M, Singh V, Dhaka P, Tomar S (2022) Virus-host protein-protein interactions as molecular drug targets for arboviral infections. Front Virol 2:70. https://doi.org/10.3389/FVIRO.2022.959586

Bhutkar M, Ruchi R, Kothiala A, Mahajan S, Waghmode B, Kumar R, Singh V, Sircar D, Kumar P, Tomar S (2023) Elucidation of antiviral mechanisms of natural therapeutic mol-

ecules Herbacetin and Caffeic acid phenethyl ester against chikungunya and dengue virus. bioRxiv:2022.05.31.494145. https://doi.org/10.1101/2022.05.31.494145
Bose ME, Littrell JC, Patzer AD, Kraft AJ, Metallo JA, Fan J, Henrickson KJ (2008) The Influenza Primer Design Resource: a new tool for translating influenza sequence data into effective diagnostics. Influenza Other Respi Viruses 2:23. https://doi.org/10.1111/J.1750-2659.2007.00031.X
Braun BA, Schein CH, Braun W (2021) DGraph clusters flaviviruses and β-coronaviruses according to their hosts, disease type, and human cell receptors. Bioinform Biol Insights 15:1–9. https://doi.org/10.1177/11779322211020316
Brusic V, Rudy G, Harrison LC (1998) MHCPEP, a database of MHC-binding peptides: update 1997. Nucleic Acids Res 26:368–371. https://doi.org/10.1093/NAR/26.1.368
Brylinski M, Skolnick J (2011) FINDSITE-metal: integrating evolutionary information and machine learning for structure-based metal binding site prediction at the proteome level. Proteins 79:735. https://doi.org/10.1002/PROT.22913
Capecchi A, Awale M, Probst D, Reymond JL (2019) PubChem and ChEMBL beyond Lipinski. Mol Inform 38. https://doi.org/10.1002/MINF.201900016
Capra JA, Laskowski RA, Thornton JM, Singh M, Funkhouser TA (2009) Predicting protein ligand binding sites by combining evolutionary sequence conservation and 3D structure. PLOS Comput Biol 5:e1000585. https://doi.org/10.1371/JOURNAL.PCBI.1000585
Chen H, Zhou HX (2005) Prediction of interface residues in protein-protein complexes by a consensus neural network method: test against NMR data. Proteins Struct Funct Genet 61:21–35. https://doi.org/10.1002/PROT.20514
Cheng X, Xiao X, Chou KC (2017) pLoc-mVirus: predict subcellular localization of multi-location virus proteins via incorporating the optimal GO information into general PseAAC. Gene 628:315–321. https://doi.org/10.1016/J.GENE.2017.07.036
Choudhary S, Malik YS, Tomar S (2020) Identification of SARS-CoV-2 cell entry inhibitors by drug repurposing using in silico structure-based virtual screening approach. Front Immunol 11:1664. https://doi.org/10.3389/fimmu.2020.01664
Choudhary S, Nehul S, Amith Kumar K, Sharma S, Rani R, Saha A, Kumar Sharma G, Tomar S, Kumar P, Kumar Sharma Senior scientist G (2022) Crystal structure and activity profiling of deubiquitinating inhibitors-bound to SARS-CoV-2 papain like protease revealed new allosteric sites for antiviral therapies. bioRxiv:2022.11.11.516107. https://doi.org/10.1101/2022.11.11.516107
Cook HV, Doncheva NT, Szklarczyk D, von Mering C, Jensen LJ (2018) Viruses.STRING: a virus-host protein-protein interaction database. Viruses 10. https://doi.org/10.3390/V10100519
Cossart P, Helenius A (2014) Endocytosis of viruses and bacteria. Cold Spring Harb Perspect Biol 6. https://doi.org/10.1101/cshperspect.a016972
Dallakyan S, Olson AJ (2015) Small-molecule library screening by docking with PyRx. Methods Mol Biol 1263:243–250. https://doi.org/10.1007/978-1-4939-2269-7_19
Darnell SJ, LeGault L, Mitchell JC (2008) KFC Server: interactive forecasting of protein interaction hot spots. Nucleic Acids Res 36:W265. https://doi.org/10.1093/NAR/GKN346
Devaurs D, Bouard L, Vaisset M, Zanon C, Al-Bluwi I, Iehl R, Siméon T, Cortés J (2013) MoMA-LigPath: a web server to simulate protein–ligand unbinding. Nucleic Acids Res 41:W297–W302. https://doi.org/10.1093/NAR/GKT380
Dhaka P, Singh A, Choudhary S, Peddinti RK, Kumar P, Sharma GK, Tomar S (2023) Mechanistic and thermodynamic characterization of antiviral inhibitors targeting nucleocapsid N-terminal domain of SARS-CoV-2. Arch Biochem Biophys 750:109820. https://doi.org/10.1016/j.abb.2023.109820
Dong J, Cao DS, Miao HY, Liu S, Deng BC, Yun YH, Wang NN, Lu AP, Bin ZW, Chen AF (2015) ChemDes: an integrated web-based platform for molecular descriptor and fingerprint computation. J Chem 7(1):1–10. https://doi.org/10.1186/s13321-015-0109-z
Douguet D (2010) e-LEA3D: a computational-aided drug design web server. Nucleic Acids Res 38:W615–W621. https://doi.org/10.1093/NAR/GKQ322

Fernandez-Recio J, Totrov M, Skorodumov C, Abagyan R (2005) Optimal docking area: a new method for predicting protein-protein interaction sites. Proteins 58:134–143. https://doi.org/10.1002/PROT.20285

Finn RD, Tate J, Mistry J, Coggill PC, Sammut SJ, Hotz HR, Ceric G, Forslund K, Eddy SR, Sonnhammer ELL, Bateman A (2008) The Pfam protein families database. Nucleic Acids Res 36:D281. https://doi.org/10.1093/NAR/GKM960

Floris M, Masciocchi J, Fanton M, Moro S (2011) Swimming into peptidomimetic chemical space using pepMMsMIMIC. Nucleic Acids Res 39. https://doi.org/10.1093/NAR/GKR287

Geoghegan JL, Duchêne S, Holmes EC (2017) Comparative analysis estimates the relative frequencies of co-divergence and cross-species transmission within viral families. PLoS Pathog 13. https://doi.org/10.1371/JOURNAL.PPAT.1006215

Gheyouche E, Launay R, Lethiec J, Labeeuw A, Roze C, Amossé A, Téletchéa S (2019) DockNmine, a web portal to assemble and analyse virtual and experimental interaction data. Int J Mol Sci 20:5062. https://doi.org/10.3390/IJMS20205062

Goyal K, Mohanty D, Mande SC (2007) PAR-3D: a server to predict protein active site residues. Nucleic Acids Res 35:W503. https://doi.org/10.1093/NAR/GKM252

Grosdidier A, Zoete V, Michielin O (2011) SwissDock, a protein-small molecule docking web service based on EADock DSS. Nucleic Acids Res 39:W270–W277. https://doi.org/10.1093/NAR/GKR366

Guedes IA, Barreto AMS, Marinho D, Krempser E, Kuenemann MA, Sperandio O, Dardenne LE, Miteva MA (2021) New machine learning and physics-based scoring functions for drug discovery. Sci Rep 111(11):1–19. https://doi.org/10.1038/s41598-021-82410-1

Gupta AK, Kumar A, Rajput A, Kaur K, Dar SA, Thakur A, Megha K, Kumar M (2020) NipahVR: a resource of multi-targeted putative therapeutics and epitopes for the Nipah virus. Database (Oxford) 2020. https://doi.org/10.1093/DATABASE/BAZ159

Hadfield J, Megill C, Bell SM, Huddleston J, Potter B, Callender C, Sagulenko P, Bedford T, Neher RA (2018) Nextstrain: real-time tracking of pathogen evolution. Bioinformatics 34:4121–4123. https://doi.org/10.1093/BIOINFORMATICS/BTY407

Heinicke S, Livstone MS, Lu C, Oughtred R, Kang F, Angiuoli SV, White O, Botstein D, Dolinski K (2007) The Princeton Protein Orthology Database (P-POD): a comparative genomics analysis tool for biologists. PLoS One 2. https://doi.org/10.1371/JOURNAL.PONE.0000766

Hiscock D, Upton C (2000) Viral Genome DataBase: storing and analyzing genes and proteins from complete viral genomes. Bioinformatics 16:484–485. https://doi.org/10.1093/BIOINFORMATICS/16.5.484

Holtzapple E, Telmer CA, Miskov-Zivanov N (2020) FLUTE: fast and reliable knowledge retrieval from biomedical literature. Database 2020:56. https://doi.org/10.1093/DATABASE/BAAA056

Hong JJ, Wu TY, Chang TY, Chen CY (2013) Viral IRES prediction system – a web server for prediction of the IRES secondary structure in silico. PLoS One 8:79288. https://doi.org/10.1371/JOURNAL.PONE.0079288

Hsu PWC, Lin LZ, Da Hsu S, Hsu JBK, Da Huang H (2007) ViTa: prediction of host microRNAs targets on viruses. Nucleic Acids Res 35:D381. https://doi.org/10.1093/NAR/GKL1009

Huang B (2009) Metapocket: a meta approach to improve protein ligand binding site prediction. OMICS J Integr Biol 13(4):325–330. https://doi.org/10.1089/omi.2009.0045

Huang B, Schroeder M (2006) LIGSITEcsc: predicting ligand binding sites using the Connolly surface and degree of conservation. BMC Struct Biol 6(1):1–11. https://doi.org/10.1186/1472-6807-6-19

Hulo C, De Castro E, Masson P, Bougueleret L, Bairoch A, Xenarios I, Le Mercier P (2011) ViralZone: a knowledge resource to understand virus diversity. Nucleic Acids Res 39. https://doi.org/10.1093/NAR/GKQ901

Irwin JJ, Shoichet BK (2005) ZINC-A free database of commercially available compounds for virtual screening. https://doi.org/10.1021/CI049714

Irwin JJ, Shoichet BK, Mysinger MM, Huang N, Colizzi F, Wassam P, Cao Y (2009) Automated docking screens: a feasibility study. J Med Chem 52(18):5712–5720. https://doi.org/10.1021/jm9006966

Jayaram B, Singh T, Mukherjee G, Mathur A, Shekhar S, Shekhar V (2012) Sanjeevini: a freely accessible web-server for target directed lead molecule discovery. BMC Bioinf 13(S17):S7. https://doi.org/10.1186/1471-2105-13-S17-S7

Kausar S, Said Khan F, Ishaq Mujeeb Ur Rehman M, Akram M, Riaz M, Rasool G, Hamid Khan A, Saleem I, Shamim S, Malik A (2021) A review: mechanism of action of antiviral drugs. Int J Immunopathol Pharmacol 35:1–12. https://doi.org/10.1177/20587384211002621

Kawabata T (2016) HOMCOS: an updated server to search and model complex 3D structures. J Struct Funct Genom 17(4):83–99. https://doi.org/10.1007/s10969-016-9208-y

Klambauer G, Wischenbart M, Mahr M, Unterthiner T, Mayr A, Hochreiter S (2015) Rchemcpp: a web service for structural analoging in ChEMBL, Drugbank and the Connectivity Map. Bioinformatics 31:3392–3394. https://doi.org/10.1093/BIOINFORMATICS/BTV373

Koes DR, Camacho CJ (2012) ZINCPharmer: pharmacophore search of the ZINC database. Nucleic Acids Res 40. https://doi.org/10.1093/NAR/GKS378

Krishnamurthy N, Brown DP, Kirshner D, Sjölander K (2006) PhyloFacts: an online structural phylogenomic encyclopedia for protein functional and structural classification. Genome Biol 7(9):1–17. https://doi.org/10.1186/gb-2006-7-9-r83

Kulkarni-Kale U, Bhosle S, Manjari GS, Kolaskar AS (2004) VirGen: a comprehensive viral genome resource. Nucleic Acids Res 32:D289. https://doi.org/10.1093/NAR/GKH098

Kumar KA, Sharma M, Dalal V, Singh V, Tomar S, Kumar P (2021a) Multifunctional inhibitors of SARS-CoV-2 by MM/PBSA, essential dynamics, and molecular dynamic investigations. J Mol Graph Model 107:107969. https://doi.org/10.1016/J.JMGM.2021.107969

Kumar R, Nehul S, Singh A, Tomar S (2021b) Identification and evaluation of antiviral potential of thymoquinone, a natural compound targeting Chikungunya virus capsid protein. Virology 561:36–46. https://doi.org/10.1016/J.VIROL.2021.05.013

Labbé CM, Rey J, Lagorce D, Vavruša M, Becot J, Sperandio O, Villoutreix BO, Tufféry P, Miteva MA (2015) MTiOpenScreen: a web server for structure-based virtual screening. Nucleic Acids Res 43:W448–W454. https://doi.org/10.1093/NAR/GKV306

Labbe CM, Pencheva T, Jereva D, Desvillechabrol D, Becot J, Villoutreix BO, Pajeva I, Miteva MA (2017) AMMOS2: a web server for protein–ligand–water complexes refinement via molecular mechanics. Nucleic Acids Res 45:W350–W355. https://doi.org/10.1093/NAR/GKX397

Laurie ATR, Jackson RM (2005) Q-SiteFinder: an energy-based method for the prediction of protein–ligand binding sites. Bioinformatics 21:1908–1916. https://doi.org/10.1093/BIOINFORMATICS/BTI315

Lawrence J (2002) DengueNet – WHO's internet based system for the global surveillance of dengue fever and dengue haemorrhagic fever. Wkly Releases 6:1883. https://doi.org/10.2807/ESW.06.39.01883-EN

Lee K, Lee M, Kim D (2017) Utilizing random Forest QSAR models with optimized parameters for target identification and its application to target-fishing server. BMC Bioinform 18:20–22. https://doi.org/10.1186/S12859-017-1960-X

Lefkowitz EJ, Dempsey DM, Hendrickson RC, Orton RJ, Siddell SG, Smith DB (2018) Virus taxonomy: the database of the International Committee on Taxonomy of Viruses (ICTV). Nucleic Acids Res 46:D708–D717. https://doi.org/10.1093/NAR/GKX932

Li H, Leung KS, Wong MH, Ballester PJ (2016) USR-VS: a web server for large-scale prospective virtual screening using ultrafast shape recognition techniques. Nucleic Acids Res 44:W436–W441. https://doi.org/10.1093/NAR/GKW320

Liang S, Zhang C, Liu S, Zhou Y (2006) Protein binding site prediction using an empirical scoring function. Nucleic Acids Res 34:3698. https://doi.org/10.1093/NAR/GKL454

Liao YC, Ko CY, Tsai MH, Lee MS, Hsiung CA (2009) ATIVS: analytical tool for influenza virus surveillance. Nucleic Acids Res 37. https://doi.org/10.1093/NAR/GKP321

Liechti R, Gleizes A, Kuznetsov D, Bougueleret L, Le Mercier P, Bairoch A, Xenarios I (2010) OpenFluDB, a database for human and animal influenza virus. Database 2010. https://doi.org/10.1093/DATABASE/BAQ004

Lim A, Zhang L (1999) WebPHYLIP: a web interface to PHYLIP. Bioinformatics 15:1068–1069. https://doi.org/10.1093/BIOINFORMATICS/15.12.1068

Lin AEJ, Guttman JA (2010) Hijacking the endocytic machinery by microbial pathogens. Protoplasma 244:75–90. https://doi.org/10.1007/S00709-010-0164-2

Liu Z, Du J, Fang J, Yin Y, Xu G, Xie L (2019) DeepScreening: a deep learning-based screening web server for accelerating drug discovery. Database 2019. https://doi.org/10.1093/DATABASE/BAZ104

London N, Miller RM, Krishnan S, Uchida K, Irwin JJ, Eidam O, Gibold L, Cimermančič P, Bonnet R, Shoichet BK, Taunton J (2014) Covalent docking of large libraries for the discovery of chemical probes. Nat Chem Biol 1012(10):1066–1072. https://doi.org/10.1038/nchembio.1666

Lu G, Rowley T, Garten R, Donis RO (2007) FluGenome: a web tool for genotyping influenza A virus. Nucleic Acids Res 35:W275. https://doi.org/10.1093/NAR/GKM365

Mackenzie JS, Drury P, Arthur RR, Ryan MJ, Grein T, Slattery R, Suri S, Domingo CT, Bejtullahu A (2014) The global outbreak alert and response network. Glob Public Health 9:1023. https://doi.org/10.1080/17441692.2014.951870

Mahajan S, Kumar R, Singh A, Pareek A, Long S, McInerney G, Tomar S (2022) Targeting host protein G3BP1 for the discovery of novel antiviral inhibitors against the Chikungunya virus. bioRxiv:2022.11.11.516135. https://doi.org/10.1101/2022.11.11.516135

Mi H, Ebert D, Muruganujan A, Mills C, Albou LP, Mushayamaha T, Thomas PD (2021) PANTHER version 16: a revised family classification, tree-based classification tool, enhancer regions and extensive API. Nucleic Acids Res 49:D394–D403. https://doi.org/10.1093/NAR/GKAA1106

Misra M, Schein CH (2007) Flavitrack: an annotated database of flavivirus sequences. Bioinformatics 23:2645–2647. https://doi.org/10.1093/BIOINFORMATICS/BTM383

Mushegian AR (2020) Are there 10^{31} virus particles on earth, or more, or fewer? J Bacteriol 202. https://doi.org/10.1128/JB.00052-20

Nagarajan N, Kingsford C (2011) GiRaF: robust, computational identification of influenza reassortments via graph mining. Nucleic Acids Res 39:e34. https://doi.org/10.1093/NAR/GKQ1232

Ofran Y, Rost B (2007) ISIS: interaction sites identified from sequence. Bioinformatics 23. https://doi.org/10.1093/BIOINFORMATICS/BTL303

Olsen LR, Zhang GL, Reinherz EL, Brusic V (2011) FLAVIdB: a data mining system for knowledge discovery in flaviviruses with direct applications in immunology and vaccinology. Immunome Res. https://pubmed.ncbi.nlm.nih.gov/25544857/

Palmeira L, Penel S, Lotteau V, Rabourdin-Combe C, Gautier C (2011) PhEVER: a database for the global exploration of virus–host evolutionary relationships. Nucleic Acids Res 39:D569. https://doi.org/10.1093/NAR/GKQ1013

Pareek A, Kumar R, Mudgal R, Neetu N, Sharma M, Kumar P, Tomar S (2022) Alphavirus antivirals targeting RNA-dependent RNA polymerase domain of nsP4 divulged using surface plasmon resonance. FEBS J 289:4901–4924. https://doi.org/10.1111/FEBS.16397

Pasznik P, Rutkowska E, Niewieczerzal S, Cielecka-Piontek J, Latek D (2019) Potential off-target effects of beta-blockers on gut hormone receptors: in silico study including GUT-DOCK—A web service for small-molecule docking. PLoS One 14:e0210705. https://doi.org/10.1371/JOURNAL.PONE.0210705

Pereira IV, Daré JK, da Cunha EFF, Freitas MP (2022) MIA-QSAR study of the structural merging of (thio)benzamide herbicides with photosynthetic system II inhibitory activities. J Biomol Struct Dyn. https://doi.org/10.1080/07391102.2022.2055649

Pickett BE, Greer DS, Zhang Y, Stewart L, Zhou L, Sun G, Gu Z, Kumar S, Zaremba S, Larsen CN, Jen W, Klem EB, Scheuermann RH (2012a) Virus pathogen Database and Analysis Resource (ViPR): a comprehensive bioinformatics Database and Analysis Resource for the Coronavirus research community. Viruses 4:3209–3226. https://doi.org/10.3390/v4113209

Pickett BE, Sadat EL, Zhang Y, Noronha JM, Squires RB, Hunt V, Liu M, Kumar S, Zaremba S, Gu Z, Zhou L, Larson CN, Dietrich J, Klem EB, Scheuermann RH (2012b) ViPR: an open bioinformatics database and analysis resource for virology research. Nucleic Acids Res 40:D593–D598. https://doi.org/10.1093/nar/gkr859

Pires DEV, Ascher DB (2016) CSM-lig: a web server for assessing and comparing protein–small molecule affinities. Nucleic Acids Res 44:W557. https://doi.org/10.1093/NAR/GKW390

Pires DEV, Veloso WNP, Myung YC, Rodrigues CHM, Silk M, Rezende PM, Silva F, Xavier JS, Velloso JPL, Da Silveira CH, Ascher DB (2020) EasyVS: a user-friendly web-based tool for molecule library selection and structure-based virtual screening. Bioinformatics 36:4200–4202. https://doi.org/10.1093/BIOINFORMATICS/BTAA480

Puranik NV, Rani R, Singh VA, Tomar S, Puntambekar HM, Srivastava P (2019) Evaluation of the antiviral potential of halogenated dihydrorugosaflavonoids and molecular modeling with nsp3 protein of chikungunya virus (chikv). ACS Omega 4(23):20335–20345. https://doi.org/10.1021/acsomega.9b02900

Qin S, Zhou HX (2007) meta-PPISP: a meta web server for protein-protein interaction site prediction. Bioinformatics 23:3386–3387. https://doi.org/10.1093/BIOINFORMATICS/BTM434

Qureshi A, Thakur N, Kumar M (2013) VIRsiRNApred: a web server for predicting inhibition efficacy of siRNAs targeting human viruses. J Transl Med 11. https://doi.org/10.1186/1479-5876-11-305

Rajput A, Kumar A, Kumar M (2018) anti-Nipah: a QSAR based prediction method to identify the inhibitors against Nipah virus. https://doi.org/10.20944/PREPRINTS201810.0103.V1

Rammensee H, Bachmann J, Emmerich NP, Bachor OA, Stevanović S (1999) SYFPEITHI: database for MHC ligands and peptide motifs. Immunogenetics 50(3-4):213–219. https://doi.org/10.1007/s002510050595

Rani R, Long S, Pareek A, Dhaka P, Singh A, Kumar P, McInerney G, Tomar S (2022) Multi-target direct-acting SARS-CoV-2 antivirals against the nucleotide-binding pockets of virus-specific proteins. Virology 577:1–15. https://doi.org/10.1016/J.VIROL.2022.08.008

Reche PA, Zhang H, Glutting JP, Reinherz EL (2005) EPIMHC: a curated database of MHC-binding peptides for customized computational vaccinology. Bioinformatics 21:2140–2141. https://doi.org/10.1093/BIOINFORMATICS/BTI269

Resch W, Zaslavsky L, Kiryutin B, Rozanov M, Bao Y, Tatusova TA (2009) Virus variation resources at the National Center for Biotechnology Information: Dengue virus. BMC Microbiol 9(1):1–7. https://doi.org/10.1186/1471-2180-9-65

Rost B, Yachdav G, Liu J (2004) The PredictProtein server. Nucleic Acids Res 32:W321–W326. https://doi.org/10.1093/NAR/GKH377

Roy A, Kucukural A, Zhang Y (2010) I-TASSER: a unified platform for automated protein structure and function prediction. Nat Protoc 54(5):725–738. https://doi.org/10.1038/nprot.2010.5

Roy A, Yang J, Zhang Y (2012) COFACTOR: an accurate comparative algorithm for structure-based protein function annotation. Nucleic Acids Res 40:W471. https://doi.org/10.1093/NAR/GKS372

Ruchi R, Singh A, Pareek A, Tomar S (2020) In silico guided drug repurposing to combat SARS-CoV-2 by targeting Mpro, the key virus specific protease. https://doi.org/10.26434/CHEMRXIV.12030345.V1

Saha S, Bhasin M, Raghava GPS (2005) Bcipep: a database of B-cell epitopes. BMC Genom 6. https://doi.org/10.1186/1471-2164-6-79

Schneidman-Duhovny D, Dror O, Inbar Y, Nussinov R, Wolfson HJ (2008) PharmaGist: a webserver for ligand-based pharmacophore detection. Nucleic Acids Res 36. https://doi.org/10.1093/NAR/GKN187

Schönbach C, Koh JLY, Flower DR, Brusic V (2005) An update on the functional molecular immunology (FIMM) database. Appl Bioinform 4:25–31. https://doi.org/10.2165/00822942-200504010-00003

Schreiber F, Patricio M, Muffato M, Pignatelli M, Bateman A (2014) TreeFam v9: a new website, more species and orthology-on-the-fly. Nucleic Acids Res 42:D922. https://doi.org/10.1093/NAR/GKT1055

Sharma R, Fatma B, Saha A, Bajpai S, Sistla S, Dash PK, Parida M, Kumar P, Tomar S (2016) Inhibition of chikungunya virus by picolinate that targets viral capsid protein. Virology 498:265–276. https://doi.org/10.1016/J.VIROL.2016.08.029

Sharma R, Kesari P, Kumar P, Tomar S (2018) Structure-function insights into chikungunya virus capsid protein: small molecules targeting capsid hydrophobic pocket. Virology 515:223–234. https://doi.org/10.1016/J.VIROL.2017.12.020

Sharma M, Mahto JK, Dhaka P, Neetu N, Tomar S, Kumar P (2022a) MD simulation and MM/PBSA identifies phytochemicals as bifunctional inhibitors of SARS-CoV-2. J Biomol Struct Dyn 40(22):12048–12061. https://doi.org/10.1080/07391102.2021.1969285

Sharma R, Shrivastava S, Singh SK, Kumar A, Singh AK, Saxena S (2022b) Deep-AVPpred: artificial intelligence driven discovery of peptide drugs for viral infections. IEEE J Biomed Heal Inform 26:5067–5074. https://doi.org/10.1109/JBHI.2021.3130825

Shu Y, McCauley J (2017) GISAID: global initiative on sharing all influenza data – from vision to reality. Eurosurveillance 22:1. https://doi.org/10.2807/1560-7917.ES.2017.22.13.30494

Soufan O, Ba-Alawi W, Magana-Mora A, Essack M, Bajic VB (2018) DPubChem: a web tool for QSAR modeling and high-throughput virtual screening. Sci Rep 81(8):1–10. https://doi.org/10.1038/s41598-018-27495-x

Squires RB, Noronha J, Hunt V, García-Sastre A, Macken C, Baumgarth N, Suarez D, Pickett BE, Zhang Y, Larsen CN, Ramsey A, Zhou L, Zaremba S, Kumar S, Deitrich J, Klem E, Scheuermann RH (2012) Influenza Research Database: an integrated bioinformatics resource for influenza research and surveillance. Influenza Other Respi Viruses 6:404. https://doi.org/10.1111/J.1750-2659.2011.00331.X

Stourac J, Vavra O, Kokkonen P, Filipovic J, Pinto G, Brezovsky J, Damborsky J, Bednar D (2019) Caver Web 1.0: identification of tunnels and channels in proteins and analysis of ligand transport. Nucleic Acids Res 47(W1):W414–W422. https://doi.org/10.1093/nar/gkz378

Sunseri J, Koes DR (2016) Pharmit: interactive exploration of chemical space. Nucleic Acids Res 44:W442–W448. https://doi.org/10.1093/NAR/GKW287

Tao A, Huang Y, Shinohara Y, Caylor ML, Pashikanti S, Xu D (2019) EzCADD: a rapid 2D/3D visualization-enabled web modeling environment for democratizing computer-aided drug design. J Chem Inf Model 59(1):18–24. https://doi.org/10.1021/acs.jcim.8b00633

Thakur N, Qureshi A, Kumar M (2012) VIRsiRNAdb: a curated database of experimentally validated viral siRNA/shRNA. Nucleic Acids Res 40. https://doi.org/10.1093/NAR/GKR1147

Tian W, Chen C, Lei X, Zhao J, Liang J (2018) CASTp 3.0: computed atlas of surface topography of proteins. Nucleic Acids Res 46:W363–W367. https://doi.org/10.1093/NAR/GKY473

Toseland CP, Clayton DJ, McSparron H, Hemsley SL, Blythe MJ, Paine K, Doytchinova IA, Guan P, Hattotuwagama CK, Flower DR (2005) AntiJen: a quantitative immunology database integrating functional, thermodynamic, kinetic, biophysical, and cellular data. Immunome Res 1:4. https://doi.org/10.1186/1745-7580-1-4

Tsai TY, Chang KW, Chen CYC (2011) IScreen: World's first cloud-computing web server for virtual screening and de novo drug design based on TCM database@Taiwan. J Comput Aided Mol Des 25(6):525–531. https://doi.org/10.1007/s10822-011-9438-9

Tuncbag N, Keskin O, Gursoy A (2010) HotPoint: hot spot prediction server for protein interfaces. Nucleic Acids Res 38:W402–W406. https://doi.org/10.1093/NAR/GKQ323

Tyagi A, Ahmed F, Thakur N, Sharma A, Raghava GPS, Kumar M (2011) HIVsirDB: a database of HIV inhibiting siRNAs. PLoS One 6:e25917. https://doi.org/10.1371/JOURNAL.PONE.0025917

Vangone A, Schaarschmidt J, Koukos P, Geng C, Citro N, Trellet ME, Xue LC, Bonvin AMJJ (2019) Large-scale prediction of binding affinity in protein–small ligand complexes: the PRODIGY-LIG web server. Bioinformatics 35:1585–1587. https://doi.org/10.1093/BIOINFORMATICS/BTY816

Vita R, Mahajan S, Overton JA, Dhanda SK, Martini S, Cantrell JR, Wheeler DK, Sette A, Peters B (2019) The immune epitope database (IEDB): 2018 update. Nucleic Acids Res 47:D339. https://doi.org/10.1093/NAR/GKY1006

Vodovar N, Goic B, Blanc H, Saleh M-C (2011) In silico reconstruction of viral genomes from small RNAs improves virus-derived small interfering RNA profiling. J Virol 85:11016–11021. https://doi.org/10.1128/JVI.05647-11

Wang J, Dokholyan NV (2019) MedusaDock 2.0: efficient and accurate protein-ligand docking with constraints. J Chem Inf Model 59(6):2509–2515. https://doi.org/10.1021/acs.jcim.8b00905

Wang S, Sundaram JP, Spiro D (2010) VIGOR, an annotation program for small viral genomes. BMC Bioinf 11(1):1–10. https://doi.org/10.1186/1471-2105-11-451

Wang J, Zhang Y, Yu Y (2015) Crescendo: A protein sequence database search engine for tandem mass spectra. J Am Soc Mass Spectrom 26(7):1077–1084. https://doi.org/10.1007/s13361-015-1120-3

Wang YL, Wang F, Shi XX, Jia CY, Wu FX, Hao GF, Yang GF (2021) Cloud 3D-QSAR: a web tool for the development of quantitative structure–activity relationship models in drug discovery. Brief Bioinform 22. https://doi.org/10.1093/BIB/BBAA276

Wass MN, Kelley LA, Sternberg MJE (2010) 3DLigandSite: predicting ligand-binding sites using similar structures. Nucleic Acids Res 38:W469. https://doi.org/10.1093/NAR/GKQ406

Watson M, Schnettler E, Kohl A (2013) viRome: an R package for the visualization and analysis of viral small RNA sequence datasets. Bioinformatics 29:1902. https://doi.org/10.1093/BIOINFORMATICS/BTT297

Weisel M, Proschak E, Schneider G (2007) PocketPicker: analysis of ligand binding-sites with shape descriptors. Chem Cent J 1(1):7. https://doi.org/10.1186/1752-153X-1-7

Xiao X, Wu ZC, Chou KC (2011) iLoc-Virus: a multi-label learning classifier for identifying the subcellular localization of virus proteins with both single and multiple sites. J Theor Biol 284:42–51. https://doi.org/10.1016/J.JTBI.2011.06.005

Yang IS, Lee JY, Lee JS, Mitchell WP, Oh HB, Kang C, Kim KH (2009) Influenza sequence and epitope database. Nucleic Acids Res 37. https://doi.org/10.1093/NAR/GKN881

Yang J, Roy A, Zhang Y (2013) Structural bioinformatics Protein-ligand binding site recognition using complementary binding-specific substructure comparison and sequence profile alignment. 29:2588–2595. https://doi.org/10.1093/bioinformatics/btt447

Yang J, Roy A, Zhang Y (n.d.) BioLiP: a semi-manually curated database for biologically relevant ligand-protein interactions. https://doi.org/10.1093/nar/gks966

Zhu H, Dennis T, Hughes J, Gifford RJ (2018) Database-integrated genome screening (DIGS): exploring genomes heuristically using sequence similarity search tools and a relational database. bioRxiv:246835. https://doi.org/10.1101/246835

Zoete V, Daina A, Bovigny C, Michielin O (2016) SwissSimilarity: a web tool for low to ultra high throughput ligand-based virtual screening. J Chem Inf Model 56:1399–1404. https://doi.org/10.1021/ACS.JCIM.6B00174

Chapter 3
Developments in Computer-Aided Drug Design for Antiviral Research

Debajit Dey

Abstract Computer-aided drug design (CADD) is a powerful tool for the rational design of new antivirals, allowing for the optimization of drug candidates based on their predicted interactions with target proteins and their expected behavior in the body. In antiviral research, CADD has greatly benefited from machine learning (ML). Large datasets of chemical compounds and their corresponding biological activity can be analyzed using ML algorithms, which enables researchers to forecast a compound's antiviral potential. As a result, the process of discovering new drugs can be substantially accelerated. This enables scientists to select which compounds to test and improve further. In order to maximize the likelihood that a possible antiviral medication would be successful in preclinical and clinical studies, ML can also be used to optimize its physicochemical properties, such as its solubility and stability. The way we identify and create new antiviral treatments may be completely changed by the use of machine learning in the field of antiviral design. In this chapter, we outline various machine learning tools that are being used in the field of computer-aided drug design with specific examples.

Keywords CADD · Antivirals · Machine learning · Deep learning · SARS-CoV-2

3.1 Introduction

Computer-aided drug design (CADD) is a rapidly growing field that utilizes advanced computational techniques to discover and design new antiviral drugs. It is an essential tool in the pharmaceutical industry because it allows researchers to virtually screen and evaluate many potential drug candidates, saving time and resources compared to traditional drug discovery methods. CADD also allows for the optimization of drug properties such as potency, selectivity, and stability and the

D. Dey (✉)
School of Medicine, University of Maryland, Baltimore, MD, USA
e-mail: ddey@som.umaryland.edu

N. Kumar et al. (eds.), *Advances in Antiviral Research*, Livestock Diseases and Management, https://doi.org/10.1007/978-981-99-9195-2_3

prediction of potential side effects. These capabilities are particularly important in developing new antiviral drugs because they can help identify the most promising candidates and avoid costly failures in the later downstream. Small molecules, peptides, and proteins, among other types of antivirals, have all been developed successfully CADD. It has also contributed significantly to the development of contemporary pharmacological research, which has led to the creation of novel therapies for a variety of viral disorders. The capacity of CADD to define binding interactions between medications and their targets at the molecular level is one of its main benefits. This enables researchers to better understand how medications function and find the most effective molecules for a particular target. (Gurung et al. 2021; Sofi et al. 2022).

One of the most common types of CADD is structure-based drug design. It is a computational approach to drug discovery that involves using computer models to predict the three-dimensional structure of a protein or other biological target. This information is then used to design or optimize molecules that are likely to bind to the target and have a desired therapeutic effect. One of the main benefits of structure-based drug design is that it allows researchers to target specific sites on a protein or other biological target, rather than relying on more general approaches that might not be as precise. Researchers can create or improve compounds that are more likely to bind to the target by understanding the precise structure of the target. For structure-based drug design, a variety of methods can be applied, including X-ray crystallography, nuclear magnetic resonance (NMR) spectroscopy, CryoEM, and homology modeling. These methods enable the three-dimensional structure of a protein or other biological target to be determined, as well as the location of specific drug-binding sites on the target. Researchers can utilize computational methods to create or optimize compounds that are likely to bind to the target and provide the intended effect once the target's structure has been identified. These techniques can consider a number of variables, including the molecule's chemical characteristics, the target site's shape and size, and interactions between the molecule and the target (Anderson 2003; van Montfort and Workman 2017).

Ligand-based drug design is a method of developing new pharmaceuticals that involve the identification and optimization of small molecules that interact with specific protein targets in the human body. This strategy has drawn a lot of interest in the field of medicinal chemistry since it enables the development of targeted medicines that may selectively modify the activity of particular proteins, possibly leading to the development of more potent and safe medications. Virtual screening, which uses computer-based algorithms to forecast the binding affinity of a chemical to a protein target, is a popular approach to ligand-based drug creation. This can be done by employing homology modeling, which uses a homologous protein as a template, or by creating 3D models of the protein based on its known structure. From a sizable database of substances, virtual screening enables the quick identification of possible ligands, which can subsequently be synthesized and tested for biological activity. The enhancement of the identified ligands' binding affinity and selectivity is a crucial component in ligand-based medication design. This can be achieved through the modification of the chemical structure of the ligand, using techniques such as rational drug design and structure-activity relationships. The optimization process can also involve the development of prodrugs, which are inactive compounds that are converted to active drugs in the body. The development of a wide variety of treatments, including small-molecule enzyme inhibitors and receptor agonists and antagonists, has proven effective using ligand-based drug design. Aspirin, a nonsteroidal anti-inflammatory medication that blocks the activity of the enzyme cyclooxygenase, and tamoxifen, a selective estrogen receptor modulator used to treat breast cancer, are two examples of medications created using this method. The potential for off-target effects and the potential for resistance arising from mutations in the target protein are drawbacks of ligand-based drug design, though. These difficulties can be overcome by finding other protein targets for drug development and utilizing several ligands in combination therapy. (Acharya et al. 2011).

Finally, there is in silico ADME (**A**bsorption, **D**istribution, **M**etabolism, and **E**xcretion) which refers to the use of computer-based simulations and modeling to predict the pharmacokinetic and pharmacodynamic properties of drug candidates. This approach has gained significant attention in the pharmaceutical industry, as it allows for the rapid and cost-effective evaluation of a large number of compounds in the early stages of drug development. One important aspect of in silico ADME is the prediction of the absorption of a drug following oral administration. This can be accomplished through the use of physiologically based pharmacokinetic (PBPK) models, which simulate the movement of the drug through the body based on its physicochemical properties and the physiological characteristics of the individual. PBPK models can be used to predict the bioavailability of a drug, as well as its distribution to different tissues and organs in the body. Predicting a drug's metabolism is a crucial component of in silico ADME/T. In silico models of drug-metabolizing enzymes can be used to forecast the creation of metabolites and the possibility of drug-drug interactions in order to achieve this. In silico models can also be used to forecast a drug's excretion, including how quickly it leaves the body and whether it has the potential to be harmful. To increase prediction accuracy and direct the optimization of medication candidates, in silico ADME can be utilized in conjunction with in vitro and in vivo research. However, it is important to note that in silico ADME has its limitations, as it relies on the accuracy of the input data and the assumptions made in the models. Therefore, it is important to validate the predictions made through in silico ADME through experimental studies (Cheng et al. 2013; Wang et al. 2015).

3.2 Machine Learning (ML) and Its Algorithms

The science of machine learning, which is expanding quickly, has the potential to completely change how we go about finding new antiviral drugs. A typical machine learning workflow consists of the following seven steps: (a) define the problem and determine the objective of the model, (b) collect and preprocess the data, (c) split data into training and testing sets, (d) choose an appropriate machine learning model and train it on the training data, (e) evaluate the model on the testing data, (f) fine-tune the model and repeat steps (d) and (e) until satisfactory performance is achieved, and (g) make predictions on new data using the final, fine-tuned model (Fig. 3.1). By leveraging the vast amounts of data generated in the pharmaceutical industry, machine learning algorithms can identify patterns and relationships that may not be immediately apparent to human analysts. This can help researchers to identify new antiviral drug targets, predict the potential effectiveness of new antivirals, and optimize drug properties such as potency and stability. Machine learning can also be used to predict the potential effectiveness of new drugs. By analyzing data on the activity of existing antiviral drugs, machine learning algorithms can identify patterns that may suggest which compounds are likely to be effective against certain targets. This can help researchers to prioritize the most promising candidates for further testing and development. Overall, the integration of machine learning into antiviral drug discovery has the potential to greatly accelerate the identification and rapid development of potent antivirals. Machine learning algorithms can be of the following types.

Supervised learning: The algorithm is trained on a labeled dataset in supervised learning, where the correct result is given for each example in the training set. Predictions on new, unobserved examples taken from the same distribution as the training set are the objective. Linear regression, logistic regression, and support vector machines are some examples of supervised learning.

Semi-supervised learning: A significant amount of unlabeled data and a small amount of labeled data are given to the algorithm during semi-supervised learning. The objective is to use the labeled examples that are already available to enhance performance on the unlabeled examples.

Unsupervised learning: The algorithm is not given labeled training examples in unsupervised learning. Instead, it must use methods like clustering or dimensionality reduction to uncover the data's underlying structure. Principal component analysis and k-means clustering are two examples of unsupervised learning.

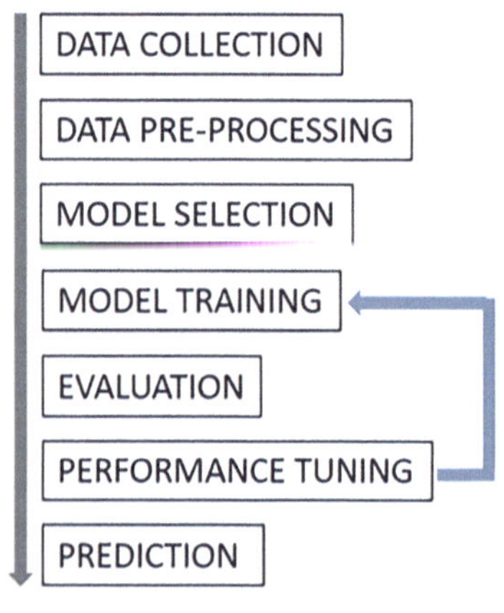

Fig. 3.1 Schematic showing the major steps of machine learning-based model building

Reinforcement learning: Here, an agent learns to interact with its surroundings in reinforcement learning to maximize a reward. The agent learns to make decisions that will maximize the cumulative benefit over time by receiving feedback in the form of rewards or penalties for its activities.

Several of the ML algorithms used in antiviral drug discovery are support vector machine (SVM), principal component analysis (PCA), linear regression, logistic regression, *k*-nearest neighbor (*k*-NN), random forest, *k*-means clustering, naïve Bayesian, AdaBoost, gradient boosting, XGBoost, and recursive partitioning (Mahesh 2020; Sarker 2021).

3.2.1 Support Vector Machine (SVM)

It is a supervised learning algorithm used for classification and regression tasks. It is an effective machine learning tool, especially when dealing with high-dimensional or nonlinearly separable data. Finding the hyperplane in a high-dimensional space that maximally divides the data points into distinct classes is the basic tenet of SVM. The support vectors, or closest data points from each class, are chosen so that the distance between the hyperplane and these points is maximized. Using a training dataset and a set of labeled examples, the SVM model is trained. The best hyperplane is then discovered by the model by maximizing the margin between the two classes using these samples. The mathematical formula for an SVM classifier can be expressed as $\hat{y} = wx + b$, where $\hat{y}$ is the predicted label, w is the weight vector, x is the feature vector, and b is the bias term. The weight vector and bias term are learned from the training data and determine the position and orientation of the hyperplane. A single data point in the high-dimensional space is represented by the feature vector, which is the input data. SVM has the ability to handle high-dimensional data successfully as one of its advantages. Furthermore, it employs the kernel approach to enhance the dimension of the data so that it might become linearly separable, making it noise-resistant and capable of handling nonlinearly separable data. SVM training can be computationally expensive, especially for large datasets. Additionally, it is sensitive to the choice of hyperparameters, which might influence the performance of the model, such as the regularization parameter and kernel type (Heikamp and Bajorath 2014; Rodriguez-Perez and Bajorath 2022).

3.2.2 Principal Component Analysis (PCA)

It is a statistical method that is frequently used to visualize complex data, uncover patterns and trends in data, and reduce the dimensionality of huge datasets. It is a potent tool that enables the most crucial data to be extracted from a dataset while minimizing noise or unimportant data. Finding a set of uncorrelated variables, referred to as principal components, that can account for the most variance in a dataset is the foundation of PCA. The linear combinations of the initial variables that best describe the variation in the data are found, and these principle components are then identified. Following key steps are involved in performing PCA. To ensure that all variables are on the same scale, the data must first be processed and scaled. The data is then split up into a number of orthogonal components, which are then rated according to how well they explain the variance in the data. The data is then displayed using a set of graphs called scree plots, which enables researchers to pinpoint the data's most crucial elements. One of PCA's key advantages is its ability to locate the underlying structure in a dataset by determining the variables that are most crucial for illuminating the data's variation. This can be especially helpful when there are numerous factors that are associated with one another and may be challenging to analyze separately. PCA is frequently used in drug discovery to examine the correlations between various factors, such as chemical compounds or biological pathways, in order to pinpoint possible drug development targets. It is also used to spot patterns and trends in therapeutic efficacy data, to spot potential off-target effects, and to depict the interactions between various chemicals in a chemical library (Jolliffe and Cadima 2016).

3.2.3 Linear Regression

It is a statistical model commonly used in machine learning for predictive modeling. It assumes that there is a linear relationship between the input variables (x) and the output variable (y). The goal of linear regression is to find the line of best fit that minimally summarizes the relationship between the variables. This is done by minimizing the sum of the squared residuals between the predicted values ($\hat{y}$) and the observed values (y). The mathematical formula for linear regression is represented as $\hat{y} = b0 + b1x$, where $b0$ and $b1$ are the intercept and slope coefficients, respectively. These coefficients are learned from the training data and can be used to make predictions on new, unseen data (Maulud and Abdulazeez 2020; Nasteski 2017).

3.2.4 Logistic Regression

It is a statistical method used to predict the probability of a binary outcome. It is a commonly used tool in the field of machine learning and is frequently used in classification tasks, such as figuring out whether or not an email is spam or whether or not a customer will leave. The main distinction between logistic regression and other regression models is that the latter uses continuous prediction while the former uses binary prediction. The logistic regression model does this by mapping the anticipated probabilities to a range between 0 and 1, using a logistic function. The logistic function is defined as $P(y) = e \wedge (b0 + b1x)/(1 + e \wedge (b0 + b1x))$, where $P(y)$

is the predicted probability of the binary outcome, *b0* is the intercept term, *b1* is the coefficient for the predictor variable *x*, and *e* is the base of the natural logarithm. The likelihood of the observed data is maximized in the logistic regression model to estimate the coefficients *b0* and *b1*. This is accomplished by the iterative gradient descent method, in which the model modifies the coefficients to reduce the discrepancy between the anticipated probability and the actual results. Comparing logistic regression to other machine learning methods, it has the benefit of being very straightforward and simple to implement. It is a preferred option in many real-world applications since it is noise-resistant and capable of handling high-dimensional data. However, logistic regression is susceptible to overfitting and may not work effectively on datasets with significant imbalances. Additionally, its capacity to simulate intricate nonlinear connections is constrained. (Dreiseitl and Ohno-Machado 2002; Ray 2019).

3.2.5 k-*Nearest Neighbor (KNN)*

It is a supervised machine learning algorithm used for both classification and regression tasks. In the absence of a specific training set, KNN is called a "lazy learner." Instead, it waits until a prediction is requested to find the nearest neighbors and make a prediction based on their labels. It calculates the separation between a given data point and every other point in the dataset. Different metrics, such as the Manhattan distance or the Euclidean distance, can be used to calculate the distance. After determining the distances, the algorithm chooses the K data points that are most nearby the given position and utilizes their labels to predict. The KNN algorithm's simplicity is one of its key benefits. It is simple to comprehend and put into practice, and it doesn't require any prior information or data assumptions. However, because the technique must determine the distance between each prediction's provided point and every other point in the dataset, it can be computationally expensive. KNN's sensitivity to the scale of the features is another drawback. The distance calculation will be dominated by one characteristic if its scale is significantly larger than that of the others, making it challenging for the other features to contribute. By normalizing the values of each feature to the same scale, feature scaling can help to mitigate this. It is often used as a baseline algorithm to compare the performance of more complex models (Taunk et al. 2019; Zhang 2016).

3.2.6 *Random Forest*

It is a supervised machine learning algorithm that can be used for both classification and regression tasks. It combines the predictions of multiple models to make a more accurate and stable prediction (ensemble learning method). Using a subset of the data that is randomly chosen, the random forest algorithm generates a collection of decision trees. The final forecast is made by taking the majority vote in the case of classification or by averaging the predictions of all the trees. The random forest algorithm's capacity to handle high-dimensional data and a large number of characteristics is one of its key advantages. Due to the incorporation of many decision trees, it also possesses a high degree of accuracy and is resistant to overfitting. However, a drawback of random forest is that it might be computationally costly, especially when there are many trees. It can also be difficult to interpret the results, as the individual decision trees are not easily accessible (Parmar et al. 2019).

3.2.7 **k-*Means Clustering***

It is a popular unsupervised machine learning algorithm that is used to group similar data points into clusters. It is a centroid-based algorithm, meaning that it uses the mean of the data points within a cluster as a representative value or centroid. The goal of *k*-means clustering is to partition a dataset into *k* clusters, where each cluster consists of data points that are similar to one another and dissimilar to data points in other clusters. To do this, the algorithm follows a simple iterative process: (a) initialize *k* centroids randomly within the data, (b) assign each data point to the closest centroid, and (c) calculate the mean of all data points assigned to each centroid and use this mean as the new centroid. Repeat steps 2 and 3 until the centroids stop changing or a maximum number of iterations is reached. One of the main advantages of *k*-means clustering is its simplicity and speed. It is relatively easy to implement and can be applied to large datasets efficiently. It is also a flexible algorithm, as the number of clusters can be specified by the user and the algorithm will automatically group the data accordingly. However, there are also some limitations to consider when using *k*-means clustering. One limitation is that the algorithm depends on the initial placement of the centroids, which can affect the final clustering results. To mitigate this issue, it is common to run the algorithm multiple times with different initializations and choose the clustering that gives the best result. Another limitation is that *k*-means clustering assumes that the clusters are spherical and equally sized, which may not always be the case in real-world data. This assumption can lead to suboptimal clustering results if the data do not conform to these assumptions (Ikotun et al. 2023).

3.2.8 Naïve Bayesian

Its classifiers are a popular machine learning algorithm that can be applied to the field of drug discovery to predict the properties and potential uses of new compounds. These classifiers are referred regarded as "naïve" because they assume that all features in a dataset are independent of one another, which may not necessarily be the case in real-world data. They rely on Bayes' theorem to make predictions based on prior information. The formula for a naïve Bayesian classifier is *P(class|data)* = [*P(data|class)** *P(class)*]/*P(data)*. Here, *P(class|data)* is the probability that a given data point belongs to a particular class, given the data. *P(data|class)* is the probability of the data given the class. *P(class)* is the probability of the class, and *P(data)* is the probability of the data. The projected class for the data point is the one with the highest probability. Because they are straightforward and efficient in a wide range of applications, naïve Bayesian classifiers are still commonly utilized. Finding molecules with the potential to be turned into powerful medicines for the treatment of diseases is the aim of drug discovery. When assessing a compound's potential, numerous elements must be taken into account, such as its chemical makeup, biological activity, and toxicity. These qualities, along with other crucial ones like how likely it is that a substance would be absorbed, distributed, metabolized, and excreted (ADME) in the body can be predicted using naïve Bayesian classifiers. In order to use naïve Bayesian classifiers for drug discovery, it is necessary to first define the event we are trying to predict (e.g., the likelihood that a compound will be effective in treating a specific disease) and the features that may be important for doing so (e.g., the compound's chemical structure). Given the values of these attributes, we next calculate the probability of each potential result (such as effective or ineffective). The prediction is made based on which result has the most likelihood. The flexibility of naïve Bayesian classifiers to handle huge datasets and a variety of features is one of its key advantages in the drug development process. They are a useful tool for researchers because they are also not computationally intensive to train and interpret. The assumption of independence between features, however, can be a substantial barrier to the development of new drugs because a compound's attributes frequently interact with one another. For instance, a compound's biological activity and toxicity may be influenced by its chemical structure. In such circumstances, a naïve Bayesian classifier's predictions might not be reliable. Naïve Bayesian classifiers are nevertheless often used in drug discovery despite this drawback due to their ease of use and versatility. They have been effectively used to forecast the ADME characteristics of substances and find possible therapeutic development leads (Lavecchia, 2015; Leung, 2007).

3.2.9 AdaBoost (Adaptive Boosting)

It is a machine learning algorithm that has been successfully applied in drug discovery. AdaBoost is a boosting algorithm, which is a type of ensemble learning method that combines the predictions of multiple weak models to get a more accurate and trustworthy prediction. AdaBoost works by weighting training data points, with misclassified data points receiving a higher weight in the subsequent iteration. This procedure is repeated until the necessary level of precision is realized or the allotted number of iterations has been used. One of AdaBoost's key advantages is that it can be used with a variety of base models, including decision trees. Decision trees are a common type of model in machine learning due to their ease of understanding and interpretation as well as their capacity to handle both continuous and categorical data. By integrating decision trees and AdaBoost, the model's accuracy, stability, and bias can all be improved. AdaBoost has been used to predict how well small molecules will interact with various target proteins in the field of drug discovery. For example, AdaBoost has been used to forecast the small-molecule binding affinity to specific proteins, which is an important component of determining the likely effectiveness of a drug. AdaBoost has also been used to predict the toxicity of small molecules in the early stages of drug development to select compounds that are most likely to be safe for human use (Ferreira and Figueiredo 2012; Mayr et al. 2014; Wang 2012).

3.2.10 Gradient Boosting

It can be applied to both classification and regression tasks. As an ensemble learning method, it integrates the predictions of various models to provide a forecast that is more precise and stable. The gradient boosting approach works by creating a series of straightforward models, each of which is trained to fix the mistakes produced by the model before it. Models are frequently decision trees, and predictions are made by adding the results of all the trees in the model. The gradient boosting algorithm's capacity to reach a high level of accuracy often beating other machine learning algorithms is one of its key features. As it integrates the predictions of various decision trees, which lowers variance, it is also resistant to overfitting. Gradient boosting has some drawbacks, though, including the potential for high computational costs, especially when there are many trees involved. Additionally, because the individual decision trees are not readily available, it can be challenging to evaluate the results (Ferreira and Figueiredo 2012; Mayr et al. 2014).

3.2.11 XGBoost (eXtreme Gradient Boosting)

It is an optimized implementation of the gradient boosting algorithm that is specifically designed for efficient and scalable training of machine learning models. It is a favored option for many applications, especially in fields like feature selection and predictive modeling. Similar to the traditional gradient boosting approach, XGBoost creates a series of straightforward models, with each model being trained to fix the mistakes produced by the one before it. Models are frequently decision trees, and predictions are made by adding the results of all the trees in the model. A set of restrictions and an objective function are components of the mathematical formula for XGBoost. In order to avoid overfitting, a regularization term is added to the loss function to create the objective function. The individual trees in the model, which are represented as decision trees, define the constraints. The objective function can be written as $Objective = \sum L(yi, \hat{y}i) + \Omega(f)$, where L is the loss function, yi is the true label, $\hat{y}i$ is the predicted label, and $\Omega(f)$ is the regularization term. The loss function measures the discrepancy between the true and predicted labels and is typically chosen based on the specific task at hand (e.g., mean squared error for regression or log loss for classification). The regularization term controls the complexity of the model and can be written as $\Omega(f) = \gamma T + 1/2\lambda \sum wi ^\wedge 2$, where γ is a parameter that controls the number of leaves in each tree, T is the total number of trees in the model, λ is a parameter that controls the amount of regularization, and wi is the weight of the ith feature. The regular gradient boosting technique is not as effective or scalable as XGBoost due to a number of improvements. The ability of XGBoost to reach a high level of accuracy—often exceeding other machine learning algorithms—is one of its key features. As it integrates the predictions of various decision trees, which lowers variance, it is also resistant to overfitting. Additionally, XGBoost is extremely effective and scalable, making it a good fit for applications with lots of features and vast amounts of data. However, one drawback of XGBoost is that because the individual decision trees are not readily available, it might be challenging to interpret the results. It is also sensitive to the scale of the features, so proper feature scaling is important to ensure that all features contribute equally to the prediction. Despite these limitations, XGBoost is a popular choice due to its high level of accuracy and efficiency (Ferreira and Figueiredo 2012; Ramdani and Furqon 2022).

3.2.12 Recursive Partitioning

It is widely used in machine learning, specifically in decision tree algorithms. It entails building a binary decision tree that may be applied to forecast outcomes for a specific data collection. The data is continually divided into smaller groups according to predetermined rules or circumstances, and this process creates the tree. The recursive partitioning forest (RPF) is a well-liked technique that makes use of recursive partitioning. RPFs are an ensemble machine learning technique that integrates numerous decision trees' predictions into a single final prediction. The final forecast is formed by averaging the predictions of all the trees in the forest, each of which has been trained on a distinct subset of the data. Compared to conventional decision trees, RPFs provide a number of benefits. The diminished chance of overfitting is a notable benefit. Overfitting is a problem with single decision trees, especially when working with large and complicated datasets. This is due to the tree's inability to generalize well to new data since it can fit the data too closely. RPFs, on the other hand, are able to improve generalization by reducing the variance of the model by averaging the predictions of various trees. The effectiveness of RPFs is another advantage. They can effectively handle massive datasets and are quite quick to train (Strobl et al. 2009).

3.3 Case Studies

3.3.1 Machine Learning-Guided Discovery of Antivirals Against Ebola Virus Disease

The Ebola virus, a member of the *Filoviridae* family, is responsible for the highly contagious and frequently fatal disease known as Ebola virus disease (EVD). Since the virus's initial discovery in 1976 in the Democratic Republic of the Congo, outbreaks have taken place in a number of African nations. Contact with the blood, saliva, organs, or other bodily fluids of infected animals or people results in the virus being transmitted. EVD has no specific cure, and it can be fatal within a short amount of time ("Ebola virus disease," 2020).

In 2017, Ekins et al. used a combination of Bayesian, support vector machines, recursive partitioning approaches, and open Bayesian software to identify three antiviral drugs against the Ebola virus. The 868 small compounds in the publicly accessible training dataset prevented the EBOV, Marburg, and Lassa viruses from replicating and entering cells in vitro. The creation of the models was based on a variety of molecular descriptors, including the number of rotatable bonds, the molecular fractional polar surface area, the maximum diameter 6 (FCFP_6)37 fingerprint of the molecules, the mass, the number of rings, and the number of aromatic rings. By applying the fivefold cross validation method, the receiver operator curve (ROC) was calculated to validate the models. Both open-source Bayesian software and Discovery Studio were used to create the models. The MicroSource spectrum, which contained 2320 small compounds, was then scored using validated models, and Quinacrine, Tilorone, and Pyronaridine were identified as potential hits with in vitro EC50 values of 350, 230, and 420 nM, respectively. The training dataset lacked any references to any of the three chemicals (Ekins et al. 2015).

In a follow-up study in 2019 using the same machine learning-based Bayesian models, Anantpadma et al. identified two potential EBOV inhibitors arterolane and lucanthone with EC_{50} values of 4.5 and 3.3 μM. Importantly, both compounds showed minimal cytotoxicity in vitro. In this study, the Bayesian models were generated using Assay Central. Both these studies prove the usefulness of ML-Bayesian models in accurate identification of effective EBOV viral inhibitors from small-molecule datasets (Anantpadma et al. 2019) (Fig. 3.2a).

3.3.2 Inhibitors Against the Novel Avian Influenza Virus A/H9N2 Using XGBoost

Avian influenza virus A/H9N2 is a subtype of influenza virus that is primarily found in birds but can also infect humans and other mammals. It is classified as a zoonotic virus, meaning that it can be transmitted between animals and humans and is characterized by its H9 hemagglutinin and N2 neuraminidase proteins. A/H9N2 viruses

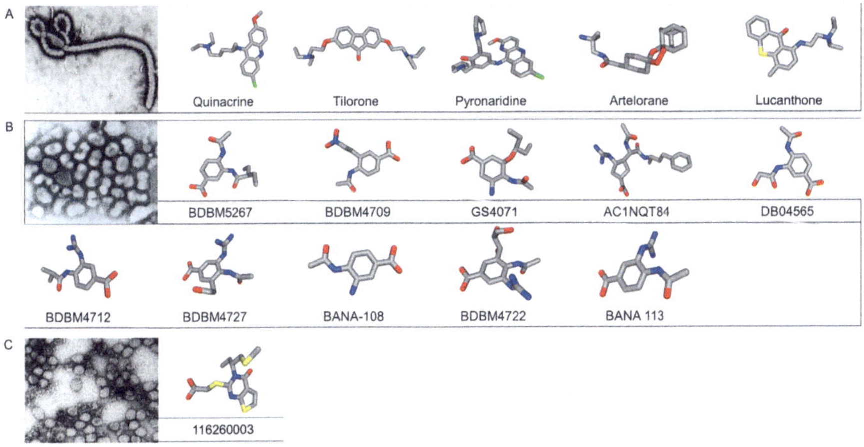

Fig. 3.2 Machine learning-based identification of antivirals against (**a**) Ebola virus, (**b**) influenza A H9N2 virus and (**c**) yellow fever virus

can cause a range of clinical signs in infected birds, ranging from mild to severe disease, and in humans, infection can cause respiratory illness with varying severity (Bi et al. 2022).

Amiroch et al. employed machine learning algorithms to identify potential inhibitors against the influenza A/H9N2 virus. In this study a total of seven machine learning algorithms were used to generate classification models. Based on parameters like ROC, accuracy, balanced accuracy, specificity, and sensitivity, the XGBoost algorithm-based model was considered to be the best. The XGBoost model had an accuracy of 0.9686, specificity of 0.9711, sensitivity of 0.96, balanced accuracy of 0.9656, and ROC of 0.9656. The other algorithms used in the study were *k*-nearest neighbor, gradient boosting, support vector machine, logistic regression, multilayer perceptron, and random forest. XGBoost identified ten compounds which gave the highest scores. Except cyclopentane amide deriv.16 g, all were reported in the literature to be influenza virus inhibitors. The others were benzoic acid inhibitor 6 or BDBM5267, benzoic acid deriv. 130 or BDBM4709, oseltamivir carboxylate or GS4071, AC1NQT84, 4-(acetylamino)-3-[(hydroxyacetyl)amino]benzoic acid or DB04565, benzoic acid deriv. 149 or BDBM4712, benzoic acid deriv. 12 or BDBM4727, 4-acetamido-3-aminobenzoic acid or BANA-108, benzoic acid deriv. 6a or BDBM4722, and BANA 113 (Amiroch et al. 2022) (Fig. 3.2b).

3.3.3 *Novel Inhibitors Against the Yellow Fever Virus*

YFV is a flavivirus and is found in tropical and subtropical regions of Africa and South America. The virus is primarily spread through the *Aedes* mosquito, which is also responsible for the transmission of other diseases such as Zika and dengue fever. Yellow fever can lead to severe illness and death, with a mortality rate of up to 50% in those who develop severe symptoms. Despite the availability of a vaccine, yellow fever remains a significant public health threat in many parts of the world, particularly in areas with high levels of poverty and inadequate access to healthcare (Oyono et al. 2022).

Gawriljuk et al. in 2021 reported the first machine learning-based identification of yellow fever virus inhibitor. The database for the training set was selected from cell culture inhibition data from ChEMBL and PubMed which consisted of 548 and 98 compounds, respectively. The final training dataset consisted of 121 molecules as active and 427 molecules as inactives, respectively. Support vector machine and Bayesian models were the top-scoring models. These models were then used to screen 7865 compounds with consensus prediction. Top five selected compounds were tested for their antiviral activity in vitro in Huh7 cells. Only the pyrazolesulfonamide derivative named 116,260,003 was able to inhibit YFV with a calculated EC_{50} of 3.2 μM in visual cytopathic assays and 3.3 μM in neutral red cytopathic assays. However, the compound was found to be somewhat cytotoxic with a calculated S.I of 7.5 and 3.3 for each corresponding assay. This study showed that ML-based models trained on nonideal datasets can reliably predict new lead compounds. However, such predictions will be reliable only within the chemical space of the training dataset (Gawriljuk et al. 2021) (Fig. 3.2c).

3.4 Deep Learning in Computer-Aided Drug Design

Artificial neural networks with numerous layers are used in deep learning, a subset of machine learning, to learn and make choices. These neural networks can evaluate and understand enormous volumes of data to carry out tasks like classification, regression, and grouping. They are motivated by the structure and operation of the human brain. Deep learning has attracted a lot of interest recently and produced outstanding outcomes in disciplines including speech recognition, natural language processing, and computer vision. Deep learning has the potential to completely change the way medications are identified and created in the area of computer-aided drug design. Antiviral drug discovery is a labor-intensive and costly process that involves the identification of new therapeutic candidates, their synthesis and optimization, and ultimately their testing in preclinical and clinical trials. Deep learning algorithms can help with this process by examining datasets to find patterns and connections that humans would not immediately see. The discovery of prospective antiviral medication candidates is one of the key applications of deep learning in drug design. This has traditionally been accomplished through a method known as virtual screening, which involves screening a large number of compounds using computer-based simulations to determine which have the most promising properties for drug development. When predicting the potential activity of new compounds, deep learning algorithms can be utilized to increase the precision and effectiveness of the process by examining vast amounts of data on the structure and function of established medications and their target proteins.

In order to maximize the synthesis of possible drug candidates, deep learning can also be applied. This entails determining the best and most efficient synthesis pathways for a particular product as well as optimizing the synthesis's environmental factors and chemical reactants. Deep learning algorithms can review data on the synthesis of existing medications and utilize this knowledge to forecast the most promising conditions and methods for synthesizing new molecules. The potential efficacy and toxicity of prospective medication candidates can also be predicted using deep learning. This is a crucial stage in the development of new drugs since it aids in the identification of candidates that are most likely to be efficient and safe for use in people. To predict efficacy and toxicity of new compounds, deep learning algorithms can examine data on the known efficacy and toxicity of medications as well as data on the structure and activity of the target proteins (Bajorath 2022).

There are several types of deep learning models, each with their own strengths and weaknesses. Some of the most common types of deep learning models include the following:

Convolutional neural networks (CNNs): These work by processing input data through multiple layers of interconnected neurons, with each layer learning to recognize different features of the input data.

Recurrent neural networks (RNNs): RNNs have the ability to process data with a temporal dimension, allowing them to take into account the context and order of the input data.

Autoencoders: These models of unsupervised learning train to encrypt and decrypt input data. They operate by first learning a condensed version of the input data that can be applied to problems like dimensionality reduction and anomaly detection.

Generative adversarial networks (GANs): A generator and a discriminator are the two neural networks that make up this kind of unsupervised learning model. While the discriminator works to separate the fake data from the real data, the generator creates artificial data. Together, the generator and discriminator networks are taught to produce data that is indistinguishable from real data and to accurately distinguish between the two.

Self-organizing maps (SOMs): These are an example of an unsupervised learning model that may be applied to tasks like dimensionality reduction and clustering. SOMs learn to map input data into a two-dimensional grid, assigning related input data points to nearby grid places.

3.4.1 Deep Learning in Identifying Inhibitors Against the SARS-CoV-2

Deep learning techniques, specifically artificial neural networks, have been successfully applied in the identification of inhibitors against SARS-2, the virus responsible for the COVID-19 pandemic. These methods involve training a model on a large dataset of chemical compounds and their corresponding properties, such as binding affinity and toxicity. The model is then able to predict the likelihood of a novel compound being an effective inhibitor against SARS-2 based on its chemical structure and other properties. This approach has the potential to significantly accelerate the drug discovery process, as it allows for the rapid screening of large numbers of compounds in silico, rather than relying on labor-intensive experimental techniques. Additionally, the use of deep learning in this context has the potential to improve the accuracy and precision of inhibitor identification, as the model is able to analyze and incorporate complex patterns and relationships in the data that may be difficult for human experts to discern. We will discuss two such reports which highlight the impact of deep learning approach for antiviral discovery against SARS-CoV-2.

A report by Deepthi et al. showed the application of a deep learning-based ensemble approach (DLEVDA) to identify potential inhibitors against SARS-CoV-2. Ensemble approach is powerful since it involves several classifiers that allow the resultant model to make more credible predictions. To build classification features, this method combines chemical structures of drugs and viral genome sequence similarities in a pairwise manner. The resultant sample is then provided as input to the first segment of DLEVDA, i.e., the convolution neural network. Here, the model learns the complex input patterns. The abstract features thus identified was used as input to train the second segment, i.e., the XGBoost classifier which can then identify potential inhibitor candidates. The fivefold cross validation of this model gives an AUC of 0.8897 with a prediction accuracy of 0.8571 and sensitivity of 0.8394. A total of 12 compounds were selected as the top hits out of which only three compounds mizoribine, N4-hydroxycytidine, and clevudine did not have any literature evidence of SARS-2 antiviral activity The top 5 compounds were identified as ribavirin, nitazoxanide, mizoribine, favipiravir, and amantadine (Deepthi et al. 2021).

Beck et al. developed a model to predict drug-target interactions based on deep learning. The model referred to as molecule transformer drug-target interaction was used to scan the SMILES input of 3410 FDA-approved drugs to identify potential inhibitors against several protein targets of SARS-2. The MT-DTI model was built based on the self-attention technique and is trained on K_i, K_d, and IC_{50} values of 1000,000,000 compounds from the Binding and Drug Target Common database. The model predictions were compared with the docking scores obtained from AutoDock Vina. The model predicted remdesivir, dolutegravir, efavirenz, boceprevir, and atazanavir as the top binders of SARS-CoV-2 proteins. Remdesivir was included in the analysis as a positive control (Beck et al. 2020).

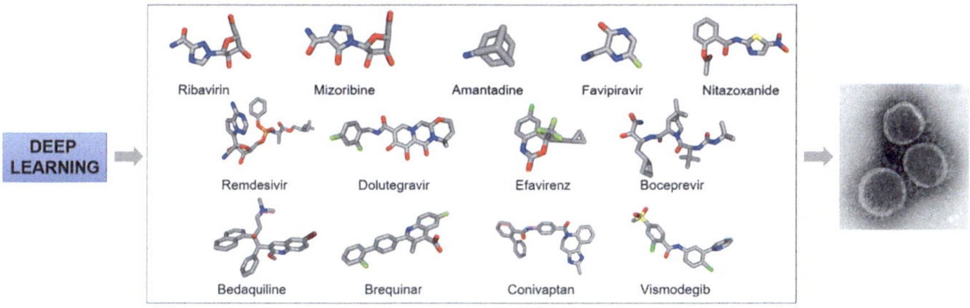

Fig. 3.3 SARS-2 antivirals identified through deep learning

On the other hand, Ke et al. adopted an approach wherein they designed and trained a deep neural network-based AI model on two different databases. The first database contained compounds known to be antivirals not only against SARS and SARS-2 but also against RNA viruses like HIV and influenza, whereas the second database consisted of viral 3C-like protease inhibitors. The predicted compounds were first validated against a feline coronavirus (feline infectious peritonitis virus) in vitro. These results were used as feedback to retrain the AI model. The DrugBank database was used for the final screening of potential compounds by the modified model. This approach identified four compounds bedaquiline, Brequinar, conivaptan, and vismodegib with antiviral activity against the feline infectious peritonitis virus as potential inhibitors of SARS-CoV-2 (Ke et al. 2020) (Fig. 3.3).

3.5 Concluding Remarks

Machine learning and deep learning have made substantial advancements in the field of computer-aided drug creation in recent years. These strategies have proven their capacity to spot patterns and linkages that were previously missed by conventional approaches by making use of intricate algorithms and large amounts of data. As a result, machine learning and deep learning have been used for a variety of tasks in drug design, such as the optimization of drug candidates and the creation of brand-new medications. Despite these successes, the use of these techniques in drug design is still in its infancy, and there remain challenges to be addressed, such as the need for large amounts of labeled data and the interpretability of the models. However, as the field continues to mature and evolve, it is likely that machine learning and deep learning will play an increasingly important role in the development of new and effective therapies.

References

Acharya C, Coop A, Polli JE, Mackerell AD Jr (2011) Recent advances in ligand-based drug design: relevance and utility of the conformationally sampled pharmacophore approach. Curr Comput Aided Drug Des 7(1):10–22. https://doi.org/10.2174/157340911793743547

Amiroch S, Irawan MI, Mukhlash I, Al Faroby MHZ, Nidom CA (2022) Machine learning for the prediction of antiviral compounds targeting avian influenza a/H9N2 viral proteins. Symmetry 14(6):1114. https://www.mdpi.com/2073-8994/14/6/1114

Anantpadma M, Lane T, Zorn KM, Lingerfelt MA, Clark AM, Freundlich JS, Davey RA, Madrid PB, Ekins S (2019) Ebola virus Bayesian machine learning models enable new in vitro leads. ACS Omega 4(1):2353–2361. https://doi.org/10.1021/acsomega.8b02948

Anderson AC (2003) The process of structure-based drug design. Chem Biol 10(9):787–797. https://doi.org/10.1016/j.chembiol.2003.09.002

Bajorath J (2022) Deep machine learning for computer-aided drug design [perspective]. Front Drug Discovery 2. https://doi.org/10.3389/fddsv.2022.829043

Beck BR, Shin B, Choi Y, Park S, Kang K (2020) Predicting commercially available antiviral drugs that may act on the novel coronavirus (SARS-CoV-2) through a drug-target interaction deep learning model. Comput Struct Biotechnol J 18:784–790. https://doi.org/10.1016/j.csbj.2020.03.025

Bi Y, Li J, Shi W (2022) The time is now: a call to contain H9N2 avian influenza viruses. Lancet Microbe 3(11):e804–e805. https://doi.org/10.1016/S2666-5247(22)00232-4

Cheng F, Li W, Liu G, Tang Y (2013) In silico ADMET prediction: recent advances, current challenges and future trends. Curr Top Med Chem 13(11):1273–1289. https://doi.org/10.2174/15680266113139990033

Deepthi K, Jereesh AS, Liu Y (2021) A deep learning ensemble approach to prioritize antiviral drugs against novel coronavirus SARS-CoV-2 for COVID-19 drug repurposing. Appl Soft Comput 113:107945. https://doi.org/10.1016/j.asoc.2021.107945

Dreiseitl S, Ohno-Machado L (2002) Logistic regression and artificial neural network classification models: a methodology review. J Biomed Inform 35(5–6):352–359. https://doi.org/10.1016/s1532-0464(03)00034-0

Ebola virus disease (2020) Nat Rev Dis Primers 6(1):14. https://doi.org/10.1038/s41572-020-0154-4

Ekins S, Freundlich JS, Clark AM, Anantpadma M, Davey RA, Madrid P (2015) Machine learning models identify molecules active against the Ebola virus in vitro. F1000Res 4:1091. https://doi.org/10.12688/f1000research.7217.3

Ferreira AJ, Figueiredo MAT (2012) Boosting algorithms: a review of methods, theory, and applications. In: Zhang C, Ma Y (eds) Ensemble machine learning: methods and applications. Springer US, pp 35–85. https://doi.org/10.1007/978-1-4419-9326-7_2

Gawriljuk VO, Foil DH, Puhl AC, Zorn KM, Lane TR, Riabova O, Makarov V, Godoy AS, Oliva G, Ekins S (2021) Development of machine learning models and the discovery of a new antiviral compound against yellow fever virus. J Chem Inf Model 61(8):3804–3813. https://doi.org/10.1021/acs.jcim.1c00460

Gurung AB, Ali MA, Lee J, Farah MA, Al-Anazi KM (2021) An updated review of computer-aided drug design and its application to COVID-19. Biomed Res Int 2021:8853056. https://doi.org/10.1155/2021/8853056

Heikamp K, Bajorath J (2014) Support vector machines for drug discovery. Expert Opin Drug Discov 9(1):93–104. https://doi.org/10.1517/17460441.2014.866943

Ikotun AM, Ezugwu AE, Abualigah L, Abuhaija B, Heming J (2023) K-means clustering algorithms: a comprehensive review, variants analysis, and advances in the era of big data. Inf Sci 622:178–210. https://doi.org/10.1016/j.ins.2022.11.139

Jolliffe IT, Cadima J (2016) Principal component analysis: a review and recent developments. Philos Transact A Math Phys Eng Sci 374(2065):20150202. https://doi.org/10.1098/rsta.2015.0202

Ke YY, Peng TT, Yeh TK, Huang WZ, Chang SE, Wu SH, Hung HC, Hsu TA, Lee SJ, Song JS, Lin WH, Chiang TJ, Lin JH, Sytwu HK, Chen CT (2020) Artificial intelligence approach

fighting COVID-19 with repurposing drugs. Biom J 43(4):355–362. https://doi.org/10.1016/j.bj.2020.05.001
Lavecchia A (2015) Machine-learning approaches in drug discovery: methods and applications. Drug Discov Today 20(3):318–331. https://doi.org/10.1016/j.drudis.2014.10.012
Leung KM (2007) Naive Bayesian classifier. In: Polytechnic University Department of Computer Science/Finance and Risk Engineering, 2007, pp 123–156
Mahesh B (2020) Machine learning algorithms-a review. Int J Sci Res 9:381–386
Maulud D, Abdulazeez AM (2020) A review on linear regression comprehensive in machine learning. J Appl Sci Technol Trends 1(4):140–147. https://doi.org/10.38094/jastt1457
Mayr A, Binder H, Gefeller O, Schmid M (2014) The evolution of boosting algorithms. Mach Learn Stat Model 53(06):419–427. https://doi.org/10.3414/ME13-01-0122
Nasteski V (2017) An overview of the supervised machine learning methods. Horizons B 4:51–62
Oyono MG, Kenmoe S, Abanda NN, Takuissu GR, Ebogo-Belobo JT, Kenfack-Momo R, Kengne-Nde C, Mbaga DS, Tchatchouang S, Kenfack-Zanguim J, Lontuo Fogang R, Zeuko'o Menkem E, Ndzie Ondigui JL, Kame-Ngasse GI, Magoudjou-Pekam JN, Bowo-Ngandji A, Nkie Esemu S, Ndip L (2022) Epidemiology of yellow fever virus in humans, arthropods, and non-human primates in sub-Saharan Africa: a systematic review and meta-analysis. PLoS Negl Trop Dis 16(7):e0010610. https://doi.org/10.1371/journal.pntd.0010610
Parmar A, Katariya R, Patel V (2019) A review on random forest: an ensemble classifier. In: Hemanth J, Fernando X, Lafata P, Baig Z (eds) International conference on intelligent data communication technologies and internet of things (ICICI) 2018, Cham
Ramdani F, Furqon M (2022) The simplicity of XGBoost algorithm versus the complexity of random forest, support vector machine, and neural networks algorithms in urban forest classification [version 1; peer review: awaiting peer review]. F1000Res 11(1069):10.12688/f1000research.124604.1
Ray S (2019) A quick review of machine learning algorithms. 2019 International conference on machine learning, big data, cloud and parallel computing (COMITCon)
Rodriguez-Perez R, Bajorath J (2022) Evolution of support vector machine and regression modeling in chemoinformatics and drug discovery. J Comput Aided Mol Des 36(5):355–362. https://doi.org/10.1007/s10822-022-00442-9
Sarker IH (2021) Machine learning: algorithms, real-world applications and research directions. SN Comput Sci 2(3):160. https://doi.org/10.1007/s42979-021-00592-x
Sofi MY, Shafi A, Masoodi KZ (2022) Chapter 22 – Introduction to computer-aided drug design. In: Sofi MY, Shafi A, Masoodi KZ (eds) Bioinformatics for everyone. Academic Press, pp 215–229. https://doi.org/10.1016/B978-0-323-91128-3.00002-1
Strobl C, Malley J, Tutz G (2009) An introduction to recursive partitioning: rationale, application, and characteristics of classification and regression trees, bagging, and random forests. Psychol Methods 14(4):323–348. https://doi.org/10.1037/a0016973
Taunk K, De S, Verma S, Swetapadma A (2019) A brief review of nearest neighbor algorithm for learning and classification. 2019 international conference on Intelligent Computing and Control Systems (ICCS), 15–17 May 2019
van Montfort RLM, Workman P (2017) Structure-based drug design: aiming for a perfect fit. Essays Biochem 61(5):431–437. https://doi.org/10.1042/EBC20170052
Wang R (2012) AdaBoost for feature selection, classification and its relation with SVM, a review. Phys Procedia 25:800–807. https://doi.org/10.1016/j.phpro.2012.03.160
Wang Y, Xing J, Xu Y, Zhou N, Peng J, Xiong Z, Liu X, Luo X, Luo C, Chen K, Zheng M, Jiang H (2015) In silico ADME/T modelling for rational drug design. Q Rev Biophys 48(4):488–515. https://doi.org/10.1017/S0033583515000190
Zhang Z (2016) Introduction to machine learning: k-nearest neighbors. Ann Transl Med 4(11):218. https://doi.org/10.21037/atm.2016.03.37

Chapter 4
Organoids in Antiviral Research: Potential and Challenges

Ashutosh Singh, Naveen Kumar, Dhruv Desai, Upasana Sahu, Sandeep Bhatia, Manoj Kumar, Bhupendra Nath Tripathi, and Aniket Sanyal

Abstract The inherent bottleneck due to divergent cell and tissue biology offered by cell culture and animal models motivated researchers to develop an alternative model system that can more precisely recapitulate the architecture and function of human and animal organs. Thanks to the extensive efforts that have led to the establishment of three-dimensional (3D) organ-like architectures called "organoids" or "mini-organs" that display the key structural, functional, and biological complexity of an organ. The organoids have been interrogated to broaden the scope of virology research, including disease modeling, personalized medicine, and drug discovery. In this chapter, we first provide fundamental protocols for culturing and passaging the organoids, as well as important points for differentiation and nutritional requirements particular to each organoid. We also discussed the outcomes of multiple antiviral investigations conducted using different organoids. Finally, the limitations of organoids in biomedical research in general and antiviral research in particular are also discussed, where we also identify critical areas for organoids improvement in the upcoming years.

Keywords Organoids · Protocols · Applications · Antivirals

A. Singh · N. Kumar (✉) · U. Sahu · S. Bhatia · M. Kumar · A. Sanyal
National Institute of High Security Animal Diseases, Bhopal, India
e-mail: naveen.kumar4@icar.gov.in

D. Desai
University of Pennsylvania, Philadelphia, PA, USA

B. N. Tripathi
Indian Council of Agricultural Research, New Delhi, India

N. Kumar et al. (eds.), *Advances in Antiviral Research*, Livestock Diseases and Management, https://doi.org/10.1007/978-981-99-9195-2_4

4.1 Introduction

Animal models and cell culture system are primarily used to study virus biology and antiviral research. However, they cannot recapitulate virus pathogenesis and drug response in humans. This is because there are extensive differences in the cell and tissue biology as well as drug metabolism in humans as compared to animals, making it very difficult to correlate the results obtained from virus pathogenesis and drug efficacy studies in experimental animal models and cell culture system to that of humans. Besides, some organizations, such as the Fund for the Replacement of Animals in Medical Experiments (FRAME) and Cruelty-Free International are constantly putting efforts to limit the animal use in biomedical research. Therefore, it became imperative to create and/or use alternative testing models that closely mimic the cellular microenvironment as that of humans, so as to promote and support biomedical research, including the drug screening, toxicity, and efficacy studies in a more humane approach.

As an alternative to animal models and cell culture system, organoids, popularly known as "mini-organs," have gained much attention globally among scientific community for the virological studies. They are basically three-dimensional (3D) self-organizing structure derived either from pluripotent embryonic stem (ES) or adult somatic stem cells (aSCs) that reflect the key structural and functional properties of organs or tissues in vivo (Clevers 2016). Although organoids and spheroids are used interchangeably sometimes due to their similar morphology and anatomical structure, there are basic differences between the two. The major distinction between the organoids and spheroids is that the former requires a scaffold-based system, while the latter can be grown on a scaffold-free system, to form a 3D structure. So, spheroids primarily consist of cell aggregates that freely float and are formed either from a single-cell type or a multicellular mixture of cells on ultralow attachment plates, whereas the organoids are derived from ES or aSCs immersed within a suitable extracellular matrix like Matrigel. The Intestinal Stem Cell Consortium (ISCC) formulated nomenclature directives in 2012: the term "organoid" refers to cultures that must contain multiple cell types like mesenchymal- and epithelial-originated cells, whereas 3D cultures containing only epithelial cell lines must be referred to as "spheroids" (Almeqdadi et al. 2019; Mustata et al. 2013). Unlike organoids, spheroids cannot self-organize or regenerate and does not mimic closely with in vivo organ-like structures (Hofer and Lutolf 2021). Therefore, organoids are much advanced 3D structures compared to spheroids.

This book chapter highlights the evolutionary history of organoids as well as their applications in biomedical research. We also discuss key considerations and basic protocols for culturing organoids, organoid-specific differentiation and nutritional requirement, and their cryopreservation and revival from the cryopreserved organoids. We also outlined various antiviral studies performed using distinct organoids along with their outcomes. Lastly, we conclude by discussing on the limitations of organoids in biomedical in general and antiviral research in particular and highlighting key priority areas for organoids improvement for the coming years.

4.2 A Brief History of Organoids

Continuous advancements in cell culture techniques have resulted in three-dimensional culture systems, giving rise to 3D organ-like in vitro models such as spheroids and organoids. For several decades, 3D cell cultures have been utilized in research. The term "organoid" was first introduced by Smith and Cochrane in 1946 when they described a cystic teratoma detected in pararenal tissue of a 2-month-old infant (Smith and Cochrane 1946). The discovery of pluripotent stem cells (PSCs) in mouse embryos in 1981 marked the beginning of stem cell research (Evans and Kaufman 1981). Later in 1998, the researchers isolated and cultivated embryonic stem cells from human blastocysts (Thomson et al. 1998). Michalopoulos et al. (2002) demonstrated that hepatocytes might phenotypically change into biliary cells in the surrounding environment of organoid cultures (Michalopoulos et al. 2002). In fact, establishing iPSCs from human and mouse fibroblasts has significantly advanced stem cell and organoid research (Takahashi et al. 2007; Takahashi and Yamanaka 2006). By demonstrating how human brain iPSCs self-organized into neuronal cells that developed polarized cortical tissues in the year 2008, the lab of Yoshiki Sasai laid the foundation for brain organoids (Eiraku et al. 2008). All of these advances from 1946 to 2008, known as the *pre-modern organoid era*, laid the foundation for modern organoid research.

The year 2009 marked the beginning of modern organoid research as a group of researchers from the Netherlands demonstrated, for the first time, the successful development of 3D intestinal organoids mimicking typical crypt-villus architecture from single adult intestinal stem cells expressing a specific leucine-rich repeat-containing G protein-coupled receptor 5 (Lgr5) (Sato et al. 2009). This was the first study to show how to grow a 3D organoid culture from a single adult stem cell (ASC). It paved the way for future organoid investigations using either ASCs or PSCs in systems such as neuroectoderm (brain and retina) and mesendoderm (liver, pancreas, stomach, lungs, and kidney). Next year, Barker and his colleagues (2010) showed that a single Lgr5 + ve cell successfully formed long-lasting organoid that resembled mature pyloric epithelium in an in vitro culture system (Barker et al. 2010). According to Eiraku et al. (2011), the dynamic and autonomous morphogenesis of the optic cup (retinal primordium) from a mouse embryonic stem cell aggregation requires an inherent self-organizing mechanism and domain-specific regulation of local epithelial features (Eiraku et al. 2011). Spence et al. (2011) developed a reliable method to control the differentiation of human pluripotent stem cells (hPSCs) into intestinal tissue by growth factor manipulations that mimic the development of the embryonic intestinal tract (Spence et al. 2011). By guiding the short-term overexpression of the transcription factors NKX2-1 and PAX8, along with the administration of recombinant human thyrotropin, Antonica et al. (2012) were able to produce functional thyroid follicular cells from mouse embryonic stem cells. These thyroid follicles restored thyroid hormone levels when implanted into hypothyroid mice (Antonica et al. 2012). Lancaster et al. (2013) created three-dimensional cerebral organoids from human pluripotent stem cells that have been shown to mimic human cortical development. They employed RNA interference and patient-specific induced pluripotent stem cells to replicate microcephaly in mice, which is notoriously difficult to replicate in mice in general (Lancaster et al. 2013).

The absence of vascularization is a major limitation in organoid development; however, Takebe et al. (2013) successfully developed a three-dimensional vascularized and functional liver from human iPSCs through the transplantation of in vitro-created liver buds. Human iPSC-derived hepatic endodermal cells were co-cultured with stromal cell populations, human mesenchymal stem cells, and human umbilical vein endothelial cells (HUVECs) to achieve this (Takebe et al. 2013). Salivary gland stem cells, according to Nanduri et al. (2014), can potentially develop into distinct ductal or lobular organoids (miniglands) and thus can be used to restore salivary gland function (Nanduri et al. 2014). In the same year, McCracken et al. (2014) and Greggio et al. (2013) demonstrated the formation of human gastric organoids and pancreatic organoids, respectively (Greggio et al. 2013; McCracken et al. 2014). Sampaziotis et al. (2015) described a method for producing cholangiocyte-like cells from human-induced pluripotent stem cells (hiPSCs), which closely resemble natural bile duct development. Furthermore, these cells can also serve as models for biliary disease and validate corresponding drugs (Sampaziotis et al. 2015). Takasato et al. (2016) described a protocol for producing kidney organoids by inducing all four progenitor populations of the kidney (renal interstitial progenitors, ureteric epithelial progenitors, endothelial progenitors, and nephron progenitors) from human pluripotent stem cells (hPSCs) (Takasato et al. 2016). D. Richards et al. (2017) devised a protocol for developing human cardiac organoids that, in terms of structure and functionality, resemble the developing heart's lumenized vascular network and have functional myocardium functions (Richards et al. 2017). Lee et al. (2020) described a composite culture condition for the growth of skin organoids having epidermis, sebaceous glands, dermis, and pigmented hair. They accomplished this by manipulating the signaling pathways of transforming growth factor (TGF) and fibroblast growth factor (FGF) to induce cranial epithelial cells and neural crest cells (Lee et al. 2020).

4.3 Diverse Applications of Organoids

Organoids offer a culturable technology that enables for manipulation of cell cultures, offering a more accurate representation of physiology and in vivo cellularity. This helps in narrowing down the gap between conventional cell culture and animal models. Table 4.1 compares the advantages and limitations of using 2D cell culture, organoids, and animal models. Overall, organoids are an excellent new technology for a variety of clinical and biological applications due to characteristics such as the capacity for long-term growth, tissue heterogeneity, and resemblance to

Table 4.1 Comparative pros and cons of the organoids to 2D cell cultures and animal models

Functions	2D cell cultures	Animal models	3D organoid cultures
High-throughput drug screening	Yes	No	Yes
Biobanking	Yes	Yes, but only at the cellular level	Yes
Patient-derived organoids (precision medicine)	No	No	Yes
Maintenance cost	Low	High	Low to medium
Recreating physiological environment	Limited	Completely physiological	Near physiological environment
Disease modeling	Poor	Good (comparable model)	Partially good
Heterogeneity	Low	High	High
Reproducibility	High	Low	Low
Vascularization	No	Yes	Partial
Manipulability	Yes	Limited	Yes

physiological 3D structure. Organoid models are widely used in disease modeling and organ development, tailored medications, pharmacology investigations, tumor modeling, and biobanking. Furthermore, human stem cell-derived organoids broaden the scope of their applications, as these organoids are being used to study genetic disorders, infectious diseases, and cancer (Fig. 4.1).

The global organoids' and spheroids' market size was valued at US$ 689.47 million in 2019, and it is expected to reach US$ 3,420.40 million by 2027 (Partners 2021). Globally, more than 70 companies are involved in organoids and models spheroids development; a few of the key players in the organoids and spheroids market are STEMCELL Technologies, Inc; 3D Biotek LLC; Cellesce Ltd.; 3D BioMatrix; Hubrecht Organoid Technology; ATCC, DefiniGEN; 3Dnamics, Inc.; PeproTech, Inc.; Corning Incorporated (Life Sciences); Merck KGaA; Organoid Therapeutics; Thermo Fisher Scientific, Inc., and others. In terms of product type, the organoid markets are divided into Intestinal, Pancreatic, Neural, Liver, Brain, Kidney, Lung, and others, of which the intestinal sector captured the largest market share (23%) in 2019. Based on the application of organoids, such as drug discovery and personalized medicine, regenerative medicine, drug toxicity and efficacy testing, developmental biology disease, and pathology of infectious disease, the developmental biology disease (29%) accounted for the largest share in 2019. In terms of source (pluripotent stem cells and organ-specific adult stem cells), the pluripotent stem cells sector secured the largest share in the global organoids markers in 2019.

The authors believe that following applications will serve as the foundation for frequent utilization of organoids in fundamental and translational research in the upcoming years.

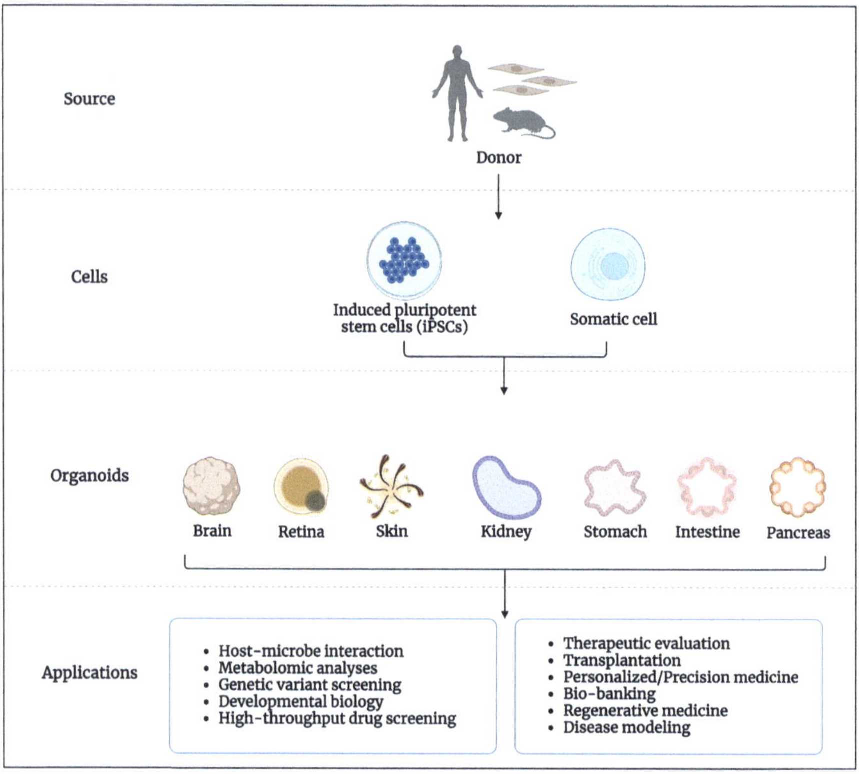

Fig. 4.1 An overview of organoids and their diverse applications

4.3.1 High-Throughput Drug Screening

In order to discover novel drugs and screen approved drug candidates (repurposing of drugs), human-derived 3D organoid models are often used as they offer opportunities for the development of personalized medicine. To achieve high-throughput screening of potential drug candidates, serum-free embryoid bodies sourced from human-induced pluripotent stem cells coupled with high-resolution imaging were employed (Durens et al. 2020). This method can compensate for experimental variability commonly observed with 3D cultures, while considerably reducing the processing time, making them an efficient tool for high-throughput drug screening.

4.3.2 Disease Modeling

The organoids can virtually recreate organ development as well as human diseases, such as genetic abnormalities, infectious diseases, and even cancers. Firstly, Dekkers et al. (2013) were successful in simulating cystic fibrosis (CF), an autosomal recessive genetic disorder caused by chloride channel dysregulation, in the human patient-derived intestinal organoids with CFTR (cystic fibrosis transmembrane conductance regulator) gene mutations (Dekkers et al. 2013). Secondly, many organoids were generated to mimic the infectious disease progression. Besides, small intestinal organoids from bovine and porcine jejunal tissues that have been created by Derricott H et al. (2019) demonstrated that these organoids are susceptible to both bacterial (*Salmonella typhimurium*) and parasitic (*Toxoplasma gondii*) infection (Derricott et al. 2019). Lastly, compared to 2D cancer cell lines, cancer-derived organoids are superior for researching tumor biology because they can maintain tumor heterogeneity. Patient-derived xenografts (PDXs), which mimic the tumor model more accurately and are also physiological, were produced, but their creation and upkeep are costly and time-consuming. Organoid technology has advanced recently, allowing for the unparalleled in vitro modeling of human diseases. As a result, several other types of cancer organoids, such as kidney (Bolck et al. 2019), brain (da Silva et al. 2018), lungs (Sachs et al. 2019), ovarian (Kopper et al. 2019), gastric (Nanki et al. 2018; Seidlitz et al. 2019), and pancreatic (Baker et al. 2016) have been produced. Using genome editing technologies like RNA interference or CRISPR-Cas9, mutations can be induced in the wild-type organoids to convert them for use in cancer modeling (Drost et al. 2015; Matano et al. 2015).

4.3.3 Precision Medicine and Biobanking

Because of organoids's long-term expansion capacity, biobanking of disease-derived organoids is now possible. These biobanks have proven to be valuable resources for therapeutic studies such as cancer classification using omics analysis and high-throughput screening for the generation of precision medicine. Some of the tumor-derived live organoid biobanks are lung (Li et al. 2020), liver (Broutier et al. 2017), breast (Sachs et al. 2018), bladder (Lee et al. 2018), gastric (Yan et al. 2018), prostate (Beshiri et al. 2018), pancreatic (Driehuis et al. 2019a, b), and glioblastoma (Jacob et al. 2020) cancer. Furthermore, patient-derived organoids (PDOs) offer invaluable opportunities for developing precision medicine due to their efficient drug screening ability. The ability to generate comparable healthy and cancerous PDOs in 3D organoids is a promising strategy for efficient drug toxicity evaluation and the likelihood of optimizing effective drug doses.

4.3.4 Regenerative Medicine

Due to the recent advances in organoid technology, characterized by expansive growth potential and stable physiological properties, PDOs may be investigated as a potential substitute for organ transplantation. Yui et al. (2012) created the first mouse-derived intestinal organoids and highlighted that mouse colonic organoids can expand and engraft into damaged mouse colons to form functional crypts (Yui et al. 2012). More importantly, when contemplating the use of organoids in the field of regenerative medicine, it is crucial to address potential safety as well as ethical implications prior to using them on patients.

4.3.5 Host-Microbe Interaction

Organoids allow researchers to study cellular and molecular pathways activated by pathogen invasions while maintaining the physiological resemblance to the tissue of origin. Organoid growth can be manipulated to produce a particular cell type, such as gastric pit cells or gastric gland cells (Bartfeld et al. 2015). These characteristics facilitate the investigation of interactions with specified target cells, such as *Salmonella* entry into M-cells, the identification of specific cells targeted by enteric viruses, or Paneth cell degranulation following infection (Bartfeld 2016).

4.3.6 Developmental Biology

Organoids provide a unique opportunity to investigate the complex process of embryogenesis at the molecular and cellular level, which could lead to the discovery of treatments for congenital defects. In terms of human embryogenesis, culturing human embryos in vitro is technically difficult and unethical to grow after the commencement of primitive streak formation (Pera 2017) (Deglincerti et al. 2016; Shahbazi et al. 2016). Human pluripotent stem cell-derived organoids are an excellent alternative for studying early-stage development processes like gastrulation and neurulation (Li et al. 2018).

4.4 Organoids: Revival, Culturing, and Cryopreservation

The basic protocols for maintaining and culturing organoids are almost similar to those employed for traditional primary or continuous cell cultures; however, it is important to note that various types of organoids require additional tissue-specific media formulations that help them in their differentiation to a particular tissue organoid as well as distinct extracellular matrices (specific matrices for a particular tissue organoid). Therefore, the researcher needs to customize the media formulations as well as extracellular matrices according to the requirement for a specific organoid culture. In this chapter, we have outlined only general and basic steps involved in the revival, culturing, and cryopreservation of organoids (Fig. 4.2).

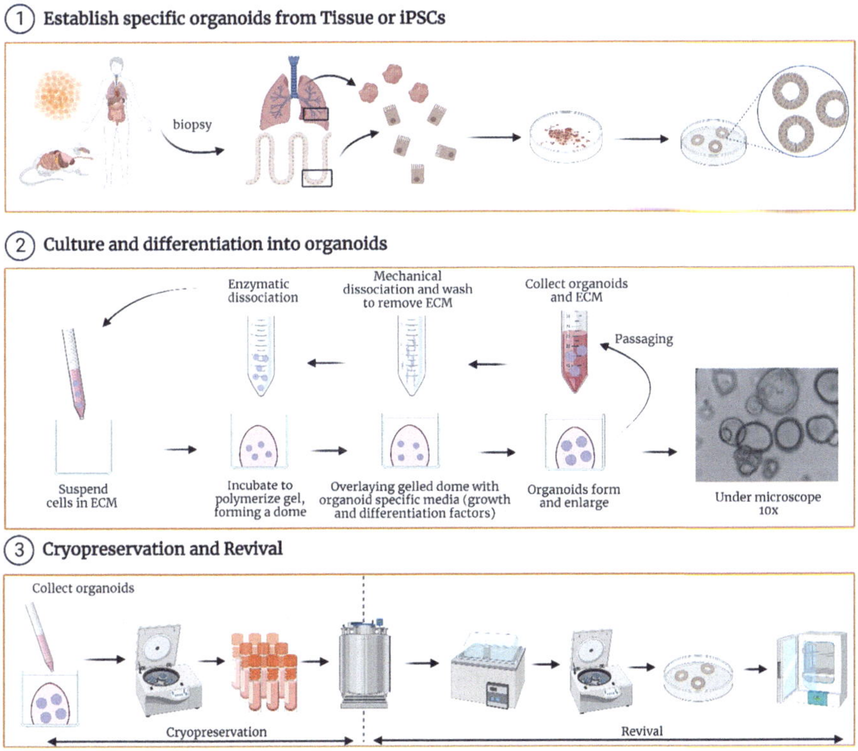

Fig. 4.2 The basic protocol for maintaining, culturing and cryopreservation of organoids

4.4.1 Revival of Cryopreserved Organoids

Organoid cultures are initiated by removing cryovials from liquid nitrogen storage and rapidly thawing them, removing the cryopreservation medium, and generating a pellet of cells. Upon being resuspended in liquid extracellular matrix (ECM), the pellet is dispersed as droplets onto tissue culture plastic, where it solidifies into a gel after incubating at 37 °C, which is further poured with warm growth medium. Within this solidified gel domes, organoids were allowed to grow and expand.

Equipment and Material required

- Cryopreserved organoids
- Basal media (HEPES buffer and L-glutamine)
- Complete organoid culture medium
- Suitable extracellular matrix (ECM)
- 70% ethanol
- Pipettors and tips

- Serological pipets
- Water bath at 37 °C
- Tabletop centrifuge
- Biosafety cabinet (BSC)
- Cell culture incubator (37 °C, 5% CO_2)
- 6-well tissue culture plates
- 15 ml conical tubes

Procedure

1. According to the requirement of different types of organoids, organoid-specific culture media (we shall be using the term "complete or growth media" for it) is required, while basal media (containing only HEPES buffer and L-Glutamine) will be used for washing purposes. (Note: Table 4.2 enlist the organoid-specific differentiation and nutritional requirement).
2. While removing cryovials from the liquid nitrogen storage, proper safety protocol should be ensured such as wearing personal protective equipment, including a protective mask or goggles, insulated gloves, and a full-sleeve lab coat. After removing, cryovial containing the organoid should immediately be thawed in a 37 °C water bath. Keep stirring the cryovial gently in the water bath, and do not submerge cryovial completely in the water. Prolonged thawing in the water bath should be avoided as it can reduce the organoid viability.
3. Transfer the cryovial to the biosafety cabinet (BSC) after surface decontaminating it with 70% ethanol.
4. Transfer all the contents of the thawed cryovial, to the tube containing 10 mL pre-warmed basal media using a 1 mL pipette, drop-wise.
5. Pellet down the organoid suspension in a centrifuge at 500 × g for 5 to 10 minutes at 25 °C temperature.
6. Remove the supernatant carefully without any disruption to the organoid pellet. If haziness is visible in basal media (supernatant), repeat step 5.
7. Resuspend the organoid pellet in 200 μL of cold ECM (thawed on ice or at 4 °C) by pipetting up and down gently to ensure complete resuspension of pellet. Gentle mixing is done to avoid bubble formation in ECM. Inoculate 2.5–5.0 $x10^5$ viable cells in 100 μL of ECM per well of a 6-well culture plate; it should be noted that 100 μL organoid suspension should be dispensed in the form of 10–15 droplets per well of the pre-warmed culture plate so that droplets make elevated domes. The same can be repeated for other wells also.
8. Cover the culture plate with a lid, and carefully invert it as soon as the inoculation is complete. Inverting the culture plate will promote dome formation as well as reduce the risk of organoids penetrating the ECM.
9. Place the culture plate in the inverted position in a cell culture incubator maintained at 37 °C and 5% CO_2. Allow the ECM to polymerize by leaving the culture plate undisturbed for 20 to 30 minutes in inverted position. Additionally, steps 7, 8, and 9 should be done quickly to avoid early polymerization of ECM.

Table 4.2 Cultural requirement of different organoids

Organoid type	Sources	Primary culture medium along with growth and differentiation factors	References
Lung	Lung alveolar epithelial cells (5×10^3) and lung fibroblasts (5×10^4)	Mix cells in the 1:1 ratio with small airway growth medium (SAGM medium, Lonza, with all additives except epinephrine) and Matrigel (Corning); ROCK inhibitor Y27632 (Sigma) was added for the first 2 days in the medium; ligand treatment of organoids [Wnt3a (200 ng/mL), Fgf7 (50 ng/mL), Fgf10 (50 ng/mL), XAV939 (10 µM), and CHIR99021 (1 µM)] were performed after the first 2 days	Zacharias et al. (2018)
Gastric	hPSCs	*Differentiation of definitive endoderm* – plating of single cells (hPSCs) in a Matrigel-coated 24-well dish, 1.5×10^5 cells per well with mTesR1 media (Stem Cell Technologies) containing ROCK inhibitor Y-27632 (10 µM; Stemgent). Next day, cells were exposed to activin A (100 ng/ml, Cell Guidance Systems), and BMP4 (50 ng/ml, R&D Systems) in RPMI 1640 media (Invitrogen) with increasing concentrations of 0%, 0.2%, and 2.0% defined FBS (Invitrogen) for 3 days *Posterior foregut spheroid formation* – culture the cells in RPMI 1640 media with 2.0% defined FBS containing growth factors such as CHIR99021 (2 µM, Stemgent), FGF4 (500 ng/ml, R&D Systems), WNT3A (500 ng/ml, R&D Systems), and NOG (200 ng/ml, R&D Systems) for 3 days; add retinoic acid (2 µM, Sigma Aldrich) on the third day *Gastric organoids* – suspend spheroids in 50 µl Matrigel (BD Biosciences), and plate in a 3D droplet. After polymerization, overlay spheroids with gut media containing advanced Dulbecco's Modified Eagle Medium (DMEM)/F12 with N2, L-glutamine, B27, HEPES (10 µM), penicillin/streptomycin, EGF (100 ng/ml, R&D Systems) (retinoic acid and NOG were added in the gut media for the first three days). Change the media after every 3–4 days	McCracken et al. (2014)
	$Lgr5^+$ stem cells	Suspend the isolated gastric gland units (100 glands) from Lgr5-EGFP-ires-CreERT2 mice in 50 µl Matrigel (BD Bioscience). After polymerization, gastric culture medium containing advanced DMEM/F12 with N2, B27, *N*-acetylcysteine (Invitrogen), gastrin (10 nM, Sigma-Aldrich), and growth factors such as EGF (50 ng/ml, PeproTech), R-spondin-1 (1 µg/ml), Noggin (100 ng/ml, PeproTech), FGF10 (100 ng/ml, PeproTech), and Wnt3A (Y-27632 was added for the first 2 days, and Exendin 4 (50 nM) was added for the differentiation of enteroendocrine lineage)	Barker et al. (2010)

(continued)

Table 4.2 (continued)

Organoid type	Sources	Primary culture medium along with growth and differentiation factors	References
Intestinal	ESCs and iPSCs	*Differentiation of definitive endoderm* – the protocol is similar to that for gastric organoids generation *Mid-hindgut spheroids* – suspend the definitive endoderm in hindgut induction medium (RPMI 1640, 2% decomplemented FBS, L-glutamine (2 mM), activin A (100 ng/ml), penicillin-streptomycin, Chiron 99021 (3 μM, Tocris), and FGF4 (500 ng/mL) for 4 days *Intestinal organoids* – plate at least 30 spheroids per well in Matrigel (BD Biosciences). After polymerization, overlay spheroids with intestinal growth media containing advanced DMEM/F-12, HEPES (15 mM), L-glutamine (2 mM), N2, B27, penicillin-streptomycin, EGF (100 ng/ml, R&D), and Noggin (100 ng/ml, R&D). Change the media twice in a week	Múnera et al. (2017)
Liver	iPSCs	Resuspend 1×10^6 human iPSC-derived hepatic cells, $0.8–1.0 \times 10^6$ human umbilical vein endothelial cells (HUVECs), and 2×10^5 human mesenchymal stem cells (MSCs) in the hepatocyte culture medium and EGM, containing oncostatin M (10 ng/ml, R&D System), dexamethasone (0.1 μM, Sigma-Aldrich), hepatocyte growth factor (20 ng/ml, PromoKine), and SingleQuots (Lonza) followed by plating on pre-solidified Matrigel. After 4–6 days of culture, cells will self-organize in 3D liver buds	Takebe et al. (2013)

Organoid type	Sources	Primary culture medium along with growth and differentiation factors	References
Pancreas	hPSCs (pancreatic progenitor)	*For differentiation of hPSCs into pancreatic progenitor* – human PSCs were grown in BE1 containing GSK3β inhibitor (3 μM, CHIR99021, Axon MedChem) and activin A (100 ng/mL). On the next day, the medium was replaced by BE1 supplemented only with 100 ng/mL activin A for additional 3 days to differentiate them into definitive endoderm. After 4 days, media were replaced with BE1 containing KFG (50 ng/mL, PeproTech) for 2 days to get gut-tube endoderm. From 6 to 10 days, gut-tube endoderm was cultured in BE3 medium supplemented with retinoic acid (2 μM, Sigma), LDN-193189 (200 nM, Sigma), SANT-1 (0.25 μM, Sigma), and PD0325901 (500 nM, Calbiochem) to differentiate into pancreatic endoderm. During days 10–14, the pancreatic endoderm was cultured into BE3 containing Indolactam V (330 nM, Stem Cell Technologies), glucose (16 mM), fibroblast growth factor (FGF10, 50 ng/mL, PeproTech), and SB431542 (10 μM, Axon MedChem) to get pancreatic progenitor *Pancreatic progenitor to 3D pancreatic organoid culture* – resuspend pancreatic progenitor in BE3 medium with ROCK inhibitor (10 μM) followed by mixing with Matrigel (1:3 ratio). After solidification, Matrigel was overlayed with BE3 medium with ROCK inhibitor (10 μM) for 4 days. On the next day, differentiation was allowed to continue in BE3 medium supplemented with FGF2 (10 ng/mL) and ROCK inhibitor (10 μM) for the next 4 days. Subsequently, pancreatic organoids were cultured in BE3 medium with FGF2 (10 ng/mL) and nicotinamide (10 mM, Sigma) and this medium was replaced after every 2–3 days (Note: BE1 medium here refers to BE1: MCDB131 (Invitrogen) supplemented with glucose (0.8 g/L, Sigma), fatty acid-free BSA (0.1%, A7030, Sigma), L-glutamine (2 mM), and sodium bicarbonate (1.174 g/L, Sigma). Besides, BE3 medium constitutes MCDB131 with sodium bicarbonate (1.754 g/L), fatty acid-free BSA (2%), glucose (0.44 g/L), L-ascorbic acid (44 mg/L), L-glutamine (2 mM), and 0.5× insulin-transferrin-selenium-ethanolamine (ITS-X)	Hohwieler et al. (2017)

(continued)

Table 4.2 (continued)

Organoid type	Sources	Primary culture medium along with growth and differentiation factors	References
Prostate	Luminal or basal- cells	After enzymatic digestion of human prostate pieces, luminal and basal cells can be sorted based on the expression of CD26 and CD49f, respectively. The prostate culture medium (50 mL) constitute advanced DMEM/F12 as basal medium supplemented with 5.0 μl A83-01 (5 mM in DMSO), 500 μl nicotinamide (1 M in PBS), 1.0 ml B27, 0.5 μl of EGF (0.5 mg/ml in PBS + 0.1% BSA), 50.0 μl R-spondin 1 (500 μg/ml in PBS + 0.1% BSA or 10% conditioned medium), 5.0 μl FGF2 (50 μg/ml in PBS + 0.1% BSA), 5.0 μl FGF10 (0.1 mg/ml in PBS + 0.1% BSA), 125.0 μl N-acetylcysteine (500 mM in PBS), 50.0 μl dihydrotestosterone (1 μM in ethanol), 50.0 μl Noggin (100 μg/ml in PBS + 0.1% BSA), 16.7 μl SB202190 (30 mM in DMSO), 5.0 μl prostaglandin E2 (10 mM in DMSO), 10 mM HEPES, penicillin/streptomycin, and GlutaMAX. (Note: Y-27632 is added to the prostate culture medium after passaging)	Drost et al. (2016)
Brain	Human embryonic stem cells (hESCs)	hESCs (6000 cells/well) were culture in cerebellar differentiation medium (growth-factor-free chemically defined medium, gfCDM) supplemented with SB431542 (10 μM, Tocris Bioscience), Y-27632 (10 μM), and insulin (7 μg/ml). On day 2, recombinant human FGF2 (50 ng/ml, R&D Systems) was added to the culture. On days 7 and 14, medium was replaced with gfCDM-supplemented insulin (one-third and full replacement, respectively). hESCs were grown up to 21 days	Muguruma et al. (2015)

10. Bring the culture plate after ECM polymerization to a BSC, and add 2 mL of complete medium maintained at room temperature in each well of the culture plate without disturbing domes. Keep the culture plate in the cell culture incubator (37 °C and 5% CO_2) for 2 to 7 days.
11. Check the growth of the organoid daily under a light microscope and change the complete medium on every alternate day.

4.4.2 Culturing and Passaging of Organoids

Similar to primary or continuous cell cultures, organoids can also be dissociated for passaging; however, this process needs to be optimized for a particular organoid culture. One must note that during the pipetting of dissociated organoids (release of organoids from domes), pre-wetting of serological pipette or plastic tips with basal medium should be ensured; otherwise dissociated organoids can stick to their walls or surfaces, resulting in reduced recovery and yield. Organoids, depending of the tissue types, will require either mechanical or enzymatic dissociation/fragmentation or both to fragment up to single-cell level or small multicellular fragments. Therefore, fragmentation process may likely vary for each organoid type. The frequency of passaging of an organoid culture generally depends on tissue type, size, and the rate of accumulation of cellular debris, and therefore, organoid-specific information may be referred or optimized, in the case of culturing of a new organoid culture. Generally, the frequency of organoid culture passaging can vary from 7 to 14 days.

Equipment and Material Required

- Similar to those required while reviving the cryopreserved organoid culture and trypsin-EDTA (organoid-specific dissociation medium are available commercially). In the absence of organoid-specific dissociation medium, trypsin-EDTA may be used.

Procedure

1. Bring the temperature of basal medium and ECM to 4 °C on ice and the complete medium to room temperature.
2. Aspirate the depleted complete medium from each well of the 6-well culture plate without disturbing the organoid domes, and discard it. During pipetting, some floating domes may hinder aspiration, and therefore, they should be carefully avoided.
3. Detach the organoid domes from the bottom of the culture plate using a cell scraper. If the cell scrapper is not available, a 1 mL pipette tip can also be used for detaching the organoid domes.
4. Add 250 to 300 μL of cold basal medium to each well. This will prevent detached organoid domes from sticking to the tips as well as ease their transfer

a new conical tube using a 1 mL pipette. While handling organoid domes, it is always recommended to use pre-wetted tips for easy transfer.

5. Transfer the detached organoid domes along with cold basal medium to a new 10 mL conical tube using a 1 mL pipette.
6. To release the organoids from ECM, add 100–200uL cold ECM and 2–3 mL organoid-specific medium in the conical tube (for each well organoids). Pipet this suspension 15–20 times up and down using a 1 mL pipette.
7. Add cold basal medium to make up the final volume of 10 mL after mechanical disruption; this will help in liquefaction of ECM and subsequent removal during the washing steps.
8. Pellet down the suspension (dissociated organoids) using a centrifuge at 500 × g for 5 to 10 minutes at 25 °C temperature.
9. Discard the supernatant carefully without any disruption to the organoid pellet. Following centrifugation, if haziness in the supernatant is visible (indicating a partial dissociation and presence of some organoids), repeat steps 6 to 9.

   (Note: After dissociating/releasing the organoid from the ECM, they have to be broken down into either smaller fragments or even single cells. Depending on the organoid type, this can be achieved by either mechanical disruption or enzymatic dissociation, while others may even require both of these methods for disruption.)
10. For gentle disruption, resuspend the pellet in 1–2 mL basal medium. Break the organoids into smaller pieces by pipetting up and down the mixture. Observe the disruption under the inverted microscope, and continue with the dissociation step to reduce the organoids to smaller fragments. Proceed to step 13.
11. For enzymatic dissociation, add room temperature dissociation buffer (2–3 mL for 1 well of 6-well cell culture plate) to the organoid pellet, and mix it using vortex. Incubate the tube in a water bath maintained at 37 °C up to 15 minutes, with vortexing periodically.
12. For organoids which may require both enzymatic dissociation and mechanical disruption, resuspend the pellet in 1–2 mL pre-warm dissociation buffer followed by pipetting up and down the mixture. This will ensure breaking the organoids into smaller pieces/fragments or single cells. It is important to observe the disruption under inverted microscope regularly. In steps 11 and 12, once the desired dissociation of organoid has been achieved, stop the dissociation process by adding 7–8 mL of basal medium. Proceed to step 13.
13. Pellet down the suspension (dissociated single cells or multicellular fragments) at 500 × g for 5 to 10 minutes at 25 °C temperature using a centrifuge.
14. Discard the supernatant carefully without disturbing the pellet.
15. Resuspend the pellet in ECM (5×10^5 cells in 100 μl of ECM per well of a 6-well plate) by pipetting it up and down. While pipetting, bubbles should be avoided in ECM. If the cell count is unknown, then split the pellet in the ratio of 1:4. For example, resuspend the pellet in 400 μl of ECM, followed by inoculating a fresh 6-well plate at 100 μl per well. It is important to note that this procedure should be completed quickly in cold ECM to avoid early polymerization.

16. Take out the pre-warmed tissue culture plate from the CO_2 incubator, and place it in the BSC. Proceed quickly to avoid cooling the culture plate, as warmth facilitates dome formation. A warm water bottle, maintained at 37 °C, can be used to place the culture plate over it while inoculating the larger container or working with ECMs that take too long to polymerize.
17. Dispense 10 to 15 µl droplets from a 100 µl organoid suspension, evenly distributed, in a well of a preheated 6-well culture plate. Repeat this for each well to be inoculated. To ensure the homogeneity of the organoid solution, pipette constantly to mix, or periodic mixing should be done in case of a larger volume of suspension, or ECM. Bubbles should be avoided in any case. Around eight to ten droplets should be dispensed quickly while making sure that droplets do not mix or touch the walls of the well. Ensure that the droplets are in the form of elevated domes.
18. Cover the plate with a lid, and carefully invert it as soon as the inoculation is complete. Inverting the culture plate can promote dome development and reduce the risk of organoids penetrating the ECM and coming into contact with the surface of the plate.
19. Place the inverted plate in a 37 °C CO_2 incubator. Allow the ECM to polymerize by leaving the plate undisturbed for 20 to 30 minutes.
20. Bring the plate to BSC and add complete medium (maintained at room temperature) without disturbing the domes. Incubate the plate in a CO_2 incubator.
21. The medium should be changed every 2 to 3 days, and the cells can be passaged every 7 to 14 days (subject to the type of organoid cultures).

4.4.3 *Cryopreservation of Organoids*

Organoids should be cryopreserved at the time of passage and can be preserved intact, in fragments, or dissociated. In order to cryopreserve intact or partially dissociated organoids, it is necessary to refer to organoid-specific information regarding cryopreservation conditions.

Equipment and Material Required

- Cell-freezing media (3dGRO™ Organoid Freeze Medium, SIGMA-SCM301, Recovery™ Cell Culture Freezing Medium, Thermo Fisher Scientific-12648010)
- Serological pipets
- Tabletop centrifuge
- Biosafety cabinet (BSC)
- Conical tubes (15 ml and 50 ml)
- Cryovials
- –80 °C freezer
- Cell scraper

Procedure

1. Collect organoids from the domes, and discard ECM by following steps 2 to 9 and 12 from the "Culturing and Passaging of Organoids" section.
2. Pool the organoid pellets from multiple wells into one conical tube, and centrifuge them down at 500 × g for 5 min at 4 °C.
3. Add 500 μl of freezing medium to 2×10^6 cells (or to 100 to 200 μl of domes collected from one to two wells of a 6-well culture plate) to resuspend the organoid pellet.
4. Transfer 0.5 ml suspension in each of the labeled cryovial. Pipetting up and down the freezing medium will ensure a homogenous mixture. Pre-wetting of the pipet with freezing medium can be done to minimize the sticking of cells.
5. Organoids should be cryopreserved at a gradual cooling rate of −1 °C/minute 1 °C/min using a rate-controlled freezer for best recovery.
6. For long-term storage, transfer vials to liquid nitrogen vapor phase.

4.5 Organoids: An Effective Tool for Antiviral Studies

Infectious agents tend to be species-specific and have limited tropism (Kumar et al. 2021a, b, 2022); this implies that not all in vivo models can be used to study all infectious diseases. Humanized rodents are a better alternative, but their widespread use is hampered by ethical concerns arising from their unchecked use (Sarkar and Heise 2019). It is difficult to efficiently mimic the pathophysiology of human disease in animal models, owing to viral pathogens having unique human tropisms and thus being unable to replicate in nonnatural hosts. An efficient model capable of replicating human physiology and predicting human disease is thus urgently required. Organoids are emerging as a valuable resource in the context of COVID-19 research (Chia et al. 2022; Clevers 2016; Sridhar et al. 2020).

4.5.1 Enteric Viruses

The first instance of how organoids enable the study of a previously unculturable virus was documented by K. Ettayebi and colleagues (2008). They hypothesized that the presence of bile is essential for strain dependent human norovirus (HuNoV) replication. They also stated that serum neutralization and inactivation techniques such as irradiation and heating can be used to eliminate HuNoV infectivity. Furthermore, Hynds et al. (2019) described a method for increasing the extended cultivation of primary human airway basal cells while preserving genomic stability through 3 T3-J2 fibroblast co-culture and ROCK inhibition. They also described the differentiation protocol as three-dimensional tracheospheres with basal, ciliated, and muco-secretory cells (Hynds et al. 2019). This enables research into lung and respiratory-related host-viral interactions, as well as antiviral studies for drug development. Kapil Saxena and colleagues created the first human intestinal enteroid (HIE) culture to demonstrate rhesus rotavirus host range restriction in vitro, correlating with in vivo results (Saxena et al. 2016). Yin et al. (2018) created an intestinal enteroid and successfully tested 6-thioguanine as a rotavirus replication inhibitor by inhibiting Rac1 activation (Yin et al. 2018). Chen et al. (2019) used DHODH enzyme inhibitors, Brequinar (BQR), and leflunomide (LFM), to deplete the pyrimidine nucleotide pool, which effectively inhibited rotavirus replication (Chen et al. 2019). They investigated gemcitabine further and discovered that it has anti-rotaviral activity by inhibiting the pyrimidine biosynthesis pathway (Chen et al. 2020). Zhao et al. produced more closely resembling human intestinal organoids after optimizing the differentiation technique. They demonstrated in their study that the optimized human intestinal organoids are significantly more sensitive to enterovirus 71 (EV-A71) than coxsackievirus A serotype 16 (CVA16) and can better mimic the native human intestinal epithelium. Their findings demonstrated that these optimized enteroids are capable of developing coronavirus and enterovirus infections and are an excellent in vitro model (Zhao et al. 2021).

4.5.2 *Zika Virus*

Emricasan (IDN-6556 or PF-03491390), a pan-caspase inhibitor, was shown by Miao Xu et al. (2016) to prevent cell death in human astrocytes and human neural progenitor cells infected with the Zika virus (ZIKV) in 2D monolayer and cerebral organoid cultures. They also discovered that niclosamide prevents ZIKV replication in 3D brain organoids (Xu et al. 2016). J. Dang and colleagues successfully infected a hESC cerebral organoid with the Zika virus and demonstrated that infection causes Toll-like receptor 3 (TLR3) activation, which results in a reduction in overall organoid size, i.e., microcephaly (Dang et al. 2016). Aleksandra Pettke et al. and colleagues (2021) successfully infected hiPSC brain organoids with ZIKV and discovered that TH6744 and TH5487 treatment can reverse ZIKV-induced neurotoxicity by impairing the formation of new viral particles late in the viral life cycle, thereby limiting ZIKV propagation (Pettke et al. 2020). Contreras D et al. (2023) investigated ZIKV-mediated ocular cell injury processes using human fetal retinal pigment epithelial cells (FRPEs), hiPSC-derived retinal stem cells (iRSCs), and retinal organoids. They discovered that 6-AZA treatment stopped virus production and apoptosis in FRPE cells immediately after ZIKV infection (Contreras et al. 2023). As demonstrated in a study by Yu W et al. (2023), desoxyrhapontigenin (DES) has anti-ZIKV activity in trophoblast cell lines, human trophoblast stem cells, and complex placental organoid models. They discovered that DES treatment reduced ZIKV load in serum and target tissue, improved ZIKV-induced pathogenic changes by targeting ZIKV E proteins, and decreased the expression of putative ZIKV receptors (Yu et al. 2023).

4.5.3 SARS-CoV-2

SARS-CoV-2 infection triggers inflammation in the surface cells of the eye, implying that the virus can enter through the eye because it expresses both the ACE2 receptor and the transmembrane serine protease 2 (TMPRSS2) (Makovoz et al. 2020). Yang L. and colleagues created hPSC-derived organoids and found that ACE2 is expressed in human pancreatic alpha and beta cells using single-cell RNA sequencing and immunostaining. They also discovered that both adult human pancreatic beta cells and hPSC-derived beta cells are prone to SARS-CoV-2 infection (Yang et al. 2020), which is consistent with a previous clinical study that found a link between COVID-19 and type 2 diabetes (Zhu et al. 2020). Jana Krüger et al. (2021) investigated antiviral drugs against SARS-CoV-2 using PSC human intestinal organoids (HIO). They discovered that remdesivir inhibits SARS-CoV-2 infection in PSC-HIOs; EK1 (spike protein inhibitor), but not famotidine inhibits SARS-CoV-2 infection in PSC-HIOs (Krüger et al. 2021). Pinar Mesci et al. (2022) investigated sofosbuvir, an FDA-approved anti-hepatitis C treatment that inhibits HCV replication by inhibiting its RNA-dependent RNA polymerase. Because SARS-CoV-2 RdRp shares a high level of sequence and structural homology with HCV, it was hypothesized that sofosbuvir could inhibit SARS-CoV-2 RdRp using iPSC brain cortical organoids. They discovered that sofosbuvir can inhibit SARS-CoV-2 replication while also preventing neuronal changes (Mesci et al. 2022).

According to the findings of Monteil V. et al., clinical-grade human recombinant soluble ACE2 (hrsACE2) was found to reduce SARS-CoV-2 infection in a dose-dependent manner. Furthermore, they demonstrated that hrsACE2 can significantly suppress the infection of human kidney and blood vascular organoids in its early stages (Monteil et al. 2020). Samuel RM and colleagues developed a screening technique to identify compounds that reduce ACE2 levels in human embryonic stem cell (hESC)-derived lung organoids and cardiac cells. They came to the conclusion that antiandrogenic drug treatment (ketoconazole, spironolactone, finasteride, and dutasteride) reduces ACE2 expression as well as TMPRSS2 expression and protects hESC-derived lung organoids from SARS-CoV-2 infection (Samuel et al. 2020). Yuling Han et al. (2021) studied SARS-CoV-2 infection and tropism using hPSC and various human organoids. They also created lung organoids and colonic organoids to study antiviral compounds. Instead of inhibiting luciferase activity, they discovered that imatinib, MPA, and QNHC specifically prevented SARS-CoV-2 entry and viral infection in hPSC-LOs; imatinib, MPA, and QNHC were also found to inhibit SARS-CoV-1 entry and viral infection in a dose-dependent manner. They demonstrated that imatinib, MPA, and QNHC inhibit SARS-CoV-2 infection in Vero E6 cells in a toxicity-independent manner, similar to hPSC-LOs and hPSC-Cos (Han et al. 2021).

Sirle Saul et al. (2021) demonstrated that lapatinib reduces both inflammation and epithelial injury in human lung organoids by inhibiting viral entry, a phenotype similar to that seen with RNAi-mediated ErbB4 suppression. As a candidate antiviral target, ErbB4 has received the least attention; they discovered that lapatinib's antiviral activity correlates with lower phospho-ErbB4 levels in vitro (Saul et al. 2021). Ebisudani T et al. (2021) investigated lopinavir, nelfinavir, and remdesivir against SARS-CoV-2 using alveolospheres. They discovered that lopinavir and nelfinavir, both protease inhibitors, significantly reduced viral titers at 1 dpi, but neither drug could terminate viral replication, whereas remdesivir demonstrated a potent antiviral effect at concentrations within the range of serum drug levels in patients (Ebisudani et al. 2021). Duarte RRR et al. (2021) created a hESC lung organoid for testing anti-SARS-CoV-2 drug. They reported that atorvastatin, which was previously predicted to bind to SARS-main CoV-2's protease (Mpro) and RNA-dependent RNA polymerase (RdRp), inhibited viral entry in a lung organoid model (Duarte et al. 2021). Jan Wysocki et al. (2021) investigated the effect of hACE2 1-168 (human-soluble ACE2 protein) on hESCs kidney organoids. With improved albumin binding affinity, hACE2 was generated and fused with an albumin binding domain (ABD) using an artificial gene encoding ABDCon. They discovered that this novel ACE2 1-618-ABD can effectively neutralize SARS-CoV-2 infectivity in human kidney organoids; whereas plasma ACE2 activity of naked ACE2 1-618 and ACE2 1-740 lasted about 8 hours, ACE2 1-618-ABD produced significant activity after 96 hours (Wysocki et al. 2021). Garcia G Jr et al. (2021) created cardiomyocytes organoids from human iPS cells and investigated berzosertib's anti-SARS-CoV-2 activity. Berzosertib, an ATR kinase inhibitor, was found to have potent antiviral activity against SARS-CoV-2 by blocking the replication cycle at the postentry step. Berzosertib also inhibited SARS-CoV-1 and the Middle East respiratory syndrome coronavirus (MERS-CoV) replication (Garcia et al. 2021).

To test the efficacy of EK1 peptide, camostat, and nafamostat against SARS-CoV-2, Shashi Kant Tiwari et al. (2021) created iPSC lung organoids and cerebral organoids. The researchers discovered that spike protein inhibitors (EK1 peptide) and TMPRSS2 inhibitors (camostat/nafamostat) prevented viral entry in LORGs (Tiwari et al. 2021). In order to test antiviral drugs that reverse SARS-CoV-2-induced cardiac injury, Mills RJ et al. (2021) created human cardiac organoids (hCOs). They discovered that bromodomain and extraterminal family inhibitors (BETi) completely prevented cardiac dysfunction and death in a mouse cytokine storm model. Furthermore, BETi reduces viral response gene transcription, ACE2 expression, and SARS-CoV-2 infection of cardiomyocytes (Mills et al. 2021). MEDS433, a novel inhibitor of human dihydroorotate dehydrogenase (hDHODH), a critical cellular enzyme in the de novo pyrimidine biosynthesis pathway, was shown to have antiviral activity against SARS-CoV and SARS-CoV-2. MEDS433 inhibited SARS-CoV-2 replication in human embryonic stem cell-derived kidney organoids, indicating that MEDS433 is still effective against SARS-CoV-2 in the kidney model (Calistri et al. 2021). Poirier et al. discovered that the antiviral Dicer (aviD), an isoform of Dicer, can activate antiviral RNAi by dicing viral double-stranded RNA, including Zika virus and SARS-CoV-2 (Poirier et al. 2021). HK Kim and colleagues used culture medium composition optimization to create an efficient human tonsil epithelial organoid that replicates in vivo characteristics such as stratified squamous epithelial layers with properly distributed tonsil biomarkers. A tonsil organoid model was used to test remdesivir's therapeutic efficacy, demonstrating its utility as a screening platform for drug candidates targeting SARS-CoV-2 (Kim et al. 2022). McNulty J et al. (2023) synthesized four novel mono-methoxy and hydroxyl substituted ring-A dihydronarciclasine derivatives, allowing the 7-hydroxyl derivative to be identified as a potent and selective antiviral agent targeting SARS-CoV-2 and HSV-1. They also reported that the concentration of this small molecule that inhibited HSV-1 infection by 50% (IC_{50}) was estimated to be 0.504 M and 0.209 M using induced pluripotent stem cells (iPCS)-derived brain organ organoids generated from two iPCS lines (McNulty et al. 2023).

4.5.4 *Other Respiratory Viruses*

Ya-Wen Chen and colleagues created lung bud organoids (LBOs) from human pluripotent stem cells. In vitro infection with the respiratory syncytial virus, which is known to cause bronchiolitis and airway blockage in newborns, caused organoid lumens to expand, detach, and shed infected cells, similar to what has been observed in human lungs (Chen et al. 2017). Using a three-dimensional organoid model of human fetal lungs, Harford et al. (2022) demonstrated that respiratory syncytial virus (RSV) can transmit from infected airways of a pregnant woman to developing fetal lungs in utero (Harford et al. 2022). A group of scientists cultured a novel intestinal organoid and demonstrated that human primary intestinal epithelial cells could serve as an alternative infection site for Middle East respiratory syndrome coronavirus (MERS-CoV) (Zhou et al. 2017).

KPY Hui et al. created a human airway organoid and concluded that the biological properties of human bronchus tissues make human airway organoids an effective tool for assessing influenza pandemic risk, as well as infection biology, pathogenesis, and drug screening (Hui et al. 2018). To study influenza virus, Zhou J. and colleagues created human ASC-derived airway organoids that morphologically and functionally resemble human airway epithelium. They also demonstrated that these organoids can detect human-infective viruses by co-evaluating H7N9/Ah and H1N1pdm strains (Zhou et al. 2018). Using a brain organoid, Xiaodong Zhang et al. (2022) investigated the effect of influenza viruses (H1N1-WSN and H3N2-HKT68), enteroviruses (EV68 and EV71), and severe fever with thrombocytopenia syndrome virus (SFTSV) on human brain development and antiviral compound. They discovered that PYC-12, an antiviral drug, could significantly reduce virus replication, apoptosis of neural stem cells (NSCs) and neurons, and inflammatory responses and restore electrophysiological function (Zhang et al. 2022).

4.5.5 *Other Viruses*

According to Holly MK and Smith JG, prototype strains (HAdV-5p and HAdV-41p) as well as clinical isolates of human adenovirus (HAdV) and enteric virus replicate efficiently in human enteroid. They also discovered that HAdV-5p is sensitive to human defensin-5 (HD5), a defense peptide expressed in the host gastrointestinal tract, whereas HAdV-41p is resistant to HD5 (Holly and Smith 2018). Porotto M and colleagues studied human parainfluenza virus (HPIV3) infection in a hPSC-derived lung bud organoid and found that pathogenesis and viral evolution in the lung organoid mirror key features of human viral infection (Porotto et al. 2019). D'Aiuto and colleagues described a differentiation procedure for creating scaffold-free, low-cost hiPSC-derived brain organoids. According to their findings, hiPSC-derived brain organoid neurons are susceptible to herpes simplex virus (HSV-1) infection, and the virus can develop latency in the neurons (D'Aiuto et al. 2019). Hans Clevers lab developed a protocol for the long-term expansion of human mucosal organoids derived from head and neck squamous cell carcinoma (HNSCC). They also validated this in vitro model for understanding mucosal pathology by demonstrating that these organoids can be infected with herpes simplex virus 1 (HSV1) and human papillomavirus (HPV16) (Driehuis et al. 2019a, b). Lohmussaar and colleagues showed that healthy and tumor-derived cervical organoid cultures closely replicate their origin tissues and are suitable for simulating sexually transmitted infections, as demonstrated by the genital virus HSV-1 (Lõhmussaar et al. 2021). Pengfei Li et al. (2022) created fetal and adult human liver organoids to test the efficacy of brequinar, homoharringtonine, and ribavirin against the hepatitis E virus (HEV). They discovered that even at very low concentrations of brequinar (0.05 and 0.5 M), significant inhibition was observed; combined homoharringtonine with ribavirin exerted a clear synergistic effect, particularly at concentrations ranging from 0 to 50 nM homoharringtonine with 0 to 50 M ribavirin (Li et al. 2022).

4.6 Challenges and Future Prospects

Organoids have proven to be an effective revolutionary tool for overcoming many of the bottlenecks associated with 2D culture systems and animal models, and they hold considerable promise for disease modeling and other virological applications. Because of their ability to self-organize and form complex structures that mimic the cellular architecture and functionality of native organs, they are an ideal candidate for expanding our knowledge of various virological studies, including antiviral drug screening; however, many technological challenges limit their widespread use in the scientific community. One of the significant bottlenecks of organoids is the absence of immune cell niches, vascular system, and microbiota, which are typically integral components of specific organs. As a result, organoids do not truly recapitulate cellular and physiological complexity of native organs or tissues. This limits their usage in modeling of antiviral responses because some of the viruses require immune cells for productive infection. To address this bottleneck, three approaches have been proposed: co-culture of epithelial organoids with tissue stromal cells, immune cells, or even the cell population from organoids of blood vessels; organ-on-a-chip, which provides a constant perfusive soluble microenvironment to organoids for maturation and vascularization; and creation of assembloids (combining organoids) to generate more complex multicellular mature and functional structures connecting multiple types of tissues.

Another obstacle is nutritional accessibility, which is especially important in iPSC-derived organoids because as organoids grow in size, the nutrient supply to cells in the organoid's center becomes constrained, resulting in cellular death. This is common in organoids that have a more compact structure, such as brain organoids. Dead cells will eventually amass in the lumens of tissue-derived organoids, forming a hollow cyst that cannot be avoided but can be addressed by mechanical fragmentation. However, because iPSC-derived organoids cannot be fragmented, techniques are being developed to circumvent nutritional accessibility. Finally, organoid models rely on nonspecific extracellular matrix such as Matrigel. Engineering methods for producing tissue-specific matrices from decellularized tissue are being researched, as is the development of synthetic matrices with more exact control over composition and stiffness.

Organoids have boosted confidence among the researchers as an alternative to the conventional models and systems, as they were lately been widely employed to study infection kinetics and antiviral drugs screening against SARS-CoV-2. Organoids are increasingly being used to test antiviral drugs against multiple viruses. We anticipate that technical advances in generating more biologically organoid models that really mirror the natural organs will broaden their applicability in infectious disease modeling, personalized therapy, virology research, and drug screening.

Acknowledgments The authors would like to thank the Indian Council of Agriculture Research, New Delhi; the National Institute of High Security Animal Diseases, Bhopal; and Science and Engineering Research Board (Grant number- CRG/2020/001274).

References

Almeqdadi M, Mana MD, Roper J, Yilmaz Ö, H. (2019) Gut organoids: mini-tissues in culture to study intestinal physiology and disease. Am J Physiol Cell Physiol 317(3):C405–c419. https://doi.org/10.1152/ajpcell.00300.2017

Antonica F, Kasprzyk DF, Opitz R, Iacovino M, Liao XH, Dumitrescu AM et al (2012) Generation of functional thyroid from embryonic stem cells. Nature 491(7422):66–71. https://doi.org/10.1038/nature11525

Baker LA, Tiriac H, Clevers H, Tuveson DA (2016) Modeling pancreatic cancer with organoids. Trends Cancer 2(4):176–190. https://doi.org/10.1016/j.trecan.2016.03.004

Barker N, Huch M, Kujala P, van de Wetering M, Snippert HJ, van Es JH et al (2010) Lgr5(+ve) stem cells drive self-renewal in the stomach and build long-lived gastric units in vitro. Cell Stem Cell 6(1):25–36. https://doi.org/10.1016/j.stem.2009.11.013

Bartfeld S (2016) Modeling infectious diseases and host-microbe interactions in gastrointestinal organoids. Dev Biol 420(2):262–270. https://doi.org/10.1016/j.ydbio.2016.09.014

Bartfeld S, Bayram T, van de Wetering M, Huch M, Begthel H, Kujala P et al (2015) In vitro expansion of human gastric epithelial stem cells and their responses to bacterial infection. Gastroenterology 148(1):126–136.e126. https://doi.org/10.1053/j.gastro.2014.09.042

Beshiri ML, Tice CM, Tran C, Nguyen HM, Sowalsky AG, Agarwal S et al (2018) A PDX/organoid biobank of advanced prostate cancers captures genomic and phenotypic heterogeneity for disease modeling and therapeutic screening. Clin Cancer Res 24(17):4332–4345. https://doi.org/10.1158/1078-0432.Ccr-18-0409

Bolck HA, Pauli C, Göbel E, Mühlbauer K, Dettwiler S, Moch H, Schraml P (2019) Cancer sample biobanking at the next level: combining tissue with living cell repositories to promote precision medicine. Front Cell Dev Biol 7:246. https://doi.org/10.3389/fcell.2019.00246

Broutier L, Mastrogiovanni G, Verstegen MM, Francies HE, Gavarró LM, Bradshaw CR et al (2017) Human primary liver cancer-derived organoid cultures for disease modeling and drug screening. Nat Med 23(12):1424–1435. https://doi.org/10.1038/nm.4438

Calistri A et al (2021) The new generation hDHODH inhibitor MEDS433 hinders the in vitro replication of SARS-CoV-2 and other human coronaviruses. 9:1731

Chen YW et al (2017) A three-dimensional model of human lung development and disease from pluripotent stem cells. Nat Cell Biol 19:542–549

Chen S et al (2019) Suppression of pyrimidine biosynthesis by targeting DHODH enzyme robustly inhibits rotavirus replication. Antiviral Res 167:35–44

Chen S et al (2020) Drug screening identifies gemcitabine inhibiting rotavirus through alteration of pyrimidine nucleotide synthesis pathway. Antiviral Res 180:104823

Chia SPS, Kong SLY, Pang JKS, Soh B-S (2022) 3D human organoids: the next "viral" model for the molecular basis of infectious diseases. Biomedicines 10:1541

Clevers H (2016) Modeling development and disease with organoids. Cell 165(7):1586–1597. https://doi.org/10.1016/j.cell.2016.05.082

Contreras D et al (2023) Differential susceptibility of fetal retinal pigment epithelial cells, hiPSC-retinal stem cells, and retinal organoids to Zika virus. Infection 15:142

D'Aiuto L et al (2019) Modeling herpes simplex virus 1 infections in human central nervous system neuronal cells using two- and three-dimensional cultures derived from induced pluripotent stem cells. J Virol 93

da Silva B, Mathew RK, Polson ES, Williams J, Wurdak H (2018) Spontaneous glioblastoma spheroid infiltration of early-stage cerebral organoids models brain tumor invasion. SLAS Discov 23(8):862–868. https://doi.org/10.1177/2472555218764623

Dang J et al (2016) Zika virus depletes neural progenitors in human cerebral organoids through activation of the innate immune receptor TLR3. Cell Stem Cell 19:258–265

Deglincerti A, Croft GF, Pietila LN, Zernicka-Goetz M, Siggia ED, Brivanlou AH (2016) Self-organization of the in vitro attached human embryo. Nature 533(7602):251–254. https://doi.org/10.1038/nature17948

Dekkers, J. F., Wiegerinck, C. L., de Jonge, H. R., Bronsveld, I., Janssens, H. M., de Winter-de Groot, K. M., … Beekman, J. M. (2013). A functional CFTR assay using primary cystic fibrosis intestinal organoids. Nat Med, 19(7), 939-945. doi:https://doi.org/10.1038/nm.3201

Derricott H, Luu L, Fong WY, Hartley CS, Johnston LJ, Armstrong SD et al (2019) Developing a 3D intestinal epithelium model for livestock species. Cell Tissue Res 375(2):409–424. https://doi.org/10.1007/s00441-018-2924-9

Driehuis E, van Hoeck A, Moore K, Kolders S, Francies HE, Gulersonmez MC et al (2019a) Pancreatic cancer organoids recapitulate disease and allow personalized drug screening. Proc Natl Acad Sci U S A 116(52):26580–26590. https://doi.org/10.1073/pnas.1911273116

Driehuis E et al (2019b) Oral mucosal organoids as a potential platform for personalized cancer therapy. Cancer Discov 9:852–871

Drost J, van Jaarsveld RH, Ponsioen B, Zimberlin C, van Boxtel R, Buijs A et al (2015) Sequential cancer mutations in cultured human intestinal stem cells. Nature 521(7550):43–47. https://doi.org/10.1038/nature14415

Drost J et al (2016) Organoid culture systems for prostate epithelial and cancer tissue. Nat Protoc 11:347–358

Duarte RRR et al (2021) Identifying FDA-approved drugs with multimodal properties against COVID-19 using a data-driven approach and a lung organoid model of SARS-CoV-2 entry. Mol Med 27:105

Durens M, Nestor J, Williams M, Herold K, Niescier RF, Lunden JW et al (2020) High-throughput screening of human induced pluripotent stem cell-derived brain organoids. J Neurosci Methods 335:108627. https://doi.org/10.1016/j.jneumeth.2020.108627

Ebisudani T et al (2021) Direct derivation of human alveolospheres for SARS-CoV-2 infection modeling and drug screening. Cell Rep 35:109218

Eiraku M, Watanabe K, Matsuo-Takasaki M, Kawada M, Yonemura S, Matsumura M et al (2008) Self-organized formation of polarized cortical tissues from ESCs and its active manipulation by extrinsic signals. Cell Stem Cell 3(5):519–532. https://doi.org/10.1016/j.stem.2008.09.002

Eiraku M, Takata N, Ishibashi H, Kawada M, Sakakura E, Okuda S et al (2011) Self-organizing optic-cup morphogenesis in three-dimensional culture. Nature 472(7341):51–56. https://doi.org/10.1038/nature09941

Evans MJ, Kaufman MH (1981) Establishment in culture of pluripotential cells from mouse embryos. Nature 292(5819):154–156. https://doi.org/10.1038/292154a0

Garcia G Jr et al (2021) Antiviral drug screen identifies DNA-damage response inhibitor as potent blocker of SARS-CoV-2 replication. Cell Rep 35:108940

Greggio C, De Franceschi F, Figueiredo-Larsen M, Gobaa S, Ranga A, Semb H et al (2013) Artificial three-dimensional niches deconstruct pancreas development in vitro. Development 140(21):4452–4462. https://doi.org/10.1242/dev.096628

Han Y et al (2021) Identification of SARS-CoV-2 inhibitors using lung and colonic organoids. Nature 589:270–275

Harford TJ et al (2022) RSV-induced changes in a 3-dimensional organoid model of human fetal lungs. PLoS One 17:e0265094

Hofer M, Lutolf MP (2021) Engineering organoids. Nat Rev Mater 6(5):402–420. https://doi.org/10.1038/s41578-021-00279-y

Hohwieler M et al (2017) Human pluripotent stem cell-derived acinar/ductal organoids generate human pancreas upon orthotopic transplantation and allow disease modelling. Gut 66:473–486

Holly MK, Smith JG (2018) Adenovirus infection of human enteroids reveals interferon sensitivity and preferential infection of goblet cells. J Virol 92

Hui KPY et al (2018) Tropism, replication competence, and innate immune responses of influenza virus: an analysis of human airway organoids and ex-vivo bronchus cultures. Lancet Respir Med 6:846–854

Hynds RE, Butler CR, Janes SM, Giangreco A (2019) Expansion of human airway basal stem cells and their differentiation as 3D tracheospheres. Methods Mol Biol 1576:43–53

Jacob F, Salinas RD, Zhang DY, Nguyen PTT, Schnoll JG, Wong SZH et al (2020) A patient-derived glioblastoma organoid model and biobank recapitulates inter- and intra-tumoral heterogeneity. Cell 180(1):188–204.e122. https://doi.org/10.1016/j.cell.2019.11.036

Kim HK et al (2022) Generation of human tonsil epithelial organoids as an ex vivo model for SARS-CoV-2 infection. Biomaterials 283:121460

Kopper O, de Witte CJ, Lõhmussaar K, Valle-Inclan JE, Hami N, Kester L et al (2019) An organoid platform for ovarian cancer captures intra- and interpatient heterogeneity. Nat Med 25(5):838–849. https://doi.org/10.1038/s41591-019-0422-6

Krüger J et al (2021) Drug inhibition of SARS-CoV-2 replication in human pluripotent stem cell-derived intestinal organoids. Cell Mol Gastroenterol Hepatol 11:935–948

Kumar N et al (2021a) Insights into the evolutionary forces that shape the codon usage in the viral genome segments encoding intrinsically disordered protein regions. Brief Bioinform 22

Kumar N et al (2021b) Evolutionary signatures governing the codon usage bias in coronaviruses and their implications for viruses infecting various bat species. Viruses 13

Kumar N et al (2022) Editorial: emerging and re-emerging viral zoonoses. Front Microbiol 13:978259

Lancaster MA, Renner M, Martin CA, Wenzel D, Bicknell LS, Hurles ME et al (2013) Cerebral organoids model human brain development and microcephaly. Nature 501(7467):373–379. https://doi.org/10.1038/nature12517

Lee SH, Hu W, Matulay JT, Silva MV, Owczarek TB, Kim K et al (2018) Tumor evolution and drug response in patient-derived organoid models of bladder cancer. Cell 173(2):515–528.e517. https://doi.org/10.1016/j.cell.2018.03.017

Lee J, Rabbani CC, Gao H, Steinhart MR, Woodruff BM, Pflum ZE et al (2020) Hair-bearing human skin generated entirely from pluripotent stem cells. Nature 582(7812):399–404. https://doi.org/10.1038/s41586-020-2352-3

Li N, Xie T, Sun Y (2018) Towards organogenesis and morphogenesis in vitro: harnessing engineered microenvironment and autonomous behaviors of pluripotent stem cells. Integ Biol 10(10):574–586. https://doi.org/10.1039/c8ib00116b

Li YF, Gao Y, Liang BW, Cao XQ, Sun ZJ, Yu JH et al (2020) Patient-derived organoids of non-small cells lung cancer and their application for drug screening. Neoplasma 67(2):430–437. https://doi.org/10.4149/neo_2020_190417N346

Li P et al (2022) Recapitulating hepatitis E virus-host interactions and facilitating antiviral drug discovery in human liver-derived organoids. Sci Adv 8:eabj5908

Lõhmussaar K et al (2021) Patient-derived organoids model cervical tissue dynamics and viral oncogenesis in cervical cancer. Cell Stem Cell 28:1380–1396.e1386

Makovoz B, Moeller R, Zebitz Eriksen A, tenOever BR, Blenkinsop TA (2020) SARS-CoV-2 infection of ocular cells from human adult donor eyes and hESC-derived eye organoids. SSRN 3650574

Matano M, Date S, Shimokawa M, Takano A, Fujii M, Ohta Y et al (2015) Modeling colorectal cancer using CRISPR-Cas9–mediated engineering of human intestinal organoids. Nat Med 21(3):256–262. https://doi.org/10.1038/nm.3802

McCracken KW, Catá EM, Crawford CM, Sinagoga KL, Schumacher M, Rockich BE et al (2014) Modelling human development and disease in pluripotent stem-cell-derived gastric organoids. Nature 516(7531):400–404. https://doi.org/10.1038/nature13863

McNulty J et al (2023) Truncated ring-A amaryllidaceae alkaloid modulates the host cell integrated stress response, exhibiting antiviral activity to HSV-1 and SARSCoV-2. Sci Rep 13:1639

Mesci P et al (2022) SARS-CoV-2 infects human brain organoids causing cell death and loss of synapses that can be rescued by treatment with Sofosbuvir. PLoS Biol 20:e3001845

Michalopoulos GK, Bowen WC, Mulè K, Lopez-Talavera JC, Mars W (2002) Hepatocytes undergo phenotypic transformation to biliary epithelium in organoid cultures. Hepatology 36(2):278–283. https://doi.org/10.1053/jhep.2002.34858

Mills RJ et al (2021) BET inhibition blocks inflammation-induced cardiac dysfunction and SARS-CoV-2 infection. Cell 184:2167–2182.e2122

Monteil V et al (2020) Inhibition of SARS-CoV-2 infections in engineered human tissues using clinical-grade soluble human ACE2. Cell 181:905–913.e907

Muguruma K, Nishiyama A, Kawakami H, Hashimoto K, Sasai Y (2015) Self-organization of polarized cerebellar tissue in 3D culture of human pluripotent stem cells. Cell Rep 10:537–550

Múnera JO et al (2017) Differentiation of human pluripotent stem cells into colonic organoids via transient activation of BMP signaling. Cell Stem Cell 21:51–64.e56

Mustata RC, Vasile G, Fernandez-Vallone V, Strollo S, Lefort A, Libert F et al (2013) Identification of Lgr5-independent spheroid-generating progenitors of the mouse fetal intestinal epithelium. Cell Rep 5(2):421–432. https://doi.org/10.1016/j.celrep.2013.09.005

Nanduri LS, Baanstra M, Faber H, Rocchi C, Zwart E, de Haan G et al (2014) Purification and ex vivo expansion of fully functional salivary gland stem cells. Stem Cell Reports 3(6):957–964. https://doi.org/10.1016/j.stemcr.2014.09.015

Nanki K, Toshimitsu K, Takano A, Fujii M, Shimokawa M, Ohta Y et al (2018) Divergent routes toward Wnt and R-spondin Niche independency during human gastric carcinogenesis. Cell 174(4):856–869.e817. https://doi.org/10.1016/j.cell.2018.07.027

Partners TI (2021) Organoids Market is exepcted to reach US$ 3,420.40 million by 2027. Retrieved from https://www.theinsightpartners.com/pr/organoids-market

Pera MF (2017) Human embryo research and the 14-day rule. Development 144(11):1923–1925. https://doi.org/10.1242/dev.151191

Pettke A et al (2020) Broadly active antiviral compounds disturb Zika virus progeny release rescuing virus-induced toxicity in brain organoids. Viruses 13

Poirier EZ et al (2021) An isoform of Dicer protects mammalian stem cells against multiple RNA viruses. Science 373:231–236

Porotto M et al (2019) Authentic modeling of human respiratory virus infection in human pluripotent stem cell-derived lung organoids. mBio 10

Richards DJ, Coyle RC, Tan Y, Jia J, Wong K, Toomer K et al (2017) Inspiration from heart development: biomimetic development of functional human cardiac organoids. Biomaterials 142:112–123. https://doi.org/10.1016/j.biomaterials.2017.07.021

Sachs N, de Ligt J, Kopper O, Gogola E, Bounova G, Weeber F et al (2018) A living biobank of breast cancer organoids captures disease heterogeneity. Cell 172(1-2):373–386.e310. https://doi.org/10.1016/j.cell.2017.11.010

Sachs N, Papaspyropoulos A, Zomer-van Ommen DD, Heo I, Böttinger L, Klay D et al (2019) Long-term expanding human airway organoids for disease modeling. Embo J 38(4). https://doi.org/10.15252/embj.2018100300

Sampaziotis F, de Brito MC, Madrigal P, Bertero A, Saeb-Parsy K, Soares FAC et al (2015) Cholangiocytes derived from human induced pluripotent stem cells for disease modeling and drug validation. Nat Biotechnol 33(8):845–852. https://doi.org/10.1038/nbt.3275

Samuel RM et al (2020) Androgen signaling regulates SARS-CoV-2 receptor levels and is associated with severe COVID-19 symptoms in men. Cell Stem Cell 27:876–889.e812

Sarkar S, Heise MT (2019) Mouse models as resources for studying infectious diseases. Clin Ther 41:1912–1922

Sato T, Vries RG, Snippert HJ, van de Wetering M, Barker N, Stange DE et al (2009) Single Lgr5 stem cells build crypt-villus structures in vitro without a mesenchymal niche. Nature 459(7244):262–265. https://doi.org/10.1038/nature07935

Saul S et al (2021) Discovery of pan-ErbB inhibitors protecting from SARS-CoV-2 replication, inflammation, and lung injury by a drug repurposing screen. bioRxiv

Saxena K et al (2016) Human intestinal enteroids: a new model to study human rotavirus infection, host restriction, and pathophysiology. J Virol 90:43–56

Seidlitz T, Merker SR, Rothe A, Zakrzewski F, von Neubeck C, Grützmann K et al (2019) Human gastric cancer modelling using organoids. Gut 68(2):207–217. https://doi.org/10.1136/gutjnl-2017-314549

Shahbazi MN, Jedrusik A, Vuoristo S, Recher G, Hupalowska A, Bolton V et al (2016) Self-organization of the human embryo in the absence of maternal tissues. Nat Cell Biol 18(6):700–708. https://doi.org/10.1038/ncb3347

Smith E, Cochrane WJ (1946) Cystic organoid teratoma: (report of a case). Can Med Assoc J 55(2):151–152

Spence JR, Mayhew CN, Rankin SA, Kuhar MF, Vallance JE, Tolle K et al (2011) Directed differentiation of human pluripotent stem cells into intestinal tissue in vitro. Nature 470(7332):105–109. https://doi.org/10.1038/nature09691

Sridhar A et al (2020) Single-Cell Transcriptomic Comparison of Human Fetal Retina, hPSC-Derived Retinal Organoids, and Long-Term Retinal Cultures. Cell Rep 30:1644–1659.e1644

Takahashi K, Yamanaka S (2006) Induction of pluripotent stem cells from mouse embryonic and adult fibroblast cultures by defined factors. Cell 126(4):663–676. https://doi.org/10.1016/j.cell.2006.07.024
Takahashi K, Tanabe K, Ohnuki M, Narita M, Ichisaka T, Tomoda K, Yamanaka S (2007) Induction of pluripotent stem cells from adult human fibroblasts by defined factors. Cell 131(5):861–872. https://doi.org/10.1016/j.cell.2007.11.019
Takasato M, Er PX, Chiu HS, Little MH (2016) Generation of kidney organoids from human pluripotent stem cells. Nat Protoc 11(9):1681–1692. https://doi.org/10.1038/nprot.2016.098
Takebe T, Sekine K, Enomura M, Koike H, Kimura M, Ogaeri T et al (2013) Vascularized and functional human liver from an iPSC-derived organ bud transplant. Nature 499(7459):481–484. https://doi.org/10.1038/nature12271
Thomson JA, Itskovitz-Eldor J, Shapiro SS, Waknitz MA, Swiergiel JJ, Marshall VS, Jones JM (1998) Embryonic stem cell lines derived from human blastocysts. Science 282(5391):1145–1147. https://doi.org/10.1126/science.282.5391.1145
Tiwari SK, Wang S, Smith D, Carlin AF, Rana TM (2021) Revealing tissue-specific SARS-CoV-2 infection and host responses using human stem cell-derived lung and cerebral organoids. Stem Cell Rep 16:437–445
Wysocki J et al (2021) A Novel soluble ACE2 variant with prolonged duration of action neutralizes SARS-CoV-2 infection in human kidney organoids. J Am Soc Nephrol 32:795–803
Xu M et al (2016) Identification of small-molecule inhibitors of Zika virus infection and induced neural cell death via a drug repurposing screen. Nat Med 22:1101–1107
Yan HHN, Siu HC, Law S, Ho SL, Yue SSK, Tsui WY et al (2018) A comprehensive human gastric cancer organoid biobank captures tumor subtype heterogeneity and enables therapeutic screening. Cell Stem Cell 23(6):882–897.e811. https://doi.org/10.1016/j.stem.2018.09.016
Yang L et al (2020) A Human pluripotent stem cell-based platform to study SARS-CoV-2 tropism and model virus infection in human cells and organoids. Cell Stem Cell 27:125–136.e127
Yin Y et al (2018) 6-Thioguanine inhibits rotavirus replication through suppression of Rac1 GDP/GTP cycling. Antiviral Res 156:92–101
Yu W et al (2023) Identification of desoxyrhapontigenin as a novel antiviral agent against congenital Zika virus infection. Antiviral Res 211:105542
Yui S, Nakamura T, Sato T, Nemoto Y, Mizutani T, Zheng X et al (2012) Functional engraftment of colon epithelium expanded in vitro from a single adult $Lgr5^+$ stem cell. Nat Med 18(4):618–623. https://doi.org/10.1038/nm.2695
Zacharias WJ et al (2018) Regeneration of the lung alveolus by an evolutionarily conserved epithelial progenitor. Nature 555:251–255
Zhang X, Lin H, Dong L, Xia Q (2022) Recapitulating influenza virus infection and facilitating antiviral and neuroprotective screening in tractable brain organoids. Theranostics 12:5317–5329
Zhao X et al (2021) Human intestinal organoids recapitulate enteric infections of enterovirus and coronavirus. Stem Cell Rep 16:493–504
Zhou J et al (2017) Human intestinal tract serves as an alternative infection route for Middle East respiratory syndrome coronavirus. Sci Adv 3:eaao4966
Zhou J et al (2018) Differentiated human airway organoids to assess infectivity of emerging influenza virus. Proc Natl Acad Sci U S A 115:6822–6827
Zhu L et al (2020) Association of blood glucose control and outcomes in patients with COVID-19 and pre-existing Type 2 diabetes. Cell Metab 31:1068–1077.e1063

Chapter 5
Metabolomics Tools in Antiviral Research

Paul S. Soma and Rushika Perera

Abstract Direct-acting antivirals (DAAs) and therapeutics must achieve two goals to effectively assist with disease mitigation. They must reduce viral replication and dissemination as well as reduce tissue damage caused by viral infection and/or the resulting immune response. They are generally "chemically modified mimics" of naturally occurring small molecules/metabolites that are intermediates or products of active biochemical pathways in the cell. However, as opposed to the naturally occurring metabolites, in a disease state, DAAs and therapeutics will divert biochemical pathways toward a more desirable outcome. Mass spectrometry (MS)-based metabolomics can detect and quantify small-molecule metabolites in cells, organs, or biofluids and provide a library of molecules that can be developed into novel chemical inhibitors or activators. However, measuring the metabolome requires consideration of various analytical challenges. Expansive chemical diversity inhibits the development of a catch-all metabolomics workflow, so various workflows must be used to measure specific molecular classes within the metabolome. A wide range of metabolite concentrations requires measurement technologies with a high dynamic range, and the temporal nature of metabolic intermediates and reactivity of certain molecular classes necessitates care with sample handling. Additionally, structural characterization and identification of metabolites are difficult due to the chemical and structural heterogeneity of the metabolome. Thus, metabolomics discovery experiments require careful consideration of study design and choice of analytical technologies with follow-up experiments for confident metabolite identification. In this chapter, we focus on the criteria that need to be considered when designing a metabolomics experiment such that quality and interpretability of data are maximized.

Keywords Metabolome · Mass spectrometry · Biochemical pathways · Antivirals · Therapeutics

---

P. S. Soma · R. Perera (✉)
Center for Vector-Borne Infectious Diseases, Department of Microbiology, Immunology, and Pathology, Colorado State University, Fort Collins, CO, USA
e-mail: Rushika.perera@colostate.edu

N. Kumar et al. (eds.), *Advances in Antiviral Research*, Livestock Diseases and Management, https://doi.org/10.1007/978-981-99-9195-2_5

5.1 Introduction

Viruses as obligate pathogens hijack host resources to obtain the raw materials needed to successfully amplify their genomes, assemble new virus particles, and transmit to a new host (Perera and Kuhn 2016). Recent technological advancements in systems biology continue to provide comprehensive overviews of how virus-host interactions occur and how they can be altered by varying environmental conditions and interventions like therapeutics and antivirals (Peng et al. 2009; Shah et al. 2022). Genomics helps map and compare the genetic composition of organisms on an evolutionary scale. Genomics also uncovers how viral genomes evolve with increasing immune pressure based on the host's competency for viral transmission. Transcriptomics provided the first insights into how host gene expression was altered by viral infection. Transcriptomics also provides avenues for differentiating gene expression activated by the host response to infection versus gene expression activated by viral proteins for the benefit of the viral life cycle. Proteomics validates gene expression changes induced by viral infection by confirming impact on translation, posttranslational processing, and protein degradation. It also provides insight into the posttranslational impact of viral infection when translation per se remains unaltered.

However, downstream of genomes, transcriptomes, and proteomes is a rich landscape of small-molecule "triggers" known as the metabolome that modulates the biochemical environment within a host. The metabolome consists of small molecules (<1500 Da) that are intermediates and products of all active biochemical pathways in a cell. Essentially, if the expression of a gene, transcript, or protein is altered following a stimulus, the outcome is reflected in downstream metabolites. Therefore, the concentration of a metabolite(s) is closely associated with a phenotype at any given time point or condition. Metabolites (e.g., lipids, sugars, amino acids, nucleotides, hormones, and other small molecules) can control intracellular events or be secreted to carry out paracrine-like signaling in neighboring cells or tissues. The metabolome can be qualitatively or quantitatively measured to provide potential for the discovery of antivirals and therapeutic agents that are, or mimic, biological molecules but function to "redirect or block" biochemical pathways such that a specific desired outcome is achieved (Tounta et al. 2021; Aderemi et al. 2021; Johnson et al. 2016).

There is increasing evidence that the metabolome can be exploited by viruses to optimize the biochemical interactions between virus and host and facilitate either persistent infection (often for the life of the host) or to replicate and be transmitted to a different host (Munger et al. 2006, 2008; Vastag et al. 2011; Perera and Kuhn 2016; Passalacqua et al. 2019; Byers et al. 2019). Therefore, the metabolome can reflect the response to and outcome of viral infection. The metabolome also reflects the physiological complexities within a host. For instance, comorbidities such as diabetes, heart disease, and metabolic syndrome that complicate the outcome of viral disease is reflected in the metabolic repertoire of the host (Fang et al. 2020; Pal et al. 2020; Kao and Huang 2021; Tong et al. 2022; Sekaran et al. 2022). Therefore, comparing the metabolome of a host with a given disease or health condition can inform on the specific cellular events related to the disease condition and inform on how to intervene and improve patient outcome. To date, the metabolome remains an "untapped vault" of chemical structures that can be exploited for development of therapeutics. This review will summarize metabolomics tools available to explore metabolic perturbations during viral infection and disease and outline how that information can be used to drive the discovery of novel antivirals.

5.2 Tools for Measuring the Metabolome

To probe a biological question, metabolomics study design includes appropriate sample handling and preparation, as well as analytical technology and data processing methods. While nuclear magnetic resonance spectroscopy is often used for metabolomics, here we will focus on mass spectrometry (MS)-based technologies. An overview of sample preparation protocols, analytical technologies, and data processing methods will be provided here as metabolomics study guidelines and recommendations have been compiled elsewhere (Alseekh et al. 2021). MS-based workflows can alleviate some of the challenges in metabolomics via a combination of analytical tools. Many methods for bulk analysis of biological samples have been developed, and MS imaging can achieve spatial analysis of metabolites on tissue. Metabolomics and lipidomics study design considerations, MS-based technologies available for experimental design, acquired metrics, and data processing protocols are summarized in Fig. 5.1.

Bulk analysis MS workflows begin with the extraction of a subset of molecular classes from a sample of biological interest (cell culture, biofluid, tissue homogenate, etc.). Stable heavy isotope-labeled internal standards are added during the extraction process to correct for losses in downstream sample preparation steps, ionization suppression, and instrument variation during analysis. After metabolite extraction the steps to analyze extracts via MS generally follow pre-ionization separation, ionization, post-ionization separation, mass analysis, and detection. Metabolite extraction and ionization dictate the types of molecules available to be analyzed by the mass spectrometer, and the pre-ionization separation must be chosen based on the molecular classes being studied. Various types of pre-ionization separation techniques and ionization sources are available, and selection depends on sample phase and molecular properties. Pre-ionization separation effectively simplifies the molecular mixture prior to analysis, resolving some isomeric or isobaric species and attenuating ionization suppression. Ionization is a prerequisite to mass analysis; to be measured, molecules must be converted into gas-phase ions. Ionization suppression occurs when different molecules within the ionization source compete for charge (King et al. 2000; Furey et al. 2013). Molecules with high ionization efficiency or high abundance can suppress ionization for other molecules

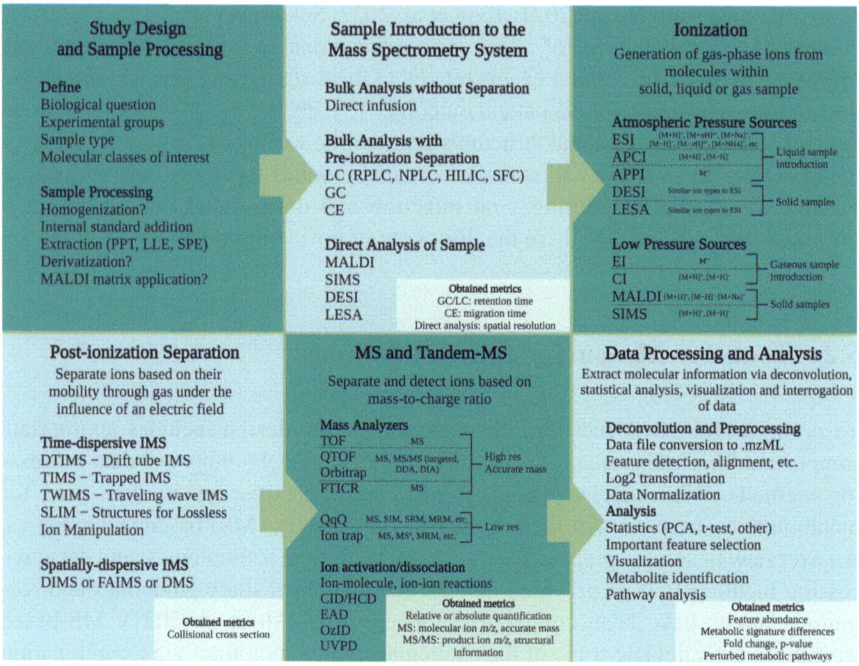

Fig. 5.1 Summary of a mass spectrometry-based metabolomics and lipidomics workflow and related technologies, metrics, and data processing options

and deteriorate detection limits. After ionization, gas-phase ions can be further separated using ion mobility spectrometry (May et al. 2017), enabling further resolution of some isomers and isobars. Isomers cannot be distinguished by MS; therefore, optimal separation prior to the MS is imperative for complex biological samples. Mass analysis involves separation of ions based on their mass-to-charge ratio (*m/z*) and subsequent detection. Because an all-encompassing metabolomics workflow does not exist, the metabolic question under study necessitates the analytical tools.

5.2.1 Sample Extraction and Internal Standards

Prior to extraction, samples must be collected, handled, and stored such that metabolism is quenched. Sample phase must be considered (e.g., biofluid, tissue), and choice of extraction type and solvent composition for metabolite extraction is critical. Common extraction types used in metabolomics are protein precipitation (PPT), liquid-liquid extraction (LLE), and solid-phase extraction (SPE). Additionally, the stability and reactivity of the molecules of interest must be considered during the sample handling and extraction process. The solvents used in any of these extraction techniques must dissolve the desired molecular classes. As a molecular class-dependent process, various extractions have been developed for polar metabolites, nonpolar lipids, and other molecular classes (Bligh and Dyer 1959; Sellick et al. 2011; Want et al. 2012; Barnes et al. 2016; Norris and Serhan 2018; Ulmer et al. 2018; Watrous et al. 2019; Liebisch et al. 2022).

Heavy isotope-labeled internal standards (IS) are added during extraction to account for incomplete extraction of metabolites, downstream sample preparation losses, ionization suppression, and analytical instrument variability (Furey et al. 2013). Commercially available heavy isotope-labeled IS's commonly contain deuterium (^{2}H), carbon-13 (^{13}C), or nitrogen-15 (^{15}N). The ideal IS has an identical molecular structure to the metabolite of interest and a higher molecular mass. Because of the heavy isotopes, the IS does not interfere with the measurement of the unlabeled endogenous metabolite because it is differentiated by *m/z* in the mass analysis step. There is potential for an IS to have isobaric interference with endogenous molecules, which cannot be distinguished by a low ("unit" mass)-resolution mass analyzer. Importantly, the extraction, separation, and ionization properties of an IS are identical to endogenous molecules. An IS mirrors the behavior of an endogenous compound throughout the metabolomics workflow and allows standardization and normalization of results.

5.2.2 *Sample Introduction Via Direct Infusion*

Biological samples can contain thousands of metabolites such that extraction results in highly complex mixtures, and, thus, the method of sample delivery to the ionization source has large implications. Some workflows forgo pre-ionization separation and use direct infusion-MS, where the total metabolite extract is consistently delivered to the ionization source. Direct infusion results in high, yet constant, ionization suppression and benefits from fast sample analysis times. However, the high ionization suppression in direct infusion-MS limits its utility for deep metabolome coverage. Since direct infusion-MS lacks pre-ionization separation, it is well complemented by post-ionization separations, such as differential ion mobility spectrometry coupled to ion trap MS (Keating and Glish 2018) or a high or ultrahigh accurate mass analyzer such as the Fourier-transform ion cyclotron resonance MS (FTICR) or Orbitrap (Marshall et al. 1998; Zubarev and Makarov 2013).

5.2.3 *Pre-Ionization Separation*

Simplification of the biological mixture via separation prior to the ionization and mass analysis steps can improve limits of detection, increase the number of metabolites detected, and reduce chimeric data by resolving isomeric or isobaric species, concentrating metabolites on the separation column prior to elution into the ionization source, and attenuating ionization suppression. Pre-ionization techniques separate gas or liquid-phase molecules based on their physicochemical properties. Separation is achieved in time, meaning molecules will traverse the chromatography column at different speeds based on their physicochemical properties and elute into the ionization source at different times. The timescale of separation is generally minutes to hours. Metabolite extracts are injected onto a pre-ionization separation platform such as liquid chromatography (LC), gas chromatography (GC), or capillary electrophoresis (CE).

The main separation mechanism of GC is based on molecular boiling point, where molecules with the lowest boiling temperature will elute from the column at the earliest time. Also, molecules have some chemical affinity to the GC column stationary phase. Separation is achieved by temporally raising the temperature of the column to gradually elute molecules with increasing boiling temperatures. GC is amenable to volatile small-molecule metabolites such as alcohols, small acids, amino acids, fatty acids, and more but commonly requires chemical derivatization to increase metabolite volatility (Fiehn 2016; Moldoveanu et al. 2018). A derivatization reaction is often specific to a certain molecular moiety (narrowing the scope of the target molecules) and requires more sample preparation steps. GC is commonly coupled to electron impact (EI) or chemical ionization (CI) sources.

LC separates molecules in the liquid phase based on their solubility in the liquid mobile phase (MP) and their chemical interactions with the solid stationary phase (SP) in the LC column. Separation of a mixture is achieved by temporally adjusting MP composition to gradually elute molecules with higher SP retention. Because of the large variety of stationary phase and mobile phase chemistries available, LC is amenable to a very wide range of molecular types, and, if necessary, the separation can be fine-tuned for challenging problems. Various broad types of LC exist, for example, reversed-phase LC (RPLC) (Uvela et al. 2019), hydrophilic interaction LC (HILIC) (Buszewski and Noga 2012), ion-exchange chromatography (IEC) (Ngere et al. 2023), and size-exclusion chromatography (SEC) (Deery et al. 2001; Ventouri et al. 2023). LC is commonly coupled to electrospray ionization (ESI), atmospheric pressure chemical ionization (APCI), or atmospheric pressure photoionization (APPI) sources.

CE separates charged molecules in solution based on their electrophoretic mobility, which dictates their migration time, through a small diameter capillary (Jorgenson and Lukacs 1983; Voeten et al. 2018). The small diameter promotes efficient heat transfer that enables application of exceptionally high separation voltages which results in highly efficient separations and fast analysis times. CE is amenable to a wide range of analytes from small-molecule metabolites, high-molecular-weight biomolecules, particles, and intact cells. CE is often coupled to ESI, but careful coupling of the high-voltage separation capillary and the ionization source must be achieved (Wu and Tang 2020).

5.2.4 *Ionization Sources*

Ionization source choice is dependent on the phase (solid, liquid, or gas) of sample being delivered to the source and the molecular classes being studied. The ionization source must convert solid, liquid, or gas-phase molecules into gas-phase ions. If used, pre-ionization separation (or direct infusion) will dictate the sample phase. Various ionization sources exist for use with either pre-ionization separation or direct infusion analysis of ranging sample types (Siuzdak 2004; Banerjee 2020). The common low-pressure sources coupled with GC are EI and CI, and common atmospheric pressure sources coupled with LC or direct infusion are ESI, APCI, and APPI. For CI, APCI, and ESI, the voltage polarity applied to the MS system will dictate whether positively or negatively charged ions are generated and transmitted into and through the MS. Based on their chemical moieties, some molecules can take on either a positive or a negative charge, while others may only ionize in one ionization mode. Additionally, the ionization sources amenable to the molecular class of interest are dependent ion molecular weight and chemical polarity (Fig. 5.2).

Each ionization source imparts a different amount of internal energy into the generated ions. Sources that impart high internal energy are referred to as "hard" sources and can fragment the ion prior to the mass analyzer. EI is a hard ionization technique that produces a radical cation, $M^{\bullet+}$, via impact of an energetic electron with a gas-phase molecule, causing ejection of an electron from that molecule. Because it is a hard source, EI produces unique fragmentation spectra for each

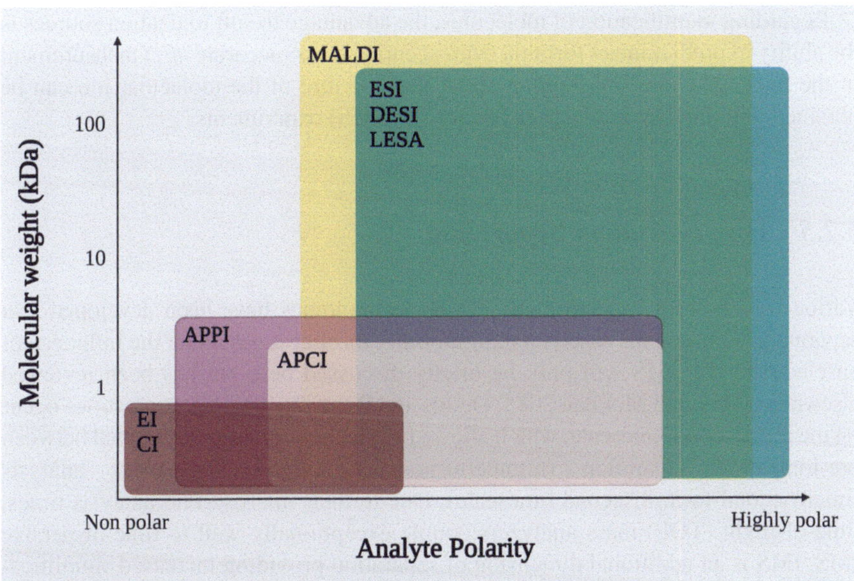

Fig. 5.2 Molecular weight and chemical polarity compatibility of various ionization sources for mass spectrometry

molecule – known as a molecular fingerprint. The EI fragmentation spectra for over 350,000 molecules have been placed into the NIST spectral library over many years, promoting confident identification of small metabolites (The National Institute of Standards and Technology 2020).

"Soft" ionization sources that do not impart high internal energy can produce the intact or "molecular" ion that is stable during transmission into the MS and mass analysis. ESI, APCI, and APPI are soft sources that are amenable to a wide range of molecular classes with some overlap. APPI is often used for very nonpolar molecules lacking heteroatoms but containing double bonds, such as polyaromatic hydrocarbons (Ghislain et al. 2012; Lung and Liu 2015). APPI produces $M^{\bullet+}$ via electron expulsion after interaction with a photon and at atmospheric pressure $M^{\bullet+}$ can abstract a hydrogen (proton) from solvent vapor to form $[M + H]^+$. APCI is amenable to small molecules with heteroatoms and forms either $[M + H]^+$ or $[M\text{-}H]^-$ by either proton transfer or abstraction via gas-phase ion-molecule reactions.

The invention of ESI revolutionized the analysis of biomolecules and is a major reason for the high versatility of modern mass spectrometry (Yamashita and Fenn 1984; Fenn et al. 1989; Cech and Enke 2001). ESI produces singly or multiply charged ions of various adducts: $[M + H]^+$, $[M + nH]^{n+}$, $[M\text{-}H]^-$, $[M\text{-}nH]^{n-}$, $[M + Na]^+$, $[M + NH4]^+$, etc. ESI allows for the analysis of small molecules ranging to very high-molecular-weight molecules (e.g., intact proteins). Upon the addition of many charges to a high-molecular-weight molecule, its *m/z* ratio is often low enough to be within the mass range of most MS systems. Multiple ionization mechanisms have been proposed for ESI, each resulting in the liberation of a solubilized molecule as a gas-phase ion from a highly charged aerosol droplet (Konermann et al. 2013).

Regarding identification of molecules, the advantage to soft ionization sources is the ability to predict intact formula from accurate mass (accurate *m/z*) measurement of the molecular ion. Information about the structure of the molecular ion can be obtained with tandem mass spectrometry (MS/MS) experiments.

5.2.5 Post-Ionization Separation

Various ion mobility spectrometry (IMS) technologies have been developed that separate gas-phase ions based on their mobility through a gas under the influence of an electric field. IMS will only be briefly discussed here but has been reviewed elsewhere (May and McLean 2015; Dodds and Baker 2019). IMS separations occur on the millisecond timescale, which allows IMS technologies to be coupled between pre-ionization separation (minute-to-hour timescale) and mass analysis (microsecond-to-millisecond timescale). Due to their microsecond analysis times, time-of-flight (TOF) mass analyzers couple exceptionally well to time-dispersive IMS. IMS is an additional dimension of separation providing increased simplification of complex biological mixtures and can be especially effective for isomer separation but, as a post-ionization separation, does not attenuate ionization suppression. Time-dispersive IMS separates ions based on their "shape-to-charge" ratio and enables the measurement of a collision cross-section (CCS) area of an ion. CCS is used as an additional piece of data for increased confidence in metabolite identification (Paglia et al. 2014; May et al. 2017; Picache et al. 2019; Baker and Bush n.d.).

5.2.6 *Mass Analysis*

A mass analyzer separates gas-phase ions based on their *m/z*, such that the electrical currents induced at the detector by different *m/z* values are differentiated. The current, or signal, for a given *m/z* is correlated to the number of ions detected and molecular abundance in the sample. Various types of mass analyzers are commercialized, each with a unique mechanism for *m/z* separation, and different timescales and modes of operation and, thus, different resulting data, metrics, and available information (Marshall et al. 1998; Hu et al. 2005; Haag 2016; Crutchfield and Clarke 2017). Trapping mass analyzers separate ions within a defined space, and beam-type mass analyzers separate ions as they traverse through space. Mass analyzer selection is imperative for collecting the correct type of data to meet the metabolomics study goals. Three common goals are: metabolite discovery, metabolite identification, and metabolite quantification.

The current standard for metabolite discovery is untargeted MS measurements using a high-resolution accurate mass (HRAM) analyzer such as time-of-flight (TOF), Orbitrap, or FTCIR. These HRAM analyzers have different mass analysis mechanisms but all allow for accurate measurement of ion monoisotopic mass (*m/z*), enable differentiation of isobars (but not isomers), and enable prediction of molecular formula and tentative metabolite identification (Schrimpe-Rutledge et al. 2016; Crutchfield and Clarke 2017; Gil-De-La-Fuente et al. 2019; Chen et al. 2021). Under typical MS (full scan) operation, all ions are transmitted into the mass analyzer, enabling detection and accurate *m/z* assignment of molecular ions. HRAM mass analyzers are often coupled with quadrupole (e.g., QTOF) or ion trap mass analyzers for MS/MS experiments that can help confirm metabolite identities. Additionally, untargeted discovery MS workflows allow for relative quantification between samples.

Though the aforementioned HRAM analyzers can be used for absolute quantification of metabolites, the current standard for absolute quantification is targeted MS measurements using a triple quadrupole (QqQ) MS system. Quadrupole mass analyzers have “unit” mass resolution and do not have accurate mass measurement capabilities – thus they are not HRAM analyzers. Quadrupoles may be operated individually for untargeted detection of all ions (MS1, full scan of the mass range) or targeted filtering of a small *m/z* window (typically 0.7, 4, or 9 Daltons wide). When coupling quadrupoles together, QqQ systems can achieve low limits of detection and high metabolite selectivity by immense reduction of chemical noise through multiple steps of mass analysis and dissociation of the ions of interest. This MS/MS technique, known as selected-reaction monitoring (SRM), and others will be discussed in the coming sections.

5.2.7 Tandem Mass Spectrometry

Tandem mass spectrometry is achieved either in time with a trapping mass analyzer or in space as ions traverse through multiple mass analyzers (Glish and Burinsky 2008). Multiple steps of mass analysis in tandem (MS/MS or MS^n) are used to aid in metabolite identification and improve selectivity for an ion of interest through pairing with gas-phase chemistry such as ion activation/dissociation, molecule-molecule, ion-molecule, or ion-ion reactions. The reaction is staged between mass analysis steps, where the first stage of mass analysis isolates the *m/z* of the *precursor ion*. The precursor ion is then the subject of a gas-phase reaction that generates one or multiple *product ions*. All product ions can either be transmitted to the detector, or another stage of mass analysis can isolate the *m/z* for a single product ion. Product ions can inform on the structure of the precursor ion.

Ion dissociation is intentionally induced within the MS system, as opposed to ion fragmentation (as described in EI ionization) that occurs during the ionization or ion transfer process. Ion activation and dissociation reactions input internal energy into the precursor ion to produce lower *m/z* product ions that retain the charge and an uncharged portion of the molecule (neutral loss) that is the complement of each product ion. Neutral losses are not detected because they no longer have a charge. Collision-induced dissociation (CID) can generate structural information for a wide range of molecules and is the most common dissociation technique. Other dissociation techniques, such as electron-activated (EAD) or ozone-induced dissociation, with different activation and dissociation chemistries, can generate orthogonal product ions that provide complementary structural information. Electron-activated dissociation techniques work well for glycan or peptide analysis but are not amenable to small-molecule metabolites (Zhao et al. 2008; Khatri et al. 2018). Ozone-induced dissociation has been used to solve double bond position for unsaturated lipid ions which is a very difficult analytical challenge (Thomas et al. 2008; Blanksby and Mitchell 2010). The utility of photon-induced dissociation, such as ultraviolet photodissociation (UVPD), has been demonstrated for various biomolecules (Brodbelt 2014; Brodbelt et al. 2020). Many gas-phase reactions have been harnessed to solve other structural elucidation challenges with MS/MS (Osburn and Ryzhov 2013; Campbell et al. 2017a, 2017b, 2018; Randolph et al. 2020; Heiles 2021).

5.2.8 MS and MS/MS Scan and Data Acquisition Modes

There are many ways to operate each type of mass analyzer to tailor the information collected in a dataset. Greater detail regarding scan types and acquisition modes are provided elsewhere (Westman-Brinkmalm and Brinkmalm 2008; Glish and Burinsky 2008; Raetz et al. 2020). For discovery metabolomics full scan (MS), mass spectra are acquired and may be supplemented by intermittent collection of

MS/MS data during analysis of a sample (Guo and Huan 2020) (most often with QTOF or Orbitrap systems). MS1 spectra include *m/z* values for all ions that are transmitted to and analyzed by the mass analyzer, allowing detection of virtually all molecules that are delivered to the ionization source and are converted to a gas-phase ion. For data-dependent acquisition (DDA), an MS spectrum is acquired from which multiple precursor ions are identified for subsequent MS/MS analysis. Two to five precursors in the MS spectrum are often selected based on highest signal. A single DDA cycle containing one MS and multiple precursor ion MS/MS scans is on the milliseconds to seconds timescale, fast enough to obtain multiple data points across a chromatographic peak that is seconds wide. DDA allows for collection of both MS spectra for discovery analysis and MS/MS spectra that provide higher selectivity and molecular structure information during the same sample analysis. Collection of MS and MS/MS data within a single sample injection saves time and sample volume because it provides potential for metabolite identification within a single experiment.

Data-independent acquisition (DIA) acquires both MS and theoretically MS/MS data for all precursor ions (Raetz et al. 2020). In DIA the precursor isolation window width is large (10 to 100 Daltons) and stepped across the entire mass range, allowing simultaneous dissociation of many *m/z* values. Because many precursor ions are selected for MS/MS at once, the product ion spectra generated are chimeric – containing structural information for many precursor ions. Hence, the data generated by DIA is highly complex and requires intensive post-processing and deconvolution (Yin et al. 2019). Because DIA collects MS/MS information for all ions, the dataset can be reanalyzed to search for metabolites of interest, whereas DDA only collects MS/MS information for a select number of precursors.

For targeted quantitative metabolomics, a QqQ mass analyzer is often used in the SRM scan mode. An SRM method is developed to target metabolites of interest and, thus, does not collect any global metabolome data. A single SRM event (called a "transition") involves selection of the precursor ion *m/z* in the first quadrupole (Q1), transmission into the collision cell (q) for CID, and then isolation of a single product ion *m/z* with the third quadrupole (Q3). Many SRM transitions can be quickly monitored for a list of metabolites – known as a multiple-reaction monitoring (MRM) method. MRM methods are often developed to quantify metabolites within metabolic pathways or specific molecular classes such as tryptophan metabolism (Whiley et al. 2019), sphingolipids (Merrill et al. 2005), oxidized lipids (Norris and Serhan 2018; Watrous et al. 2019; Hinz et al. 2019; Li et al. 2019), etc.

5.2.9 Mass Spectrometry Imaging

Mass spectrometry imaging (MSI) enables mapping of molecular spatial distribution on the surface of a complex sample such as biological tissues (Amstalden van Hove et al. 2010; Buchberger et al. 2018). MSI promotes detection of a wide variety of molecular classes and does not require chemical labeling. After tissue sectioning and preparation, a sample grid is defined, and mass spectra are acquired to characterize the chemical components of each defined pixel. MSI requires direct analysis of a solid sample; thus, the ionization technique must ablate or desorb and ionize molecules from the surface. Ionization techniques for surface analysis can be laser (matrix-assisted laser desorption ionization (MALDI)) (Cornett et al. 2007), particle or ion beam (secondary ion mass spectrometry (SIMS)) (Van Vaeck et al. 1999), liquid microjunction (liquid extraction surface analysis (LESA), MasSpec Pen) (Sarsby et al. 2015; Zhang et al. 2017), or electrospray based (desorption electrospray ionization (DESI)) (Eberlin 2014). MSI workflows commonly utilize HRAM mass analyzers because of the sample molecular complexity. Mapping molecular spatial distribution of diseased tissue can help identify molecular effectors of pathology. Thus, MSI has become a powerful tool for discovery of disease-related biomolecules.

5.2.10 Metabolite Annotation and Identification

Analogous to trilateration of GPS coordinates with satellite data, the goal of metabolite identification is to pinpoint metabolite structure using data from multiple MS-based technologies. Thus, unambiguous metabolite structural elucidation and identification may require multiple experiments, and there are multiple defined levels of confidence for metabolite identification (Schrimpe-Rutledge et al. 2016). In an untargeted metabolomics dataset, molecular features are initially defined by their chromatographic retention time (RT) and accurate mass (*m/z*) value. Molecular formulas can be proposed using accurate mass for tentative metabolite identification but lack structural information. Other types of data used to improve metabolite identification confidence are RT, CCS, molecular ion *m/z*, isotope distribution (Böcker et al. 2009; Dührkop et al. 2014), complementary ion adducts (from ESI process), and MS/MS product ion spectra. All these metrics are ideally measured for a given metabolite in the biological sample and compared to those collected for a pure synthetic standard of that metabolite. Matching metrics to a synthetic standard yields a validated metabolite identity. Additionally, MS/MS product ion spectra can be matched to published or in-house libraries (The National Institute of Standards and Technology 2020; mzCloud Mass Spectral Database 2022).

5.2.11 Metabolomics Data Processing and Analysis

An LC-MS-based discovery metabolomics study generates large datasets embedded with signals for many metabolites. Multiple previous publications cover metabolomics data processing and analysis (Uppal et al. 2016; Gardinassi et al. 2017; Gassiot et al. 2022; Parker et al. 2023). To unlock molecular information, LC-MS data can be deconvoluted using feature extraction algorithms such as XCMS or MZmine3 that detect and align retention time of chromatographic peaks across multiple samples (Smith et al. 2006; Myers et al. 2017; Schmid et al. 2023). Feature detection results in a peak table containing features (each defined by chromatographic RT and *m/z* value) and their abundance in each sample represented by chromatographic peak areas. Peak areas are generally normalized and log-transformed and then used for downstream statistical analyses and metabolic pathway analyses.

Multivariate analysis, such as principal component analysis (PCA), is used to broadly determine whether the measured metabolic signature is different between treatment groups (e.g., healthy vs. unhealthy). Pairwise statistical analysis can be used to calculate fold change and p value between two treatment groups for each feature. Feature level fold change information can be used to determine features that have high differential abundance between treatment groups. Volcano plots are effective for visualization of feature differential abundance between treatment groups, enabling pinpointing of features of interest. Additionally, feature selection methodologies can be employed to pinpoint potentially important metabolites within a dataset (Antonelli et al. 2019).

Metabolic pathway analysis is best used as a hypothesis generation tool. The results of a pairwise statistical comparison for each feature including RT, *m/z*, t score, and p value can be fed into the *mummichog* algorithm from MetaboAnalyst (Li et al. 2013). The *mummichog* algorithm tentatively identifies metabolites based on *m/z* and maps them to known metabolic pathways. When multiple metabolites are mapped to a single pathway, total pathway enrichment is calculated. Many pathways may be proposed by pathway analysis, but the results must be validated by confident metabolite identification to confirm dysregulation of the pathways of interest.

5.3 Applications of Metabolomics Tools in the Development of Antivirals and Therapeutics

Several studies that have used metabolomics/lipidomics methods to profile biomarkers of viral diseases have identified common and unique metabolite signatures associated with each disease (Cui et al. 2013; Cui et al. 2016; Gale et al. 2017, 2020; Kyle et al. 2019; Lewis et al. 2022; Rabinowitz et al. 2023). These signatures highlight metabolic pathways that are elevated or decreased during infection and disease onset and provide a potential entry point for early therapeutic intervention. Metabolite signatures can also be used to develop host-targeted therapies for synergy with DAAs.

DAAs target viral proteins encoded in the genome. Metabolomics studies on cultured human fibroblasts infected with influenza virus identified

N-acetylneuraminic acid as the most significantly increased metabolite in infected cells compared to uninfected cells. This metabolite is cleaved from viral glycoproteins by the action of the neuraminidase enzyme encoded in the viral genome. Oseltamivir (Tamiflu) was developed as an inhibitor of influenza virus neuraminidase providing an example where results of metabolomics studies are directly linked to the activity of enzymes and the development of inhibitors of those enzymes (Munger et al. 2006, 2008; Davies 2010; Schmitz and Swaminathan 2022). In another example, metabolomics studies of herpes simplex virus type-1 (HSV-1)-infected cells showed an increase in deoxypyrimidines such as dTTP and dTMP (Munger et al. 2006, 2008; Vastag et al. 2011; Rabinowitz et al. 2023). Inhibitors such as aciclovir, the leading antiviral against HSV-1, target the viral thymidine kinase to reduce these deoxypyrimidines (Richards et al. 1983).

The efficacy of DAAs can also be enhanced by host-targeted therapeutics. A key observation from the SARS-CoV-2 pandemic is the existence of comorbidities that exacerbate the pathology of COVID-19 (Singh et al. 2020; Hua et al. 2021). Many of these comorbidities such as obesity and diabetes are directly linked to imbalances within the human body surrounding glucose and lipid metabolism (Goodpaster and Sparks 2017; Athyros et al. 2018). These metabolic pathways are a direct target of many viruses that strive to transform the host environment for their replicative advantage. Altered glucose and lipid metabolism has been linked to the replication mechanisms of various global pathogens including flaviviruses like Zika virus (ZIKV), dengue viruses (DENVs) (Perera and Kuhn 2016; Martín-Acebes et al. 2019), retroviruses such as HIV (Waters and Hsue 2019; Lagathu et al. 2019), and coronaviruses SARS-CoV-1 and MERS-CoV (Wu et al. 2017; Yan et al. 2019). This link between pathogenicity and metabolism is partially due to these viruses being obligate parasites that need to hijack metabolic pathways for viral entry, genome translation, replication, virus assembly, and egress. During this process they also steal host-derived lipid membranes as a structural component of the virus particle. All these virus-host interactions increase energy demands on the cell causing infection-driven upregulation of both glucose and lipid metabolism associated with glycolysis and mitochondrial respiration (Perera and Kuhn 2016). Additionally, these metabolic pathways also influence the immune response to infection, adding to the resulting pathology (O'Neill et al. 2016; Hotamisligil 2017). Therefore, when the human body is already altered in metabolic state due to diabetes, metabolic syndrome, or obesity, it essentially has upregulated the very metabolic intermediates necessary to upstart viral replication providing a "proviral" metabolic environment. Hence, any therapeutic strategy needs to consider a combined approach that addresses this "proviral metabolic state" in the host in conjunction with direct targeting of viral enzymatic functions.

Studies have demonstrated superiority of combinatorial therapeutics (Simon and Ho 2003; Lin et al. 2016; Shyr et al. 2021). Antiviral development for human immunodeficiency virus (HIV) provides a perfect example. Drugs developed early in the HIV epidemics targeted individual viral proteins to block steps in the viral life cycle. These drugs resulted in partially reconstituting the immune response, decreasing mortality, and reducing opportunistic infections and related malignancies. However, with time drug resistance mutations arose driving a loss in clinical benefit. These patients were then treated with a drug that targeted a different viral protein, once again driving the emergence of drug-resistant viral variants. However, groundbreaking discoveries by Dr. David Ho demonstrated that while it took only days to create resistance mutations against a single drug, it took months to create resistance against a double-drug cocktail and thousands of years for a three-drug cocktail (Simon and Ho 2003). These discoveries of the power of combination therapeutics have saved millions of lives. The currently used combinations against HIV are successful in driving viral amplification below detection thresholds, making HIV-acquired immunodeficiency syndrome (AIDS) a chronic, manageable disease with almost no accumulation of drug-resistant mutations. Combination therapeutics is also now applied to hepatitis C virus (HCV) infections with triple-drug cocktails that are effective against all known genotypes of the virus and driving a cure in weeks (Boesecke and Rockstroh 2015; De Nicola and Colombo 2016).

However, the concept of combining direct-acting antivirals with host-targeted therapeutics is in its infancy. For example, nitazoxanide (2-(acetyloxy)-N-(5-nitro-2-thiazolyl) benzamide) was discovered as a first-in-class broad-spectrum inhibitor against a plethora of viruses including influenza viruses, human immunodeficiency virus, hepatitis B and C, rotavirus, parainfluenza, coronavirus, respiratory syncytial virus, norovirus, as well as flaviviruses such as dengue, yellow fever, and Japanese encephalitis viruses. It has a broad mechanism of action but in mammalian hosts disrupts metabolic pathways such as autophagy, the unfolded protein response, mitochondrial uncoupling, AKT/mTOR, AMPK, and Wnt/β-catenin signaling pathways (Rossignol 2014). It has been shown that nitazoxanide can enhance the activity of (synergize with) inhibitors against influenza virus neuraminidase such as oseltamivir achieving greater antiviral efficacy (Rossignol 2014). It can also synergize with ribavirin in the inhibition of SARS-CoV-2 (Lian et al. 2020). Alisporivir is a cyclophilin inhibitor that has been used in combination with Ribavirin to enhance the efficacy of HCV genotype 2 or 3 inhibition (Pawlotsky et al. 2015).

Host-directed therapies for acute viral infection have the potential to help reduce resistance to direct-acting antivirals targeted against viral proteins (Crouchet et al. 2018; Cakir et al. 2021). When host-targeted therapies are given in combination with direct-acting antivirals, synergy between the combined antivirals is optimized to achieve efficacy at much lower doses than required when using single antivirals alone (Prichard and Shipman 1990). Reduced dose and short treatment time frame for these acute viral infections ensure fewer side effects and toxicity compared to when prolonged host therapies are used for chronic conditions like cancer, diabetes, heart disease, and metabolic syndrome. Metabolomics can increase our ability to identify host-targeted therapies by providing insight into the specific pathways perturbed by virus infection and exposing metabolic vulnerabilities that can be exploited to gain a health advantage.

5.4 Conclusions

Metabolomics involves interrogating the active biochemical environment of a system via measurement of various classes of small molecules. Metabolism is driven by upstream gene expression and enzyme activity, rendering it a complex system of biochemical reactions that, when measured in a thoughtful and controlled manner, is embedded with rich biological information. The metabolome is dynamic, responds to external stimuli, and is commonly accepted as a close effector of phenotype. As molecular phenotype effectors, metabolites provide biochemical information with greater temporal resolution than other "omics" studies such as genomics, transcriptomics, and proteomics. Therefore, the metabolome provides an accessible library of small molecules that can be exploited as antivirals or therapeutics to alter virus-host interactions such that more positive outcomes are achieved for the host. Although measuring the metabolome can provide great advantages, various analytical challenges exist in metabolomics that do not exist for other "omics" fields. Accordingly, careful design of metabolomics experiments is needed to ensure acquisition of meaningful data.

Acknowledgments The authors acknowledge Oshani C. Ratnayake for critical reading of the manuscript.

Funding: PS was supported by a NIH/NIAID R01AI151166 to RP.

Author contributions: PS and RP wrote the manuscript.

Conflicts of interest: The authors declare no conflicts of interest.

References

Aderemi AV, Ayeleso AO, Oyedapo OO, Mukwevho E (2021) Metabolomics: a scoping review of its role as a tool for disease biomarker discovery in selected non-communicable diseases. Meta 11:418. https://doi.org/10.3390/metabo11070418

Alseekh S, Aharoni A, Brotman Y, Contrepois K, D'Auria J, Ewald JC, Ewald J, Fraser PD, Giavalisco P, Hall RD, Heinemann M, Link H, Luo J, Neumann S, Nielsen J, Perez de Souza L, Saito K, Sauer U, Schroeder FC, Schuster S, Siuzdak G, Skirycz A, Sumner LW, Snyder MP, Tang H, Tohge T, Wang Y, Wen W, Wu S, Xu G, Zamboni N, Fernie AR (2021) Mass spectrometry-based metabolomics: a guide for annotation, quantification and best reporting practices. Nat Methods 18:747–756. https://doi.org/10.1038/s41592-021-01197-1

Amstalden van Hove ER, Smith DF, Heeren RMA (2010) A concise review of mass spectrometry imaging. J Chromatogr A 1217:3946–3954. https://doi.org/10.1016/j.chroma.2010.01.033

Antonelli J, Claggett BL, Henglin M, Kim A, Ovsak G, Kim N, Deng K, Rao K, Tyagi O, Watrous JD, Lagerborg KA, Hushcha PV, Demler OV, Mora S, Niiranen TJ, Pereira AC, Jain M, Cheng S (2019) Statistical workflow for feature selection in human metabolomics data. Metabolites 9:143. https://doi.org/10.3390/METABO9070143

Athyros VG, Doumas M, Imprialos KP, Stavropoulos K, Georgianou E, Katsimardou A, Karagiannis A (2018) Diabetes and lipid metabolism. Hormones 17:61–67

Baker ES, Bush MF (n.d.) Collision cross section database. http://depts.washington.edu/bushlab/ccsdatabase/. Accessed 11 Apr 2023

Banerjee S (2020) Empowering clinical diagnostics with mass spectrometry. ACS Omega 5:2041–2048. https://doi.org/10.1021/acsomega.9b03764

Barnes S, Benton HP, Casazza K, Cooper SJ, Cui X, Du X, Engler J, Kabarowski JH, Li S, Pathmasiri W, Prasain JK, Renfrow MB, Tiwari HK (2016) Training in metabolomics research. I. Designing the experiment, collecting and extracting samples and generating metabolomics data. J Mass Spectrom 51:461–475. https://doi.org/10.1002/JMS.3782

Blanksby SJ, Mitchell TW (2010) Advances in mass spectrometry for lipidomics. 3:433–465. https://doi.org/10.1146/ANNUREV.ANCHEM.111808.073705

Bligh EG, Dyer WJ (1959) A rapid method of total lipid extraction and purification. Can J Biochem Physiol 37:911–917. https://doi.org/10.1139/o59-099

Böcker S, Letzel MC, Lipták Z, Pervukhin A (2009) SIRIUS: decomposing isotope patterns for metabolite identification. Bioinformatics 25:218–224. https://doi.org/10.1093/bioinformatics/btn603

Boesecke C, Rockstroh JK (2015) Treatment of chronic HCV genotype 1 coinfection. Curr HIV/AIDS Rep 12:326–335. https://doi.org/10.1007/s11904-015-0278-4

Brodbelt JS (2014) Photodissociation mass spectrometry: new tools for characterization of biological molecules. Chem Soc Rev 43:2757–2783. https://doi.org/10.1039/c3cs60444f

Brodbelt JS, Morrison LJ, Santos I (2020) Ultraviolet Photodissociation mass spectrometry for analysis of biological molecules. Chem Rev 120:3328–3380. https://doi.org/10.1021/acs.chemrev.9b00440

Buchberger AR, DeLaney K, Johnson J, Li L (2018) Mass spectrometry imaging: a review of emerging advancements and future insights. Anal Chem 90:240–265. https://doi.org/10.1021/acs.analchem.7b04733

Buszewski B, Noga S (2012) Hydrophilic interaction liquid chromatography (HILIC)-a powerful separation technique. Anal Bioanal Chem 402:231–247. https://doi.org/10.1007/S00216-011-5308-5/TABLES/3

Byers N, Fleshman A, Perera R, Molins C (2019) Metabolomic insights into human Arboviral infections: dengue, chikungunya, and Zika viruses. Viruses 11:225. https://doi.org/10.3390/v11030225

Cakir M, Obernier K, Forget A, Krogan NJ (2021) Target discovery for host-directed antiviral therapies: application of proteomics approaches. mSystems 6. https://doi.org/10.1128/mSystems.00388-21

Campbell MT, Chen D, Glish GL (2017a) Identifying the D-Pentoses using water adduction to lithium Cationized molecule. J Am Soc Mass Spectrom 28:1420–1424. https://doi.org/10.1007/s13361-017-1656-5

Campbell MT, Chen D, Wallbillich NJ, Glish GL (2017b) Distinguishing biologically relevant hexoses by water adduction to the lithium-Cationized molecule. Anal Chem 89:10504–10510. https://doi.org/10.1021/acs.analchem.7b02647

Campbell MT, Chen D, Glish GL (2018) Distinguishing linkage position and Anomeric configuration of glucose-glucose disaccharides by water adduction to Lithiated molecules. Anal Chem 90:2048–2054. https://doi.org/10.1021/acs.analchem.7b04162

Cech NB, Enke CG (2001) Practical implications of some recent studies in electrospray ionization fundamentals. Mass Spectrom Rev 20:362–387. https://doi.org/10.1002/MAS.10008

Chen L, Lu W, Wang L, Xing X, Chen Z, Teng X, Zeng X, Muscarella AD, Shen Y, Cowan A, McReynolds MR, Kennedy BJ, Lato AM, Campagna SR, Singh M, Rabinowitz JD (2021) Metabolite discovery through global annotation of untargeted metabolomics data. Nat Methods 18:1377–1385. https://doi.org/10.1038/s41592-021-01303-3

Cornett DS, Reyzer ML, Chaurand P, Caprioli RM (2007) MALDI imaging mass spectrometry: molecular snapshots of biochemical systems. Nat Methods 4:828–833. https://doi.org/10.1038/nmeth1094

Crouchet E, Wrensch F, Schuster C, Zeisel MB, Baumert TF (2018) Host-targeting therapies for hepatitis C virus infection: current developments and future applications. Ther Adv Gastroenterol 11:175628481875948. https://doi.org/10.1177/1756284818759483

Crutchfield CA, Clarke W (2017) High resolution accurate mass (HRAM) mass spectrometry. Mass Spectrometry for the Clinical Laboratory, pp 247–259. https://doi.org/10.1016/B978-0-12-800871-3.00012-2
Cui L, Lee YH, Kumar Y, Xu F, Lu K, Ooi EE, Tannenbaum SR, Ong CN (2013) Serum metabolome and Lipidome changes in adult patients with primary dengue infection. PLoS Negl Trop Dis 7. https://doi.org/10.1371/journal.pntd.0002373
Cui L, Lee YH, Thein TL, Fang J, Pang J, Ooi EE, Leo YS, Ong CN, Tannenbaum SR (2016) Serum metabolomics reveals serotonin as a predictor of severe dengue in the early phase of dengue fever. PLoS Negl Trop Dis 10. https://doi.org/10.1371/journal.pntd.0004607
Davies BE (2010) Pharmacokinetics of oseltamivir: an oral antiviral for the treatment and prophylaxis of influenza in diverse populations. J Antimicrob Chemother 65(Suppl):2
De Nicola S, Colombo M (2016) The eradication of HCV. Minerva Gastroenterol Dietol 62:63–75
Deery MJ, Stimson E, Chappell CG (2001) Size exclusion chromatography/mass spectrometry applied to the analysis of polysaccharides. Rapid Commun Mass Spectrom 15:2273–2283. https://doi.org/10.1002/RCM.458
Dodds JN, Baker ES (2019) Ion mobility spectrometry: fundamental concepts, instrumentation, applications, and the road ahead. J Am Soc Mass Spectrom 30:2185–2195. https://doi.org/10.1007/s13361-019-02288-2
Dührkop K, Hufsky F, Böcker S (2014) Molecular formula identification using isotope pattern analysis and calculation of fragmentation trees. Mass Spectrometry 3:S0037–S0037. https://doi.org/10.5702/massspectrometry.s0037
Eberlin LS (2014) DESI-MS imaging of lipids and metabolites from biological samples. Methods Mol Biol 1198:299–311. https://doi.org/10.1007/978-1-4939-1258-2_20/COVER
Fang X, Li S, Yu H, Wang P, Zhang Y, Chen Z, Li Y, Cheng L, Li W, Jia H, Ma X (2020) Epidemiological, comorbidity factors with severity and prognosis of COVID-19: a systematic review and meta-analysis. Aging 12:12493–12503. https://doi.org/10.18632/aging.103579
Fenn JB, Mann M, Meng CK, Wong SF, Whitehouse CM (1989) Electrospray ionization for mass spectrometry of large biomolecules. Science 1979(246):64–71. https://doi.org/10.1126/science.2675315
Fiehn O (2016) Metabolomics by gas chromatography–mass spectrometry: combined targeted and untargeted profiling. Curr Protoc Mol Biol 114:30.4.1–30.4.32. https://doi.org/10.1002/0471142727.MB3004S114
Furey A, Moriarty M, Bane V, Kinsella B, Lehane M (2013) Ion suppression; a critical review on causes, evaluation, prevention and applications. Talanta 115:104–122. https://doi.org/10.1016/j.talanta.2013.03.048
Gale TV, Horton TM, Grant DS, Garry RF (2017) Metabolomics analyses identify platelet activating factors and heme breakdown products as Lassa fever biomarkers. PLoS Negl Trop Dis 11. https://doi.org/10.1371/journal.pntd.0005943
Gale TV, Schieffelin JS, Branco LM, Garry RF, Grant DS (2020) Elevated l-threonine is a biomarker for Lassa fever and Ebola. Virol J 17. https://doi.org/10.1186/s12985-020-01459-y
Gardinassi LG, Xia J, Safo SE, Li S (2017) Bioinformatics tools for the interpretation of metabolomics data. Curr Pharmacol Rep 3:374–383. https://doi.org/10.1007/s40495-017-0107-0
Gassiot C, Tesio Torta F, Burla BJ, Hoffmann N, Mayer G, Has C, Kopczynski D, Al Machot F, Schwudke D, Ahrends R, Marcus K, Eisenacher M, Turewicz M (2022) A current encyclopedia of bioinformatics tools, data formats and resources for mass spectrometry Lipidomics. Metabolites 12:584. https://doi.org/10.3390/METABO12070584
Ghislain T, Faure P, Michels R (2012) Detection and monitoring of PAH and oxy-PAHs by high resolution mass spectrometry: comparison of ESI, APCI and APPI source detection. J Am Soc Mass Spectrom 23:530–536. https://doi.org/10.1007/s13361-011-0304-8
Gil-De-La-Fuente A, Godzien J, Saugar S, Garcia-Carmona R, Badran H, Wishart DS, Barbas C, Otero A (2019) CEU mass mediator 3.0: a metabolite annotation tool. J Proteome Res 18:797–802. https://doi.org/10.1021/acs.jproteome.8b00720

Glish GL, Burinsky DJ (2008) Hybrid mass spectrometers for tandem mass spectrometry. J Am Soc Mass Spectrom 19:161–172. https://doi.org/10.1016/J.JASMS.2007.11.013
Goodpaster BH, Sparks LM (2017) Metabolic flexibility in health and disease. Cell Metab 25:1027–1036
Guo J, Huan T (2020) Comparison of full-scan, data-dependent, and data-independent acquisition modes in liquid chromatography-mass spectrometry based untargeted metabolomics. Anal Chem 92:8072–8080. https://doi.org/10.1021/acs.analchem.9b05135
Haag AM (2016) Mass analyzers and mass spectrometers. Adv Exp Med Biol 919:157–169. https://doi.org/10.1007/978-3-319-41448-5_7
Heiles S (2021) Advanced tandem mass spectrometry in metabolomics and lipidomics—methods and applications. Anal Bioanal Chem 413:5927–5948. https://doi.org/10.1007/s00216-021-03425-1
Hinz C, Liggi S, Mocciaro G, Jung S, Induruwa I, Pereira M, Bryant CE, Meckelmann SW, O'Donnell VB, Farndale RW, Fjeldsted J, Griffin JL (2019) A comprehensive UHPLC ion mobility quadrupole time-of-flight method for profiling and quantification of eicosanoids, other Oxylipins, and fatty acids. Anal Chem 91:8025–8035. https://doi.org/10.1021/acs.analchem.8b04615
Hotamisligil GS (2017) Foundations of Immunometabolism and implications for metabolic health and disease. Immunity 47:406–420. https://doi.org/10.1016/j.immuni.2017.08.009
Hu Q, Noll RJ, Li H, Makarov A, Hardman M, Cooks RG (2005) The Orbitrap: a new mass spectrometer. J Mass Spectrom 40:430–443. https://doi.org/10.1002/JMS.856
Hua S, Yang Y, Zou D, Li J, Yan K, Xu Y, Jiang X, Rong X, Ye D (2021) COVID-19 and metabolic comorbidities: an update on emerging evidences for optimal therapies. Biomed Pharmacother 140
Johnson CH, Ivanisevic J, Siuzdak G (2016) Metabolomics: beyond biomarkers and towards mechanisms. Nat Rev Mol Cell Biol 17:451–459. https://doi.org/10.1038/nrm.2016.25
Jorgenson JW, Lukacs KDA (1983) Capillary zone electrophoresis. Science 1979(222):266–272. https://doi.org/10.1126/SCIENCE.6623076
Kao T-W, Huang C-C (2021) Recent Progress in metabolic syndrome research and therapeutics. Int J Mol Sci 22:6862. https://doi.org/10.3390/ijms22136862
Keating JE, Glish GL (2018) Dual emitter Nano-electrospray ionization coupled to differential ion mobility spectrometry-mass spectrometry for shotgun Lipidomics. Anal Chem 90:9117–9124. https://doi.org/10.1021/acs.analchem.8b01528
Khatri K, Pu Y, Klein JA, Wei J, Costello CE, Lin C, Zaia J (2018) Comparison of collisional and electron-based dissociation modes for middle-down analysis of multiply glycosylated peptides. J Am Soc Mass Spectrom 29:1075–1085. https://doi.org/10.1007/s13361-018-1909-y
King R, Bonfiglio R, Fernandez-Metzler C, Miller-Stein C, Olah T (2000) Mechanistic investigation of ionization suppression in electrospray ionization. J Am Soc Mass Spectrom 11:942–950. https://doi.org/10.1016/S1044-0305(00)00163-X
Konermann L, Ahadi E, Rodriguez AD, Vahidi S (2013) Unraveling the mechanism of electrospray ionization. Anal Chem 85:2–9. https://doi.org/10.1021/ac302789c
Kyle JE, Burnum-Johnson KE, Wendler JP, Eisfeld AJ, Halfmann PJ, Watanabe T, Sahr F, Smith RD, Kawaoka Y, Waters KM, Metz TO (2019) Plasma lipidome reveals critical illness and recovery from human Ebola virus disease. Proc Natl Acad Sci U S A 116:3919–3928. https://doi.org/10.1073/pnas.1815356116
Lagathu C, Béréziat V, Gorwood J, Fellahi S, Bastard J-P, Vigouroux C, Boccara F, Capeau J (2019) Metabolic complications affecting adipose tissue, lipid and glucose metabolism associated with HIV antiretroviral treatment. Expert Opin Drug Saf 18:829–840. https://doi.org/10.1080/14740338.2019.1644317
Lewis HM, Liu Y, Frampas CF, Longman K, Spick M, Stewart A, Sinclair E, Kasar N, Greener D, Whetton AD, Barran PE, Chen T, Dunn-Walters D, Skene DJ, Bailey MJ (2022) Metabolomics markers of COVID-19 are dependent on collection wave. Meta 12. https://doi.org/10.3390/metabo12080713

Li L, Zhong S, Shen X, Li Q, Xu W, Tao Y, Yin H (2019) Recent development on liquid chromatography-mass spectrometry analysis of oxidized lipids. Free Radic Biol Med 144:16–34. https://doi.org/10.1016/J.FREERADBIOMED.2019.06.006

Li S, Park Y, Duraisingham S, Strobel FH, Khan N (2013) Predicting network activity from high throughput metabolomics. PLoS Comput Biol 9:1003123. https://doi.org/10.1371/journal.pcbi.1003123

Lian E, McAlister C, Ramirez G, Chernoff DN, Went G, Hoopes J, Perera R (2020) Triple Combination Nitazoxanide, Ribavirin, and Hydroxychloroquine results in the multiplicative reduction of in vitro SARS-CoV-2 viral replication. bioRxiv. https://doi.org/10.1101/2020.11.25.399055

Liebisch G, Höring M, Stieglmeier C, Schnabel K, Hallmark T, Ekroos K, Burkhardt R (2022) Benchmarking one-phase lipid extractions for plasma Lipidomics. Anal Chem 94:12292–12296. https://doi.org/10.1021/ACS.ANALCHEM.2C02117/ASSET/IMAGES/LARGE/AC2C02117_0003.JPEG

Lin B, He S, Yim HJ, Liang TJ, Hu Z (2016) Evaluation of antiviral drug synergy in an infectious HCV system. Antivir Ther 21:595–603. https://doi.org/10.3851/IMP3044

Lung SCC, Liu CH (2015) Fast analysis of 29 polycyclic aromatic hydrocarbons (PAHs) and nitro-PAHs with ultra-high performance liquid chromatography-atmospheric pressure photoionization-tandem mass spectrometry. Sci Rep 5:1–13. https://doi.org/10.1038/srep12992

Marshall AG, Hendrickson CL, Jackson GS (1998) Fourier transform ion cyclotron resonance mass spectrometry: a primer. Mass Spectrom Rev 17:1–35. https://doi.org/10.1002/(SICI)1098-2787(1998)17:1<1::AID-MAS1>3.0.CO;2-K

Martín-Acebes MA, de Oya NJ, Saiz JC (2019) Lipid metabolism as a source of druggable targets for antiviral discovery against zika and other flaviviruses. Pharmaceuticals 12. https://doi.org/10.3390/ph12020097

May JC, McLean JA (2015) Ion mobility-mass spectrometry: time-dispersive instrumentation. Anal Chem 87:1422–1436. https://doi.org/10.1021/ac504720m

May JC, Morris CB, McLean JA (2017) Ion mobility collision cross section compendium. Anal Chem 89:1032–1044. https://doi.org/10.1021/acs.analchem.6b04905

Merrill AH, Sullards MC, Allegood JC, Kelly S, Wang E (2005) Sphingolipidomics: high-throughput, structure-specific, and quantitative analysis of sphingolipids by liquid chromatography tandem mass spectrometry. Methods 36:207–224. https://doi.org/10.1016/j.ymeth.2005.01.009

Moldoveanu SC, David V, Moldoveanu SC, David V (2018) Derivatization methods in GC and GC/MS. Gas chromatography – derivatization, sample preparation, application. https://doi.org/10.5772/INTECHOPEN.81954

Munger J, Bajad SU, Coller HA, Shenk T, Rabinowitz JD (2006) Dynamics of the cellular metabolome during human cytomegalovirus infection. PLoS Pathog 2:1165–1175. https://doi.org/10.1371/journal.ppat.0020132

Munger J, Bennett BD, Parikh A, Feng XJ, McArdle J, Rabitz HA, Shenk T, Rabinowitz JD (2008) Systems-level metabolic flux profiling identifies fatty acid synthesis as a target for antiviral therapy. Nat Biotechnol 26:1179–1186. https://doi.org/10.1038/nbt.1500

Myers OD, Sumner SJ, Li S, Barnes S, Du X (2017) Detailed investigation and comparison of the XCMS and MZmine 2 chromatogram construction and chromatographic peak detection methods for preprocessing mass spectrometry metabolomics data. Anal Chem 89:8689–8695. https://doi.org/10.1021/acs.analchem.7b01069

mzCloud – Advanced Mass Spectral Database. https://www.mzcloud.org/. Accessed 13 Nov 2022

Ngere JB, Ebrahimi KH, Williams R, Pires E, Walsby-Tickle J, McCullagh JSO (2023) Ion-exchange chromatography coupled to mass spectrometry in life science, environmental, and medical research. Anal Chem 95:152–166. https://doi.org/10.1021/ACS.ANALCHEM.2C04298/ASSET/IMAGES/LARGE/AC2C04298_0005.JPEG

Norris PC, Serhan CN (2018) Metabololipidomic profiling of functional immunoresolvent clusters and eicosanoids in mammalian tissues. Biochem Biophys Res Commun 504:553–561. https://doi.org/10.1016/j.bbrc.2018.03.037

O'Neill LAJ, Kishton RJ, Rathmell J (2016) A guide to immunometabolism for immunologists. Nat Rev Immunol 16:553–565. https://doi.org/10.1038/nri.2016.70

Osburn S, Ryzhov V (2013) Ion-molecule reactions: analytical and structural tool. Anal Chem 85:769–778. https://doi.org/10.1021/AC302920A/ASSET/IMAGES/LARGE/AC-2012-02920A_0004.JPEG

Paglia G, Williams JP, Menikarachchi L, Thompson JW, Tyldesley-Worster R, Halldórsson S, Rolfsson O, Moseley A, Grant D, Langridge J, Palsson BO, Astarita G (2014) Ion mobility derived collision cross sections to support metabolomics applications. Anal Chem 86:3985–3993. https://doi.org/10.1021/ac500405x

Pal A, Gowdy KM, Oestreich KJ, Beck M, Shaikh SR (2020) Obesity-driven deficiencies of specialized pro-resolving mediators May drive adverse outcomes during SARS-CoV-2 infection. Front Immunol 11. https://doi.org/10.3389/fimmu.2020.01997

Parker EJ, Billane KC, Parker EJ, Billane KC, Austen N, Cotton A, George RM, Hopkins D, Lake JA, Pitman JK, Prout JN, Walker HJ, Williams A, Cameron DD (2023) Untangling the complexities of processing and analysis for untargeted LC-MS data using open-source tools. Metabolites 13:463. https://doi.org/10.3390/METABO13040463

Passalacqua KD, Purdy JG, Wobus CE (2019) The inert meets the living: the expanding view of metabolic alterations during viral pathogenesis. PLoS Pathog 15:1–6. https://doi.org/10.1371/journal.ppat.1007830

Pawlotsky J, Flisiak R, Sarin SK, Rasenack J, Piratvisuth T, Chuang W, Peng C, Foster GR, Shah S, Wedemeyer H, Hézode C, Zhang W, Wong KA, Li B, Avila C, Naoumov NV (2015) Alisporivir plus ribavirin, interferon free or in combination with pegylated interferon, for hepatitis C virus genotype 2 or 3 infection. Hepatology 62:1013–1023. https://doi.org/10.1002/hep.27960

Peng X, Chan EY, Li Y, Diamond DL, Korth MJ, Katze MG (2009) Virus–host interactions: from systems biology to translational research. Curr Opin Microbiol 12:432–438. https://doi.org/10.1016/j.mib.2009.06.003

Perera R, Kuhn RJ (2016) Host metabolism and its contribution in Flavivirus biogenesis. In: Duane G, Nikos V (eds) Arboviruses: molecular biology, evolution and control. Caister Academic Press, University of Texas Medical Branch/Emerging Infectious Diseases Program, Duke-NUS Graduate Medical School, Galveston/Singapore, pp 45–60

Picache JA, Rose BS, Balinski A, Leaptrot KL, Sherrod SD, May JC, McLean JA (2019) Collision cross section compendium to annotate and predict multi-omic compound identities. Chem Sci 10:983–993. https://doi.org/10.1039/c8sc04396e

Prichard MN, Shipman C (1990) A three-dimensional model to analyze drug-drug interactions. Antivir Res 14:181–205. https://doi.org/10.1016/0166-3542(90)90001-N

Rabinowitz JD, Purdy JG, Vastag L, Shenk T, Koyuncu E (2023) Metabolomics in drug target discovery. Cold Spring Harb Symp Quant Biol. https://doi.org/10.1101/sqb.2011.76.010694

Raetz M, Bonner R, Hopfgartner G (2020) SWATH-MS for metabolomics and lipidomics: critical aspects of qualitative and quantitative analysis. Metabolomics 16:71. https://doi.org/10.1007/s11306-020-01692-0

Randolph CE, Blanksby SJ, McLuckey SA (2020) Toward complete structure elucidation of Glycerophospholipids in the gas phase through charge inversion ion/ion chemistry. Anal Chem 92:1219–1227. https://doi.org/10.1021/ACS.ANALCHEM.9B04376/ASSET/IMAGES/LARGE/AC9B04376_0004.JPEG

Richards DM, Carmine AA, Brogden RN et al (1983) Acyclovir. Drugs 26:378–438. https://doi.org/10.2165/00003495-198326050-00002

Rossignol JF (2014) Nitazoxanide: a first-in-class broad-spectrum antiviral agent. Antivir Res 110:94–103

Sarsby J, Griffiths RL, Race AM, Bunch J, Randall EC, Creese AJ, Cooper HJ (2015) Liquid extraction surface analysis mass spectrometry coupled with field asymmetric waveform ion mobility spectrometry for analysis of intact proteins from biological substrates. Anal Chem 87:6794–6800. https://doi.org/10.1021/acs.analchem.5b01151

Schmid R, Heuckeroth S, Korf A, Smirnov A, Myers O, Dyrlund TS, Bushuiev R, Murray KJ, Hoffmann N, Lu M, Sarvepalli A, Zhang Z, Fleischauer M, Dührkop K, Wesner M, Hoogstra SJ, Rudt E, Mokshyna O, Brungs C, Ponomarov K, Mutabdžija L, Damiani T, Pudney CJ, Earll M, Helmer PO, Fallon TR, Schulze T, Rivas-Ubach A, Bilbao A, Richter H, Nothias LF, Wang M, Orešič M, Weng JK, Böcker S, Jeibmann A, Hayen H, Karst U, Dorrestein PC, Petras D, Du X, Pluskal T (2023) Integrative analysis of multimodal mass spectrometry data in MZmine 3. Nat Biotechnol 1–3. https://doi.org/10.1038/s41587-023-01690-2

Schmitz U, Swaminathan S (2022) Discovery and development of oseltamivir at Gilead Sciences. Antivir Ther 27

Schrimpe-Rutledge AC, Codreanu SG, Sherrod SD, McLean JA (2016) Untargeted metabolomics strategies—challenges and emerging directions. J Am Soc Mass Spectrom 27:1897–1905. https://doi.org/10.1007/s13361-016-1469-y

Sekaran SD, Liew ZM, Yam HC, Raju CS (2022) The association between diabetes and obesity with dengue infections. Diabetol Metab Syndr 14:101. https://doi.org/10.1186/s13098-022-00870-5

Sellick CA, Hansen R, Stephens GM, Goodacre R, Dickson AJ (2011) Metabolite extraction from suspension-cultured mammalian cells for global metabolite profiling. Nat Protocol 6:1241–1249. https://doi.org/10.1038/NPROT.2011.366

Shah PS, Beesabathuni NS, Fishburn AT, Kenaston MW, Minami SA, Pham OH, Tucker I (2022) Systems biology of virus-host protein interactions: from hypothesis generation to mechanisms of replication and pathogenesis. Annu Rev Virol 9:397–415. https://doi.org/10.1146/annurev-virology-100520-011851

Shyr ZA, Cheng Y-S, Lo DC, Zheng W (2021) Drug combination therapy for emerging viral diseases. Drug Discov Today 26:2367–2376. https://doi.org/10.1016/j.drudis.2021.05.008

Simon V, Ho DD (2003) HIV-1 dynamics in vivo: implications for therapy. Nat Rev Microbiol 1:181–190. https://doi.org/10.1038/nrmicro772

Singh AK, Gupta R, Ghosh A, Misra A (2020) Diabetes in COVID-19: prevalence, pathophysiology, prognosis and practical considerations. Diabetes Metab Syndr 14:303–310. https://doi.org/10.1016/j.dsx.2020.04.004

Siuzdak G (2004) An introduction to mass spectrometry ionization: an excerpt from the expanding role of mass spectrometry in biotechnology, 2nd edn. MCC Press, San Diego, 2005. JALA J Assoc Lab Autom 9:50–63. https://doi.org/10.1016/j.jala.2004.01.004

Smith CA, Want EJ, O'maille G, Abagyan R, Siuzdak G (2006) XCMS: processing mass spectrometry data for metabolite profiling using nonlinear peak alignment, matching, and identification. https://doi.org/10.1021/ac051437y

The National Institute of Standards and Technology (2020) NIST20: updates to the NIST tandem and electron ionization spectral libraries

Thomas MC, Mitchell TW, Harman DG, Deeley JM, Nealon JR, Blanksby SJ (2008) Ozone-induced dissociation: elucidation of double bond position within mass-selected lipid ions. Anal Chem 80:303–311. https://doi.org/10.1021/AC7017684/SUPPL_FILE/AC7017684-FILE002.PDF

Tong L, Xiao X, Li M, Fang S, Ma E, Yu X, Zhu Y, Wu C, Tian D, Yang F, Sun J, Qu J, Zheng N, Liao S, Tai W, Feng S, Zhang L, Li Y, Wang L, Han X, Sun S, Yang L, Zhong H, Zhao J, Liu W, Liu X, Wang P, Li L, Zhao G, Zhang R, Cheng G (2022) A glucose-like metabolite deficient in diabetes inhibits cellular entry of SARS-CoV-2. Nat Metab 4:547–558. https://doi.org/10.1038/s42255-022-00567-z

Tounta V, Liu Y, Cheyne A, Larrouy-Maumus G (2021) Metabolomics in infectious diseases and drug discovery. Mol Omics 17:376–393. https://doi.org/10.1039/D1MO00017A

Ulmer CZ, Jones CM, Yost RA, Garrett TJ, Bowden JA (2018) Optimization of Folch, Bligh-Dyer, and Matyash sample-to-extraction solvent ratios for human plasma-based lipidomics studies. Anal Chim Acta 1037:351–357. https://doi.org/10.1016/j.aca.2018.08.004

Uppal K, Walker DI, Liu K, Li S, Go YM, Jones DP (2016) Computational metabolomics: a framework for the million metabolome. Chem Res Toxicol 29:1956–1975. https://doi.org/10.1021/acs.chemrestox.6b00179

Uvela PŽ, Skoczylas M, Liu JJ, Bą T, Kaliszan R, Wong MW, Buszewski B (2019) Column characterization and selection systems in reversed-phase high-performance liquid chromatography. Chem Rev. https://doi.org/10.1021/acs.chemrev.8b00246

Van Vaeck L, Adriaens A, Gijbels R (1999) Static secondary ion mass spectrometry: (S-SIMS) part 1. Methodology and structural interpretation. Mass Spectrom Rev 18:1–47. https://doi.org/10.1002/(SICI)1098-2787(1999)18:1<1::AID-MAS1>3.0.CO;2-W

Vastag L, Koyuncu E, Grady SL, Shenk TE, Rabinowitz JD (2011) Divergent effects of human cytomegalovirus and herpes simplex virus-1 on cellular metabolism. PLoS Pathog 7. https://doi.org/10.1371/journal.ppat.1002124

Ventouri IK, Malheiro DBA, Voeten RLC, Kok S, Honing M, Somsen GW, Haselberg R (2023) Probing protein denaturation during size-exclusion chromatography using native mass spectrometry. Anal Chem 21:7. https://doi.org/10.1021/acs.analchem.9b04961

Voeten RLC, Ventouri IK, Haselberg R, Somsen GW (2018) Capillary electrophoresis: trends and recent advances. Anal Chem 90:1464–1481. https://doi.org/10.1021/ACS.ANALCHEM.8B00015/ASSET/IMAGES/LARGE/AC-2018-000153_0010.JPEG

Want EJ, Masson P, Michopoulos F, Wilson ID, Theodoridis G, Plumb RS, Shockcor J, Loftus N, Holmes E, Nicholson JK (2012) Global metabolic profiling of animal and human tissues via UPLC-MS. Nature Protocols 8:17–32. 2013 8:1. https://doi.org/10.1038/NPROT.2012.135

Waters DD, Hsue PY (2019) Lipid abnormalities in persons living with HIV infection. Can J Cardiol 35:249–259. https://doi.org/10.1016/j.cjca.2018.11.005

Watrous JD, Niiranen TJ, Lagerborg KA, Henglin M, Xu YJ, Rong J, Sharma S, Vasan RS, Larson MG, Armando A, Mora S, Quehenberger O, Dennis EA, Cheng S, Jain M (2019) Directed non-targeted mass spectrometry and chemical networking for discovery of eicosanoids and related Oxylipins. Cell Chem Biol 26:433–442.e4. https://doi.org/10.1016/J.CHEMBIOL.2018.11.015

Westman-Brinkmalm A, Brinkmalm G (2008) Tandem mass spectrometry. Mass spectrometry: instrumentation, interpretation, and applications, pp 89–103. https://doi.org/10.1002/9780470395813.ch3

Whiley L, Nye LC, Grant I, Andreas N, Chappell KE, Sarafian MH, Misra R, Plumb RS, Lewis MR, Nicholson JK, Holmes E, Swann JR, Wilson ID (2019) Ultrahigh-performance liquid chromatography tandem mass spectrometry with electrospray ionization quantification of tryptophan metabolites and markers of gut health in serum and plasma – application to clinical and epidemiology cohorts. Anal Chem 91:5207–5216. https://doi.org/10.1021/ACS.ANALCHEM.8B05884/SUPPL_FILE/AC8B05884_SI_001.PDF

Wu H, Tang K (2020) Highly sensitive and robust capillary electrophoresis-electrospray ionization-mass spectrometry: interfaces, preconcentration techniques and applications. Rev Anal Chem 39:45–55. https://doi.org/10.1515/revac-2020-0112

Wu Q, Zhou L, Sun X, Yan Z, Hu C, Wu J, Xu L, Li X, Liu H, Yin P, Li K, Zhao J, Li Y, Wang X, Li Y, Zhang Q, Xu G, Chen H (2017) Altered lipid metabolism in recovered SARS patients twelve years after infection. Sci Rep 7:9110. https://doi.org/10.1038/s41598-017-09536-z

Yamashita M, Fenn JB (1984) Electrospray ion source. Another variation on the free-jet theme. J Phys Chem 88:4451–4459. https://doi.org/10.1021/J150664A002/ASSET/J150664A002.FP.PNG_V03

Yan B, Chu H, Yang D, Sze K-H, Lai P-M, Yuan S, Shuai H, Wang Y, Kao RY-T, Chan JF-W, Yuen K-Y (2019) Characterization of the Lipidomic profile of human coronavirus-infected cells: implications for lipid metabolism remodeling upon coronavirus replication. Viruses 11:73. https://doi.org/10.3390/v11010073

Yin Y, Wang R, Cai Y, Wang Z, Zhu ZJ (2019) DecoMetDIA: deconvolution of multiplexed MS/MS spectra for metabolite identification in SWATH-MS-based untargeted metabolomics. Anal Chem 91:11897–11904. https://doi.org/10.1021/acs.analchem.9b02655

Zhang J, Rector J, Lin JQ, Young JH, Sans M, Katta N, Giese N, Yu W, Nagi C, Suliburk J, Liu J, Bensussan A, Dehoog RJ, Garza KY, Ludolph B, Sorace AG, Syed A, Zahedivash A, Milner TE, Eberlin LS (2017) Nondestructive tissue analysis for ex vivo and in vivo cancer diagnosis using a handheld mass spectrometry system. Sci Transl Med 9. https://doi.org/10.1126/scitranslmed.aan3968

Zhao C, Xie B, Chan SY, Costello CE, O'Connor PB (2008) Collisionally activated dissociation and electron capture dissociation provide complementary structural information for branched Permethylated oligosaccharides. J Am Soc Mass Spectrom 19:138–150. https://doi.org/10.1016/j.jasms.2007.10.022

Zubarev RA, Makarov A (2013) Orbitrap mass spectrometry. Anal Chem 85:5288–5296. https://doi.org/10.1021/ac4001223

Chapter 6
Epigenetic- and Epitranscriptomic-Targeted Reprogramming: Novel Targets for the Development of Broad-Spectrum Antivirals

Naveen Kumar, Ashutosh Singh, Upasana Sahu, Dhruv Desai, Manoj Kumar, Sandeep Bhatia, and Aniket Sanyal

Abstract Epigenetic and epitranscriptomic modifications have emerged as promising targets for the development of broad-spectrum antiviral therapies. These modifications involve alterations to DNA, histones, and RNA molecules, which can regulate gene expression and cellular processes. By manipulating these modifications, it is possible to modulate host–virus interactions, disrupt viral replication cycles, and enhance antiviral immune responses. This chapter explores the potential of epigenetic- and epitranscriptomic-targeted reprogramming as a novel approach for the development of broad-spectrum antivirals. We discuss the impact of epigenetic modifications, such as DNA methylation and histone modifications, on viral gene expression, latency, and immune evasion, as well as viral RNA modifications that influence its stability, translation, and protein synthesis. Specific targets within the epigenetic and epitranscriptomic landscape that have shown promise in inhibiting viral replication and reactivating latent viruses are described. Furthermore, the challenges and opportunities associated with these novel therapeutic strategies, including drug specificity, off-target effects, and safety considerations are discussed. A deeper comprehension of the intricate interactions among host factors, viral genomes, and epigenetic/epitranscriptomic machinery will be crucial for harnessing the full potential of epigenetic- and epitranscriptomic-targeted reprogramming in the development of broad-spectrum antivirals. Largely, this chapter sheds light on the exciting frontier of epigenetic- and epitranscriptomic-based approaches and their potential to revolutionize antiviral therapy.

Keywords Epigenetics · Epitranscriptomics · Viruses · Drugs

N. Kumar (✉) · A. Singh · U. Sahu · M. Kumar · S. Bhatia · A. Sanyal
National Institute of High Security Animal Diseases, Bhopal, India
e-mail: naveen.kumar4@icar.gov.in

D. Desai
University of Pennsylvania, Philadelphia, PA, USA

N. Kumar et al. (eds.), *Advances in Antiviral Research*, Livestock Diseases and Management, https://doi.org/10.1007/978-981-99-9195-2_6

6.1 Introduction

Epigenetic and epitranscriptomic modifications are essential regulatory mechanisms that control gene expression and cellular processes. They involve modifications to DNA, histones, and RNA molecules that can influence gene activity and protein synthesis. Recently, there has been growing interest in leveraging these modifications as novel targets for the development of broad-spectrum antiviral therapies. Epigenetic- and epitranscriptomic-targeted reprogramming offers unique opportunities to modulate host–virus interactions, disrupt viral replication cycles and enhance antiviral immune responses.

Epigenetic modifications encompassing DNA methylation and histone modifications can affect viral gene expression, latency, and immune evasion (Tsai and Cullen 2020; Wang et al. 2021). Targeting these modifications with specific inhibitors or activators can reactivate dormant viruses, suppress viral replication, or restore antiviral immune responses. Moreover, the emerging field of epitranscriptomics focuses on modifications of RNA molecules, including RNA methylation, which can impact RNA stability, translation, and protein synthesis (Tsai and Cullen 2020). Modulating these RNA modifications holds promise for interfering with viral RNA functions and inhibiting viral replication.

By targeting epigenetic and epitranscriptomic modifications, broad-spectrum antiviral strategies can be developed that target multiple viruses and potentially overcome the limitations of traditional antiviral therapies. Unlike direct-acting antivirals that target viral proteins, epigenetic and epitranscriptomic modifications are predominantly host-directed, allowing for the manipulation of cellular processes that are hijacked by viruses. This approach can minimize the emergence of drug resistance and provide a broader spectrum of activity against diverse viral families (Kumar et al. 2020).

Several epigenetic and epitranscriptomic targets have already been identified and studied for their potential antiviral effects. For example, DNA methyltransferase (DNMT) and histone deacetylase inhibitors have demonstrated promise in reactivating latent viruses, inhibiting viral replication, and enhancing antiviral immune responses (Kauder et al. 2009; Tsai et al. 2011; Diamantopoulos et al. 2012; Rawson et al. 2016; Lin et al. 2022). Similarly, modulating RNA modifications encompassing N6-methyladenosine (m^6A) or 5-methylcytosine (m^5C) can interfere with viral RNA stability, translation, and viral protein production (Huggins et al. 1999; Bray et al. 2000; Durbin et al. 2016; Kennedy et al. 2016; Lichinchi et al. 2016; Courtney et al. 2017; Ye et al. 2017; Tsai et al. 2018; Hao et al. 2019; Lu et al. 2020; Nance and Meier 2021; Starostina et al. 2021). The development of broad-spectrum antivirals targeting epigenetic and epitranscriptomic modifications is an exciting frontier in antiviral research. However, challenges such as optimizing drug specificity, minimizing off-target effects, and ensuring the safety and efficacy of these novel therapies remain. Further understanding of the complex interactions among host factors, viral genomes, and epigenetic/epitranscriptomic machinery is needed to harness the full potential of these approaches.

In conclusion, epigenetic- and epitranscriptomic-targeted reprogramming provides novel avenues for the development of broad-spectrum antiviral therapies. By manipulating host cellular processes and regulatory mechanisms, these approaches have the potential to disrupt viral replication cycles, reactivate latent viruses, and enhance antiviral immune responses. Continued research and advancements in this field may lead to the development of innovative treatments that can combat a wide range of viral infections.

6.2 Epigenetic Modifications

Epigenetic modifications refer to heritable alterations in gene expression that do not entail changes to the DNA sequence itself. These modifications hold the power to regulate gene activity by modifying the DNA structure or influencing the packaging of DNA within chromatin complexes. Vital for normal cellular functions, development, and disease processes, including viral infections, epigenetic modifications play a crucial role. Within the realm of viruses, various types of epigenetic modifications can occur.

6.2.1 DNA Methylation

DNA methylation, involving the addition or removal of methyl groups from cytosine residues in DNA, represents the predominant epigenetic modifications in the genome, playing a pivotal role in the regulation of genomic function. The most prevalent form of DNA methylation occurs at the fifth position of cytosine-phosphate-guanine (CpG) dinucleotides, known as m^5C. CpG sites, known as CpG islands, serve as major targets for DNA methylation (Baubec and Schübeler 2014). Additionally, methylation events can take place on cytosine residues without neighboring guanine, especially in differentiated tissues (Schultz et al. 2015). CpG island methylation is frequently linked to the transcriptional repression of repetitive sequences and the silencing of imprinted gene alleles. Promoters and enhancers of transcribed genes typically exhibit lower CpG methylation levels, while the gene body can display higher levels (Lister et al. 2009). Based on CpG content and surrounding genomic characteristics, methylated cytosines are categorized into distinct regions, namely CpG islands (2%), shores (4%), shelves (6%), and open sea/others (88%) (Rechache et al. 2012).

DNA methylation is an enduring modification that is heritably propagated through cell divisions. It employs diverse mechanisms to modulate gene expression. Methylated DNA molecules can attract methyl-CpG-binding domain (MBD) proteins, which subsequently facilitate the recruitment of histone-modifying and chromatin-remodeling complexes to the methylated sites (Stirzaker et al. 2017). Additionally, DNA methylation can also impede the binding of DNA-binding proteins to their target sites, inhibiting the transcription of genes (Johnson et al. 2014b). Proteins that interact with methylated DNA sites or MBD proteins, acting as "readers," provide functional information associated with DNA modification. However, DNA methylation can be passively erased during DNA replication when the modification is not restored, while active removal of DNA methylation, acting as "erasers," can also occur (Andrews et al. 2016).

The transfer of methyl groups from S-adenosyl methionine (SAM) to DNA is carried out by a class of enzymes known as DNMTs. The DNMT family consists of five members: DNMT1, DNMT2, DNMT3a, DNMT3b, and DNMT3L. Among them, methyltransferase activity is present in DNMT1, DNMT3a, and DNMT3b. While DNMT1 acts as a maintenance enzyme, specifically methylating hemimethylated CpG dinucleotides following DNA replication during the S phase of the cell cycle, DNMT3a and DNMT3b function as catalytically active members (Jones and Liang 2009). The balance between DNA demethylation and methylation is regulated by DNA demethyltransferases (DeMTs) and DNMTs, respectively. Collectively, the combined roles of DNMTs and DeMTs are essential for preserving the DNA methylation patterns during cell cycle control, apoptosis, DNA damage response, immune recognition, and cellular self-renewal.

6.2.2 *Histone Modifications*

Proteins called histones wrap DNA into the small chromatin structure. The fundamental building block of chromatin, known as a nucleosome, is composed of DNA encased in an octamer of the four histone proteins (H2A, H2B, H3, H4), which are joined by histone protein H1 (Kaplan et al. 2009). Various chemical modifications can occur on the amino acid residues of histones, including acetylation, methylation, phosphorylation, ubiquitination, and more. These modifications can alter the chromatin structure (also known as chromatin remodeling) and accessibility of DNA, thereby influencing gene expression. Viruses can modulate histone modifications to either promote or suppress viral gene expression, evading host immune responses or establishing latency. The common histone modifications are discussed further.

6.2.2.1 Histone Acetylation

With the help of histone acetyltransferases (HATs), histone acetylation entails the addition of an acetyl group from acetyl CoA to the lysine residues on histone tails. This modification is generally associated with gene activation (Sterner and Berger 2000). The positive charge on lysine residues is neutralized by acetylation, which lowers the affinity of histones for DNA. Because of the relaxed chromatin structure, transcription factors and other regulatory proteins have easier access to the DNA, promoting gene transcription. However, deacetylation of histone, which is commonly, carried out by histone deacetylases (HDACs) leads to gene repression. Therefore, a dynamic balance between HATs and HDACs regulates the chromatin structure and gene expression (Yang and Seto 2007).

6.2.2.2 Histone Methylation

Histone methyltransferases (HMTs) add a methyl group to the lysine or arginine residues of histones (H3, and H4) in the process known as histone methylation. Depending on the particular residue and level of methylation, unlike acetylation, methylation can have both activating and repressive effects on gene expression without altering the charge on the histone protein. For instance, trimethylations of histone H3 lysine 4 (H3K4me3), H3K79me3; monomethylations of histone 3 lysine 9 (H3K9me1), H3K27me1, H3K79me1; and monomethylations of histone 4 lysine 20 (H4K20me1) are associated with active gene transcription, while trimethylations of histone H3 lysine 27 (H3K27me3) and H3K9me3 are linked to gene repression (Barski et al. 2007; Wesche et al. 2017). Therefore, different histone modifications created by HMTs and removed by histone demethylases (HDMs) control the expression of certain genes (Van Rechem and Whetstine 2014).

6.2.2.3 Histone Phosphorylation

Histone phosphorylation is the process of adding phosphate groups to serine, threonine, or tyrosine residues on histone tails, with the help of various kinases and removed by phosphatases. The phosphorylated sites vary on different core histone proteins, such as Ser1 gets phosphorylated on H2A and H4, Ser14 and Ser32 on H2B, and Ser10 and Ser28 on H3 (Galasinski et al. 2002). This modification can regulate gene expression by influencing chromatin condensation, transcription factor binding, and recruitment of other chromatin-modifying enzymes. Phosphorylation of histones is often associated with changes in gene expression in response to cellular signaling pathways and environmental cues (Curradi et al. 2002; Si et al. 2010; Rossetto et al. 2012).

6.2.2.4 Histone Ubiquitination

Histone ubiquitination is the process by which ubiquitin molecules bind to histone lysine residues. Depending on the precise position and degree of ubiquitination, this modification can have diverse effects on the expression of genes (Weake and Workman 2008). For instance, monoubiquitination of histone H2B at lysine 120 (H2BK120ub) is associated with gene activation, while ubiquitination of histone H2A at lysine 119 (H2AK119ub) is linked to gene repression.

6.2.2.5 Histone Sumoylation

Small ubiquitin-like modifier proteins are added to histone lysine or arginine residues during the process of histone sumoylation. By changing the chromatin structure and recruiting transcriptional regulators, sumoylation can control gene expression. Numerous biological activities, such as DNA repair, chromosome segregation, and transcriptional regulation, depend on it (Hendriks et al. 2014).

All these histone modifications can act individually or in combination to create a "histone code," which helps determine the functional state of chromatin and regulates gene expression patterns. The interpretation of the histone code by various proteins and complexes is crucial for maintaining cellular homeostasis and responding to environmental cues. Dysregulation of histone modifications has been implicated in numerous diseases, including cancer, neurodegenerative disorders, and immune dysfunctions. It's important to note that the specific epigenetic modifications and their roles can vary depending on the virus and the infected host. The field of viral epigenetics is a thriving one with continuing investigations uncovering various processes by which viruses manipulate host epigenetic machinery for their benefit.

6.3 Epigenetic Repression and Strategies Evolved by Viruses to Avoid it

When the viral DNA enters the host cells, it is sensed by the cells as foreign DNA, and certain genes or regions of the viral genome are silenced or suppressed through epigenetic modifications. This suppression or repression can occur via a variety of pathways, including DNA methylation, histone modifications, and noncoding RNA (ncRNA) activity. However, viruses have evolved strategies to avoid or counteract epigenetic repression to facilitate their replication and evade host immune responses.

One example of a cellular defense mechanism against viral infections is the formation of promyelocytic leukemia nuclear bodies (PML-NBs). PML-NBs are dynamic subnuclear structures that contain a variety of proteins, such as PML, Daxx, Sp100, and ATRX, involved in antiviral responses. These structures are formed upon activation of the host innate immune system, particularly in response to interferon signaling (Knipe 2015). Viruses have evolved mechanisms to disrupt or avoid the formation and function of PML-NBs. One strategy employed by viruses is the expression of viral proteins that can inhibit the assembly or induce the dispersal of PML-NBs. For example, upon entering host cells, the histone chaperone complex HIRA, along with repressive H3K9me3 marks and PML-NB components (Daxx and ATRX), marks the herpes simplex virus 1 (HSV-1) DNA. Initially, HSV-1 employs its viral protein- 16 (VP16) protein to recruit host-cell factor 1 and octamer-binding factor, which leads to the recruitment of HDMs lysine-specific demethylase 1 and Jumonji domain 2 proteins. These proteins finally remove H3K9me3 repressive marks from viral immediate early promoters (Cliffe and Knipe 2008; Oh and Fraser 2008; Liang et al. 2009; Placek et al. 2009; Liang et al. 2013; Cohen et al. 2018). Furthermore, the VP16 complex recruits the methyltransferases Set1 and MLL1, which results in the formation of H3K4me3 activation marks on histone H3 linked to viral DNA (Narayanan et al. 2007). This process promotes the development of the immediate early viral protein ICP0, which then in turn targets and degrades PML-NBs, counteracting the antiviral effects of these structures. Other viruses, such as human cytomegalovirus and Epstein–Barr virus (EBV), also produce proteins (pp71 and BNRF1, respectively) that specifically act on PML-NBs to counteract their antiviral functions (Cantrell and Bresnahan 2006; Lukashchuk et al. 2008; Tsai et al. 2011).

In addition to PML-NBs, another component of the host innate immune response that is involved in the epigenetic repression of viral DNA is the interferon-inducible protein 16 (IFI16) (Orzalli et al. 2013; Diner et al. 2016). IFI16 is a DNA sensor that detects viral DNA in the cytoplasm and nucleus of infected cells and has been documented to interact with PML-NBs (Diner et al. 2015). Upon binding to viral DNA, IFI16 recruits other proteins to form a complex that leads to the repression of viral gene expression through epigenetic modifications. However, viruses have evolved mechanisms to evade or counteract the actions of IFI16. For example, human papillomavirus (HPV) encodes viral proteins, such as E7, that can interact with IFI16 and inhibit its function. By blocking the recognition of viral DNA by IFI16, HPV can avoid the epigenetic repression induced by IFI16 and continue its replication and persistence within host cells (Song et al. 2020). Additionally, HSV-1 has been demonstrated to degrade IFI16 (Johnson et al. 2014a).

Overall, viruses have evolved multiple strategies to avoid or counteract epigenetic repression mechanisms, such as the formation of PML-NBs and the actions of IFI16. By disrupting these host defense mechanisms, viruses can establish a favorable environment for their replication and immune evasion. Understanding these viral evasion strategies is crucial for the development of effective antiviral therapies that can restore or enhance the host's ability to repress viral gene expression through epigenetic modifications.

6.4 Antiviral Drugs Targeting Epigenetic Modifications

These drugs offer a relatively new and promising approach in the field of virology. Viruses, like all other organisms, rely on host cellular machinery to replicate and survive. They can exploit host epigenetic mechanisms to establish a favorable environment for their replication or to evade the immune response. By targeting these epigenetic modifications, antiviral drugs aim to disrupt viral replication, suppress viral gene expression, and enhance host defense mechanisms. The epigenetic drugs can broadly be classified into two groups namely, DNA methylating agents and histone deacetylation inhibitors (HDACis).

6.4.1 Histone Deacetylation Inhibitors

HDACis are another class of epigenetic drugs that have been explored for antiviral therapy. HDACis, including vorinostat and romidepsin, promote histone hyperacetylation, leading to transcriptional activation of silenced genes, modulation of immune responses, and inhibition of viral replication (Fig. 6.1a). HDACis have shown antiviral activity against various viruses, including herpesviruses, HPV, HIV, hepatitis C virus (HCV), and respiratory syncytial virus (RSV). However, the optimal dosing, treatment duration, and potential for viral resistance require further investigation. Following are the HDACis that have been used against several viruses.

6.4.1.1 Vorinostat

Vorinostat (also known as suberoylanilide hydroxamic acid, SAHA) is a well-studied broad-spectrum HDACi that has demonstrated antiviral activity against several viruses. It inhibits the activity of histone deacetylases, leading to hyperacetylation of histones and other proteins, which leads to transcriptional activation of silenced genes and modulation of immune responses (Fig. 6.1a). For instance, Banerjee et al. (2018) demonstrated that vorinostat treatment significantly reduced human papillomavirus type 18 DNA amplification by modulating the chromatin structure and also enhanced cell apoptosis by suppressing the DNA repair mechanism and improving the expression of the proapoptotic protein Bim (Banerjee et al. 2018). Furthermore, both SAHA and trichostatin A (another HDACi,) have been demonstrated to efficiently limit the replication of RSV. They achieve this by increasing the activity of interferon-α-related signaling pathways, lowering RSV-induced airway inflammation and oxidative stress (Feng et al. 2016). Shapira et al. (2016) demonstrated that vorinostat and trichostatin A decreased viral replication by increasing the production of host protein ND10, which is known to suppress viral replication (Shapira et al. 2016). Additionally, SAHA or RGFP966 (also an HDACi) treatment has been found to suppress HCV infection. This is achieved by increasing the expression of liver-expressed antimicrobial peptide 1 and decreasing the expression of apolipoprotein-A1 (Zhou et al. 2018). Furthermore, vorinostat has been shown to inhibit the DNA replication of human adenoviruses (HAdVs), HAdV-4, HAdV-5, and HAdV-7 (Saha and Parks 2019).

6.4.1.2 Romidepsin (Istodax)

In a similar vein to vorinostat, romidepsin is an HDACi approved by the Food and Drug Administration (FDA) that has shown antiviral activities against various viruses. It induces hyperacetylation of histones, which reactivates silenced genes and modulates viral gene expression. The most effective antiviral strategy for latency-producing viruses is kick-and-kill strategy, which means to reactivate them

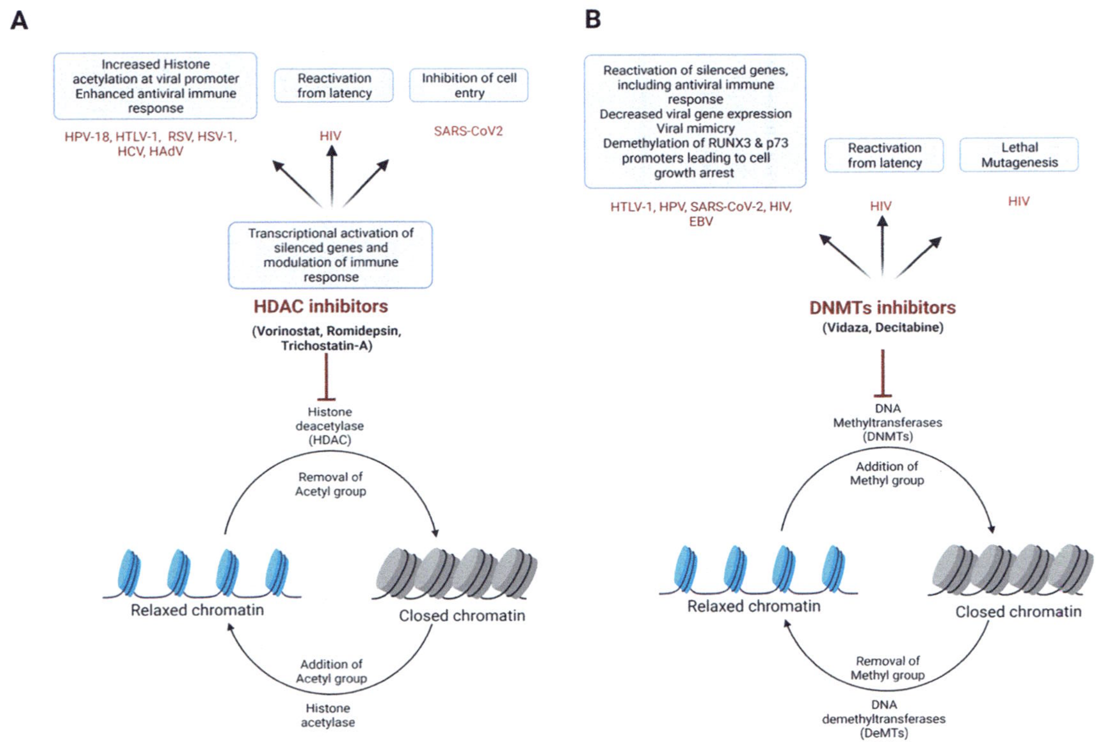

Fig. 6.1 Antiviral drugs reprogramming the epigenetic modifications to disrupt viral replication, suppress viral gene expression, and enhance host defense mechanisms. (**a**) Histone deacetylation inhibitors (HDACis) employ multiple mechanisms for the inhibition of many viruses by promoting histone hyperacetylation; (**b**) DNA methyltransferase (DNMT) inhibitors incorporate into DNA during replication and inhibit the activity of DNA demethyltransferases (DeMTs), and therefore, they result in DNA hypomethylation and reactivation of silenced genes, including those involved in antiviral immune responses

from latency followed by killing the infected cells by the cytotoxic T cells. This strategy has recently been used for human T-cell leukemia virus type 1 (HTLV-1), where romidepsin and panobinostat (another HDACi) treatment enhanced the expression of tax protein, an immunodominant target of cytotoxic T cells, by histone acetylation, resulting in clearance of latently infected cells (Schnell et al. 2022).

Furthermore, six HDACis were discovered to block SARS-CoV-2 cell entry in a SARS-CoV-2-S pseudovirus cell entry assay, including romidepsin, panobinostat, givinostat hydrochloride monohydrate, CAY10603, and sirtinol (Fig. 6.1a) (Saha and Parks 2019). An important therapeutic strategy for human immunodeficiency virus type 1 (HIV-1) infection has been to reactivate the virus from latency and then eliminate it using either a vaccine (Mothe et al. 2020) or monoclonal antibodies with a CD4-binding site, 3BNC117 (Gunst et al. 2022), in combination with romidepsin. Romidepsin treatment increases histone acetylation, cell-associated HIV-1 RNA, and T-cell activation, thereby limiting long-term HIV-1 persistence (Mothe et al. 2020). These findings emphasize the potential of HDACis, such as vorinostat and romidepsin, in the treatment of viral infections by modulating histone acetylation and gene expression. However, additional research is required to optimize dosing and treatment duration, and evaluate potential off-target effects for their safe and effective use in clinical settings.

6.4.2 DNA Methylating Agents

Although FDA has licensed DNA methylating drugs (also known as DNMT inhibitors), such as 5-azacytidine (vidaza) and 5-aza-2′-deoxycytidine (decitabine) for use in myelodysplastic syndrome, they have also demonstrated potential in combating viral infections. Both of these drugs are nucleoside analogs of cytidine that function as a DNMT inhibitor. They incorporate into DNA during replication and inhibit the activity of DNMTs. By doing so, it leads to DNA hypomethylation and reactivation of silenced genes, including those involved in antiviral immune responses (Fig. 6.1b). Additionally, they can also block cytosine methylation at m^5C in SARS-CoV-2 RNA by removing RNA methyltransferases. Studies have demonstrated the efficacy of DNA methylating agents against viral pathogens, such as human immunodeficiency virus (HIV), HTLV-1, hepatitis B virus, and SARS-CoV-2. However, the use of DNA methylating agents in viral infections requires careful consideration due to potential off-target effects and impacts on cellular DNA methylation patterns.

6.4.2.1 5-Azacytidine (Vidaza)

Several studies have utilized 5-azacytidine, known as vidaza, either to reactivate viruses from latency (Kauder et al. 2009) or to inhibit virus replication (Diamantopoulos et al. 2012; Rawson et al. 2016; Lin et al. 2022). Kauder et al.

(2009) conducted a study demonstrating that treating latently infected cells with 5-azacytidine can reactivate HIV-1 from its dormant state. This reactivation resulted in enhanced viral gene expression and infectious virus particle formation (Kauder et al. 2009). Rawson et al. (2016) revealed that 5-azacytidine can cause deadly mutagenesis in HIV-1 by generating G-to-C transversions during the process of reverse transcription (Fig. 6.1b) (Rawson et al. 2016). Interestingly, 5-azacytidine has also been successfully used to treat myelodysplastic syndrome associated with HTLV-1 and autoimmune manifestations (Diamantopoulos et al. 2012). Furthermore, a recent study by Lin et al. (2022) demonstrated that 5-azacytidine incorporates into SARS-CoV-2 RNA and exhibits antiviral activity by inhibiting cytosine methylation at m^5C (Lin et al. 2022).

6.4.2.2 5-Aza-2′-Deoxycytidine (Decitabine)

Decitabine, commonly known as 5-Aza-2′-deoxycytidine, is a cytidine nucleoside analog that inhibits DNMT. When incorporated into DNA during replication, it induces DNA hypomethylation, resulting in the reactivation of genes. Stich et al. (2017) conducted a study showing that decitabine suppressed the replication of HPV in cervical cancer cells by downregulating the expression of oncogenes, E6 and E7, and thereby, it blocked the tumor growth of HPV-associated cancer (Stich et al. 2017). Another approach for treating HPV-associated cancer was recently demonstrated in which decitabine treatment induced the production of cytoplasmic double-stranded RNA, also known as viral mimicry, resulting in the activation of IFN-related genes. This led to the sensitization of infected cells to immune-mediated clearance (Alexandraki and Strati 2022) (Fig. 6.1b).

Decitabine was also used for the treatment of EBV-associated malignancies, like gastric cancer (Nakamura et al. 2017). The antitumor activity was achieved by demethylating the promoters of RUNX3 and p73, which increased the transcription of p21, resulting in the arrest of tumor cell growth and apoptosis. Similarly, it can also sensitize the tumor cells to cytotoxic T cells by enhancing the expression of Epstein–Barr nuclear antigen (EBNA1), which otherwise escapes detection from cytotoxic T cells, leading to the regression of EBV-Burkitt lymphoma (Dalton et al. 2020).

Additionally, decitabine has been found to improve the efficacy of existing antiviral drugs such as gemcitabine and resveratrol for the treatment of HIV-1 infection (Rawson et al. 2013).

These studies highlight the potential of DNA methylating agents, such as 5-azacytidine and decitabine, in viral infection treatment by reversing DNA methylation-mediated gene silencing. However, for the safe and effective use of these drugs in the context of viral infections, the optimal dosing, treatment duration, and potential off-target effects must be carefully considered.

6.5 Epitranscriptomic Modifications

With over 100 distinct covalent modifications on RNA transcripts in mammalian cells have been identified—collectively referred to as epitranscriptomic modifications—that occur posttranscriptionally in ncRNAs, including transfer RNA. However, at least 10 of these modifications have been reported to occur in mammalian mRNA. Nevertheless, we will specifically concentrate on four epitranscriptomic modifications that have demonstrated their influence on viral gene expression, facilitation of various stages in the viral replication cycle, and evasion from host antiviral immunity—m^6A, m^5C, *N*4-acetylcytidine, and 2′O-methylation of the ribose moiety of four ribonucleosides (Am, Cm, Gm, and Um, collectively referred to as Nm) (Fig. 6.2). The study of chemical modifications in viral RNA that can regulate viral gene expression and replication is known as viral epitranscriptomics. These chemical modifications occur through a process known as RNA epigenetics, where specific chemical tags are added to RNA molecules, altering their structure and function. By studying the modifications present in viral RNA, researchers are deciphering the mechanisms by which these modifications influence viral replication, immune evasion, and pathogenesis. This knowledge is being utilized to develop therapeutic strategies targeting specific viral RNA modifications.

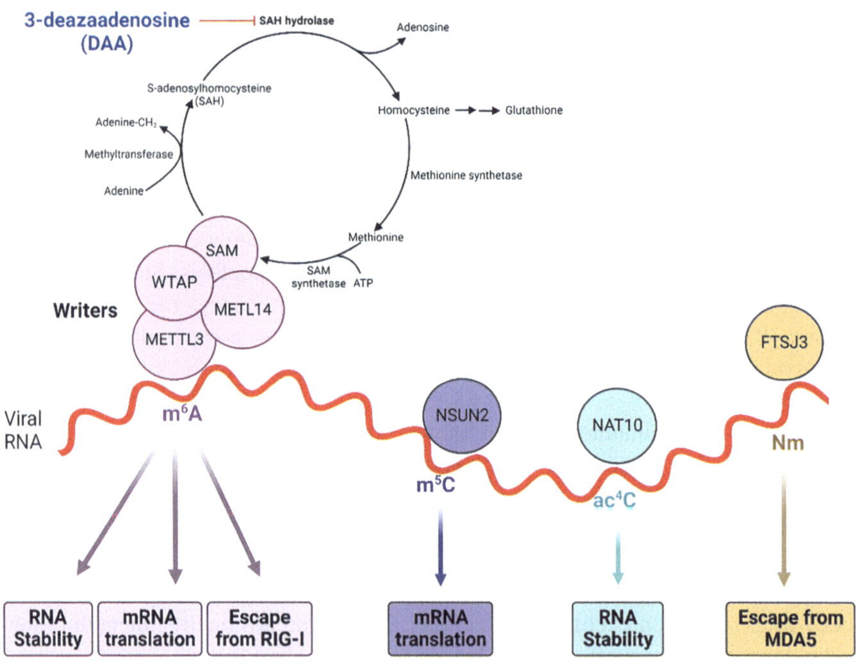

Fig. 6.2 Epitranscriptomic modifications deposited on the viral RNA promoting viral replication and escape from the host innate immune response and mechanism of action of 3-deazaadenosine (an inhibitor of m^6A modification)

Additionally, researchers are exploring the possibility of directly targeting viral RNA modifications with therapeutic drugs. Small molecules or nucleotide analogs can be designed to specifically recognize and bind to modified sites in viral RNA, thereby interfering with viral replication and gene expression. The field of viral epitranscriptomics is gaining attention among antiviral researchers due to its potential implications in antiviral therapy and the development of novel drugs. The four epitranscriptomic modifications found in viral RNA are detailed in the following sections.

6.5.1 N6-Methyladenosine (m^6A)

It is a common epitranscriptomic modification reported on eukaryotic cellular mRNA as well as on viral RNA transcripts (Meyer and Jaffrey 2014; Roundtree et al. 2017). This modification is added to the viral transcript by an enzyme, methyltransferase-like 3—known as the m^6A "writer"—in conjugation with several cofactors, including METTL14, Wilms tumor 1-associated protein, and methyl donor SAM (Fig. 6.2) (Meyer and Jaffrey 2014; Roundtree et al. 2017). The m^6A writer complex is predominantly nuclear, and once, this modification is deposited on viral transcript, m^6A residues are detected by five m^6A "readers," including the predominantly nuclear protein YTHDC1, and the cytoplasmic factors, YTHDC2, YTHDF1, YTHDF2, and YTHDF3. These five m^6A readers carry a YTH-domain that directly binds to m^6A residues, facilitating the posttranscriptional function of m^6A. This modification is proposed to be reversible, i.e., it can be removed in response to certain signals with the help of two m^6A "erasers" (or demethylases), including AlkB homolog 5 (ALKBH5) and fat mass and obesity-associated protein (FTO) (Darnell et al. 2018; Zhao et al. 2018). Several studies using knockdown or overexpression of one or more m^6A writers, readers, or erasers have demonstrated that m^6A residues promote viral replication by modulating different phases of the viral replication cycle. This modulation was achieved by promoting stability of viral mRNA or mRNA translation or by escaping recognition from the host innate immune RNA sensor, retinoic acid-inducible gene I (RIG-I), for a range of RNA (HIV-1, influenza A virus, RSV, enterovirus 71, HCV, human metapneumovirus, Zika virus, dengue virus, West Nile virus, and yellow fever virus) and DNA (polyomavirus SV40) viruses (Kennedy et al. 2016; Lichinchi et al. 2016; Courtney et al. 2017; Ye et al. 2017; Tsai et al. 2018; Hao et al. 2019; Xue et al. 2019; Lu et al. 2020).

6.5.2 5-Methylcytidine (m^5C)

It is another common epitranscriptomic modification found on eukaryotic cellular mRNA, but not so common across the different viral family's transcripts (Courtney et al. 2019a, b). In mammals, C5 methylation of cytosines on mRNA is specifically carried out by two methyltransferases, namely, DNMT homolog DNMT2 and that from the NOL1/NOP2/SUN domain (NSUN) family. In particular, of seven types of NSUN methyltransferases, NSUN2 catalyzes the addition of the majority of m^5C residues to eukaryotic as well as viral mRNAs (Fig. 6.2). However, the m^5C reader and eraser for viral mRNA have not been identified yet. A study on blocking the NSUN2 expression revealed a peculiar decrease in the efficiency of translation for both HIV-1 and cellular transcripts, indicating that the addition of m^5C appears to enhance viral replication and gene expression by acting primarily at the mRNA translation level (Courtney et al. 2019b).

6.5.3 N4-Acetylcytidine

This modification is highly conserved across a spectrum of RNAs from archaea to humans. It is primarily involved in the correct reading of codons during the translation process (and thereby, enhances translation efficiency) and improves cellular mRNA stability (Jin et al. 2020). The acetylation of cytidine is catalyzed by an ac^4C writer, i.e., N-acetyltransferase 10 (NAT10) (Fig. 6.2). The ac^4C reader and eraser have not been characterized so far. Inhibition of NAT10 in HIV-1 and enterovirus 71 in separate studies resulted in a decrease in viral gene expression and replication (Tsai et al. 2020; Hao et al. 2022). This modification, primarily detected in the untranslated region of the virus, improved viral translation by enhancing mRNA stability.

6.5.4 2′O-Methylation

The addition of a methyl group to the 2′-hydroxyl (2′-OH) group of the ribose sugar in four ribonucleosides (Am, Cm, Gm, and Um, collectively referred to as Nm) is known as 2′O-methylation modification. The nucleolar protein FTSJ3, also known as 2′-O-ribose RNA methyltransferase, adds this modification (Nm) on retroviral transcripts (Fig. 6.2) (Ringeard et al. 2019). However, this modification did not seem to affect the gene expression of HIV-1, as observed in a previous study. The 2′O-methylation modification in viral RNA can contribute to immune evasion strategies employed by viruses. In mammalian cells, the host immune system recognizes viral RNA molecules through pattern recognition receptors (PRRs), such as RIG-I and melanoma differentiation-associated gene 5 (MDA5). These receptors detect viral RNA with specific features, including the absence of 2′O-methylation. By adding 2′O-methyl groups to their RNA, viruses can evade detection by PRRs, suppressing the host immune response and facilitating viral replication. In a recent study, when HIV-1 was grown in the presence of reduced levels of FTSJ3, writer for this modification, the virus strongly activated MDA5 (Ringeard et al. 2019). Other studies have also demonstrated that epitranscriptomic RNA modifications, such as

Nm and pseudouridine, can dampen innate immune responses to transfected mRNA molecules (Robbins et al. 2007; Anderson et al. 2010; Durbin et al. 2016). Therefore, it is essential to investigate whether other viruses utilize this epitranscriptomic modification, to evade detection by host innate immune factors. Importantly, developing approaches to prevent 2′O-methylation in viral RNA could enhance immune recognition of the virus and promote antiviral immune responses. Further research is required to unravel the mechanisms and consequences of 2′O-methylation in viral RNA.

6.6 Methods to Map the Epitranscriptomic Modifications

Mapping epitranscriptomic modifications involves identifying the locations and patterns of chemical modifications within RNA molecules. Several methods have been developed to map these modifications (Table 6.1). Below are a few commonly used techniques:

(a) *Antibody-based methods*: Antibody-based methods utilize antibodies that specifically recognize and bind to specific RNA modifications. For instance, the use of antibodies against specific modified nucleotides, such as m6A or m^5C, can be employed to map these modifications. Methylated RNA immunoprecipitation sequencing and m6A-seq are examples of immunoprecipitation-based techniques that combine immunoprecipitation with high-

Table 6.1 Pros and cons of the techniques employed for mapping the epitranscriptomic modifications

Methods	Application	Disadvantages
Methylated RNA immunoprecipitation sequencing (meRIP-seq)	Used for mapping m^6A and m^5C sites on mRNA	Provide information about a broad region that may have one or more m^6A and m^5C sites
Individual-nucleotide-resolution cross-linking immunoprecipitation (iCLIP)	Allow mapping of modifications (m^6A and m^5C) at single nucleotide level	–
Bisulfite sequencing	Used to map m^5C residues on mRNA (single nucleotide level)	Can result in incorrect identification of RNA stems
High-performance liquid chromatography–tandem mass spectrometry (HPLC–MS/MS)	Identification and quantification of epitranscriptomic modifications	Do not provide information on the exact location of modifications
RiboMeth-seq	Allow quantification and individual nucleotide level mapping of 2′O-methylation	Large amounts of input RNA
Direct RNA nanopore sequencing (DRS)	Allow single-nucleotide level mapping of epitranscriptomic modifications	Large amounts of input RNA

throughput sequencing to find modified RNA locations (Dominissini et al. 2012; Meyer et al. 2012). The miCLIP or m6A individual-nucleotide-resolution cross-linking and immunoprecipitation (iCLIP) is an improved version of these methods which entails RNA fragmentation and subsequent reaction with anti-m6A antibodies to produce cross-links under UV light (Linder et al. 2015). Following the binding of RNA complexes to m6A antibodies, protein A/G-affinity purification is used to separate them. The complexes are then resolved by SDS-PAGE and transferred onto a nitrocellulose membrane. The RNA is released from the membrane using proteinase K and subsequently subjected to reverse transcription to produce a cDNA library, employing the iCLIP protocol (Hafner et al. 2010). The resultant cDNA is then amplified through PCR, and then it is subjected to NGS protocols for sequencing.

(b) *Chemical-based methods*: Chemical-based methods involve the chemical labeling or conversion of modified nucleotides, allowing for their detection and mapping. For instance, bisulfite sequencing is commonly used to map RNA modifications like m^5C. Unmethylated cytosines undergo conversion to uracils during the bisulfite process, but methylated cytosines do not. By comparing the sequencing data before and after bisulfite treatment, the modified cytosine sites can be identified (Schaefer et al. 2009). Another chemical method is RiboMeth-seq, which is used for the quantification and individual nucleotide level mapping of 2′O-methylation (Krogh et al. 2017). This method merges alkaline fragmentation with a unique library construction protocol that relies on 5′-OH and 2′,3′ cyclic phosphate ends. This preparation method is employed to ready RNA for sequencing. By examining the read-ends of library fragments, the RNA molecules can be accurately mapped with nucleotide precision, allowing for the determination of the proportion of molecules methylated at the 2′-O-Me sites.

(c) *Antibody-free method*: Deamination adjacent to RNA modification target (DART)-seq is a technique that does not require antibodies to identify m^6A sites. The cytidine deaminase APOBEC1 and the m^6A-binding YTH domain are fused in DART-seq. When APOBEC1-YTH is expressed in cells, C-to-U deamination occurs close to m^6A residues. These deaminated sites can then be detected through standard RNA-seq analysis (Meyer 2019).

(d) *Mass spectrometry-based methods*: Mass spectrometry techniques such as high-resolution ion mobility spectrometry-mass spectrometry can be employed to directly analyze and quantify modified nucleotides. These methods involve isolating and fragmenting RNA molecules into individual nucleotides, followed by mass spectrometry analysis of the resulting nucleotides. By comparing the mass spectra of modified and unmodified nucleotides, the presence and locations of the modifications can be determined (Kenderdine et al. 2020).

(e) *Nanopore sequencing*: Third-generation sequencing technology known as "nanopore sequencing" allows for the detection of individual RNA molecules as they move through a nanopore. Oxford Nanopore Technologies created the direct RNA nanopore sequencing platform (Loman et al. 2015). This platform makes use of a motor protein to thread a single RNA molecule through an

E. coli CsgG-derived protein nanopore sensor that is integrated into a synthetic hydrophobic membrane. The sensor detects variations in an ionic current as the RNA molecule traverses the nanopore. By analyzing alterations in the current and the duration the molecule spends in the pore, valuable data can be obtained to distinguish between unmodified and modified nucleotides.

Notably, each method has its benefits and drawbacks, and the choice of technique depends on the specific RNA modification being studied and the research goals (Table 6.1). Additionally, these methods are continually evolving, and new techniques are being developed to improve the accuracy and resolution of epitranscriptomic mapping.

6.7 Epitranscriptomic Modifications as a Target for Antiviral Therapy

The epitranscriptomic modifications identified in diverse virus families and their primary roles in enhancing viral gene expression, replication, and escape from the host's innate immune response offer themselves as possible targets for designing of antivirals. However, at the same time, these modifications also frequently detected on cellular mRNA do offer challenges of targeting specifically viral RNA. The nonspecific targeting of drugs to cellular mRNA may lead to disruption of the physiological function of cellular protein, and therefore, it may lead to toxicity. Antiviral strategies for targeting epitranscriptomic modifications are explained further.

6.7.1 Inhibition of Viral Gene Expression

Epitranscriptomic modifications are essential for controlling viral gene expression. For example, modifications like m^6A promote viral mRNA stability and translation efficiency, as well as assist in escaping recognition from the host's innate immune system. The precise balance of viral gene expression can be upset by aiming the enzymes that add or remove these changes. For instance, 3-deazaadenosine (DAA) is the only epitranscriptomic modification inhibitor documented to exhibit broad-spectrum antiviral activity against many viruses, including RSV, Rous sarcoma virus, vesicular stomatitis virus, HIV-1, parainfluenza virus, reovirus, and measles (Bader et al. 1978; de Clercq and Montgomery 1983; Wyde et al. 1990; Mayers et al. 1995; Gordon et al. 2003).

The methyl donor SAM, which is a crucial component of m^6A "writer" for adding m^6A on viral mRNA, is hydrolyzed by S-adenosylhomocysteine (SAC) hydrolase. DAA can inhibit this enzymatic step, depleting SAM and inhibiting m^6A addition on viral as well as cellular mRNA (Fig. 6.2). However, antiviral activity achieved in this way can also lead to cellular toxicity. Fortunately, several studies have already demonstrated that broad-spectrum antiviral activity of DAA can be achieved at a concentration, that is nontoxic to the cell cultures (Bader et al. 1978; de Clercq and Montgomery 1983; Wyde et al. 1990; Mayers et al. 1995; Gordon et al. 2003). Moreover, DAA demonstrated remarkable efficacy in blocking the replication of RSV in cotton rats (Wyde et al. 1990) and Ebola virus in mice (Huggins et al. 1999; Bray et al. 2000) without causing any noticeable toxicity. Nevertheless, inhibitors that specifically block epitranscriptomic modifications deposited on the viral RNA would not only be tolerated by the host cells but also pave the way for the creation of broad-spectrum antiviral drugs in the future.

6.7.2 *Modulation of Viral Immune Evasion*

Epitranscriptomic modifications can impact how the immune system detects the viral RNA. Some modifications, such as m^6A and 2′O-methylation, can mask viral RNA from immune surveillance, allowing the virus to evade the immune response (Durbin et al. 2016; Kennedy et al. 2016; Lichinchi et al. 2016; Courtney et al. 2017; Ye et al. 2017; Tsai et al. 2018; Hao et al. 2019; Ringeard et al. 2019; Xue et al. 2019; Lu et al. 2020). By targeting these modifications, it may be possible to enhance the immune recognition of viral RNA, triggering a more robust antiviral immune response. Strategies could involve inhibiting the enzymes responsible for the modifications or designing modified RNA analogs that competitively bind to viral RNA, preventing its immune evasion.

6.7.3 *RNA-Based Vaccines and Therapies*

Epitranscriptomic modifications can be utilized in the development of RNA-based vaccines and therapies. By incorporating specific modifications into viral RNA sequences, it may be possible to enhance the stability, translational efficiency, and immunogenicity of RNA-based vaccines. For instance, modifications like m^6A or pseudouridine (Ψ) have been shown to improve the efficacy and immunogenicity of the influenza vaccine (Starostina et al. 2021). Additionally, targeted modifications, such as pseudouridine (Ψ), can be introduced into therapeutic RNA molecules, specifically in mRNA vaccine, which was shown to enhance translational efficiency by inhibiting the activation of innate immune sensors like Toll-like receptors (TLR) (Nance and Meier 2021).

To completely comprehend the intricacies and specificities of these modifications in various viral infections, it is crucial to keep in mind that the development of antiviral therapies targeting epitranscriptomic modifications is still in its early stages. However, the ability to specifically modulate viral RNA modifications offers hope for the future development of novel antiviral strategies.

6.8 Challenges and Future Prospects of Epigenetic and Epitranscriptomic Drugs

The major challenges that hamper the clinical use of epigenetic and epitranscriptomic drugs against viruses are: (a) Off-target effects: Epigenetic drugs can affect the DNA methylation or histone acetylation status at unintended genomic regions, leading to alterations in gene expression patterns unrelated to the therapeutic target. These off-target effects can result in unpredictable biological consequences and potential toxicity; (b) Nonspecific cytotoxicity: High doses of epigenetic drugs can induce nonspecific cytotoxicity, causing damage to normal cells and tissues. This can lead to adverse effects on various organs and systems in the body.

Despite the challenges associated with toxicity, these drugs hold significant promise as therapeutic agents. The future prospects of epigenetic drugs include (i) Targeted delivery systems: Developing targeted delivery systems, such as nanoparticle-based drug delivery, can enhance the selective uptake of epigenetic drugs by targeted cells, minimizing off-target effects and reducing toxicity to normal tissues; (ii) Combination therapies: The combination of epigenetic drugs with other antivirals, vaccines, or immunotherapies holds great potential to enhance treatment outcomes, improve response rates, and reduce toxicity; (iii) Personalized medicine approaches: Advancements in genomic profiling and epigenetic biomarkers can help identify patient-specific molecular alterations and guide the selection of appropriate epigenetic drugs. This personalized approach can optimize treatment response and minimize toxicity, especially tumor-associated DNA viruses; (iv) Epigenetic/epitranscriptomic editing technologies: Emerging technologies, such as CRISPR-based epigenome editing, offer the potential to precisely modify epigenetic/epitranscriptomic marks and gene expression patterns. These technologies may provide new avenues for targeted and reversible modulation of the epigenome/epitranscriptome with reduced off-target effects and toxicity.

6.9 Conclusion

Epigenetic- and epitranscriptomic-targeted reprogramming provides novel avenues for the development of broad-spectrum antiviral therapies. By manipulating host cellular processes and regulatory mechanisms, these approaches have the potential to disrupt viral replication cycles, reactivate latent viruses, and enhance antiviral immune responses. However, the toxicity of epigenetic/epitranscriptomic drugs is a crucial issue that needs to be carefully addressed. Ongoing research and development efforts aim to optimize dosing strategies, explore combination therapies, develop targeted delivery systems, and improve the specificity of epigenetic drugs. Continued research and advancements in this field may lead to the discovery of novel therapies that can combat a wide range of viral infections.

Acknowledgments The authors would like to thank the Indian Council of Agriculture Research-National Institute of High Security Animal Diseases, and Science and Engineering Research Board (Grant number- CRG/2020/001274).

References

Alexandraki A, Strati K (2022) Decitabine treatment induces a viral mimicry response in cervical cancer cells and further sensitizes cells to chemotherapy. Int J Mol Sci 23:14042

Anderson BR, Muramatsu H, Nallagatla SR, Bevilacqua PC, Sansing LH, Weissman D, Karikó K (2010) Incorporation of pseudouridine into mRNA enhances translation by diminishing PKR activation. Nucleic Acids Res 38:5884–5892

Andrews FH, Strahl BD, Kutateladze TG (2016) Insights into newly discovered marks and readers of epigenetic information. Nat Chem Biol 12:662–668

Bader JP, Brown NR, Chiang PK, Cantoni GL (1978) 3-Deazaadenosine, an inhibitor of adenosylhomocysteine hydrolase, inhibits reproduction of Rous sarcoma virus and transformation of chick embryo cells. Virology 89:494–505

Banerjee NS, Moore DW, Broker TR, Chow LT (2018) Vorinostat, a pan-HDAC inhibitor, abrogates productive HPV-18 DNA amplification. Proc Natl Acad Sci U S A 115:E11138–e11147

Barski A, Cuddapah S, Cui K, Roh TY, Schones DE, Wang Z, Wei G, Chepelev I, Zhao K (2007) High-resolution profiling of histone methylations in the human genome. Cell 129:823–837

Baubec T, Schübeler D (2014) Genomic patterns and context specific interpretation of DNA methylation. Curr Opin Genet Dev 25:85–92

Bray M, Driscoll J, Huggins JW (2000) Treatment of lethal Ebola virus infection in mice with a single dose of an S-adenosyl-L-homocysteine hydrolase inhibitor. Antivir Res 45:135–147

Cantrell SR, Bresnahan WA (2006) Human cytomegalovirus (HCMV) UL82 gene product (pp71) relieves hDaxx-mediated repression of HCMV replication. J Virol 80:6188–6191

Cliffe AR, Knipe DM (2008) Herpes simplex virus ICP0 promotes both histone removal and acetylation on viral DNA during lytic infection. J Virol 82:12030–12038

Cohen C, Corpet A, Roubille S, Maroui MA, Poccardi N, Rousseau A, Kleijwegt C, Binda O, Texier P, Sawtell N, Labetoulle M, Lomonte P (2018) Promyelocytic leukemia (PML) nuclear bodies (NBs) induce latent/quiescent HSV-1 genomes chromatinization through a PML NB/Histone H3.3/H3.3 Chaperone Axis. PLoS Pathog 14:e1007313

Courtney DG, Kennedy EM, Dumm RE, Bogerd HP, Tsai K, Heaton NS, Cullen BR (2017) Epitranscriptomic enhancement of influenza a virus gene expression and replication. Cell Host Microbe 22:377–386.e5

Courtney DG, Chalem A, Bogerd HP, Law BA, Kennedy EM, Holley CL, Cullen BR (2019a) Extensive epitranscriptomic methylation of A and C residues on murine Leukemia virus transcripts enhances viral gene expression. mBio (3):10, e01209–e01219

Courtney DG, Tsai K, Bogerd HP, Kennedy EM, Law BA, Emery A, Swanstrom R, Holley CL, Cullen BR (2019b) Epitranscriptomic addition of m(5)C to HIV-1 transcripts regulates viral gene expression. Cell Host Microbe 26:217–227.e6

Curradi M, Izzo A, Badaracco G, Landsberger N (2002) Molecular mechanisms of gene silencing mediated by DNA methylation. Mol Cell Biol 22:3157–3173

Dalton T, Doubrovina E, Pankov D, Reynolds R, Scholze H, Selvakumar A, Vizconde T, Savalia B, Dyomin V, Weigel C, Oakes CC, Alonso A, Elemento O, Pan H, Phillip JM, O'Reilly RJ, Gewurz BE, Cesarman E, Giulino-Roth L (2020) Epigenetic reprogramming sensitizes immunologically silent EBV+ lymphomas to virus-directed immunotherapy. Blood 135:1870–1881

Darnell RB, Ke S, Darnell JE Jr (2018) Pre-mRNA processing includes N(6) methylation of adenosine residues that are retained in mRNA exons and the fallacy of "RNA epigenetics". RNA 24:262–267

de Clercq E, Montgomery JA (1983) Broad-spectrum antiviral activity of the carbocyclic analog of 3-deazaadenosine. Antivir Res 3:17–24

Diamantopoulos PT, Michael M, Benopoulou O, Bazanis E, Tzeletas G, Meletis J, Vayopoulos G, Viniou NA (2012) Antiretroviral activity of 5-azacytidine during treatment of a HTLV-1 positive myelodysplastic syndrome with autoimmune manifestations. Virol J 9:1

Diner BA, Lum KK, Javitt A, Cristea IM (2015) Interactions of the antiviral factor interferon gamma-inducible protein 16 (IFI16) mediate immune signaling and Herpes Simplex Virus-1 immunosuppression. Mol Cell Proteomics 14:2341–2356

Diner BA, Lum KK, Toettcher JE, Cristea IM (2016) Viral DNA sensors IFI16 and cyclic GMP-AMP synthase possess distinct functions in regulating viral gene expression, immune defenses, and apoptotic responses during herpesvirus infection. mBio 7:e01553–e01516

Dominissini D, Moshitch-Moshkovitz S, Schwartz S, Salmon-Divon M, Ungar L, Osenberg S, Cesarkas K, Jacob-Hirsch J, Amariglio N, Kupiec M, Sorek R, Rechavi G (2012) Topology of the human and mouse m6A RNA methylomes revealed by m6A-seq. Nature 485:201–206

Durbin AF, Wang C, Marcotrigiano J, Gehrke L (2016) RNAs containing modified nucleotides fail to trigger RIG-I conformational changes for innate immune signaling. mBio 7:e00833–e00816

Feng Q, Su Z, Song S, Xu H, Zhang B, Yi L, Tian M, Wang H (2016) Histone deacetylase inhibitors suppress RSV infection and alleviate virus-induced airway inflammation. Int J Mol Med 38:812–822

Galasinski SC, Louie DF, Gloor KK, Resing KA, Ahn NG (2002) Global regulation of post-translational modifications on core histones. J Biol Chem 277:2579–2588

Gordon RK, Ginalski K, Rudnicki WR, Rychlewski L, Pankaskie MC, Bujnicki JM, Chiang PK (2003) Anti-HIV-1 activity of 3-deaza-adenosine analogs. Inhibition of S-adenosylhomocysteine hydrolase and nucleotide congeners. Eur J Biochem 270:3507–3517

Gunst JD, Pahus MH, Rosás-Umbert M, Lu IN, Benfield T, Nielsen H, Johansen IS, Mohey R, Østergaard L, Klastrup V, Khan M, Schleimann MH, Olesen R, Støvring H, Denton PW, Kinloch NN, Copertino DC, Ward AR, Alberto WDC, Nielsen SD, Puertas MC, Ramos V, Reeves JD, Petropoulos CJ, Martinez-Picado J, Brumme ZL, Jones RB, Fox J, Tolstrup M, Nussenzweig MC, Caskey M, Fidler S, Søgaard OS (2022) Early intervention with 3BNC117 and romidepsin at antiretroviral treatment initiation in people with HIV-1: a phase 1b/2a, randomized trial. Nat Med 28:2424–2435

Hafner M, Landthaler M, Burger L, Khorshid M, Hausser J, Berninger P, Rothballer A, Ascano M Jr, Jungkamp AC, Munschauer M, Ulrich A, Wardle GS, Dewell S, Zavolan M, Tuschl T (2010) Transcriptome-wide identification of RNA-binding protein and microRNA target sites by PAR-CLIP. Cell 141:129–141

Hao H, Hao S, Chen H, Chen Z, Zhang Y, Wang J, Wang H, Zhang B, Qiu J, Deng F, Guan W (2019) N6-methyladenosine modification and METTL3 modulate enterovirus 71 replication. Nucleic Acids Res 47:362–374

Hao H, Liu W, Miao Y, Ma L, Yu B, Liu L, Yang C, Zhang K, Chen Z, Yang J, Zheng Z, Zhang B, Deng F, Gong P, Yuan J, Hu Z, Guan W (2022) N4-acetylcytidine regulates the replication and pathogenicity of enterovirus 71. Nucleic Acids Res 50:9339–9354

Hendriks IA, D'Souza RC, Yang B, Verlaan-de Vries M, Mann M, Vertegaal AC (2014) Uncovering global SUMOylation signaling networks in a site-specific manner. Nat Struct Mol Biol 21:927–936

Huggins J, Zhang ZX, Bray M (1999) Antiviral drug therapy of filovirus infections: S-adenosylhomocysteine hydrolase inhibitors inhibit Ebola virus in vitro and in a lethal mouse model. J Infect Dis 179(Suppl 1):S240–S247

Jin G, Xu M, Zou M, Duan S (2020) The processing, gene regulation, biological functions, and clinical relevance of N4-Acetylcytidine on RNA: a systematic review. Mol Ther Nucleic Acids 20:13–24

Johnson KE, Bottero V, Flaherty S, Dutta S, Singh VV, Chandran B (2014a) IFI16 restricts HSV-1 replication by accumulating on the HSV-1 genome, repressing HSV-1 gene expression, and directly or indirectly modulating histone modifications. PLoS Pathog 10:e1004503

Johnson LM, Du J, Hale CJ, Bischof S, Feng S, Chodavarapu RK, Zhong X, Marson G, Pellegrini M, Segal DJ, Patel DJ, Jacobsen SE (2014b) SRA- and SET-domain-containing proteins link RNA polymerase V occupancy to DNA methylation. Nature 507:124–128

Jones PA, Liang G (2009) Rethinking how DNA methylation patterns are maintained. Nat Rev Genet 10:805–811

Kaplan N, Moore IK, Fondufe-Mittendorf Y, Gossett AJ, Tillo D, Field Y, LeProust EM, Hughes TR, Lieb JD, Widom J, Segal E (2009) The DNA-encoded nucleosome organization of a eukaryotic genome. Nature 458:362–366

Kauder SE, Bosque A, Lindqvist A, Planelles V, Verdin E (2009) Epigenetic regulation of HIV-1 latency by cytosine methylation. PLoS Pathog 5:e1000495

Kenderdine T, Nemati R, Baker A, Palmer M, Ujma J, FitzGibbon M, Deng L, Royzen M, Langridge J, Fabris D (2020) High-resolution ion mobility spectrometry-mass spectrometry of isomeric/isobaric ribonucleotide variants. J Mass Spectrom 55:e4465

Kennedy EM, Bogerd HP, Kornepati AV, Kang D, Ghoshal D, Marshall JB, Poling BC, Tsai K, Gokhale NS, Horner SM, Cullen BR (2016) Posttranscriptional m(6)A editing of HIV-1 mRNAs enhances viral gene expression. Cell Host Microbe 19:675–685

Knipe DM (2015) Nuclear sensing of viral DNA, epigenetic regulation of herpes simplex virus infection, and innate immunity. Virology 479–480:153–159

Krogh N, Birkedal U, Nielsen H (2017) RiboMeth-seq: profiling of 2′-O-me in RNA. Methods Mol Biol 1562:189–209

Kumar N, Sharma S, Kumar R, Tripathi BN, Barua S, Ly H, Rouse BT (2020) Host-directed antiviral therapy. Clin Microbiol Rev 33:e00168–e00119

Liang Y, Vogel JL, Narayanan A, Peng H, Kristie TM (2009) Inhibition of the histone demethylase LSD1 blocks alpha-herpesvirus lytic replication and reactivation from latency. Nat Med 15:1312–1317

Liang Y, Vogel JL, Arbuckle JH, Rai G, Jadhav A, Simeonov A, Maloney DJ, Kristie TM (2013) Targeting the JMJD2 histone demethylases to epigenetically control herpesvirus infection and reactivation from latency. Sci Transl Med 5:167ra5

Lichinchi G, Gao S, Saletore Y, Gonzalez GM, Bansal V, Wang Y, Mason CE, Rana TM (2016) Dynamics of the human and viral m(6)A RNA methylomes during HIV-1 infection of T cells. Nat Microbiol 1:16011

Lin X, Ke X, Jian X, Xia L, Yang Y, Zhang T, Xiong H, Zhao B, Liu W, Chen Q, Tang C (2022) Azacytidine targeting SARS-CoV-2 viral RNA as a potential treatment for COVID-19. Sci Bull (Beijing) 67:1022–1025

Linder B, Grozhik AV, Olarerin-George AO, Meydan C, Mason CE, Jaffrey SR (2015) Single-nucleotide-resolution mapping of m6A and m6Am throughout the transcriptome. Nat Methods 12:767–772

Lister R, Pelizzola M, Dowen RH, Hawkins RD, Hon G, Tonti-Filippini J, Nery JR, Lee L, Ye Z, Ngo QM, Edsall L, Antosiewicz-Bourget J, Stewart R, Ruotti V, Millar AH, Thomson JA, Ren B, Ecker JR (2009) Human DNA methylomes at base resolution show widespread epigenomic differences. Nature 462:315–322

Loman NJ, Quick J, Simpson JT (2015) A complete bacterial genome assembled de novo using only nanopore sequencing data. Nat Methods 12:733–735

Lu M, Zhang Z, Xue M, Zhao BS, Harder O, Li A, Liang X, Gao TZ, Xu Y, Zhou J, Feng Z, Niewiesk S, Peeples ME, He C, Li J (2020) N(6)-methyladenosine modification enables viral RNA to escape recognition by RNA sensor RIG-I. Nat Microbiol 5:584–598

Lukashchuk V, McFarlane S, Everett RD, Preston CM (2008) Human cytomegalovirus protein pp71 displaces the chromatin-associated factor ATRX from nuclear domain 10 at early stages of infection. J Virol 82:12543–12554

Mayers DL, Mikovits JA, Joshi B, Hewlett IK, Estrada JS, Wolfe AD, Garcia GE, Doctor BP, Burke DS, Gordon RK et al (1995) Anti-human immunodeficiency virus 1 (HIV-1) activities of 3-deazaadenosine analogs: increased potency against 3′-azido-3′-deoxythymidine-resistant HIV-1 strains. Proc Natl Acad Sci U S A 92:215–219

Meyer KD (2019) DART-seq: an antibody-free method for global m(6)a detection. Nat Methods 16:1275–1280

Meyer KD, Jaffrey SR (2014) The dynamic epitranscriptome: N6-methyladenosine and gene expression control. Nat Rev Mol Cell Biol 15:313–326

Meyer KD, Saletore Y, Zumbo P, Elemento O, Mason CE, Jaffrey SR (2012) Comprehensive analysis of mRNA methylation reveals enrichment in 3′ UTRs and near stop codons. Cell 149:1635–1646

Mothe B, Rosás-Umbert M, Coll P, Manzardo C, Puertas MC, Morón-López S, Llano A, Miranda C, Cedeño S, López M, Alarcón-Soto Y, Melis GG, Langohr K, Barriocanal AM, Toro J, Ruiz I, Rovira C, Carrillo A, Meulbroek M, Crook A, Wee EG, Miró JM, Clotet B, Valle M, Martinez-Picado J, Hanke T, Brander C, Moltó J (2020) HIVconsv vaccines and Romidepsin in early-treated HIV-1-infected individuals: safety, immunogenicity and effect on the viral reservoir (study BCN02). Front Immunol 11:823

Nakamura M, Nishikawa J, Saito M, Sakai K, Sasaki S, Hashimoto S, Okamoto T, Suehiro Y, Yamasaki T, Sakaida I (2017) Decitabine inhibits tumor cell proliferation and up-regulates e-cadherin expression in Epstein-Barr virus-associated gastric cancer. J Med Virol 89:508–517

Nance KD, Meier JL (2021) Modifications in an emergency: the role of N1-Methylpseudouridine in COVID-19 vaccines. ACS Central Science 7:748–756

Narayanan A, Ruyechan WT, Kristie TM (2007) The coactivator host cell factor-1 mediates Set1 and MLL1 H3K4 trimethylation at herpesvirus immediate early promoters for initiation of infection. Proc Natl Acad Sci U S A 104:10835–10840

Oh J, Fraser NW (2008) Temporal association of the herpes simplex virus genome with histone proteins during a lytic infection. J Virol 82:3530–3537

Orzalli MH, Conwell SE, Berrios C, DeCaprio JA, Knipe DM (2013) Nuclear interferon-inducible protein 16 promotes silencing of herpesviral and transfected DNA. Proc Natl Acad Sci U S A 110:E4492–E4501

Placek BJ, Huang J, Kent JR, Dorsey J, Rice L, Fraser NW, Berger SL (2009) The histone variant H3.3 regulates gene expression during lytic infection with herpes simplex virus type 1. J Virol 83:1416–1421

Rawson JM, Heineman RH, Beach LB, Martin JL, Schnettler EK, Dapp MJ, Patterson SE, Mansky LM (2013) 5,6-Dihydro-5-aza-2′-deoxycytidine potentiates the anti-HIV-1 activity of ribonucleotide reductase inhibitors. Bioorg Med Chem 21:7222–7228

Rawson JM, Daly MB, Xie J, Clouser CL, Landman SR, Reilly CS, Bonnac L, Kim B, Patterson SE, Mansky LM (2016) 5-Azacytidine enhances the mutagenesis of HIV-1 by reduction to 5-Aza-2′-deoxycytidine. Antimicrob Agents Chemother 60:2318–2325

Rechache NS, Wang Y, Stevenson HS, Killian JK, Edelman DC, Merino M, Zhang L, Nilubol N, Stratakis CA, Meltzer PS, Kebebew E (2012) DNA methylation profiling identifies global methylation differences and markers of adrenocortical tumors. J Clin Endocrinol Metab 97:E1004–E1013

Ringeard M, Marchand V, Decroly E, Motorin Y, Bennasser Y (2019) FTSJ3 is an RNA 2′-O-methyltransferase recruited by HIV to avoid innate immune sensing. Nature 565:500–504

Robbins M, Judge A, Liang L, McClintock K, Yaworski E, MacLachlan I (2007) 2′-O-methyl-modified RNAs act as TLR7 antagonists. Mol Ther 15:1663–1669

Rossetto D, Avvakumov N, Côté J (2012) Histone phosphorylation: a chromatin modification involved in diverse nuclear events. Epigenetics 7:1098–1108

Roundtree IA, Evans ME, Pan T, He C (2017) Dynamic RNA modifications in gene expression regulation. Cell 169:1187–1200

Saha B, Parks RJ (2019) Histone deacetylase inhibitor Suberoylanilide Hydroxamic acid suppresses human adenovirus gene expression and replication. J Virol 93

Schaefer M, Pollex T, Hanna K, Lyko F (2009) RNA cytosine methylation analysis by bisulfite sequencing. Nucleic Acids Res 37:e12

Schnell AP, Kohrt S, Aristodemou A, Taylor GP, Bangham CRM, Thoma-Kress AK (2022) HDAC inhibitors Panobinostat and Romidepsin enhance tax transcription in HTLV-1-infected cell lines and freshly isolated patients' T-cells. Front Immunol 13:978800. Published 2022 Aug 16. https://doi.org/10.3389/fimmu.2022.978800

Schultz MD, He Y, Whitaker JW, Hariharan M, Mukamel EA, Leung D, Rajagopal N, Nery JR, Urich MA, Chen H, Lin S, Lin Y, Jung I, Schmitt AD, Selvaraj S, Ren B, Sejnowski TJ, Wang W, Ecker JR (2015) Human body epigenome maps reveal noncanonical DNA methylation variation. Nature 523:212–216

Shapira L, Ralph M, Tomer E, Cohen S, Kobiler O (2016) Histone deacetylase inhibitors reduce the number of herpes simplex Virus-1 genomes initiating expression in individual cells. Front Microbiol 7:1970

Si J, Boumber YA, Shu J, Qin T, Ahmed S, He R, Jelinek J, Issa JP (2010) Chromatin remodeling is required for gene reactivation after decitabine-mediated DNA hypomethylation. Cancer Res 70:6968–6977

Song Y, Wu X, Xu Y, Zhu J, Li J, Zou Z, Chen L, Zhang B, Hua C, Rui H, Zheng Q, Zhou Q, Wang Q, Cheng H (2020) HPV E7 inhibits cell pyroptosis by promoting TRIM21-mediated degradation and ubiquitination of the IFI16 inflammasome. Int J Biol Sci 16:2924–2937

Starostina EV, Sharabrin SV, Antropov DN, Stepanov GA, Shevelev GY, Lemza AE, Rudometov AP, Borgoyakova MB, Rudometova NB, Marchenko VY, Danilchenko NV, Chikaev AN, Bazhan SI, Ilyichev AA, Karpenko LI (2021) Construction and immunogenicity of modified mRNA-vaccine variants encoding influenza virus antigens. Vaccines (Basel) 9:452

Sterner DE, Berger SL (2000) Acetylation of histones and transcription-related factors. Microbiol Mol Biol Rev 64:435–459

Stich M, Ganss L, Puschhof J, Prigge ES, Reuschenbach M, Guiterrez A, Vinokurova S, von Knebel Doeberitz M (2017) 5-aza-2′-deoxycytidine (DAC) treatment downregulates the HPV E6 and E7 oncogene expression and blocks neoplastic growth of HPV-associated cancer cells. Oncotarget 8:52104–52117

Stirzaker C, Song JZ, Ng W, Du Q, Armstrong NJ, Locke WJ, Statham AL, French H, Pidsley R, Valdes-Mora F, Zotenko E, Clark SJ (2017) Methyl-CpG-binding protein MBD2 plays a key role in maintenance and spread of DNA methylation at CpG islands and shores in cancer. Oncogene 36:1328–1338

Tsai K, Cullen BR (2020) Epigenetic and epitranscriptomic regulation of viral replication. Nat Rev Microbiol 18:559–570

Tsai K, Thikmyanova N, Wojcechowskyj JA, Delecluse HJ, Lieberman PM (2011) EBV tegument protein BNRF1 disrupts DAXX-ATRX to activate viral early gene transcription. PLoS Pathog 7:e1002376

Tsai K, Courtney DG, Cullen BR (2018) Addition of m6A to SV40 late mRNAs enhances viral structural gene expression and replication. PLoS Pathog 14:e1006919

Tsai K, Jaguva Vasudevan AA, Martinez Campos C, Emery A, Swanstrom R, Cullen BR (2020) Acetylation of cytidine residues boosts HIV-1 gene expression by increasing viral RNA stability. Cell Host Microbe 28:306–312.e6

Van Rechem C, Whetstine JR (2014) Examining the impact of gene variants on histone lysine methylation. Biochim Biophys Acta 1839:1463–1476

Wang X, Xia H, Liu S, Cao L, You F (2021) Epigenetic regulation in antiviral innate immunity. Eur J Immunol 51:1641–1651

Weake VM, Workman JL (2008) Histone ubiquitination: triggering gene activity. Mol Cell 29:653–663

Wesche J, Kühn S, Kessler BM, Salton M, Wolf A (2017) Protein arginine methylation: a prominent modification and its demethylation. Cell Mol Life Sci 74:3305–3315

Wyde PR, Ambrose MW, Meyer HL, Zolinski CL, Gilbert BE (1990) Evaluation of the toxicity and antiviral activity of carbocyclic 3-deazaadenosine against respiratory syncytial and parainfluenza type 3 viruses in tissue culture and in cotton rats. Antivir Res 14:215–225

Xue M, Zhao BS, Zhang Z, Lu M, Harder O, Chen P, Lu Z, Li A, Ma Y, Xu Y, Liang X, Zhou J, Niewiesk S, Peeples ME, He C, Li J (2019) Viral N(6)-methyladenosine upregulates replication and pathogenesis of human respiratory syncytial virus. Nat Commun 10:4595

Yang XJ, Seto E (2007) HATs and HDACs: from structure, function and regulation to novel strategies for therapy and prevention. Oncogene 26:5310–5318

Ye F, Chen ER, Nilsen TW (2017) Kaposi's sarcoma-associated herpesvirus utilizes and manipulates RNA N(6)-adenosine methylation to promote lytic replication. J Virol 91:e00466–e00417

Zhao BS, Nachtergaele S, Roundtree IA, He C (2018) Our views of dynamic N(6)-methyladenosine RNA methylation. RNA 24:268–272

Zhou Y, Wang Q, Yang Q, Tang J, Xu C, Gai D, Chen X, Chen J (2018) Histone deacetylase 3 inhibitor suppresses hepatitis C virus replication by regulating Apo-A1 and LEAP-1 expression. Virol Sin 33:418–428

Chapter 7
Application of Cryoelectron Microscopy in Antiviral Research

Pankhuri Narula, Milan Kumar Lokshman, and Manidipa Banerjee

Abstract In the last decade, single-particle reconstruction from cryoelectron micrographs has been established as a powerful technique for generating high-resolution structures of biological macromolecules. In comparison to conventional methods such as X-ray crystallography and nuclear magnetic resonance, cryogenic electron microscopy (cryoEM) does not require large quantities of starting material or crystallization of biomolecules. The preservation of biomolecules at near-native states and the ability to characterize large complexes, transition states, and heterogeneous and dynamic systems make this method valuable for effective drug development against viral diseases. While there are challenges associated with the utilization of cryoEM as a high-throughput method for drug discovery, the issues are not insurmountable in the near future. The establishment of antiviral discovery pipelines is a crucial scientific activity in the current scenario, given the human health crisis and economic hardships precipitated by the recent pandemic and strong prospects for the emergence or re-emergence of epidemic or pandemic viral strains. The prioritization of high-resolution structure determination of viral proteins and protein complexes is essential as it will aid in drug discovery. In this chapter, we discuss the advantages of cryoEM over other structural biology methods, followed by prevailing drug discovery techniques, the current and potential contribution of cryoEM to virus-specific structure determination and identification and validation of drug candidates, and the future prospects of the technique. Although efforts have been made to include as many recent studies in this area as possible, the vastness of the topic has precluded the inclusion of all such studies.

Keywords Cryoelectron microscopy · Structure-based drug discovery · Fragment-based drug discovery · Virus-receptor complex · Neutralizing antibody

---

Pankhuri Narula and Milan Kumar Lokshman contributed equally.

P. Narula · M. K. Lokshman · M. Banerjee (✉)
Kusuma School of Biological Sciences, Indian Institute of Technology Delhi, New Delhi, India
e-mail: mbanerjee@bioschool.iitd.ac.in

N. Kumar et al. (eds.), *Advances in Antiviral Research*, Livestock Diseases and Management, https://doi.org/10.1007/978-981-99-9195-2_7

7.1 Introduction

The emergence and reemergence of pathogenic viruses can have devastating effects on human health on a global scale. The last century has seen numerous viral epidemics or pandemics of different intensities, including the Spanish flu outbreak in 1918 (Martini et al. 2019), the HIV/AIDS outbreak in the early 1980s (Beyrer 2021), SARS-CoV in 2002 (Zhu et al. 2020), H1N1 strain influenza in 2009 (Sullivan et al. 2010), MERS-CoV in 2012 (Al Hajjar et al. 2013), Ebola in 2014–2016 (Kamorudeen et al. 2020), and the SARS-CoV2 pandemic in 2019–2022 (Hu et al. 2020). The clinical manifestations of viral diseases range from mild or chronic symptoms to acute and life-threatening conditions (Piret and Boivin 2021). In addition to adversely affecting human health, the annual burden of viral diseases results in severe social and economic impact globally including disruption in healthcare facilities, collapse of financial systems, and prevalence of mental health issues. Therefore, it is essential to devise control, prevention, and treatment strategies against current as well as potentially re-emerging viral pathogens (Moeti et al. 2022).

Viruses are dynamic biological entities capable of withstanding extreme environmental conditions, assisted replication in the host cells, and rapid mutations for adapting to different hosts. While virus families differ immensely based on their morphology, type of genetic material, and mode of replication, a base virion unit typically consists of a nucleoprotein core surrounded by a protein capsid, which can, in some cases, be enveloped by a host-derived lipid layer (Mateu 2013). Each infectious cycle typically initiates with the binding of infectious virus particles with cognate host cell receptors, followed by internalization into cellular compartments like endosomes, cellular membrane penetration, and release of the genome. Genome replication and viral protein translation follow, which is facilitated by virus-triggered impairment of the host's immune responses. The last step involves the assembly of new infectious virions released to surrounding cells and tissues to propagate the infection. An infection cycle therefore consists of several discrete stages such as receptor binding, membrane disruption, disassembly, replication, translation, assembly, and egress; and the viral components crucial for these steps can potentially be targeted for generating antivirals (Jones et al. 2021). However, the stages to be targeted for antiviral development are carefully chosen for each virus family to minimize any adverse effect on the host cell, since the viral life cycle is closely intertwined with cellular pathways and processes at the molecular level. At this point, the majority of the approved antivirals target the viral replication machinery, followed by inhibitors of the initial host–virus interaction and egress/release pathways. Many challenges in antiviral development remain including high rates of resistance, host cell cytotoxicity, high mutation rates in viral genome, and targeted application (Adamson et al. 2021). The strategies and methods utilized for antiviral drug development are detailed in the following sections (Dai et al. 2020).

7.1.1 Structure-Based Drug Design (SBDD)

Generating high-resolution structural information of viral proteins, or host–viral protein complexes, and utilizing the knowledge of active site architecture or interface chemistry to design small molecules for inhibition of the activity/process (Bajad et al. 2021).

7.1.2 In Silico Screening

Employing a computational pipeline to screen a large number of compounds/small molecule libraries and predict their interaction with viral targets, which are eventually validated experimentally (Murgueitio et al. 2012). This is typically utilized in conjunction with SBDD as mentioned in Sect. 7.1.1.

7.1.3 Combinatorial Approach

Combining different antiviral drugs, with diverse mechanisms of action, to achieve a significant synergistic antiviral activity (Hoopes et al. 2011).

7.1.4 Fragment-Based Drug Design (FBDD)

Identification of small molecules that have low binding affinities to a viral target, followed by linking several such molecules or expanding them to generate a usable drug with high affinity to the target (Mallakuntla et al. 2022).

7.1.5 Proteolysis-Targeting Chimera

This approach involves targeted protein degradation using a proteolysis-targeting chimera (PROTAC), which consists of a heterobifunctional molecule binding to a target of interest as well as an E3 ubiquitin ligase. Simultaneous binding of a target protein and the ligase by PROTAC results in the ubiquitination and degradation of the target protein (Békés et al. 2022).

7.2 Modes for Generation of High-Resolution Structural Information

As the previous section highlights, the elucidation of high-resolution three-dimensional (3D) structures of biological macromolecules is crucial for providing a template for the design of potential drug candidates. A detailed understanding of biomolecular structures allows prediction of their functional roles within an organism, interaction partners, and disorders associated with mutations or malfunctions while assisting in the development of suitable therapeutics against molecules that require to be inhibited for the benefit of the organism. Analyzing molecular structures also results in the exposition of evolutionary pathways and suggests how biomolecules, such as virus-encoded, receptor-binding glycoproteins, have evolved in their structural and ultimately functional aspects. Given that detailed structural characterization of biomolecules can have a strong impact in a variety of domains, there have been continuous advancements in the techniques for investigating structures. The earliest efforts date back to 1912 with the discovery of X-ray diffraction by Max von Laue, Friedrich, and Knipping, which resulted in the structural resolution of a crystallized molecule (Thomas 2012). This was followed by the refinement of the technique by the father–son duo William Henry Bragg and William Lawrence Bragg in the early twentieth century, leading to the birth of modern X-ray crystallography (Thomas 2012). The first protein structure at atomic resolution was resolved by John Kendrew in 1957, resulting in a revolution in structural analysis of biomolecules using this method (de Chadarevian 2018). Nuclear magnetic resonance (NMR), a technique used for studying the structures of small- to medium-sized molecules in solution, was first demonstrated by Felix Bloch and Edward Mills Purcell in 1946, who shared the 1952 Nobel Prize in Physics for their work (Shampo et al. 2012). Electron microscopy first came into the picture in 1931, with the development of the first electron microscope by Ernst Ruska and Max Knoll, for which they were awarded the Nobel Prize in Physics in 1986. A strong impetus to apply electron microscopy methods to biological samples resulted in the birth of cryogenic electron microscopy (cryoEM) in the early 1970s when Jacques Dubochet developed vitrification methods to reduce radiation damage and to preserve molecules in near-native states (Cressey and Callaway 2017). Joachim Frank, Marin van Heel, and others made significant advancements in the development of methods for single-particle reconstruction (SPR), in which the two-dimensional (2D) images of molecules at different orientations are computationally connected to construct a 3D model of the structure (Callaway 2020). In the initial decades after its discovery, cryoEM applied only to relatively larger macromolecules like viruses and ribosomes; and the resolutions were limited. However, in the last decade, the technique has advanced exponentially with advancements in instrumentation, such as the development of the direct detector by Richard Henderson and others, as well as in computational methods of reconstruction (Nogales and Zhang 2016). The considerable improvement in resolution and the breaking of the size barrier of cryoEM has had a profound impact on structural biology, which was recognized by a Nobel

Prize in Chemistry to Dubochet, Frank, and Henderson in 2017 (Cressey and Callaway 2017). CryoEM has become an essential tool for studying the intricacies of molecular structures. Further advancement in the field, such as in situ structural determination of biomolecular complexes in their native environment within cells, is ongoing (Kaplan et al. 2021). The current pipeline for cryoEM and SPR is a systematic multistep process. The criteria for quality control at each step are detailed below (Cheng 2018; Weissenberger et al. 2021).

7.2.1 Sample Preparation

Purified protein or protein complexes, generated from recombinant expression systems, are primarily utilized for image data collection in cryoEM. Assessment of a few important factors is essential for obtaining high-quality images for reconstruction:

(i) *Sample purity and homogeneity*: Highly pure, homogeneous samples with limited impurities will lead to clear images and better interpretation of 2D classes during reconstruction.
(ii) *Sample concentration:* An optimized concentration should result in images with well-separated particles in sufficient quantities. Highly concentrated samples will lead to particle aggregation and background noise.
(iii) *Buffer conditions*: The buffer composition, particularly pH, concentration of salt, and of additives such as glycerol, detergent, etc., can significantly impact the stability and conformation of the sample, as well as vitreous ice formation. Buffer conditions are empirically judged for each sample; however, preknowledge of appropriate buffers for specific samples is typically utilized for standardization.
(iv) *Sample stability*: The effect of environmental factors like temperature, pH, and the presence of additives has to be carefully considered to ensure the stability of complexes on the grids.

7.2.2 Grid Preparation

The choice of grids depends on the types of samples. Typically, holey carbon or quantifoil grids of different hole sizes are used, with or without additional thin coatings of carbon or gold. The grids are glow-discharged or plasma-cleaned before the adsorption of the sample, to render them hydrophilic; however, this is also sample-dependent, and the parameters are determined empirically. An ideal cryoEM grid is expected to have an optimal distribution of particles in all orientations to achieve a successful reconstruction.

7.2.3 Sample Freezing

Sample freezing or vitrification leads to the immobilization of particles undergoing Brownian motion in solution in their near-native conformation and random orientation. After the adsorption of the sample onto the grid, the excess solution is blotted away with filter paper to ensure thin ice generation, and the grid is plunge-frozen within a very short timescale to generate vitreous ice. During vitrification, samples are rapidly frozen in liquid ethane, which also prevents the formation of crystalline ice. Usually, vitrification is carried out in automated setups where factors like temperature, humidity, and blotting parameters are optimized empirically. A major problem during freezing is the migration of the particles to the air–water interface, which should be minimized to prevent particle aggregation, denaturation, and preferred orientation issues which can drastically reduce the quality of reconstruction (Hands-Portman and Bakker 2022). Reducing the time required for freezing and altering the chemistry of the buffer using additives can be beneficial in avoiding problems associated with AWI interactions (Li et al. 2021). The vitrified grids can be utilized immediately for imaging or stored in liquid nitrogen for later use.

7.2.4 CryoEM Imaging and Acquisition

For imaging, vitrified grids are loaded onto a cryo-holder, and instrument conditions are optimized, followed by the setting up of automated data collection parameters to acquire images from different areas of the grid. Typically, the grid is scanned at low magnification to identify areas that have optimal ice and sample distribution for data collection. Correct sample morphology is inferred from the preknowledge of similar samples. It is essential to set up parameters such as magnification and electron dosage accurately for each sample to ensure optimal data collection with minimal radiation damage.

7.2.5 Image Processing, Reconstruction, and Refinement

The image-processing pipeline involves motion correction to eliminate induced movement, contrast transfer function correction to restore distortions in images, followed by picking of particles from the acquired images and clustering of particle images in similar orientations to achieve well-distributed 2D classes that are combined to generate 3D densities. Many thousands of images in each orientation have to be assembled to improve the signal-to-noise ratio of particles imaged at low electron doses. Repeated refinement of the angular assignments over multiple cycles leads to improvement in resolution. Several pipelines for SPR have been developed over the years including EMAN (Ludtke et al. 1999), RELION (Scheres 2012), and CryoSPARC (Punjani et al. 2017), among others.

7.3 Comparison of CryoEM with Other Structural Techniques: Unique Advantages

As evident from Sect. 7.1, targeted drug discovery requires high-resolution structures of biological macromolecules. The vast majority of biomolecular structures deposited in the Protein Data Bank (PDB) have been resolved using X-ray crystallography. X-ray diffraction studies on single crystals are known to provide atomic-level information; however, the generation of well-diffracting single crystals is frequently a challenge. Crystallization requires highly concentrated and pure recombinant proteins that can be difficult to produce, particularly in the case of large multiprotein complexes, posttranslationally modified proteins, or those with membrane-binding domains. The 3D structures generated by X-ray crystallography represent static states of biomolecules associated with crystal contacts (Smyth and Martin 2000). On the other hand, NMR provides information about the dynamic state of biomolecules. However, this method is severely limited by the size of the biomolecule and does not provide structural information about large molecular weight molecules (Ardenkjaer-Larsen et al. 2015).

CryoEM offers a variety of advantages over other structure determination techniques. This method requires neither crystallization of biomolecules nor highly concentrated protein, which makes it a more amenable technique for studying targets like membrane proteins, which are difficult to generate in recombinant form. Some degree of heterogeneity in the sample does not preclude 3D structure determination, and since the particles are quickly frozen, they are expected to be preserved in near-native, dynamic states. The sample requirement for cryoEM is significantly less compared to that for either X-ray crystallography or NMR, and the time required for grid condition optimization, data collection, and reconstructions compares favorably with the other methods, provided microscope time and computational infrastructures are readily available. CryoEM is a relatively young technique compared to the other two methods, and several aspects of instrumentation and computational processes are still undergoing refinement. Along with the improvement in reconstruction methods, there is significant emphasis on automating sample preparation methods to make this step more effective and less dependent on user skills (Lyumkis 2019). More automation in the pipeline is expected to make the method amenable to high-throughput applications. The cost, infrastructure, and specialized knowledge and expertise required for cryoEM makes it imperative to explore avenues for automating the pipeline, to ensure high returns from each such setup.

7.4 Examples of Application of CryoEM in Drug Development

Despite being a relatively new technique, cryoEM has had several major contributions in unraveling the structural basis of human diseases. During the recent SARS-CoV-2 pandemic, several 3D structures of the spike protein in pre- and postfusion conformations—the major antigenic determinant of the virus—were determined using cryoEM (Gui et al. 2017). This method also resulted in structural elucidation of the binding parameters of spike protein to its receptor angiotensin-converting enzyme 2 (ACE2) (Song et al. 2018; Zhou et al. 2020), as well as neutralizing antibodies. This plethora of information, generated very rapidly in the early months of the pandemic, was crucial in designing antiviral strategies including drug development, monoclonal antibody-based therapeutics, and vaccine candidates. This part will be covered in detail in later sections.

CryoEM-based strategies have been employed to obtain detailed structural information of signaling or apoptotic protein complexes crucial in cancer biology. For example, the 3D structure determination of the protein complex mammalian target of rapamycin complex 1 has provided insights into the architecture and regulation of the complex, facilitating the development of targeted therapies (Chen et al. 2018). CryoEM-based methods have been utilized to resolve the structure of the BRCA1-BARD1 complex, which is involved in DNA repair. Mutations in this complex are associated with increased susceptibility to breast and ovarian cancers (Dai et al. 2021); therefore, high-resolution structural information can be harnessed for developing targeted strategies for mitigation or cure. A few notable studies have highlighted the use of cryoEM in the area of neurodegenerative disease biology. Structures of amyloid-beta (Aβ) fibrils and tau protein filaments (Kollmer et al. 2019; Fitzpatrick et al. 2017), which are key pathological hallmarks of Alzheimer's disease, have been elucidated using cryoEM which has advanced our understanding of disease mechanisms. In the case of Parkinson's disease, the detailed architecture of alpha-synuclein fibrils, as well as those of distinct strains associated with different clinical subtypes of the disease, has been elucidated using cryoEM (Guerrero-Ferreira et al. 2018). This method has also been extensively utilized to study the structures of protein aggregates, such as superoxide dismutase 1 and C9orf72 dipeptide repeat proteins (Wang et al. 2022), associated with the pathophysiological manifestations of amyotrophic lateral sclerosis. In this Chapter, we will focus on the current and future applications of CryoEM toward the development of antiviral strategies.

7.5 Role of CryoEM in Viral Structure Elucidation and Antiviral Development

CryoEM has been extensively utilized to resolve high-resolution structures of whole virus capsids, viral and host–viral protein complexes, as well as transition states during dynamic processes such as capsid disassembly. Such structural information of druggable targets is crucial for antiviral drug development as represented in Fig. 7.1. Some category-wise examples are detailed below:

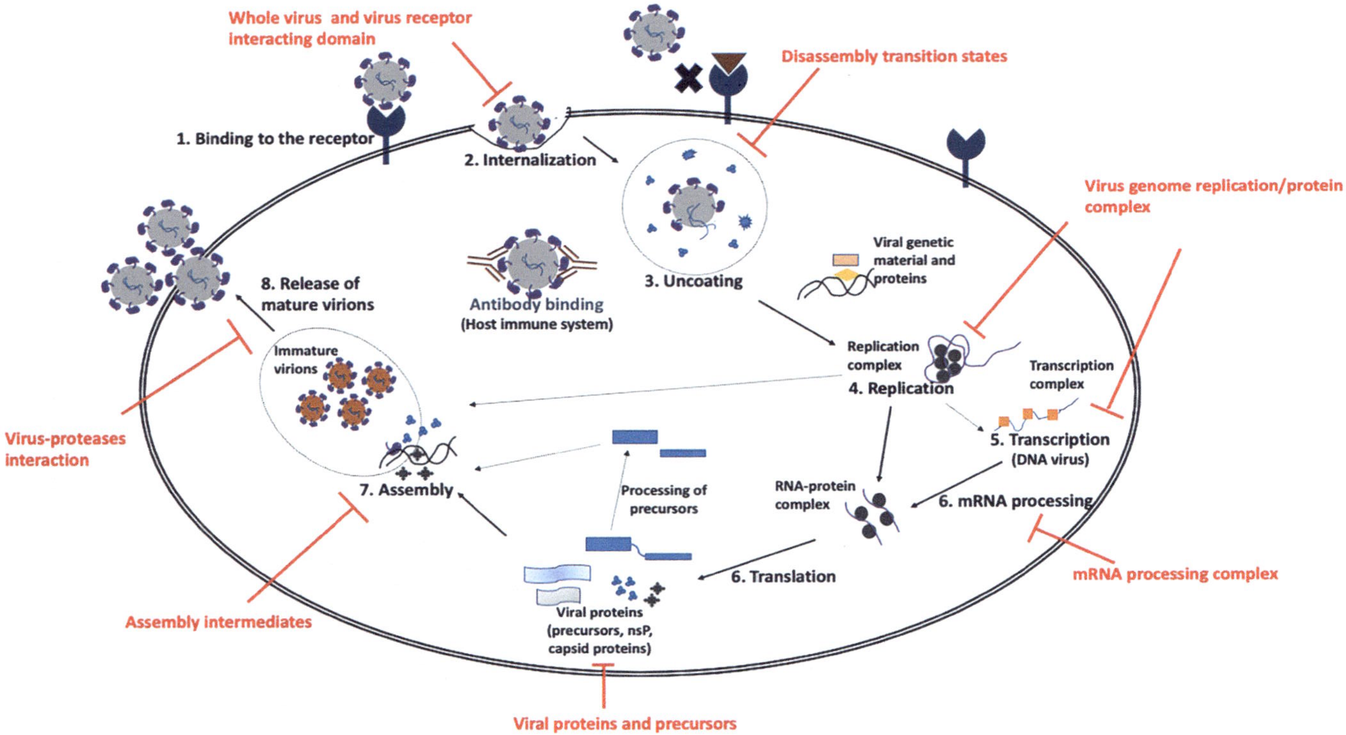

Fig. 7.1 *Targets for antiviral development in a typical virus life cycle*. The schematic depicts the discrete steps in virus–host interaction and the virus-encoded components that are critical for executing these steps

7.5.1 Whole Virus Capsids

CryoEM is an important technique for the 3D reconstruction of whole viral capsids and can provide important insights into assembly, stability, and maturation processes. Virus capsids are uniquely suited to cryoEM-based structural analysis because of their large size which can impede crystallization. The symmetry inherent in icosahedral capsids also makes it possible to resolve structures with limited data at high resolutions. One of the first 3D reconstructions from 2D images was that of the tomato bushy stunt virus—an icosahedral plant virus (Johnson and Olson 2021). A medium-resolution structure of adenovirus marked a high point in the cryoEM field in 1984 (Adrian et al. 1984), which was followed by the first subnanometer structure—that of hepatitis B virus at a resolution of 7.4 Å in 1997 (Böttcher et al. 1997). The structures of native Coxsackievirus A9 in 2023 (Fig. 7.2) and human enterovirus D-68 at 2.9 Å (Domanska et al. 2022) and 2.0 Å (EMD-25765), respectively, are some of the highest resolution virus structures resolved using cryoEM to date. The 3D structures of whole viruses at reasonably high resolutions provide detailed information about the arrangement of viral proteins, nucleic acids, and additional associated components like antibodies, receptors, or small molecules. Such information can provide crucial insights into the overall organization and assembly of the virus, conformation of antigenic sites, and conformational transitions during host interaction. The structural comparison of strains and variants can provide critical insights into viral evolution, the key variations that can be responsible for virulence, transmissibility, immune escape, and drug resistance. Some examples of virus capsid structures obtained using cryoEM are included in Table 7.1.

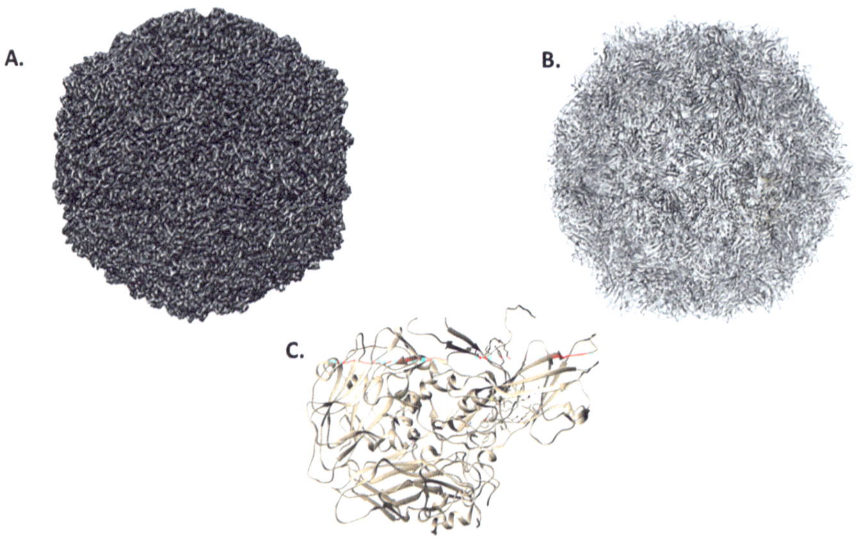

Fig. 7.2 *A 2.9-Å cryogenic electron microscopic structure of native Coxsackievirus A9 [EMD-15634]*. (**a**) Isosurface representation of the whole virus. (**b**) Molecular surface representation, shown in ribbon view. (**c**) The icosahedral asymmetric unit

Table 7.1 Cryogenic electron microscopic structures of whole viruses

EMD/PDB ID	Virus name	Strain	Resolution
EMD-10103	Murine norovirus	MNV-1	3.1 Å
EMD-33403	Scylla serrata reovirus	SZ-2007	3.1 Å
EMD-15859	*Rosellinia necatrix* megabirnavirus 1/W779	W779	3.3 Å
EMD-25765	Human enterovirus D-68	US/MO/14–18,947	2.0 Å
EMD-15634	Coxsackievirus A9	Griggs	2.9 Å
EMD-14195	Saccharomyces cerevisiae virus L-BCLa	–	3.7 Å
EMD-27559	Chikungunya virus	s27	6.75 Å
EMD-13022	Enterovirus 70	J670/71	2.63 Å
EMD-33502	Echovirus 18	HeB15–254	3.37 Å
EMD-32550	Adeno-associated virus 9	–	3.87 Å
EMD-14512	Tick-borne encephalitis virus	Kuutsalo-14	3.3 Å
EMD-32102	Infectious bursal disease virus	Gx	3.8 Å
EMD-25905	W6 possum enterovirus (EV-F4)	–	2.96 Å
EMD-12171	Human rhinovirus 14	1059	2.6 Å
EMD-22961	Mayaro virus	IQT4235	4.4 Å
EMD-12182	Kashmir bee virus	–	2.8 Å
EMD-30375	Apple-latent spherical virus	–	2.87 Å
EMD-22645	Murine polyomavirus	A2	3.9 Å
EMD-11370	Dengue virus 2	16,681	3.1 Å
EMD-20595	*Bombyx mori* cytoplasmic polyhedrosis virus	–	3.7 Å
EMD-9204	Cucumber leaf spot virus	–	3.2 Å
EMD-9205	Red clover necrotic mosaic virus	–	2.9 Å
EMD-0054	*Feline calicivirus*	F9	3.0 Å

Further, the transient molecular states generated during viral assembly and disassembly are best investigated using cryoEM. Imaging viral protein complexes or capsids in these dynamic stages of the life cycle can provide insights into the nature of capsid protein—protein and protein–nucleic acid interaction. CryoEM structures of disassembling nonenveloped viruses have been investigated to identify the sites for genomic RNA release (Bostina et al. 2011; Harutyunyan et al. 2014; Buchta et al. 2019; Azad and Banerjee 2019). Structures of such dynamic states at the molecular level can be utilized to understand the physical and chemical basis of disassembly—a crucial target step for preventing viral infections.

7.5.2 Virus–Receptor Complexes and Blocking the Early Stages of Entry

Structural information of viruses complex with the host receptor is of great importance in elucidating the mechanistic details of the initial stages of virus entry. Such information can provide critical insights about the specificity of binding interactions, as well as conformational changes and rearrangements in the virus during

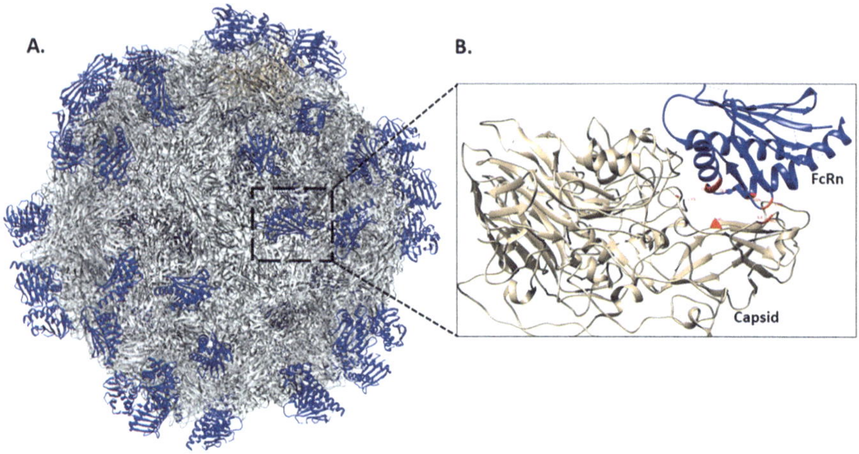

Fig. 7.3 *Echovirus 18 (E18) bound with its receptor FcRn [EMD-33499].* (**a**) A 3.09-Å cryogenic electron microscopic structure of E18 bound with its receptor FcRn (shown in blue) at 3.09 Å. (**b**) Enlarged view of the capsid region bound to FcRn. Interacting regions are shown in red

attachment and entry. Molecular-level knowledge of receptor recognition can assist in elucidating virus tropism and host range. By understanding the structural features and binding sites involved in receptor recognition, inhibitors or neutralizing antibodies that block viral entry and prevent infection can be designed. The first report of the structure of a virus–receptor complex resolved by cryoEM was published in 1994, which showed two amino-terminal, immunoglobulin-like domains of the intercellular adhesion molecule-1 (ICAM-1) bound to human rhinovirus 16. This study showed that ICAM-1 binds into a 12 Å-deep "canyon" on the surface of the virus capsid (Rossmann et al. 1994). Other cryoEM studies that showed the interaction of host receptors with virus capsid include the 3D structure of Coxsackievirus B3 in complex with its full-length receptor—the human coxsackievirus adenovirus receptor (CAR), and also the D1D2 fragment of CAR (He et al. 2001), the association of adenovirus fibers with the receptor (Wu et al. 2003), among others. Such studies have been crucial in elucidating the structural determinants of host–virus attachment and downstream processes, such as the requirement of fiber flexibility for adenovirus-receptor binding, and a two-in-one attachment-uncoating process triggered by the binding of the human neonatal Fc receptor (FcRn) receptor to echovirus 18 (Fig. 7.3) (Chen et al. 2022a). Some of the examples are listed in Table 7.2.

An early example of cryoEM-based antiviral drug discovery is against members of the Picornaviridae family. These viruses contain a pseudo T = 3 icosahedral structure generated from 60 copies of each of the four structural proteins—VP1, VP2, VP3, and VP4. Destabilization of the icosahedron is necessary for the release of RNA inside the host cell. Structural analysis of members of the picornavirus family such as poliovirus or enterovirus 71 revealed a hydrophobic pocket near the receptor-binding canyon generated by VP1, that could be targeted to block genome release. CryoEM studies revealed that an inhibitor, NLD-22, shows a strong

Table 7.2 Cryogenic electron microscopic structures of altered virus capsids

EMDB/PDB ID	Virus name	State observed	Unique features
EMD-15692	Coxsackievirus A9 (CVA9)	CVA9 primed for uncoating	Expanded
EMD-15076	Coxsackievirus A9 (CVA9)	CVA9 primed for uncoating	Expanded
EMD-13125	Enterovirus 70	Altered particle	Uneven genomic density
EMD-13126	Enterovirus 70	Empty particle	State after genome release
EMD-33505	Echovirus 18	Receptor FcRn bound state at pH 5.5	Pocket factor lost, capsid destabilized, uncoating initiated
EMD-24204/24200	Eastern equine encephalitis virus	Prefusion states 1 and 2	Structural rearrangements of E1 and E2 glycoprotein, essential for membrane fusion
EMD-32551	Adeno-associated virus 9	Adeno-associated virus receptor (AAVR) bound state	Receptor bound to threefold protrusions and the region between three- and fivefold axes.
EMD-12594	Human rhinovirus 14	Receptor ICAM-1-bound-activated state	C terminal of VP4 translocated, primed for genome release
EMD-12595	Human rhinovirus 14	Receptor ICAM-1-bound state	Lacks genome, empty particle
EMD-12173	Kashmir bee virus	Activated particle	Reduction in interpentamer contacts
EMD-12154	Kashmir bee virus	Empty particle	Postgenome release state
EMD-9730	Flock house virus	Eluted particle	Increased diameter, stain permeable
EMD-9732	Flock house virus	Puffed particle	Particle in the process of genome release
EMD-20469	Poliovirus type 1	135S particle	Expanded capsid with openings at twofold and quasi-threefold axes
EMD-4115	Israeli acute bee paralysis virus (IAPV)	Empty IAPV particles	Detachment of VP4, partial loss of structure at VP2 N-terminus, loss of RNA

association with the hydrophobic pocket of VP1 in the EV 71 capsid and blocks viral infection (Zhang et al. 2020). Similarly, antivirals WIN51711 (Fig. 7.4) and pleconaril bind to the hydrophobic VP1 pocket of EV70 and block viral infection (Füzik et al. 2022). A nanobody, NB-5829, raised against murine norovirus, was shown to bind with the P-dimer in the main capsid protein VP1 and cause a 42° rotation that interferes with the receptor attachment and blocks cell entry (Füzik et al. 2022).

During the recent pandemic, two cryoEM structures of the spike protein from SARS-CoV-2 were reported at 3.5 Å–2.8 Å in March 2022 (Wrapp et al. 2020; Walls et al. 2020), almost coinciding with the declaration of a pandemic by the World Health Organization. These early structures, obtained at unprecedented speed, were crucial to establish the receptor-binding mechanism, immunogenicity,

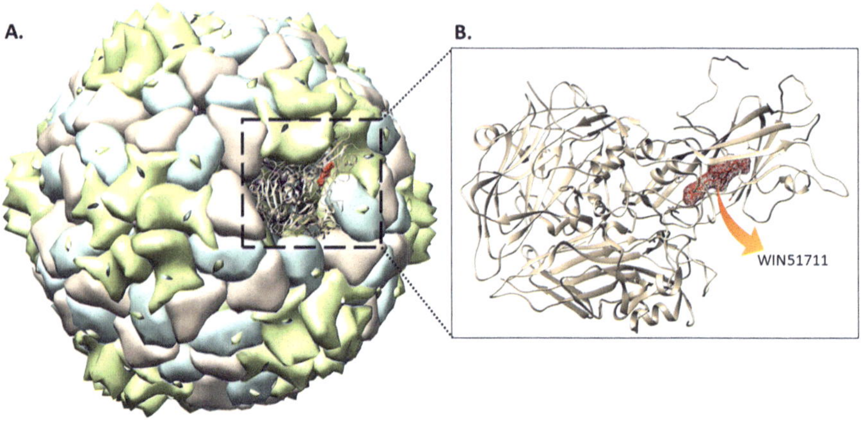

Fig. 7.4 *Human enterovirus 70 (EV70) in complex with antiviral WIN51711.* (**a**) Cartoon representation of EV70 bound with inhibitor WIN51711. One icosahedral asymmetric unit bound with WIN51711 shown in ribbon view. (**b**) Enlarged view of the capsid-WIN51711 complex. Density of WIN51711 is shown in red

and immune escape potential of the virus. Subsequent studies resolved cryoEM reconstructions of spike 2 in different conformational states, pre- and postfusion, as well as in complex with its main receptor ACE2. Together, these studies have been crucial for the design of peptides or small molecules for disrupting the association and preventing virus entry (Gui et al. 2017). Several types of inhibitors have been designed and tested, including a synthetic helix-hairpin peptide that acts as a protein–protein interaction (PPI) inhibitor (Khatri et al. 2022) shown in Fig. 7.5. This peptide arrests the spike protein in dimeric form (2:2 oligomeric association) and thus prevents the association of spike to ACE2. CryoEM has also been an essential technique to understand the association of ACE2 with spike components from different SARS-CoV-2 strains, as well as the interaction of spike with neutralizing antibodies. The latter group of studies has been crucial in designing antibody-based therapeutics as well as vaccine candidates (Wec et al. 2020; Cameroni et al. 2021; Zhu et al. 2023). An example of antibody-based COVID therapy is the ambavirumab/romisevirumab combination approved for usage by the State Food and Drug Administration of China. While ambavirumab directly inhibits spike–ACE2 binding, romisevirumab inhibits viral membrane fusion by binding to a different site of the spike. Since the neutralizing antibodies do not compete against each other for target binding, their joint efficacy is enhanced, even against strains such as omicron. A recent study takes this approach a bit further to combine cryoEM with artificial intelligence-based techniques and next-generation sequencing (NGS). The cryoEM densities of neutralizing antibodies from patient samples are directly assigned using bioinformatic approaches to get sequences, and this is combined with NGS to

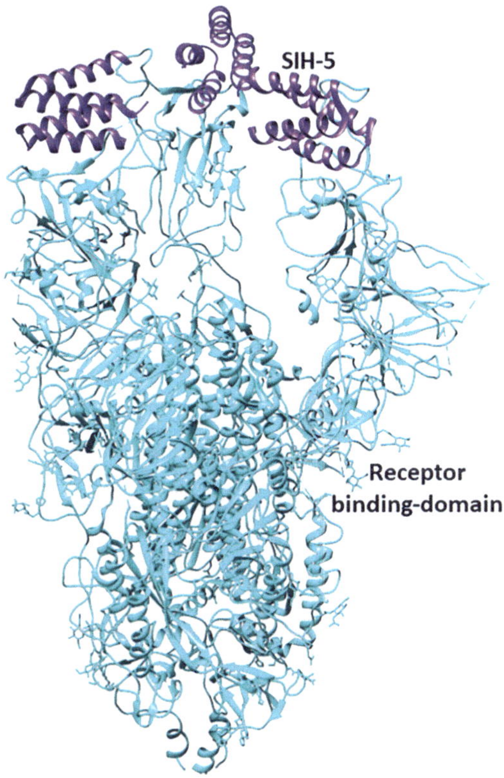

Fig. 7.5 *Spike receptor binding domain (RBD)-inhibitor complex resolved by cryogenic electron microcopy (cryoEM).* A 4.47-Å cryoEM structure of RBD from the spike protein of SARS-CoV2 in complex with synthetic inhibitor peptide SIH-5 (EMD-33042)

specifically identify monoclonals from polyclonal population and synthesize them for reiterative vaccine design (Antanasijevic et al. 2022). Some of the examples of virus–antibody interactions are provided in Table 7.3.

CryoEM reconstructions of virus–antibody complexes have been utilized for deciphering several antibody-mediated neutralization mechanisms against numerous other viruses such as dengue virus (Dowd et al. 2023; Sharma et al. 2021), Zika virus (Singh et al. 2022), hepatitis A virus (Cao et al. 2019), etc. An interesting example includes antibody design against the Eastern equine encephalitis virus (EEEV). During cellular membrane fusion orchestrated by EEEV, the viral surface glycoprotein E2 dissociates from E1, by a rotation of 60°, resulting in the exposure of the B domain of E2. This conformational change is required for successful fusion. A cryoEM reconstruction of EEEV with the neutralizing antibody Fab22 shows that the antibody binds to the E2-B domain and thus stabilizes the E2–E1 interaction and inhibits fusion, resulting in reduced infectivity (Chen et al. 2022b). Thus, the cryoEM reconstruction has been crucial for understanding the importance of the E2-B domain, and effective inhibitors against E2 could probably be potential antivirals against EEEV.

Table 7.3 Cryogenic electron microscopic structures of antibody/inhibitor complexed virus

EMDB/PDB ID	Name of the virus	Antibody/ inhibitor	Resolution	Site of interaction/ function
EMD-26947	Eastern equine encephalitis virus	Fab fragment of IgG-106	5.2 Å	Overlapping epitope on E2 protein
EMD-24202	Eastern equine encephalitis virus	Antigen binding fragment Fab22	8.1 Å	Blocks membrane fusion
EMD-29021	Dengue virus (DENV2)	prM13 antibody	9.8 Å	prM
EMD-29020	Dengue virus (DENV2)	prM12 antibody	10.2 Å	prM
EMD-25606	Zika virus	IgM antibody DH1017 fab fragment	5.26 Å	Envelope dimer epitope within domain II
EMD-13127	Enterovirus 70	Pleconaril	2.31 Å	VP1 pocket
EMD-13128	Enterovirus 70	WIN51711	2.74 Å	VP1 pocket
EMD-0943	Enterovirus 71	NLD-22	3.264 Å	VP1 pocket
EMD-30465	Dengue virus (DENV2)	EDE1 C10	3.5 Å	Envelope (E) protein dimer
EMD-10600	Murine norovirus	Neutralizing nanobody NB-5829 and Mg^{2+}	4.5 Å	Binds with the P dimers and blocks receptor binding.
EMD-32914	Chikungunya virus nonstructural protein 1 (nsP1)	Inhibitor FHA	2.62 Å	SAM/SAH binding site, blocks methyltransferase activity
EMD 9827,9828,9829, 9830	Hepatitis A virus	Neutralizing antibody F4, F6, F7, F9, respectively	3.9 Å, 3.68 Å, 3.05 Å, 3.79 Å, respectively	Heavy chains bind to BC and EF loop of VP3, while light chains bind to the BC loop of VP2 and the BC loop of VP3.
EMD-14240	Influenza virus A/H7N9 polymerase	Nucleotide analog T1106	2.58 Å	Stalled and backtracked by five nucleotides of the replication complex
EMD-28756	SARS-CoV2 omicron BA.1 spike protein	Fab of RBD-2b antibody 1C3	3.4 Å	RBD
EMD-34688	SARS-CoV2 spike trimer	Rabbit monoclonal antibody (RmAb) 9H1 fab	3.59 Å	Receptor binding motif (RBM) of RBD
EMD-33042	SARS-CoV2 RBD trimer	Synthetic peptide SIH5	4.47 Å	Arrests spike protein in dimeric form

7.5.3 Nonstructural Viral Proteins and Protein Complexes

Nonstructural proteins and protein complexes, which are encoded by the viral genome but are not part of the virus capsid, are critical for virus propagation in infected cells. Such components, namely polymerases, replication and transcription complexes, proteases, ion channel-forming proteins, etc., are required for viral genome amplification, viral protein production, and assembly, for mitigating the host's immune responses, and generally for maintaining an environment favorable for virus replication with minimal host interference (Yadav et al. 2021). The importance of these crucial components in virus propagation makes them excellent targets for antiviral drug development, provided there is limited similarity and minimal cross-reactivity with host components. The last aspect requires careful consideration since molecular mimicry of host components by coevolving pathogens can be a major concern. Thus, it becomes important to obtain and compare molecular architecture and binding scaffolds of crucial druggable targets of viral origin with analogous host proteins to rule out potential cross-reactivity of inhibitors. Further, generating and crystallizing large viral complexes such as replication complexes, which may contain multiple proteins with membrane-spanning domains and post-translational modification such as glycosylation, is a challenging process; however, in the majority of such cases, structural details of the entire complex, the mode of PPI, and the chemistry of the binding surfaces, may be crucial for effective drug design. It has been possible to generate cryoEM-based structural details of such complexes with limited, or heterogeneous, material at near-atomic resolution essential for structure-based drug designing. Some examples are as follows.

Nonstructural protein 1 (nsP1) of the chikungunya virus (Fig. 7.6) is required for viral RNA capping and membrane association. A cryoEM-based reconstruction of nsP1 resulted in 3D structures of the nsP1 pore complexed with cofactors and substrates such as SAM, SAH, m7GTP, etc., needed for the RNA capping reaction, at near-atomic (2.5 Å–2.8 Å) resolution (Jones et al. 2023). The availability of these structural details made it possible to check the potency and binding parameters of a potential inhibitor, 6′-β-fluoro-homoaristeromycin (FHA), against the RNA-capping activity of nsP1. The cryoEM reconstruction of nsP1-FHA conjugate revealed the mechanism of capping inhibition by FHA which involved blocking the binding of SAM to nsP1 (Jones et al. 2023). Similar high-resolution structures of replication complex of several viruses have been reported (Kouba et al. 2023; Peng et al. 2022; Tan et al. 2022). A cryoEM reconstruction of the influenza virus replication complex after consecutive incorporation of T1106 (a nucleoside analog) shows backtracking and termination of the replication complex (Kouba et al. 2023). CryoEM structures of the SARS-CoV2 RNA replication machinery captured during RNA translocation depict that the nsp12-nsp7-nsp8 complex undergoes conformational rearrangements for RNA binding, and that polymerase activity is strictly dependent on the availability of Mn^{2+} (Wang et al. 2020). Most importantly, the similarity of the polymerase active site to the corresponding sites in viruses such as HCV and poliovirus suggested that known nucleoside analogs such as remdesivir may act as

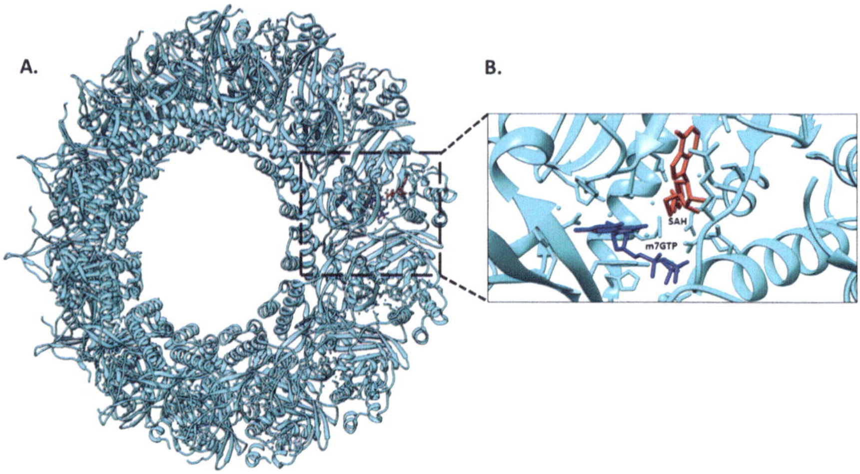

Fig. 7.6 *RNA-capping complex of chikungunya virus [EMD-15555].* (**a**) A 2.7-Å cryogenic electron microscopic structure of nonstructural protein 1 capping pore in complex with S-adenosyl homocysteine (SAH) and m7GTP. (**b**) Enlarged view of the capping pore complex showing the binding of SAH (red) and m7GTP (blue)

potent inhibitors of SARS-CoV-2 (Yin et al. 2020; Fertig et al. 2021). CryoEM structures of the polymerase in the presence of inhibitors such as favipiravir and suramin (Naydenova et al. 2021; Yin et al. 2021) have provided important information about inhibition mechanisms, which is invaluable for the next-generation drug design.

Apart from replicases, other nonstructural components important for virus propagation include ion channel proteins and proteases. Many virus families encode ion channel proteins called viroporins, which assist in virus assembly and budding. The structure of the viroporin E (envelope protein) from SARS-CoV-2, in conjunction with its host binding partner PALS1, was resolved using cryoEM in 2021 (Chai et al. 2021). The structure of the ORF3a ion channel from SARS-CoV-2 embedded in lipid nanodiscs has also been resolved at 2.1 Å (Kern et al. 2021). As these components are involved in modulating the host inflammatory responses against viral infection, an understanding of potential binding surfaces is invaluable for drug discovery. The structure of the main protease of SARS-CoV-2, in conjunction with the substrate, has provided new insights into the molecular mechanism of viral polyprotein processing, which will also be important for the design of more effective inhibitors (Narwal et al. 2023).

Thus, structural information about the binding modes of ligands, cofactors, and nucleic acid to such large complexes is capable of providing unique insights. Dynamic transition states of such protein complexes are more promising palettes for structure-based drug designing. Some examples of cryoEM structures of nonstructural viral proteins and complexes are provided in Table 7.4.

Table 7.4 Cryogenic electron microscopic structures of viral proteins/protein complexes

EMDB/ PDB ID	Virus name	Protein complex/state visualized	Component(s)	Resolution
EMD 31183	Mammalian reovirus	Initial transcription complex	mRNA, RNA polymerase, and cofactor μ2 complex	3.3 Å
EMD-15554	Chikungunya virus	Capping pore complex	nsP1 capping pore with m7GTP and SAH ligands	2.7 Å
EMD-15555	Chikungunya virus	Capping pore complex	nsp1 capping pore with SAM	2.8 Å
EMD-31582	Chikungunya virus	Capped viral RNA	nsP1 with m7GpppAU-10RNA	2.51 Å
EMD-33591	Chikungunya virus	Replication complex	nsP1, 2, 4, Zn^{2+}, GTP	2.8 Å
EMD-20581, 20582, 20585, 20586, 20587	*Bombyx mori* cypovirus 1	Sequential stages of RNA transcription: Quiescent, initiation, abortive, early elongation, and elongation	Template RNA, nontemplate RNA, transcript, RdRp	2.8–3.5 Å
EMD-28162, 28200	SARS-CoV2	Stepwise polyprotein cleavage reaction	M^{pro} in complex with nsp7–10 of polyprotein substrate pp1a	2.49 and 3.36 Å
EMD-14144	Influenza virus	Polymerase elongation complex	Stalled polymerase in conjunction with template and antiviral pseudobase T1106	2.48 Å
EMD-34731	Monkeypox virus	DNA polymerase holoenzyme	DNA, DNA polymerase, processivity factor component a20	2.8 Å
EMD-13476	Rotavirus A	NSP2-RNP complex	NSP2, ribonucleoprotein	3.1 Å
EMD-30275, 30252	SARS-CoV2	RdRp pre- and posttranslocated catalytic complex	nsp12-nsp7-nsp8, RNA, remdesivir	2.93 and 3.26 Å
EMD-31977	SARS-CoV2	Spike protein conformation 1	M protein dimer, YN7756_1 Fab	4.3 Å

7.6 Future Prospects of CryoEM in Accelerating Antiviral Drug Discovery and Potential Challenges

Over the last four decades, cryoEM has undergone a remarkable evolution from being a niche technique primarily used by a small subset of research labs to being a predominant structural biology method. This "resolution revolution" is due to a combination of various factors, including but not limited to (a) improvement in instrumentation capabilities such as the development of direct electron detectors; (b) increased automation for data collection; and (c) accelerated methods for data processing. Atomic-level information about the 3D structures of biological macromolecules involved in disease pathways is essential for SBDD and validation, and the reduced sample requirement for obtaining such information by cryoEM makes it a highly applicable method in drug discovery. A review of literature shows numerous efforts to develop inhibitors against critical stages of virus propagation—including virus entry, replication and assembly, and egress. CryoEM reconstructions of virus-encoded protein complexes, which are critical for specific stages in virus propagation, can provide detailed structural insights into potential interaction sites for inhibitors, including small molecules and therapeutic antibodies. Virus capsids contain vulnerable regions essential for receptor binding and metastability of the capsid required for the disassembly and release of the genome. Identifying these sites and entry- or disassembly-related conformational changes are possible by cryoEM. Antivirals could be targeted against those specific sites, and effective binding of those antivirals and subsequent conformational changes could also be visualized by cryoEM. The power of cryoEM as a drug discovery tool became particularly evident during the recent SARS-CoV-2 pandemic, where this technique was extensively utilized to obtain structural information about virus protein structures, mode of receptor, and neutralizing antibody binding and the effect of mutations. This information was utilized fairly quickly in the design of small molecule, peptidomimetic, and antibody drugs for blocking virus attachment, entry, and replication, as well as in the design of vaccine candidates. It is expected that the adoption of cryoEM to understand human disease phenomena and for the development of mitigation strategies, particularly strategies against viral diseases, will expand even further as the technology becomes more accessible and easier to use.

A concern for the utilization of cryoEM for SBDD and FBDD lies in the effective setting up of a robust pipeline, which allows quick generation of inhibitor-bound structures at reasonably high resolutions of 2.0–2.5 Å. This step, which is required for lead optimization, is still dependent on X-ray crystallography as a gold standard. However, increased automation in the pipeline will be essential in overcoming such barriers and establishing cryoEM as the primary method for antiviral drug discovery.

Acknowledgments This work was supported by the Department of Science and Technology India, Grant number SPF/2021/000115 to MB. MKL thanks the Department of Biotechnology for a Junior Research Fellowship. PN thanks to Department of Biotechnology for DBT-BioCARe project (Project Number: BT/PR50894/BIC/101/1317/2023) to support this work.

References

Adamson CS, Chibale K, Goss RJM, Jaspars M, Newman DJ, Dorrington RA (2021) Antiviral drug discovery: preparing for the next pandemic. Chem Soc Rev 50(6):3647–3655. https://doi.org/10.1039/D0CS01118E

Adrian M, Dubochet J, Lepault J, McDowall AW (1984) Cryo-electron microscopy of viruses. Nature 308(5954):32–36. https://doi.org/10.1038/308032a0

Al Hajjar S, Memish ZA, McIntosh K (2013) Middle East respiratory syndrome coronavirus (MERS-CoV): a perpetual challenge. Ann Saudi Med 33(5):427–436. https://doi.org/10.5144/0256-4947.2013.427

Antanasijevic A, Bowman CA, Kirchdoerfer RN, Cottrell CA, Ozorowski G et al (2022) From structure to sequence: antibody discovery using cryoEM. Sci Adv 8(3):eabk2039. https://doi.org/10.1126/sciadv.abk2039

Ardenkjaer-Larsen J-H, Boebinger GS, Comment A, Duckett S, Edison AS, Engelke F et al (2015) Facing and overcoming sensitivity challenges in biomolecular NMR spectroscopy. Angew Chem Int Ed Engl 54(32):9162–9185. https://doi.org/10.1002/anie.201410653

Azad K, Banerjee M (2019) Structural dynamics of nonenveloped virus disassembly intermediates. J Virol 93(22). https://doi.org/10.1128/JVI.01115-19

Bajad NG, Rayala S, Gutti G, Sharma A, Singh M, Kumar A, Singh SK (2021) Systematic review on role of structure based drug design (SBDD) in the identification of anti-viral leads against SARS-Cov-2. Curr Res Pharmacol Drug Discov 2:100026. https://doi.org/10.1016/j.crphar.2021.100026

Békés M, Langley DR, Crews CM (2022) PROTAC targeted protein degraders: the past is prologue. Nat Rev Drug Discov 21(3):181–200. https://doi.org/10.1038/s41573-021-00371-6

Beyrer C (2021) A pandemic anniversary: 40 years of HIV/AIDS. Lancet (London, England) 397(10290):2142–2143. https://doi.org/10.1016/S0140-6736(21)01167-3

Bostina M, Levy H, Filman DJ, Hogle JM (2011) Poliovirus RNA is released from the capsid near a twofold symmetry axis. J Virol 85(2):776–783. https://doi.org/10.1128/JVI.00531-10

Böttcher B, Wynne SA, Crowther RA (1997) Determination of the fold of the core protein of hepatitis B virus by electron cryomicroscopy. Nature 386(6620):88–91. https://doi.org/10.1038/386088a0

Buchta D, Füzik T, Hrebík D, Levdansky Y, Sukeník L, Mukhamedova L et al (2019) Enterovirus particles expel capsid pentamers to enable genome release. Nat Commun 10(1):1138. https://doi.org/10.1038/s41467-019-09132-x

Callaway E (2020) Revolutionary cryo-EM is taking over structural biology. Nature 578(7794):201–201. https://doi.org/10.1038/d41586-020-00341-9

Cameroni E, Bowen JE, Rosen LE, Saliba C, Zepeda SK, Culap K et al (2021) Broadly neutralizing antibodies overcome SARS-CoV-2 Omicron antigenic shift. Nature 602(7898):664–670. https://doi.org/10.1038/s41586-021-04386-2

Cao L, Liu P, Yang P, Gao Q, Li H, Sun Y et al (2019) Structural basis for neutralization of hepatitis a virus informs a rational design of highly potent inhibitors. PLoS Biol 17(4):e3000229. https://doi.org/10.1371/journal.pbio.3000229

Chai J, Cai Y, Pang C, Wang L, McSweeney S, Shanklin J, Liu Q (2021) Structural basis for SARS-CoV-2 envelope protein recognition of human cell junction protein PALS1. Nat Commun 12(1):3433. https://doi.org/10.1038/s41467-021-23533-x

Chen X, Liu M, Tian Y, Li J, Qi Y et al (2018) Cryo-EM structure of human mTOR complex 2. Cell Res 28(5):518–528. https://doi.org/10.1038/s41422-018-0029-3

Chen X, Qu X, Liu C, Zhang Y, Zhang G, Han P et al (2022a) Human FcRn is a two-in-one attachment-uncoating receptor for echovirus 18. MBio 13(4). https://doi.org/10.1128/mbio.01166-22

Chen C-L, Klose T, Sun C, Kim AS, Buda G, Rossmann MG et al (2022b) Cryo-EM structures of alphavirus conformational intermediates in low pH-triggered prefusion states. Proc Natl Acad Sci U S A 119(30):e2114119119. https://doi.org/10.1073/pnas.2114119119

Cheng Y (2018) Single-particle cryo-EM-how did it get here and where will it go. Sci (New York, N.Y.) 361(6405):876–880. https://doi.org/10.1126/science.aat4346

Cressey D, Callaway E (2017) Cryo-electron microscopy wins chemistry Nobel. Nature 550(7675):167–167. https://doi.org/10.1038/nature.2017.22738

Dai W, Zhang B, Jiang X-M, Su H, Li J, Zhao Y et al (2020) Structure-based design of antiviral drug candidates targeting the SARS-CoV-2 main protease. Science 368(6497):1331–1335. https://doi.org/10.1126/science.abb4489

Dai L, Dai Y, Han J, Huang Y, Wang L, Huang J, Zhou Z (2021) Structural insight into BRCA1-BARD1 complex recruitment to damaged chromatin. Mol Cell 81(13):2765–2777.e6. https://doi.org/10.1016/j.molcel.2021.05.010

de Chadarevian S (2018) John Kendrew and myoglobin: protein structure determination in the 1950s. Protein Sci Publ Protein Soc 27(6):1136–1143. https://doi.org/10.1002/pro.3417

Domanska A, Plavec Z, Ruokolainen V, Löflund B, Marjomäki V, Butcher SJ (2022) Structural studies reveal that endosomal cations promote formation of infectious coxsackievirus A9 A-particles, facilitating RNA and VP4 release. J Virol 96(24). https://doi.org/10.1128/jvi.01367-22

K A. Dowd, ., Sirohi, D., S D. Speer, VanBlargan, L. A., Chen, R. E., Mukherjee, S., et al 2023. prM-reactive antibodies reveal a role for partially mature virions in dengue virus pathogenesis. Proc Natl Acad Sci, 120(3): e2218899120. https://doi.org/10.1073/pnas.2218899120

Fertig TE, Chitoiu L, Terinte-Balcan G, Peteu V, Marta D, Gherghiceanu M (2021) The atomic portrait of SARS-CoV-2 as captured by cryo-electron microscopy. J Cell Mol Med 26(1):25–34. https://doi.org/10.1111/jcmm.17103

Fitzpatrick AWP, Falcon B, He S, Murzin AG, Murshudov G, Garringer HJ et al (2017) Cryo-EM structures of tau filaments from Alzheimer's disease. Nature 547(7662):185–190. https://doi.org/10.1038/nature23002

Füzik T, Moravcová J, Kalynych S, Plevka P (2022) Structure of human enterovirus 70 and its inhibition by capsid-binding compounds. J Virol 96(17). https://doi.org/10.1128/jvi.00604-22

Guerrero-Ferreira R, Taylor NM, Mona D, Ringler P, Lauer ME, Riek R et al (2018) Cryo-EM structure of alpha-synuclein fibrils. elife 7. https://doi.org/10.7554/eLife.36402

Gui M, Song W, Zhou H, Xu J, Chen S, Xiang Y, Wang X (2017) Cryo-electron microscopy structures of the SARS-CoV spike glycoprotein reveal a prerequisite conformational state for receptor binding. Cell Res 27(1):119–129. https://doi.org/10.1038/cr.2016.152

Hands-Portman I, Bakker SE (2022) Customising the plunge-freezing workflow for challenging conditions. Faraday Discuss 240(0):44–54. https://doi.org/10.1039/D2FD00060A. PMID: 35913403; PMCID: PMC9641998

Harutyunyan S, Kowalski H, Blaas D (2014) The rhinovirus subviral a-particle exposes 3′-terminal sequences of its genomic RNA. J Virol 88(11):6307–6317. https://doi.org/10.1128/JVI.00539-14

He Y, Chipman PR, Howitt J, Bator CM, Whitt MA, Baker TS et al (2001) Interaction of coxsackievirus B3 with the full length coxsackievirus-adenovirus receptor. Nat Struct Biol 8(10):874–878. https://doi.org/10.1038/nsb1001-874

Hoopes JD, Driebe EM, Kelley E, Engelthaler DM, Keim PS, Perelson AS et al (2011) Triple combination antiviral drug (TCAD) composed of amantadine, Oseltamivir, and ribavirin impedes the selection of drug-resistant influenza A virus. PLoS One 6(12):e29778. https://doi.org/10.1371/journal.pone.0029778

Hu B, Guo H, Zhou P, Shi Z-L (2020) Characteristics of SARS-CoV-2 and COVID-19. Nat Rev Microbiol 19(3):141–154. https://doi.org/10.1038/s41579-020-00459-7

Johnson JE, Olson AJ (2021) Icosahedral virus structures and the protein data bank. J Biol Chem 296:100554. https://doi.org/10.1016/j.jbc.2021.100554

Jones JE, Le Sage V, Lakdawala SS (2021) Viral and host heterogeneity and their effects on the viral life cycle. Nat Rev Microbiol 19(4):272–282. https://doi.org/10.1038/s41579-020-00449-9

Jones R, Hons M, Rabah N, Zamarreño N, Arranz R, Reguera J (2023) Structural basis and dynamics of chikungunya alphavirus RNA capping by nsP1 capping pores. Proc Natl Acad Sci 120(12):e2213934120. https://doi.org/10.1073/pnas.2213934120

Kamorudeen RT, Adedokun KA, Olarinmoye AO (2020) Ebola outbreak in West Africa, 2014–2016: epidemic timeline, differential diagnoses, determining factors, and lessons for future response. J Infect Public Health 13(7):956–962. https://doi.org/10.1016/j.jiph.2020.03.014

Kaplan M, Nicolas WJ, Zhao W, Carter SD, Metskas LA, Chreifi G et al (2021) In situ imaging and structure determination of biomolecular complexes using electron Cryo-tomography. Methods Mol Biol (Clifton, N.J.) 2215:83–111. https://doi.org/10.1007/978-1-0716-0966-8_4

Kern DM, Sorum B, Mali SS, Hoel CM, Sridharan S, Remis JP et al (2021) Cryo-EM structure of SARS-CoV-2 ORF3a in lipid nanodiscs. Nat Struct Mol Biol 28(7):573–582. https://doi.org/10.1038/s41594-021-00619-0

Khatri B, Pramanick I, Malladi SK, Rajmani RS, Kumar S, Ghosh P et al (2022) A dimeric proteomimetic prevents SARS-CoV-2 infection by dimerizing the spike protein. Nat Chem Biol 18(10):1046–1055. https://doi.org/10.1038/s41589-022-01060-0

Kollmer M, Close W, Funk L, Rasmussen J, Bsoul A, Schierhorn A et al (2019) Cryo-EM structure and polymorphism of Aβ amyloid fibrils purified from Alzheimer's brain tissue. Nat Commun 10(1):4760. https://doi.org/10.1038/s41467-019-12683-8

Kouba T, Dubankova A, Drncova P, Donati E, Vidossich P, Speranzini V et al (2023) Direct observation of backtracking by influenza A and B polymerases upon consecutive incorporation of the nucleoside analog T1106. Cell Rep 42(1):111901. https://doi.org/10.1016/j.celrep.2022.111901

Li B, Zhu D, Shi H, Zhang X (2021) Effect of charge on protein preferred orientation at the air–water interface in cryo-electron microscopy. J Struct Biol 213(4):107783. https://doi.org/10.1016/J.JSB.2021.107783

Ludtke SJ, Baldwin PR, Chiu W (1999) EMAN: semiautomated software for high-resolution single-particle reconstructions. J Struct Biol 128(1):82–97. https://doi.org/10.1006/JSBI.1999.4174

Lyumkis D (2019) Challenges and opportunities in cryo-EM single-particle analysis. J Biol Chem 294(13):5181–5197. https://doi.org/10.1074/jbc.REV118.005602

Mallakuntla MK, Togre NS, Santos DB, Tiwari S (2022) Implications of fragment-based drug discovery in tuberculosis and HIV. Pharmaceuticals 15(11). https://doi.org/10.3390/ph15111415

Martini M, Gazzaniga V, Bragazzi NL, Barberis I (2019) The Spanish influenza pandemic: a lesson from history 100 years after 1918. J Prev Med Hyg 60(1):E64–E67. https://doi.org/10.15167/2421-4248/jpmh2019.60.1.1205

Mateu MG (2013) Introduction: the structural basis of virus function. Subcell Biochem 68:3–51. https://doi.org/10.1007/978-94-007-6552-8_1

Moeti M, Gao GF, Herrman H (2022) Global pandemic perspectives: public health, mental health, and lessons for the future. Lancet (London, England) 400(10353):e3–e7. https://doi.org/10.1016/S0140-6736(22)01328-9

Murgueitio MS, Bermudez M, Mortier J, Wolber G (2012) In silico virtual screening approaches for anti-viral drug discovery. Drug Discov Today Technol 9(3):e219–e225. https://doi.org/10.1016/j.ddtec.2012.07.009

Narwal M, Armache J-P, Edwards TJ, Murakami KS (2023) SARS-CoV-2 polyprotein substrate regulates the stepwise Mpro cleavage reaction. J Biol Chem 299(5):104697. https://doi.org/10.1016/j.jbc.2023.104697

Naydenova K, Muir KW, Wu L-F, Zhang Z, Coscia F, Peet MJ et al (2021) Structure of the SARS-CoV-2 RNA-dependent RNA polymerase in the presence of favipiravir-RTP. Proc Natl Acad Sci 118(7):e2021946118. https://doi.org/10.1073/pnas.2021946118

Nogales E, Zhang R (2016) Visualizing microtubule structural transitions and interactions with associated proteins. Curr Opin Struct Biol 37:90–96. https://doi.org/10.1016/j.sbi.2015.12.009

Peng Q, Xie Y, Kuai L, Wang H, Qi J, Gao GF, Shi Y (2022) Structure of monkeypox virus DNA polymerase holoenzyme. Science 379(6627):100–105. https://doi.org/10.1126/science.ade6360

Piret J, Boivin G (2021) Pandemics throughout history. Front Microbiol 11:3594. https://doi.org/10.3389/fmicb.2020.631736

Punjani A, Rubinstein JL, Fleet DJ, Brubaker MA (2017) cryoSPARC: algorithms for rapid unsupervised cryo-EM structure determination. Nat Methods 14(3):290–296. https://doi.org/10.1038/nmeth.4169

Rossmann MG, Olson NH, Kolatkar PR, Oliveira MA, Cheng RH, Greve JM et al (1994) Crystallographic and cryo EM analysis of virion-receptor interactions. Arch Virol Suppl 9:531–541. https://doi.org/10.1007/978-3-7091-9326-6_51

Scheres SHW (2012) RELION: implementation of a Bayesian approach to cryo-EM structure determination. J Struct Biol 180(3):519–530. https://doi.org/10.1016/J.JSB.2012.09.006
Shampo MA, Kyle RA, Steensma DP (2012) Richard Ernst—nobel prize for nuclear magnetic resonance spectroscopy. Mayo Clin Proc 87(12):e109. https://doi.org/10.1016/j.mayocp.2012.01.023
Sharma A, Zhang X, Dejnirattisai W, Dai X, Gong D, Wongwiwat W et al (2021) The epitope arrangement on flavivirus particles contributes to Mab C10's extraordinary neutralization breadth across Zika and dengue viruses. Cell 184(25):6052–6066.e18. https://doi.org/10.1016/j.cell.2021.11.010
Singh T, Hwang K-K, Miller AS, Jones RL, Lopez CA, Dulson SJ et al (2022) A Zika virus-specific IgM elicited in pregnancy exhibits ultrapotent neutralization. Cell 185(25):4826–4840.e17. https://doi.org/10.1016/j.cell.2022.10.023
Smyth MS, Martin JH (2000) X ray crystallography. Mol Pathol MP 53(1):8–14. https://doi.org/10.1136/mp.53.1.8
Song W, Gui M, Wang X, Xiang Y (2018) Cryo-EM structure of the SARS coronavirus spike glycoprotein in complex with its host cell receptor ACE2. PLoS Pathog 14(8):e1007236. https://doi.org/10.1371/journal.ppat.1007236
Sullivan SJ, Jacobson RM, Dowdle WR, Poland GA (2010) 2009 H1N1 influenza. Mayo Clin Proc 85(1):64–76. https://doi.org/10.4065/mcp.2009.0588
Tan YB, Chmielewski D, Law MCY, Zhang K, He Y, Chen M et al (2022) Molecular architecture of the chikungunya virus replication complex. Sci Adv 8(48):eadd2536. https://doi.org/10.1126/sciadv.add2536
Thomas JM (2012) The birth of X-ray crystallography. Nature 491(7423):186–187. https://doi.org/10.1038/491186a
Walls AC, Park Y-J, Tortorici MA, Wall A, McGuire AT, Veesler D (2020) Structure, function, and antigenicity of the SARS-CoV-2 spike glycoprotein. Cell 181(2):281–292.e6. https://doi.org/10.1016/j.cell.2020.02.058
Wang Q, Wu J, Wang H, Gao Y, Liu Q, Mu A et al (2020) Structural basis for RNA replication by the SARS-CoV-2 polymerase. Cell 182(2):417–428.e13. https://doi.org/10.1016/j.cell.2020.05.034
Wang L-Q, Ma Y, Yuan H-Y, Zhao K, Zhang M-Y, Wang Q et al (2022) Cryo-EM structure of an amyloid fibril formed by full-length human SOD1 reveals its conformational conversion. Nat Commun 13(1):3491. https://doi.org/10.1038/s41467-022-31240-4
Wec AZ, Wrapp D, Herbert AS, Maurer DP, Haslwanter D, Sakharkar M et al (2020) Broad neutralization of SARS-related viruses by human monoclonal antibodies. Science 369(6504):731–736. https://doi.org/10.1126/science.abc7424
Weissenberger G, Henderikx RJM, Peters PJ (2021) Understanding the invisible hands of sample preparation for cryo-EM. Nat Methods 18(5):463–471. https://doi.org/10.1038/s41592-021-01130-6
Wrapp D, Wang N, Corbett KS, Goldsmith JA, Hsieh C-L, Abiona O et al (2020) Cryo-EM structure of the 2019-nCoV spike in the prefusion conformation. Science (New York, N.Y.) 367(6483):1260–1263. https://doi.org/10.1126/science.abb2507
Wu E, Pache L, Von Seggern DJ, Mullen T-M, Mikyas Y, Stewart PL, Nemerow GR (2003) Flexibility of the adenovirus fiber is required for efficient receptor interaction. J Virol 77(13):7225–7235. https://doi.org/10.1128/jvi.77.13.7225-7235.2003
Yadav R, Chaudhary JK, Jain N, Chaudhary PK, Khanra S, Dhamija P et al (2021) Role of structural and non-structural proteins and therapeutic targets of SARS-CoV-2 for COVID-19. Cell 10(4). https://doi.org/10.3390/cells10040821
Yin W, Mao C, Luan X, Shen D-D, Shen Q, Su H et al (2020) Structural basis for inhibition of the RNA-dependent RNA polymerase from SARS-CoV-2 by remdesivir. Science (New York, N.Y.) 368(6498):1499–1504. https://doi.org/10.1126/science.abc1560

Yin W, Luan X, Li Z, Zhou Z, Wang Q, Gao M et al (2021) Structural basis for inhibition of the SARS-CoV-2 RNA polymerase by suramin. Nat Struct Mol Biol 28(3):319–325. https://doi.org/10.1038/s41594-021-00570-0

Zhang M, Wang Y, He W, Sun Y, Guo Y, Zhong W et al (2020) Design, synthesis, and evaluation of novel enterovirus 71 inhibitors as therapeutic drug leads for the treatment of human hand, foot, and mouth disease. J Med Chem 63(3):1233–1244. https://doi.org/10.1021/acs.jmedchem.9b01414

Zhou T, Tsybovsky Y, Gorman J, Rapp M, Cerutti G, Chuang G-Y et al (2020) Cryo-EM structures of SARS-CoV-2 spike without and with ACE2 reveal a pH-dependent switch to mediate endosomal positioning of receptor-binding domains. Cell Host Microbe 28(6):867–879.e5. https://doi.org/10.1016/j.chom.2020.11.004

Zhu Z, Lian X, Su X, Wu W, Marraro GA, Zeng Y (2020) From SARS and MERS to COVID-19: a brief summary and comparison of severe acute respiratory infections caused by three highly pathogenic human coronaviruses. Respir Res 21(1):224. https://doi.org/10.1186/s12931-020-01479-w

Zhu K-F, Yuan C, Du Y-M, Sun K-L, Zhang X-K, Vogel H et al (2023) Applications and prospects of cryo-EM in drug discovery. Mil Med Res 10(1):10. https://doi.org/10.1186/s40779-023-00446-y

Chapter 8
Host-Directed Antiviral Therapy

Joyshikh Sonowal, Rajib Deb, S. Arutkumaran, Gyanendra Singh Sanger, Swaraj Rajkhowa, Seema Rani Pegu, and Vivek Kumar Gupta

Abstract Most of the antiviral drugs developed so far have been specifically designed to target particular pathogens and their unique viral pathways. However, due to their pathogen-specific nature, these drugs may not effectively combat a wide range of viruses. This necessitates the development of broad-spectrum antivirals capable of tackling multiple viral infections. Host-directed antiviral therapy (HDAT) presents a promising treatment option for viral infections. HDAT has the potential to be effective against various viruses, including those that mutate quickly or show resistance to traditional antiviral medications, as it focuses on host factors rather than the virus itself. By boosting the immune response or modifying cellular processes, HDAT aims to create an unfavorable environment for viral replication. The efficacy of HDAT and vaccines depends on the specific virus and the characteristics of the individual undergoing treatment. HDAT offers new opportunities for antiviral drug development by targeting host-specific vulnerabilities. The article also discusses different types of HDAT, such as immunomodulatory therapies, cell-based therapies, therapies that target host factors, repurposed drugs, and natural products. Ultimately, the choice of therapy depends on the virus's properties, the interaction between the virus and the host, and the host's health status.

Keywords HDAT · Host factors · Host-directed therapies · Immunomodulatory therapies · Cell-based therapies · Host factor targeting therapies · Repurposed drugs · Natural products

J. Sonowal · R. Deb (✉) · G. S. Sanger · S. Rajkhowa · S. R. Pegu · V. K. Gupta
ICAR-National Research Centre on Pig, Guwahati, Assam, India
e-mail: Rajib.Deb@icar.gov.in

S. Arutkumaran
ICAR-Indian Veterinary Research Institute, Bareilly, Uttar Pradesh, India

N. Kumar et al. (eds.), *Advances in Antiviral Research*, Livestock Diseases and Management, https://doi.org/10.1007/978-981-99-9195-2_8

8.1 Introduction

Throughout the twentieth and twenty-first centuries, humanity has experienced several devastating viral outbreaks, including smallpox caused by the Variola virus (https://www.who.int/), the H1N1 influenza virus that caused the Spanish flu pandemic in 1918–1919 (https://www.cdc.gov/ 2019a), Polio viral disease that peaked in the 1950s (Concepcion et al. 2021), the human immunodeficiency virus (HIV)/AIDS pandemic that emerged in the early 1980s and still lacks a cure (Eisinger and Fauci 2018), the rapid spread of severe acute respiratory syndrome (SARS) in 2003 (Christian et al. 2004), several outbreaks of Ebola viral disease in West Africa between 2014 and 2016 (https://www.cdc.gov/ 2019b), and the COVID-19 pandemic caused by the SARS-CoV-2 virus (Piret and Boivin 2021). As of April 2023, the COVID-19 pandemic has resulted in far-reaching social and economic impacts, with over 200 million individuals infected and more than 4 million deaths recorded worldwide. The development of multiple vaccines and therapeutics has been a key response to this crisis. Vaccines and antiviral drugs are designed to directly act against the virus causing the infection. Vaccines stimulate the immune system of the host, enabling it to recognize and effectively respond to a particular virus. While vaccines can be highly effective at preventing infection and reducing the severity of disease, they are often specific to a particular strain or variant of the virus. The strategies of the direct approach of using vaccines and other antiviral medications to combat viral infections frequently encounter setbacks when the virus mutates or develops resistance against the drug.

Host-directed antiviral therapy (HDAT) involves targeting host factors such as cellular pathways or immune responses that the virus uses to replicate and cause disease. By targeting the host rather than the virus itself, HDAT can potentially be effective against a broad range of viruses, including those that mutate quickly or that are resistant to traditional antiviral drugs. This approach was first used to control a herpes virus infection and is currently used to control HIV-1 infection (Kumar et al. 2020). The effectiveness of host-directed antiviral therapy and vaccines depends on the virus and the individual receiving the treatment. Factors, such as the virus's characteristics, how it interacts with the host, the accessibility of effective treatments, and the individual's medical history and state, will ultimately determine the treatment option. Several types of host-directed antiviral therapies are discussed in the following sections.

8.1.1 Immunomodulatory Therapies

This therapy aims to modulate the immune response of the host to control viral infections. Some examples of immunomodulatory therapies in HDAT include:

(a) Interferons: Host cells produce natural proteins in response to viral infections, known as interferons (IFNs), which activate the immune system and hinder viral replication. To enhance the effectiveness of IFN therapy, various technological modifications, such as the addition of ribavirin and conjugation of IFN with polyethylene glycol (PegIFN), have been made. PegIFN has a longer half-life and is less susceptible to enzymatic degradation, although it still has some side effects. However, patient unresponsiveness and the production of anti-IFN antibodies remain challenging with this approach (Kumar et al. 2020).

(b) Immune checkpoint inhibitors: Immune checkpoint inhibitors are medications designed to counteract the function of molecules on immune cells that restrict the immune response, resulting in enhanced immune activity against viruses. The most significant immune checkpoints are PD1 and CTLA-4, which are expressed on immune cells as membrane-bound molecules. Immune checkpoint inhibitors work by blocking the binding of checkpoint molecules to their ligands, which restores T cell activity, strengthens immune response, and improves the ability to fight off viral infections (Gao et al. 2019).

(c) Immunoglobulins: These are proteins that are produced by the immune system in response to viral infections. They can be administered as a therapy to provide passive immunity against viral infections (Neurath 2008).

(d) Toll-like receptor (TLR) agonists: TLR agonists are molecules that activate TLRs found on immune cells, which detect specific pathogen-associated molecular patterns (PAMPs). When activated, TLRs can strengthen the immune response against viruses. Some TLRs are located on the cell surface and constantly monitor extracellular stimuli to distinguish between harmless environmental molecules and PAMPs, maintaining a balance in the immune system. Others are located inside cells and detect pathogen-related DNA or RNA structures (Girkin et al. 2022).

(e) Cytokine inhibitors: These are drugs that can block the activity of cytokines, which are molecules that are produced by the immune system in response to viral infections. By blocking cytokines, cytokine inhibitors can reduce the inflammation and tissue damage caused by viral infections (Arend 2002).

(f) Steroids (Glucocorticosteroids): These are drugs that can suppress the immune system and reduce inflammation. They may be used in severe cases of viral infections to prevent tissue damage caused by an overactive immune response (Zurier and Weissmann 1973).

(g) Janus kinase (JAK) inhibitors: The virus replication cycle can be influenced by several kinases at each stage, and these kinases can be blocked by small molecule chemical inhibitors (Kumar et al. 2020). These are drugs that can block the activity of JAK enzymes, which are involved in the signaling pathways of many cytokines. By blocking JAK enzymes, JAK inhibitors can reduce inflammation and the immune response against viruses.

(h) Immunostimulatory oligonucleotides: These are short DNA or RNA sequences that can activate the immune system by stimulating toll-like receptors (TLRs) or other immune receptors. They can enhance the immune response against viruses. Antisense nucleic acid sequences can act as blockers of viral gene expression and can able to disrupt the transcription, splicing, and translation. Back-splicing of certain pre-mRNA exons gives rise to circRNAs, which is a type of noncoding RNA that is present in all eukaryotes, and can also be used for therapeutic purposes (Oberemok et al. 2023).

(i) Colony-stimulating factors: These are proteins that can stimulate the production and differentiation of immune cells, such as white blood cells. They can enhance the immune response against viruses.
(j) Monoclonal antibodies (mAbs) against immune cells: These are antibodies that can specifically target and block the activity of certain immune cells, such as T cells or B cells. By blocking these cells, mAbs can modulate the immune response against viruses.
(k) Immunomodulatory peptides: These are short chains of amino acids that can modulate the immune response by interacting with immune cells or other molecules involved in the immune response. They can enhance or suppress the immune response against viruses.
(l) Nucleoside analogs: These are drugs that can interfere with viral replication by mimicking the structure of nucleotides, which are the building blocks of DNA and RNA.
(m) Nitric oxide (NO) donors: These are molecules that can increase the production of NO, which is a signaling molecule that plays a role in the immune response against viral infections.
(n) Antioxidants: These are compounds that can reduce oxidative stress and inflammation, which can be beneficial in reducing tissue damage caused by viral infections and modulating the immune response.
(o) Probiotics: These are live bacteria that can interact with the immune system and enhance the immune response against viral infections.

These immunomodulatory therapies have shown promise in treating viral infections by modulating the host immune response.

8.1.2 Cell-Based Therapies

These use cells to attack the virus or promote tissue repair in the infected individual. This therapy involves the transplantation of cells into the patient to aid in the eradication of the virus or restore normal immune function (Aly 2020; Musiał-Wysocka et al. 2019). The cell-based therapy options for HDAT include mesenchymal stem cell (MSC) therapy, natural killer (NK) cell therapy, T-cell therapy, chimeric antigen receptor (CAR) T-cell therapy, regulatory T-cell therapy, decoy receptor therapy, and so on. These therapies have shown potential in treating viral infections by targeting various aspects of the host immune response and promoting tissue repair.

(a) T-cell therapy—it utilizes T cells, a specialized immune cell that can identify and eliminate infected cells. However, T cells can have both beneficial and detrimental effects on the course of COVID-19. Several approaches have been proposed to enhance the antiviral T-cell response against SARS-CoV-2, including adoptive T-cell therapies, vaccines that activate T-cell responses, recombinant cytokines, Th1 activators, and Th17 blockers. There is also a possibility of using immune checkpoint inhibitors either alone or in conjunction with anti-inflammatory drugs to boost the antiviral T-cell response (Toor et al. 2021).

(b) Dendritic cell therapy—dendritic cells are a type of immune cell that can stimulate an immune response against viruses. Dendritic cell therapy involves the collection and manipulation of dendritic cells, which are then infused back into the patient to stimulate an immune response against the virus (Wack 2021; Gardner et al. 2020).

(c) CAR T-cell therapy—this involves the genetic modification of T cells to express a chimeric antigen receptor, which allows them to specifically target infected cells. CAR T-cell therapy has been investigated for its potential in treating viral infections such as hepatitis B and C and HIV (Wang et al. 2020; Alfageme-Abello et al. 2021).

(d) MSC therapy—It has been studied as a potential therapy for viral infections such as influenza, herpes, and coronavirus (Sadeghi et al. 2020; Shaw et al. 2021; Alagesan et al. 2022). MSCs are a type of stem cell found in bone marrow and other tissues, and they can differentiate into various cell types and modulate the immune response. In viral infections, MSCs can help to reduce inflammation, promote tissue repair, and enhance the immune response against the virus. They do this by releasing factors that inhibit the virus's replication, stimulate the production of antiviral cytokines, and suppress the activity of immune cells that exacerbate inflammation (Vishnupriya et al. 2021).

(e) NK cell therapy—which involves the use of cells that are part of the innate immune system and can kill infected cells. NK cells have been shown to have activity against a variety of viruses, including SARS-CoV-2, flavivirus, influenza, HIV, and hepatitis C (Björkström et al. 2022; Smith et al. 2023). NK cell therapy has the potential to be used as a standalone therapy or in combination with other antiviral treatments.

(f) Decoy receptor therapy—this involves using decoy receptors that can bind to the virus and prevent it from infecting host cells (Arimori et al. 2022).

It's worth noting that many of these cell-based therapies are still in the experimental stages and have not yet been approved for clinical use. Further research is needed to determine their safety and effectiveness in treating viral infections in humans. However, cell-based therapies have their limitations and potential risks. These include the potential for adverse immune reactions, the need for specialized facilities and equipment for cell processing and storage, and the potential for the cells to be rejected by the recipient's immune system.

8.1.3 Host Factor Targeting Therapies

These target cellular factors that the virus uses to replicate or infect cells. Host factor targeting therapies (HFTs) aim to interfere with the host factors that viruses depend on for their replication and spread. Classical antiviral therapy inhibits the biological activities of viral structural proteins (capsid, nucleocapsid, envelope, etc.), nonstructural proteins, and replication enzymes (RNA methyltransferase, capping enzyme, protease, RNA-dependent RNA polymerase, helicase, etc.). A different strategy to control viral infections involves designing molecules that specifically target the host pathways utilized by viruses for replication. These pathways include host metabolic pathways such as glucose, lipid, and polyamine metabolism, glycosylation processes, inflammatory cascades, the ubiquitin-proteasome system, and programmed ribosomal frameshifting, etc. Some of the methods used for identifying the host-viral interactions are RNA interference (RNAi)-based methods, haploid cell screening, drug combination approach, genomic approaches, transcriptome and proteomic analysis of virus-infected cells, and CRISPR/Cas9 screens (Mahajan et al. 2021).

RNAi is a type of HFT that uses small RNA molecules to selectively target and degrade viral RNA or host RNA that is essential for viral replication. This approach entails the selective knockdown of the host gene (mRNA), resulting in the loss of function. RNAi can be employed to induce knockdown by transiently suppressing the expression of gene of interest (GoI) through small interfering RNA transfection or by achieving long-term suppression through the integration of GoI into the host genome using short hairpin RNAs (Badia et al. 2022). RNAi-based therapies can be designed to target specific viral genes or host factors that are required for viral replication. Other examples of HFTs include small molecule inhibitors that target host enzymes or proteins that are critical for viral replication or entry into host cells. For example, some HFTs target proteases that are involved in the processing of viral proteins, or receptors on the host cell surface that are used by the viruses to enter the cell. Another approach is to use mAbs that target specific viral proteins or host factors that are involved in viral replication or immune evasion. Antiviral mAbs are immunoglobulins with a defined specificity and a single isotype. Through their antigen-binding fragment, these antibodies exhibit therapeutic effects and are utilized to target specific diseases (Mahajan et al. 2021). These antibodies can neutralize the virus or enhance the immune response against the virus. In addition, CRISPR/Cas9 is a powerful gene editing tool that has the potential to be used in HDAT HFTs. One potential application of CRISPR/Cas9 in HDAT HFTs is to directly target genes that are essential for viral replication, such as viral entry receptors, viral polymerases, or other viral proteins. Using CRISPR/Cas9 to disable these genes, the virus may be unable to replicate and cause infection.

Another potential application of CRISPR/Cas9 in HDAT HFTs is to target host genes that are involved in the immune response to viral infections. For example, CRISPR/Cas9 could be used to disrupt genes that encode immune checkpoint proteins or cytokines that are involved in the regulation of the immune response. By disabling these genes, the immune response may be enhanced, leading to more effective clearance of the virus. CRISPR tools are used to target the gene that encodes for the C-C chemokine receptor type 5 (CCR5) receptor, which is used by HIV to enter and infect cells. By deleting or modifying the CCR5 gene using CRISPR, researchers have been able to block HIV infection in laboratory experiments. In addition, CRISPR/Cas9 can be used to engineer immune cells, such as T cells or natural killer cells, to enhance their antiviral activity. For example, CRISPR/Cas9 could be used to modify T cells to express CARs that specifically recognize viral antigens. This would allow the engineered T cells to selectively target and kill infected cells while leaving uninfected cells unharmed.

Other examples of successful host-directed therapies (HDT) include the application of CCR5 antagonists against HIV, the use of cyclosporine for influenza A virus, and the utilization of mAbs targeting claudin-1 and occludin for hepatitis C virus (HCV) (Badia et al. 2022). One advantageous aspect of this approach is its ability to provide broad-spectrum inhibitory activity against multiple viruses, offering a wider range of effectiveness. Furthermore, the evolutionary conservation of host factors contributes to a higher threshold for the emergence of drug resistance, adding another significant advantage to this approach. Viral evasion of HDT would face a substantial genetic barrier when attempting to develop resistant strains. Moreover, the use of drugs targeting host pathways brings another advantage, as many of these medications are already FDA-approved for treating various diseases. This allows for their prompt utilization in the treatment of viral infections, without the need for extensive regulatory processes.

One drawback of this approach is that the identification and development of HDTs necessitate a thorough understanding of virus–host interactions and their biological implications for viral replication. This requires extensive research and knowledge to determine which host factors play a substantial role in viral replication and can be effectively targeted for pharmacological inhibition. Additionally, it is crucial to ensure that the modulation of these host factors does not interfere with critical endogenous molecular functions, as only those factors meeting these criteria are suitable for HDT-based interventions.

8.1.4 Repurposed Drugs

These are the class of drugs that were originally developed to treat a specific disease or condition but have been found to have therapeutic potential for other diseases or conditions, including viral infections. Given below are some examples of repurposed drugs that have been investigated as HDAT for viral infections.

(a) Remdesivir: It is an antiviral drug that was initially developed to treat Ebola virus disease. It has been repurposed for the treatment of COVID-19 and has been shown to reduce the duration of hospitalization and improve clinical outcomes in hospitalized patients with severe COVID-19.

(b) Chloroquine/hydroxychloroquine: They are drugs that were originally developed to treat malaria and autoimmune diseases such as lupus and rheumatoid arthritis. They have been investigated as potential treatments for COVID-19, but their efficacy and safety are still a matter of debate.
(c) Interferons: They are a class of proteins that were initially developed as treatments for certain types of cancer and viral infections such as hepatitis B and C. They have been repurposed for the treatment of COVID-19 and have shown promise in reducing the severity of the disease.
(d) Ivermectin: It is an antiparasitic drug that has been used for decades to treat diseases such as river blindness and scabies. It has been repurposed for the treatment of COVID-19 and has been shown to reduce the duration of hospitalization and improve clinical outcomes in some studies, but further research is needed to establish its efficacy and safety.
(e) Oseltamivir: It is an antiviral drug that was originally developed to treat influenza virus infections. It has been investigated as a potential treatment for COVID-19, but its efficacy in treating this disease is still uncertain.
(f) Ribavirin: It is an antiviral drug that was initially developed to treat HCV infections. It has been repurposed for the treatment of COVID-19 and has been used in combination with other drugs such as interferon and remdesivir.
(g) Favipiravir: It is an antiviral drug that was initially developed to treat influenza virus infections. It has been repurposed for the treatment of COVID-19 and has been shown to reduce the time to viral clearance and improve clinical outcomes in some studies.

Repurposed drugs have the advantage of being already approved for use in humans and having established safety profiles. However, their efficacy in treating viral infections may vary depending on the virus and the stage of the disease.

8.1.5 Natural Products

These are compounds derived from plants, animals, or microorganisms that have antiviral properties, such as flavonoids or alkaloids (Lin et al. 2014; El Sayed 2000). Following are some examples of natural products that have been investigated as potential HDAT for viral infections:

(a) Flavonoids: They are a class of compounds that are found in many fruits, vegetables, and herbs. They have been shown to have antiviral properties against a variety of viruses, including influenza, HIV, and herpes simplex virus.
(b) Alkaloids: They are a class of nitrogen-containing compounds that are found in many plants. They have been shown to have antiviral properties against a variety of viruses, including influenza, hepatitis B, and herpes simplex virus.
(c) Polyphenols: They are a class of compounds that are found in many fruits, vegetables, and herbs. They have been shown to have antiviral properties against a variety of viruses, including influenza, HIV, and herpes simplex virus.

(d) Essential oils: They are aromatic compounds that are found in many plants. They have been shown to have antiviral properties against a variety of viruses, including influenza, herpes simplex virus, and respiratory syncytial virus.
(e) Probiotics: They are live microorganisms that are found in many foods, such as yogurt and kefir. They have been shown to have antiviral properties against a variety of viruses, including influenza and rotavirus.
(f) Mushroom extracts: Certain mushrooms, such as shiitake and reishi, contain compounds that have been shown to have antiviral properties against a variety of viruses, including influenza and hepatitis B.

While natural products have the advantage of being derived from natural sources and are generally considered safe, their effectiveness against viruses may vary depending on the specific virus and the individual receiving the treatment.

The antiviral agents that are commonly utilized to combat significant viruses are thoroughly listed in Table 8.1. Additionally, the host target of each antiviral agent is provided in Table 8.1, sourced from standard reference articles.

8.2 Conclusion

The HDAT holds great promise as a viable strategy to combat emerging viral diseases. With the lack of authorized medications for emerging and re-emerging viruses such as influenza and COVID-19, it is imperative to develop effective therapies to mitigate their global threat. By targeting host factors that viruses exploit for their replication and survival, host-directed therapies offer a broad-spectrum approach that can potentially overcome the limitations of traditional antiviral therapies. Such host-directed antivirals are also likely to offer a higher barrier to the development and selection of drug-resistant mutations. With the global threat caused by several emerging viral agents, efficient therapy against those is urgently necessary.

Table 8.1 Antiviral and antifungal agents used against important viruses (Kumar et al. 2020)

Virus	Antiviral agent	Host target
Dengue virus	Fenretinide (4-HPR)	Activator of retinoid receptors
	Brefeldin A	ADP-ribosylation factor
	UV-4B	ER glycosylation pathway
	FR180204, Ag-126	ERK1/2
	Difluoromethylornithine, diethylnorspermine	Host polyamine synthesis
Ebola virus	Emetine, cephaeline	Lysosomal function
	MLS000394177, MLS000733230, MLS000730532	Macropinocytic uptake
Foot and mouth disease virus	Apigenin	hnRNP A1 and A2
Hepatitis B virus	APG-1387	cIAP2
Hepatitis C virus	HL05100P2, cyclosporine, NIM-811, CRV431, CMX157	Cyclophilin
	Inarigivir (SB 9200)	RIG-I agonist
	AGK2	SirT2
	Phorbol myristate acetate	Synthesis of NF-κB-mediated antiviral protein
	EYP001	Synthetic farnesoid X receptor
	Rintatolimod (Ampligen)	TLR3 agonist
	GS9620, RO6864018, RO7020531, AL-034, imiquimod	TLR7 agonist
	GS9688	TLR8 agonist
	IFN-α, PegIFN-α, Alferon N	TNF-α-mediated antiviral activity
	Sunitinib	AAK1
	HL05100P2, Cyclosporine, NIM-811, CRV431, CMX157, DEBIO-025	Cyclophilin
	PF4620110, LCQ908	DGAT1
	Erlotinib	EGRF and GAK
	TVB-2640	FAS
	Statins	HMG-CoA reductase
	IFN-α-2b, INF-alfacon-1, PegIFN-α-2b Plus ribavirin, PegIFN-α-2b	IFN mediated
	Ribavirin	Multiple modes of action
	Rintatolimod (Ampligen)	TLR3 agonist
	Curcumin, EF-24, FLLL32	Ubiquitin-proteasome, PKC, NF-KB, Akt

(continued)

Table 8.1 (continued)

Virus	Antiviral agent	Host target
Human immune deficiency virus	STI-571, Gleevec, imatinib, nilotinib, dasatinib	Abl family protein kinases
	Maraviroc	Blocks gp120 and CCR5 interaction
	Fattiviracin	Internalization factors
	LJ001	Modulation of membrane fluidity
	Rintatolimod (Ampligen)	TLR3 agonist
	CL097	TLR7/8 agonist
	Genistein	Tyrosine kinase
Herpes simplex virus-1	PD-0332991	CDK4/6
	Dynasore	Dynamin
	U0126, Cl-1040 (PD184352)	MEK1/2
	CGP57380	MNK1
	AG879	NGFR
	Tyrphostin A9	PDGFR
	SD-29	RACK1
	TG100572	Src family kinases
Influenza virus	MK2206	Akt
	U18666A	Annexin A6
	Statins	Anti-inflammatory and immunomodulatory effects
	Verapamil, chlorpromazine	Calcium channel blocker
	Concanamycin, saliphenylhalamide	Cellular vacuolar ATPases
	Clypearin, corilagin, TG003	CLK1
	Silvestrol, pateamine	eIF4A
	UV-4B	ER glycosylation pathway
	Mycophenolic acid, ribavirin	IMPDH
	Ivermectin	Importins
	COX-2 depletion	Induction of type I IFN-mediated antiviral immunity
	AS601245, SP600125	JNK
	U0126, Cl-1040 (PD184352)	MEK1/2
	LJ001	Modulation of membrane fluidity
	AG879	NGFR
	Tyrphostin A9	PDGFR
	Aprotinin, camostat	Protease inhibitor
	Vemurafenib	Raf
	Glycyrrhizin	Regulation of endocytic uptake
	DAS181	Sialic acid receptor
	SIP agonist	Suppression of virus-induced cytokine storm
	PPAR agonist	Suppression of virus-induced cytokine storm and associated lethality in mice
	Verdinexor, DP2392-E10, leptomycin B	XPO1
	Clonidine	α2-Adrenergic receptors

(continued)

Table 8.1 (continued)

Virus	Antiviral agent	Host target
Japanese encephalitis virus	Quercetin	TNF-α-mediated antiviral activity
Lentivirus	FR180204, Ag-126	ERK1/2
Newcastle disease virus	Emetine, cephaeline	Lysosomal function
	Thapsigargin	SERCA
Poxvirus	STI-571, Gleevec, imatinib, nilotinib, dasatinib	Abl family protein kinases
	Gefitinib (Iressa)	EGFR
	CGP57380	MNK1
	LJ001	Modulation of membrane fluidity
	CI1033	ErbB1(EGFR)-mediated antiviral immunity
Zika virus	Fenretinide (4-HPR)	Activator of retinoid receptors
	R448, cabozantinib, nanchangmycin	AXL kinase
	Hippuristanol, silvestrol	eIF4A
	Suramin	Glycosylation (secretory pathway)
	Difluoromethylornithine, diethylnorspermine	Host polyamine synthesis
	Mycophenolic acid, ribavirin	IMPDH
	Daptomycin	Modulation of late endosomal function
	Bortezomib	Proteasome function
	Obatoclax, chloroquine, bafilomycin A1, ammonium chloride, Arbidol (umifenovir), chlorpromazine, niclosamide, mefloquine HCl	Regulation of endosomal pH
	Tenovin-1	SirT1 and SirT2

References

Alagesan S, Brady J, Byrnes D, Fandiño J, Masterson C, McCarthy S, Laffey J, O'Toole D (2022) Enhancement strategies for mesenchymal stem cells and related therapies. Stem Cell Res Ther 13:75

Alfageme-Abello O, Porret R, Perreau M, Perez L, Muller YDJCO i H, AIDS (2021) Chimeric antigen receptor T-cell therapy for HIV cure. Curr Opin HIV AIDS 16:88–97

Aly RM (2020) Current state of stem cell-based therapies: an overview. Stem Cell Investig 7:8

Arend WP (2002) The mode of action of cytokine inhibitors. J Rheumatol Suppl 65:16–21

Arimori T, Ikemura N, Okamoto T, Takagi J, Standley DM, Hoshino A (2022) Engineering ACE2 decoy receptors to combat viral escapability. Trends Pharmacol Sci 43:838–851

Badia R, Garcia-Vidal E, Ballana EJFV d f (2022) Viral-host dependency factors as therapeutic targets to overcome antiviral drug-resistance: a focus on innate immune modulation. Front Virol 2

Björkström NK, Strunz B, Ljunggren H-G (2022) Natural killer cells in antiviral immunity. Nat Rev Immunol 22:112–123

Christian MD, Poutanen SM, Loutfy MR, Muller MP, Low DE (2004) Severe acute respiratory syndrome. Clin Infect Dis 38:1420–1427

Concepcion, F. E., L.-G. Ruth and S. Tom, 2021. Epidemiology and prevention of vaccine-preventable diseases: poliomyelitis
Eisinger RW, Fauci AS (2018) Ending the HIV/AIDS pandemic. Emerg Infect Dis 24:413–416
El Sayed KAJS (2000) Natural products as antiviral agents. Stud Nat Prod Chem 24:473–572
Gao P, Lazare C, Cao C, Meng Y, Wu P, Zhi W, Lin S, Wei J, Huang X, Xi LJJ o h (2019) Immune checkpoint inhibitors in the treatment of virus-associated cancers. J Hematol Oncol 12:1–10
Gardner A, de Mingo Pulido Á, Ruffell BJF i I (2020) Dendritic cells and their role in immunotherapy. Front Immunol 11:924
Girkin JL, Maltby S, Bartlett NWJERR (2022) Toll-like receptor-agonist-based therapies for respiratory viral diseases: thinking outside the cell. Eur Respir Rev 31:210274
https://www.cdc.gov/ (2019a) 1918 pandemic (H1N1 virus)
https://www.cdc.gov/ (2019b) Ebola Disease: 2014–2016 Ebola Outbreak in West Africa
https://www.who.int/, Smallpox
Kumar N, Sharma S, Kumar R, Tripathi BN, Barua S, Ly H, Rouse BT (2020) Host-directed antiviral therapy. Clin Microbiol Rev 33
Lin L-T, Hsu W-C, Lin C-CJJ o t, c. medicine (2014) Antiviral natural products and herbal medicines. J Tradit Complement Med 4:24–35
Mahajan S, Choudhary S, Kumar P, Tomar S (2021) Antiviral strategies targeting host factors and mechanisms obliging +ssRNA viral pathogens. Bioorg Med Chem 46:116356–116356
Musiał-Wysocka A, Kot M, Majka MJC t (2019) The pros and cons of mesenchymal stem cell-based therapies. Cell Transplant 28:801–812
Neurath AJE o v (2008) Immune response to viruses: antibody-mediated immunity. Ency Virol:56
Oberemok VV, Andreeva OA, Alieva EEJIJ o MS (2023) DNA oligonucleotides as antivirals and vaccine constituents against SARS coronaviruses: a prospective tool for immune system tuning. Int J Mol Sci 24:1553
Piret J, Boivin GJF i m (2021) Pandemics throughout history. Front Microbiol 11:631736
Sadeghi S, Soudi S, Shafiee A, Hashemi SM (2020) Mesenchymal stem cell therapies for COVID-19. Current status and mechanism of action. Life Sci 262:118493
Shaw TD, Krasnodembskaya AD, Schroeder GN, Zumla A, Maeurer M, O'Kane CM (2021) Mesenchymal stromal cells: an antimicrobial and host-directed therapy for complex infectious diseases. Clin Microbiol Rev 34:e0006421
Smith DM, Schafer JR, Tullius B, Witkam L, Paust SJSTM (2023) Natural killer cells for antiviral therapy. Sci Transl Med 15:eabl5278
Toor SM, Saleh R, Sasidharan Nair V, Taha RZ, Elkord EJI (2021) T-cell responses and therapies against SARS-CoV-2 infection. Immunology 162:30–43
Vishnupriya M, Naveenkumar M, Manjima K, Sooryasree N, Saranya T, Ramya S, Winster SH, Paulpandi M, Balachandar V, Arul NJERMPS (2021) Post-COVID pulmonary fibrosis: therapeutic efficacy using with mesenchymal stem cells–how the lung heals. Eur Rev Med Pharmacol Sci 25:2748–2751
Wack A (2021) Monocyte and dendritic cell defects in COVID-19. Nat Cell Biol 23:445–447
Wang Y, Liu Y, Tan X, Pan B, Ge J, Qi K, Cheng H, Cao J, Shi M, Yan Z, Qiao J, Jing G, Wang X, Sang W, Xia R, Zhang X, Li Z, Gale RP, Zheng J, Zhu F, Xu K (2020) Safety and efficacy of chimeric antigen receptor (CAR)-T-cell therapy in persons with advanced B-cell cancers and hepatitis B virus-infection. Leukemia 34:2704–2707
Zurier RB, Weissmann G (1973) Anti-immunologic and anti-inflammatory effects of steroid therapy. Med Clin N Am 57:1295–1307

Chapter 9
Role of Herbal Formulations in Antiviral Therapy - an Overview

Amruta Prabhakar Padakanti, Sakshi Arjun, and Naveen Chella

Abstract The incidence and prevalence of various viral infections have risen tremendously and remain a life-threatening risk to people over the past century. Viral outbreaks and reemerging of viral infection are constantly challenging the global healthcare system. To combat crises, vaccines, antibiotics, and broad-spectrum antiviral agents are used and have often shown limited efficacy with unsatisfactory clinical outcomes. The world is looking for a promising alternative medicine that will be more effective with high safety profile. Traditional medicine gained much interest recently with the increase in people's usage after the recent outbreak of COVID-19. In the current chapter, we have tried to cover the role of herbal drugs and their respective formulations in treating various viral diseases. A brief introduction to viral classification, epidemiology, and types of viral infections was provided to understand the topic and its relevance better. Various plant-based materials, including extracts and pure active constituents that are studied by multiple researchers in treating different viral infections, were reported. A discussion on the antiviral formulations currently under various phases of clinical trials, e.g., Kovir, Xagrotin, COVIDEX™, Septilin®, and Sho-Saiko-to (SST) and patents related to herbal antiviral molecules were also discussed with their merits. A comprehensive list of antiviral medications was given for the readers' reference. Finally, the regulatory guidelines related to herbal products and regulatory bodies of India and other countries were enlisted for the complete understanding of the research stage to the marketing of herbal systems with particular emphasis on viral therapy.

Keywords Antiviral herbal medicines · Viral load · Antiviral therapy · Clinical trials · Patent · Herbal regulatory guidelines

---

Amruta Prabhakar Padakanti and Sakshi Arjun contributed equally with all other contributors.

A. P. Padakanti · S. Arjun · N. Chella (✉)
Department of Pharmaceutical Technology (Formulations), National Institute of Pharmaceutical Education and Research, (NIPER) Guwahati, Guwahati, India
e-mail: naveen@niperguwahati.in

N. Kumar et al. (eds.), *Advances in Antiviral Research*, Livestock Diseases and Management, https://doi.org/10.1007/978-981-99-9195-2_9

9.1 Introduction

Illness caused by viruses continues to be one of the major contributors to morbidity and millions of deaths throughout the globe. Outbreaks of influenza, AIDS, COVID-19, and reemerging new variants are dangerous infections and challenges to health sectors (Feehan and Apostolopoulos 2020). A virus is a tiny microbe that propagates and replicates only inside the host body and is responsible for infections and diseases. Their size usually measures between 20 nm to micron and is enclosed with a protective protein coat that comprises a nucleic acid segment (Van Regenmortel 2007). Viruses are classified broadly based on morphology, genome configuration, and mode of replication. According to literature studies (Xu et al. 2011), the Holmes classification, Lwoff, Home, and Tournier (LHT) classification, Baltimore classification, and ICTV classification are the four primary classification systems of viruses, and each has a unique evolutionary background (Fig. 9.1).

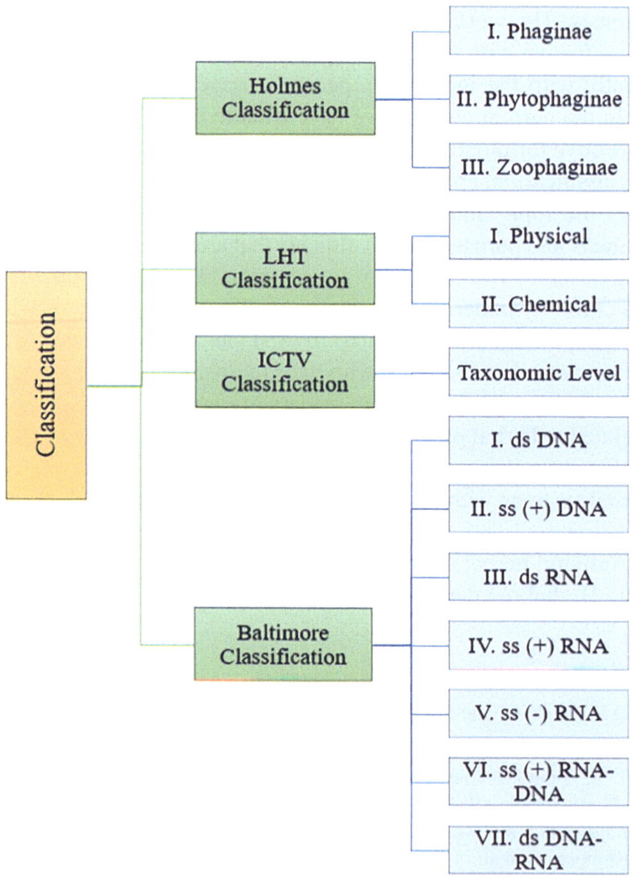

Fig. 9.1 Virus according to Holmes classification, LHT classification, Baltimore classification, and ICTV classification

History of virus taxonomy and early developments of virus classifications system described systemically by *Jens H Kuhn*. In 1948, Francis O. Holmes proposed the first comprehensive classification system based on host tropism. Holmes categorized viruses into three groups: Phaginae, which targets bacteria; Phytophaginae, which attacks plants; Zoophaginae, which infects animals. However, most virologists rejected this system as Holmes excluded the morphological features of viruses (Kuhn 2012). After that, emphasis was given to LHT classification This system was approved in 1962 and is grouped based on the physical (type of symmetry as helical or icosahedral or other morphological types) and chemical (presence of DNA or RNA) properties of viruses (Gibbs 2020) properties. These classification systems have additional flaws and are unsatisfactory since it is often challenging to identify and distinguish virus morphology under the microscope and many viruses that cause the same infection or have a similar appearance. In 1971, David Baltimore outlined the classification of viruses into seven groups according to the modes of genomic replication and type of nucleic acid content. Baltimore classification (BC) gained more attention and remains highly useful for researchers (Koonin et al. 2021). Baltimore's seven classes are described below:

1. **BC Group I:** Viruses consisting of double-stranded (ds) DNA as genomic content (e.g., Adenoviridae, Poxviridae).
2. **BC Group II:** Viruses consisting of positive single-stranded (ss) DNA as genomic content (e.g., Parvoviridae).
3. **BC Group III:** Viruses composed of double-stranded (ss) RNA as genomic content (e.g., Reoviridae).
4. **BC Group IV:** Viruses consisting of positive single-stranded (ss) RNA as genomic content (e.g., Togaviridae).
5. **BC Group V:** Viruses with negative single-stranded (ss) RNA as genomic content (e.g., Rabdoviridae).
6. **BC Group VI:** Viruses with reverse transcribing RNA, consisting of positive single-stranded genomic content that replicates through DNA intermediates during the life cycle (e.g., Retroviridae).
7. **BC Group VII:** This group was later added and is composed of reverse-transcribing DNA consisting of double-stranded genomic content that replicates through RNA intermediates during the life cycle (e.g., Hepadnaviridae).

Later, the virologist addressed initiating an official viral taxonomy in the virology domain. The International Committee on Nomenclature of Viruses (ICNV) was started in 1966 by the International Congress for Microbiology and the International Association of Microbiological Societies (IAMS) (Hull and Rima 2020). The primary objective was to organize, refine, and establish taxonomy of viruses that would be globally recognized and widely accepted. The ICNV was renamed officially into the International Committee on Virus Taxonomy (ICTV) in 1974 (Kuhn 2012). The universal viral classification developed a hierarchical order as species, subgenus, genus, subfamily, family, suborder, order, class, phylum, kingdom, and realm (Lefkowitz et al. 2018). Virus taxonomy starts with the lowest suffix (species) taxonomic rank to the highest (Realm) taxonomic category.

Emerging viral diseases and unexpected outbreaks continue to disrupt the health-care sector, and the difficulty in diagnosis further remains disease management complicated (Mourya et al. 2019). The risk of virus infection that affects the occurrence, cause, and intensity can be determined by the characteristic nature of the virus, the host, and the host population. In addition, behavioral, environmental, and ecological aspects influence the virus spreading from one host to many (Pekar et al. 2022). Epidemiology has come to have a significant role that aims to integrate these aspects into a single, comprehensive data that systematically collects, assembles, evaluates, and provides information in the health sector. Epidemiology is the quantitative measurement study of the disease frequency, distribution, transmission, and control of diseases in the population (Lin et al. 1999, 2018).

Epidemiology of viruses is the study of determinants, dynamics, and distribution of viral infection and disease in the population. It is critical to differentiate between infection and disease because infection represents the introduction and proliferation of a virus within the host body (Monto 2002). In contrast, disease indicates the host's response to infection is adequate to produce a specific recognizable pattern of clinical symptoms. Quantitative assessments of infections and diseases assist in understanding the infection's general nature and alter and direct disease control (Zardini et al. 2021). Parameters used to describe and compare the intensity of disease among various populations are represented in the form of rate and proportions, including incidence, prevalence, morbidity rate, and mortality rate. The incidence rate, also called the "attack rate," is the number of new infections that emerge in people over time (Ng et al. 2021). The prevalence rate is the number of existing disease cases in the population to the size of the population during a specific period. The morbidity rate measures the number of people affected and developed clinical signs for a particular disease. In contrast, the mortality rate measures the number of deaths from the disease at a specific time (You et al. 2020). These rates may be influenced by various attributes that differentiate one rate from another, including age, gender, genetic makeup, immunological condition, nutritional condition, behavior, medical care, and patient management parameters (Hassan et al. 2020).

In epidemiology, endemic, epidemic, and pandemic are used to describe the disease states that affect the human population. Epidemiologists study and investigate the origins of an epidemic and monitor their spread and management to control and prevent the diseases. An epidemiological investigation is designed to be carried out by case-control, cohort, cross-sectional, and long-term herd methods that provide a conceptual framework for understanding the correlation between risk factors and disease incidence and prevalence (Funk et al. 2009).

Virus spread can happen horizontally or vertically. Vertical transmission is the transfer from the mother to the offspring, and horizontal transmission is between animals in the population. Most transmissions spread through horizontal virus transmission (Stewart et al. 2005) which can happen through direct touch, indirect contact, or a shared vehicle. It can also be airborne, zoonotic, vector-borne, or iatrogenic. Some viruses spread naturally through various modalities, whereas others do so just once. The transmitted virus affects specific human tissue and develops clinical symptoms called viral syndrome (Zheng et al. 2022). The most reported human viral syndromes include respiratory tract, skin, gastroenteritis, liver, and ocular infections.

9.1.1 Respiratory Tract Syndrome

The virus that causes localized illness in the respiratory system may be released as airborne particles (aerosols) in saliva produced by speaking or coughing. Large aerosols (> 5 microns) deposited in the nasopharyngeal region can evoke upper respiratory tract infections (URTI) (Eccles 2002). The most common virus associated with URI is Respiratory Syncytial Virus (RSV) (Welliver 2003), Influenza virus (IV) (Nishimura et al. 2005), Parainfluenza Virus (PIV) (Lau et al. 2005), Rhinovirus (RV) (Granados et al. 2017), Enterovirus (EV) (Imamura and Oshitani 2015), Coronavirus (CoV) (Gralinski and Baric 2015), Bocavirus (BoV) (Akturk et al. 2015), and Coxsackievirus (CAV) (Nekoua and Hober 2022). Smaller aerosols (< 5 microns) can reach and deposit in the alveolar lumen (deeper in the lung), causing infection in the lower respiratory system. Influenza Virus (IV), Human metapneumovirus (HMPV), and Severe Acute Respiratory Syndrome (SARS) are the primary virus responsible for lower respiratory tract infections (Wang et al. 2021). Upper and lower respiratory tracts develop common symptoms like cold, runny nose, cough, wheezing, pneumonia, shortness of breath, and asthma (Woodhead 2005).

9.1.2 Gastroenteritis Viral Syndrome

Gastroenteritis is a viral infection or inflammation that affects the stomach epithelium by pathogens. A stomachache, nausea, vomiting, and diarrhea are all symptoms of gastroenteritis. Acute gastroenteritis is one of the most prevalent illnesses affecting children and continues to be a significant cause of the morbidity and death of children all over the globe (Blacklow and Greenberg 1991). Rotaviruses (RV) (Blacklow and Greenberg 1991), enteric adenoviruses (EAdV), Astrovirus (AstV), and Norovirus are the most common cause of viral gastroenteritis (Borrows and Turner 2014).

9.1.3 Central Nervous System Viral Syndrome

When pathogens invade the cerebral, this may sometimes result in ictogenic and epileptogenic brain changes. This is especially true if the invasion includes destroying the blood-brain barrier (BBB). Inflammation in Central nervous system (CNS) infection produces aseptic meningitis, encephalitis, and meningoencephalitis

(Orvedahl et al. 2010). Viral pathogens responsible for developing CNS infection include Rhabdovirus, Orthomyxovirus, Retrovirus, Flavivirus, Herpesvirus (HSV), Measles, Rabies, and EVs. In most cases, the first symptom to appear is fever, followed by agitation, an inability or reluctance to eat, migraines, neck discomfort, and occasionally convulsions (Kramer-Hämmerle et al. 2005).

9.1.4 Ocular Viral Syndrome

Herpes simplex virus (HSV) (Newman and Gooding 2013), varicella-zoster virus (VZV) (Zaal et al. 2001), cytomegalovirus (CMV) (Joye and Gonzales 2018), adenovirus (Garcia-Zalisnak et al. 2018), and vaccinia virus (VACV) (Montgomery et al. 2011) are the major prominent viral species that may affect the eye causing eye infections. Critical indications of eye infections comprise redness, itching, sensitivity to light, and swelling of the cornea, inflammation in the eye.

9.1.5 Hepatitis Viral Syndrome

Hepatitis is a scientific term that refers to the inflammation of the liver. Viral diseases are generally controlled through various strategies, including sanitization using disinfectants and hygiene, elimination through arthropod vectors, maintenance of quarantine, vaccination, and drug-based therapies. Hepatic inflammation is transmitted through the fecal-oral route, water, or food (Aggarwal 2011). The most common types of liver infection include Hepatitis A, Hepatitis B, Hepatitis C, Hepatitis delta virus (HDV), and Hepatitis E virus (HEV). Hepatitis can cause flu, tiredness, a lack of appetite, dizziness, vomiting, stomach pain, joint pain, jaundice, and pain in the liver (Jacobson et al. 2010)

9.1.6 Cutaneous Viral Syndrome

Virus infection that impacts deep skin tissue may develop warts, rashes, swelling, inflammation, edema, pus, and itching on the skin. Varicella Zoster virus (VZV), Human herpesvirus (HSV), Cowpox virus (CPXV), smallpox, measles, Coxsackie A virus (CAV), Human papillomavirus (HPV), Poxvirus, and rabies are few viral species that cause skin infections (Perry and Halsey 2004).

9.1.7 Genitourinary Viral Syndrome

Several virus pathogens occur in the genital tract and, through genital secretions, develop a genitourinary infection. Common pathogens that cause genitourinary infection include (HSV, HPV, HIV, and Trichomonas vaginalis, Treponema pallidum (Ferreira et al. 2007; Masarani et al. 2008).

Viral diseases are generally controlled through various strategies. Primary measures include sanitization by disinfectants and hygiene, elimination through arthropod vectors, and maintaining quarantine contributed significantly (Teshome 2019). Secondary control measures that successfully manage viral infections include using vaccines, antibiotics, broad-spectrum antiviral agents, and chemotherapeutics (Fraser et al. 2004). Unfortunately, after COVID-19, many viral infections emerged as variants or resistant to current measures, thus failing to reach global demand by currently available control therapies or lack of novel discovery (Field 2001). Scientists worldwide immediately began squeezing their brains for a solution to decode the infection, find its source, and develop countermeasures. The globe was set on high alert to prevent the virus from spreading further by using herbal medicine as an alternative measure.

9.2 Evolving and Reemerging Herbal Medicine

Active molecules derived from animals, plants, or other natural sources were extensively used, practiced for medicinal purposes for thousands of years, and reported in ancient Indian, Chinese, Egyptian, Greek, Roman, and Japanese documents (Halberstein 2005). Rigveda, Atharva Veda, Charaka Samhita, and Sushruta Samhita are a few traditional Indian books on herbal medicine (Jaiswal and Williams 2017). Plant-based therapies are the oldest used in various civilizations and are still the most widespread worldwide. An active substance derived from crude plant parts that have therapeutic potential, either raw or processed material, using one or a combination of plants to treat, prevent, cure diseases, or promote wellness, is known as herbal medicine or phytochemical (Calixto 2000). The crude plant parts are extracted, separated, concentrated, and purified to produce a single active molecule essential for therapeutic activity (Sasidharan et al. 2010). Generally believed that morphine was the first isolated active ingredient, as it was discovered accidentally by Freidrich Wilhelm Adam Serturner from crude opium (Brook et al. 2017). In the early nineteenth century, scientists began to isolate and modify pharmacologically active phytochemicals into synthetic molecules. In 1897, the first synthetic drug, aspirin, produced from the willow plant, was marketed widely for its analgesics and antipyretic activity (Mahdi 2010). This discovery sparked the use of synthetic drugs due to their onset of action and began allopathic drug development by pharmaceutical industries to treat complex diseases (Jones 2011). Medical community experts and pharmaceutical companies were not ready to implement herbal medicine as they claimed herbs lack its quality, safety, and efficacy requirements, lack of knowledge, complex nature, poor standardization, nomenclature errors, and difficulties in identifying active substances (Carmona and Soares Pereira 2013). Consequently, synthetic drug-based therapies developed a considerable gap between herbal medicine and healthcare.

Currently, herbal medicines are regaining importance and increasing tremendously as an alternative medicine to synthetic therapies (Lu and Lu 2019). Due to numerous advantages of these products, mainly being safe, non-toxic, have fewer side effects, have excellent compatibility, and are available at reasonable rates (Ekor 2014). According to WHO, 80% global population regularly uses herbal medicine as primary healthcare. About 40–50% population in India still prefers to use herbal medicine as primary care (Sen and Chakraborty 2017). There was also increased knowledge or scientific evidence on the activity of herbal products.

9.2.1 Antiviral Medicinal Plants and Phytochemicals

Most herbal compounds, either as an extract or purified active constituent, were identified and reported in the literature as antiviral medicine. The mechanism behind their antiviral activity was established for a few medicinal plants and is still unclear for many herbal compounds. Based on the literature search, the following compounds were found to be consumed more for their antiviral properties.

Glycyrrhizin (glycyrrhizic acid) is a triterpene glycoside and is the *Glycyrrhiza* species (licorice) root's most investigated bioactive compound. This compound has several reported antiviral activities. Glycyrrhizin is active against coronavirus, enterovirus 71, coxsackievirus A16, dengue virus, Epstein-Barr virus, hepatitis c virus, HIV, influenza A, Japanese encephalitis virus, porcine reproductive and respiratory syndrome virus, respiratory syncytial virus, rotavirus, and VZV (Baba and Shigeta 1987; Ito et al. 1988; Badam 1997; Lin 2003; Wolkerstorfer et al. 2009; Alfajaro et al. 2012; Feng Yeh et al. 2013; Wang et al. 2013; Adianti et al. 2014; Duan et al. 2015; Baltina et al. 2019; Banerjee et al. 2023). Glycyrrhizin has activity against coronavirus by increasing the production of nitric oxide in macrophages, acting on pathways of cell signaling and transcription, lipid-bilayer membrane alteration, and by ACE2 enzyme binding (Banerjee et al. 2023). Still, the mechanism of action remains unclear for remaining viral activities. Identification of the antiviral activity and its mechanism of action to inhibit the virus from herbal medicine or existing ancient literature is needed. Hence, we summarized the herbal medicines against various viral pathogens (Table 9.1).

Table 9.1 Detailed list of herbal compounds used in various viral therapy

S. No.	Natural product evaluated	Action studied for	Proposed mechanism	References
1	Ethanolic and aqueous extracts of the Andrographis polyherbal formulation (Vathasura kudineer)	Chikungunya virus	Inhibition of interleukin-1 beta, pro-inflammatory cytokines, and tumor necrosis factor-alpha production by suppressing the nitric oxide, prostaglandin E2, and inflammatory mediators.	(Hasan et al. 2023)
2	Ethanolic extract of Nilavembu Kudineer	Chikungunya virus, dengue	Unclear	(Jain et al. 2020)
3	Saikosaponins B2 from Bupleurum spp., Heteromorpha spp. and Scrophularia scorodonia	Coronavirus	Intervention in the viral attachment, early-stage viral replication, and penetration.	(Cheng et al. 2006b)
4	Lycorine from Lycoris radiata	Coronavirus	Unclear	(Li et al. 2005)
5	Sinigrin, indigo, beta-sitosterol from Isatis indigotica root extract	Coronavirus	Inhibit protease of coronavirus that mediates the proteolytic processing.	(Lin et al. 2005)
6	Amentoflavone from ethanolic extract of Torreya nucifera leaves	Coronavirus	Inhibit protease of coronavirus	(Ryu et al. 2010)
7	Myricetin, scutellarein from Scutettaria baicalensis	Coronavirus	Inhibition of the helicase protein of SARS-CoV	(Yu et al. 2012)
8	Houttuynia cordata water extract	Coronavirus	Inhibition of SARS-CoV RNA polymerase and protease.	(Lau et al. 2008)
9	Neobavaisoflavone, Bavachinin, 4′-O-methylbavachalcone, Isobavachalcone, Corylifol A, and Psoralidin from ethanolic extract of Psoralea corylifolia seeds	Coronavirus	Inhibit virus replication of SARS by inhibiting papain-like protease enzyme	(Kim et al. 2014)
10	Spirooliganone B extracted from Illicium oligandrum roots	Influenza virus A and coxsackie virus B3 and	Unclear	(Ma et al. 2013)
11	Ursolic acid from Ocimum basilicum (OB) aqueous and ethanolic extract	Coxsackievirus B1, and enterovirus 71	Inhibit infection process and replication	(Chiang et al. 2005)

(continued)

Table 9.1 (continued)

S. No.	Natural product evaluated	Action studied for	Proposed mechanism	References
12	Raoulic acid from Raoulia australis	Coxsackie B3, B4 virus, enterovirus 71, human rhinovirus 2 and 3	Unclear	(Choi et al. 2009)
13	Baicalein from root of Scutellaria baicalensis	Dengue virus	Inhibit adsorption and intracellular replication of the dengue virus	(Zandi et al. 2012)
14	Quercetin	Dengue virus	Inhibit dengue virus replication	(Zandi et al. 2011)
15	Narasin	Dengue virus	Blocks dengue virus protein expression	(Low et al. 2011)
16	Marine seaweed extract of Padina gymnospora, Canistrocarpus cervicornis, Caulerpa racemose, Palisada perforata	Dengue virus	Inhibition of early-stage cycle of virus infection	(Koishi et al. 2012)
17	Gallic acid extracted from flowers of Woodfordia fruticosa	Enterovirus	Unclear	(Choi et al. 2010)
18	Tea polyphenols-epigallocatechin gallate and gallocatechin gallate	Enterovirus	Inhibit enterovirus 71 (EV71) replication	(Ho et al. 2009)
19	Isochlorogenic acid A extracted from Laggera alata	Hepatitis B virus	Unclear	(Hao et al. 2012)
20	Piperine extracted from Piper longum	Hepatitis B virus	Unclear	(Jiang et al. 2013)
21	Dehydrocheilanthifoline extracted from Corydalis saxicola	Hepatitis B virus	Inhibit the replication	(Zeng et al. 2013)
22	Saikosaponins C from Bupleurum species	Hepatitis B virus	Unclear	(Chiang et al. 2003)
23	Polygonum cuspidatum Ethanolic extract	Hepatitis B virus	Unclear	(Chang et al. 2005)
24	Curcumin	Hepatitis B virus	Inhibition of HBV replication and gene expression	(Mouler Rechtman et al. 2010)
25	Tellimagrandin I isolated from methanolic extract from Rosa rugosa	Hepatitis c virus	By inhibiting invasion	(Tamura et al. 2010)
26	Ladanein isolated from Marrubium peregrinum	Hepatitis c virus	Inhibit a post-attachment entry, resistant to an entry inhibitor.	(Haid et al. 2012)
27	(−)-epigallocatechin-3-gallate flavonoid extracted from green tea	Hepatitis c virus	Inhibit early stage of the viral life cycle and virus entry	(Calland et al. 2012)

(continued)

Table 9.1 (continued)

S. No.	Natural product evaluated	Action studied for	Proposed mechanism	References
28	Curcumin	Hepatitis c virus	Inhibited virus replication	(Kim et al. 2010)
29	Silybin A, silybin B, and isosilybin A, isosilybin B derived from Silybum marianum	Hepatitis c virus	Induction of Stat1 phosphorylation	(Polyak et al. 2007)
30	Punicalagin and chebulagic acid isolated from Terminalia chebula	Herpes simplex virus 1	Inhibit glycoprotein-glycosaminoglycan interactions to inhibit cell-to-cell spread and entry of herpes simplex virus 1	(Lin et al. 2011)
31	Proanthocyanidin from Myrothamnus flabellifolia	Herpes simplex virus type 1	Blocking cell surface viral attachment and inhibit virus penetration into the host cell.	(Gescher et al. 2011b)
32	Blackberry extract	Herpes simplex virus type 1	Inhibit viral replication	(Danaher et al. 2011)
33	Aqueous extract from Rhododendron ferrugineum	Herpes simplex virus type 1	Inhibit of viral penetration and adsorption	(Gescher et al. 2011a)
34	Houttuynoids A-E isolated from Houttuynia cordata	Herpes simplex virus 1	Unclear	(Chen et al. 2012)
35	Meliacine isolated from Melia azedarach	Herpes simplex virus 2	Increase level of IFN-gamma and TNF-alpha	(Petrera and Coto 2009)
36	Excoecarianin isolated from the acetone extract of P. Urinaria plant	Herpes simplex virus 2	Inactivate virus particles to prevent viral infection	(Cheng et al. 2011)
37	Hippomanin A from acetone extract of Phyllanthus urinaria	Herpes simplex virus 2	Unclear	(Yang et al. 2007a)
38	Ent-epiafzelechin-(4alpha-- > 8)-epiafzelechin (EEE) extracted from the fresh leaves of Cassia javanica	Herpes simplex virus 2	Inhibit replication	(Cheng et al. 2006a)
39	1,3,4,6-tetra-O-galloyl-beta--D-glucose and Geraniin extracted from Phyllanthus urinaria	Herpes simplex virus 1 and 2	Unclear	(Yang et al. 2007b)

(continued)

Table 9.1 (continued)

S. No.	Natural product evaluated	Action studied for	Proposed mechanism	References
40	Punicalagin and Chebulagic acid and extracted Terminalia chebula	Broad-spectrum antiviral activity for human hepatitis C virus, cytomegalovirus, measles virus, dengue virus, and respiratory syncytial virus.	Inhibit interactions of viral glycoprotein with glycosaminoglycans.	(Lin et al. 2013)
41	Tricyclic coumarin GUT-70	Human immunodeficiency virus	Inhibit replication	(Kudo et al. 2013)
42	Artemisia annua tea infusion	Human immunodeficiency virus	Unclear	(Lubbe et al. 2012)
43	Homoisoflavonoids, sappanone A isolated from heartwood of Caesalpinia sappan	Influenza virus	Inhibit neuraminidase	(Jeong et al. 2012)
44	Xanthones from polygala karensium	Influenza virus	Inhibit neuraminidase	(Dao et al. 2012)
45	Chalcones isolated from the acetone extract of Glycyrrhiza inflata.	Influenza virus	Inhibit neuraminidase	(Dao et al. 2011)
46	Aqueous extracts from dandelion	Influenza virus	Inhibition of virus replication.	(He et al. 2011)
47	Pelargonium sidoides root extract	Influenza virus	Early stage of infection inhibition, impaired neuraminidase activity, and viral hemagglutination	(Theisen and Muller 2012)
48	Extract of Sambucus nigra L.	Influenza virus	Inhibit propagation	(Krawitz et al. 2011)
49	Robustaflavone, Amentoflavone, and agathisflavone isolated from Rhus succedanea and Garcinia multiflora	Influenza A and influenza B viruses	Unclear	(Lin et al. 1999)
50	Aqueous and ethanol extracts of Cajanus cajan	Measles virus	Unclear	(Nwodo et al. 2011)
51	Zanthoxylum chalybeum and Warburgia ugandensis extract	Measles virus	Unclear	(Olilaa et al. 2002)
52	Spicebush	Measles virus	Unclear	(McWhorter 1996)

(continued)

Table 9.1 (continued)

S. No.	Natural product evaluated	Action studied for	Proposed mechanism	References
53	Resveratrol (trans-3,4,5-trihydroxystilbene)	Respiratory syncytial virus	Decrease gamma interferon levels associated with virus-mediated airway inflammation, which is mediated via toll-like receptor 3 signaling	(Zang et al. 2011)
54	Cimicifugin isolated from Cimicifuga foetida	Respiratory syncytial virus	Inhibit viral attachment and viral internalization	(Wang et al. 2012a)
55	Four C-glycosides extracted from Lophatherum gracile leaves	Respiratory syncytial virus	Unclear	(Wang et al. 2012b)
56	Genkwanol B, C and stelleranol extracted from radix Wikstroemiae	Respiratory syncytial virus	Unclear	(Huang et al. 2010)
57	Uncinoside A and B isolated from the whole dried plants of Selaginella uncinata	Respiratory syncytial virus	Unclear	(Ma et al. 2003)

9.3 Research on Antiviral Formulations

Herbal compounds can be processed into a suitable dosage and are intended for delivery to targeted sites of action within the body. Dosage forms can be categorized into three main classes liquid, solid, or semisolid herbal products processed with or without excipients (Byeon et al. 2019). Herbal liquid preparations are intended for internal use. The most commonly available herbal liquid preparations include tea, decoctions, infusions, syrup, emulsions, glycerites, and oxymels (Doris Kumadoh 2017). Herbal compounds or plant crude are suspended in hot boiling water for a specific time and filtered to prepare herbal tea, decoctions, and infusions. The use of herbal Unani decoctions (Fazil and Nikhat 2022), Chinese infusions (Li et al. 2013), and herbal tea (Scarmozzino and Visioli 2020) during the COVID-19 pandemic was suggested by AYUSH for mild illnesses in residential quarantine and as a self-care drink for healthy persons (Ghosh et al. 2020).

Additionally, the effective use of herbal syrup containing *IVY Leaf* dry extract to treat cough in children was reported (Olszanecka-Glinianowicz et al. 2020). Solid dosage is widely available in the form of herbal powders, herbal mixtures, granules, pills, lozenges, tablets, and capsules (Doris Kumadoh 2017). Yinqiao herbal powder, which has significant anti-inflammatory and antiviral effects, is a traditional Chinese medicine approach for treating COVID-19 in China (Zhang and Li 2022). Capsules are oral solid dosage forms that typically encapsulate the herbal compounds in a gelatin container. Herbal products are widely preferred to formulate in solid dosage forms, as it is easy to produce on a large scale and have more extended stability. Most commonly available semisolid herbal dosage forms like creams, pastes, and ointments are intended for topical application (Maver et al. 2015).

However, conventional herbal therapies are poorly soluble and thus have a major limitation in terms of poor bioavailability, high systemic clearance, and poor efficacy. Currently, researchers and manufacturers focus on using novel delivery systems to overcome the problems associated with conventional delivery systems. These novel delivery systems offer many advantages, including improved solubility and bioavailability, sustained drug delivery, and protection of active constituents from physical or chemical degradation. Hence, developing active herbal components into appropriate nanomedicines (liposomes, nanoemulsions, nanospheres, nano micelles, solid lipid nanoparticles, and nanogels, etc.) to deliver in a sustained manner is the current research emphasis (Ansari et al. 2012). Nanoparticles are colloidal submicron entities with an average diameter of 1 to 100 nm. Herbal extract or active components are dissolved, entrapped, or encapsulated into a matrix (Huang and Chang 2009; Nooreen et al. 2018). Generally, polymeric, metallic, and inorganic matrix systems are widely used. Table 9.2 indicates a few dosage forms (conventional and novel drug delivery systems) of antiviral herbal formulations that are currently under research.

9.4 Antiviral Herbal Formulations in Clinical Trials

Comprehensive research study on antiviral herbal formulations has led to the development of various new formulations for managing viral diseases. The below (Fig. 9.2) represents different products currently under clinical research and their current status. All data collected and summarized from *https://clinicaltrials.gov*. This section describes herbal formulations with their composition, route of administration, and indications.

Phase III

The formulations that entered phase III or completed phase III trials are mentioned under this heading.

9.4.1 Kovir (TD0069) Capsule (Registered Number: NCT05044650)

Kovir (TD0069) is a hard gelatin capsule made of polyherbal medicine according to the traditional Ren Shen Bai Du San formula to treat COVID-19 in Vietnam. "Ren Shen Bai Du San" is a conventional polyherbal composition used to treat acute plaque. During the pandemic, Vietnamese Ministry of Health officially recommended this to prevent and treat COVID-19 (Loc et al. 2022a). One Kovir hard capsule consists of 12 herbal mixture compositions, including *Bupleurum chinense*

Table 9.2 Novel herbal formulations on viral therapies

Sr. N	Type of formulation	Natural product evaluated	Action studied for and proposed mechanism	References
1	Powder	Carica papaya leaves extract	Chikungunya virus	(Patil et al. 2022a)
2	Herbomineral pills	KanthaRasaVillai	Hepatitis C virus	(Ranjitsingh et al. 2022)
3	Cyclodextrin conjugate	Conjugate of glycyrrhetinic acid and β-cyclodextrin	Influenza A	(Liang et al. 2019)
4	Nano-formulation	Linga chenduram	Inhibition of the hepatitis C virus replication	(Al-Ansari et al. 2021)
5	Nanoemulsion	Cuphea ignea extract	SARS-CoV-2	(Mahmoud et al. 2021)
6	Cerium oxide nanoparticles (particle size- 14 nm)	Fruit extract of Hyphaene thebaica	Poliovirus	(Mohamed et al. 2020)
7	Chitosan nanoparticles (particle size- 29–39.5 nm)	Curcumin	Inhibition of hepatitis C viral entry and replication	(Loutfy et al. 2020)
8	Silver nanoparticles (particle size- 68.08, 28.38, 37.10 nm)	Andrographis paniculata, Phyllanthus niruri, and Tinospora cordifolia	Chikungunya virus	(Sharma et al. 2019)
9	Silver nanoparticles and supercritical fluid formulation (particle size- 76.27 nm)	Carica papaya leaves extract	Dengue virus type 2	(Patil et al. 2022)
10	Silver nanoparticles (particle size- 50 nm)	Catechin	Inhibition of Japanese encephalitis virus attachment and post-infection spread.	(Chowdhury et al. 2016)
11	Silver nanoparticles (particle size- 42 nm)	Cinnamomum cassia	H7N3 influenza A virus	(Fatima et al. 2016)
12	Silver nanoparticles (particle size- 13.69 nm)	Curcumin	Inactivating respiratory syncytial virus directly	(Yang et al. 2016)
13	Silver nanoparticles	Lampranthus coccineus and Malephora lutea	Herpes simplex virus-1, hepatitis A, Coxsackie B4 virus	(Haggag et al. 2019)
14	Silver nanoparticles (particle size- 5–15 nm)	Panax ginseng root extract	Influenza A virus	(Sreekanth et al. 2018)

(continued)

Table 9.2 (continued)

Sr. N	Type of formulation	Natural product evaluated	Action studied for and proposed mechanism	References
15	Gold nanoparticles (particle size- 11 nm)	Garlic extract	Inhibit measles virus infection by blocking viral particles directly	(Meléndez-Villanueva et al. 2019)
16	Cationic carbon dots (particle size- 4.8 nm)	Curcumin	Coronavirus replication suppression and stimulation of proinflammatory cytokines and interferon-stimulating genes production	(Ting et al. 2018)

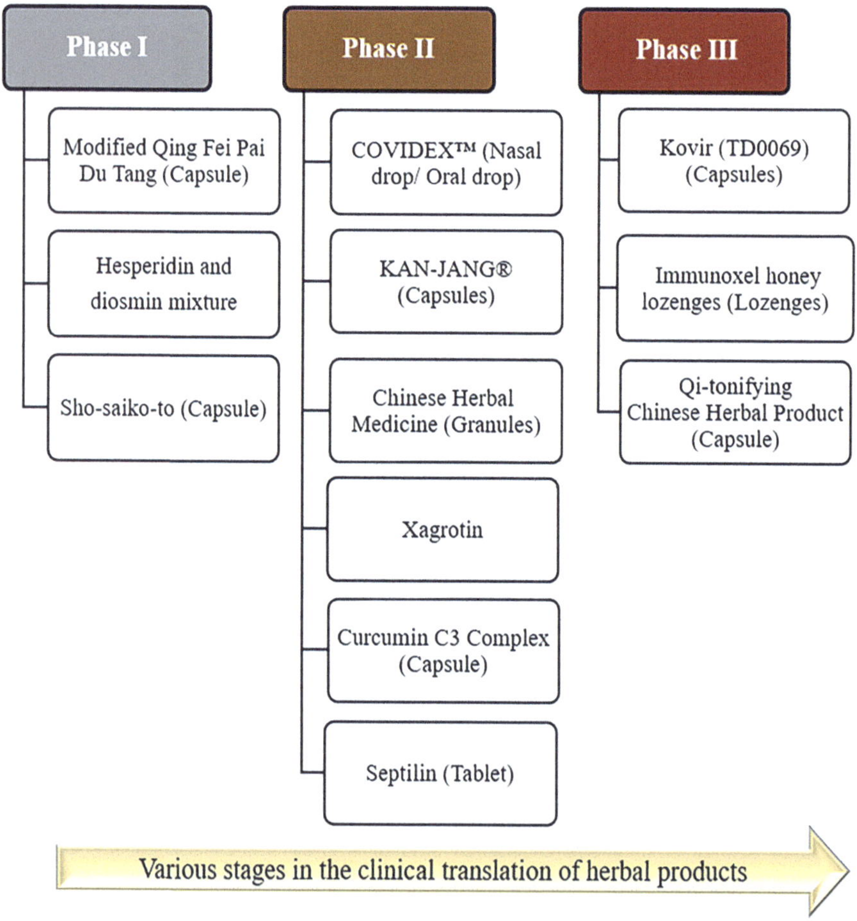

Fig. 9.2 Herbal formulations present under various stages of clinical trial studies

DC and, Nannf, Poria cocos (Schw.) Wolf, Peucedanum praeruptorum Dunn, Glycyrrhiza glabra L, Codonopsis pilosula (Franch., Platycodon grandiflorum (Jacq.) A. DC, Ligusticum wallichii Franch, Citrus aurantium L, Notopterygium incisum Ting ex H. T. Chang, Angelica pubescens Maxim, Zingiber officinale Rose, Mentha arvensis L. It is grouped into four groups Chief (Jun), Deputy (Chen), Helper (Zuo), and Envoy (Shi) aspects. The "chief" aspect majorly prevents chills, dizziness, uncomfortable throat, and muscular discomfort. The Deputy aspect improves the activity of the "chief" components. In contrast, the "Assistant" and "Envoy" components balance the function and can accelerate the body's healing process. The Kovir hard capsule at present is in the phase III clinical trial.

Each capsule comprises 600 mg of finely grounded 12 herbs and 200 mg of excipients such as maize Aerosil, calcium carbonate, magnesium stearate, talc, starch, and sodium benzoate. A randomized study, double-blind, and placebo-controlled clinical trial was conducted to examine the safety and effectiveness of the oral dose of Kovir hard capsule in managing mild COVID-19 subjects. The Sao Thai Duong Joint-stock company sponsored this clinical trial study. The investigation was conducted in SARS-CoV-2 positive individuals of Vietnam aged 18 to 65 with mild or moderate symptoms. Sixty-six subjects were categorized into two team groups, in which 34 were nominated to take Kovir capsules, and the remaining 32 participants received an empty capsule. Both Subjects received either capsules or the same amount of placebo as three capsules three times a day for a maximum of 14 days. Capsule potency was characterized by the intention to treat the database on viral clearance and symptom improvement. The safety studies of the Kovir capsule described the treatment dataset on adverse events. The study observed viral RNA load and viral clearance decreased over time. The Kovir experimental group exhibited more rapid recovery and reduced symptom grades than the placebo. In the entire study, Kovir individuals did not experience any adverse events. Thus, it concluded capsules are safer to use and can improve symptoms in COVID-19 patients (Loc et al. 2022b).

9.4.2 Immunoxel Honey Lozenges (Registered Number: NCT01061593)

Immunoxel is a polyherbal product with immunostimulatory activity. The Ministry of Health of Ukraine approved it as a dietary supplement in 2006. It is the most recognized polyherbal medicine in Ukraine, manufactured into a suitable dosage form by Ekomed company, and used widely to treat chronic bacterial and viral infections (tuberculosis or HIV) (Efremenko et al. 2012). Immunoxel is composed of hydroalcoholic extracts of various herbal into one blend in suitable proportions, including *Aloe arborescens, Glycyrrhiza glabra, Rosa canina, Oreganum majorana, Thymus vulgaris, Salvia officinalis, Foeniculum vulgare, Taraxacum officinale, Urtica dioica, Plantago major, Calendula officinalis, Echinacea purpurea,*

Polygonum aviculare, Artemisia sp., *Achillea millefolium, Centaurium erythraea, Inula helenium, Potentilla erecta, Gnaphalium uliginosum, Bidens tripartite, Rhodiola rosea, Inonotus obliquus, Inonotus obliquus, Hippophae rhamnoides, and Juniperus communis* (Arjanova et al. 2009). Further, with the support of Island Abbey Food Ltd., designed for oral delivery sublingual honey lozenges using Immunoxel as the active ingredient. Each lozenge contains 300 μL Immunoxel dose. The dosage form proved stable for 3 years at room temperature. A phase III clinical trial is completed to investigate and confirm the clinical benefit of Immunoxel sublingual honey lozenges.

Phase III investigation was carried out for 30 days by random, double-blind, placebo-controlled study with patients diagnosed with pulmonary tuberculosis in Mongolia and Ukraine. The study was conducted with a different group of patients diagnosed with pulmonary tuberculosis, including drug-susceptible tuberculosis (DS-TB), extensively drug-resistant tuberculosis (XRD-TB), multidrug resistance tuberculosis (MDR-TB), and tuberculosis and HIV coinfection (TB-HIV). These patients were randomly categorized into two groups: 137 patients were assigned group one and administered a once-in-a-day sublingual Immunoxel honey lozenge. In contrast, the remaining 132 patients were treated with placebo lozenges and regular antituberculosis therapy (ATT). Evaluation of the study was done by identification of mycobacterium tuberculosis in a sputum smear using the microscopic examination. HIV infection was characterized by a count of CD4 T cells.

A negative sputum smear was observed in 87 patients out of 132 (65.9%) who received Immunoxel honey lozenge, whereas only 32 out of 127 (25.2%) were in the placebo group. In the immunotherapy group, it was reported that the overall body mass improved by 2 kg, whereas, in the blank group, the increase in weight was only 0.6 kg. Absolute CD4 lymphocyte cell count was improved notably within individuals who were treated with Immunoxel lozenges from 309 ± 156 to 417 ± 159 (p 0.0001), but it still dropped in control patients, from 306 ± 228 to 271 ± 211 ($p = 0.53$). ESR and leukocyte counts confirmed that Immunoxel lozenges had a significant anti-inflammatory impact. The study demonstrated a positive outcome of Immunoxel honey lozenges being safe to use, as it has not produced any adverse effects, are effective against virus and tuberculosis infections, and are easy to administer (Batbold et al. 2017).

9.4.3 Qi-Tonifying Chinese Herbal Products (Registered Number: NCT01631032)

Qi-tonifying is a traditional Chinese herbal formula used clinically to control and treat various diseases and promotes overall human strength. Bu-Zhong-Yi-Qi-Tang (BZYQT) is a well-known qi-tonifying herbal product proven for its immune-modulating activity in treating allergic rhinitis or other complex conditions (Liu et al. 2019). The composition and type of herbs used in the Bu-Zhong-Yi-Qi-Tang

(BZYQT) formula are not disclosed. A capsule was developed with 500 mg of Bu-Zhong-Yi-Qi-Tang (BZYQT) and a few excipients.

The double-blinded, randomized, and placebo-controlled experimental design was carried out with 120 enrolled intermittent allergic rhinitis patients aged between 20 to 50 years. In group one, 60 individuals were assigned to receive a BZYQT capsule 3 gm (6 pieces of 500 mg capsule) three times a day. In group two, the remaining 60 individuals were assigned to receive a placebo capsule with a similar quantity to the experiment group for 3 months. The study was evaluated once a month, and data were collected statistically by the scoring system for the intensity of allergic rhinitis symptoms and the visual analog scale for fatigue severity. Additionally, several parameters on an immunological response, like the amount of IgE, CD4/CD8, T lymphocytes, cytokines, PGE2, LTC4, and COX, were also screened. This study was completed successfully and concluded with a positive outcome.

Phase II

In this section, case studies currently under Phase II clinical trial studies are provided.

9.4.4 Covidex™ (Registered Number: NCT05228626)

In 2021, SARS-CoV-2 remained as the pandemic and mortality rates drastically increased worldwide. Available treatments and protective vaccines were insufficient to overcome the pandemic. Researchers and scientists were urgently rushed to find solutions to avoid viral spread to prevent and cure illness. Based on reports released by the Uganda Ministry of Health, in the year 2021, Uganda battled its worst COVID-19 epidemic with mild symptoms. Professor Patrick Engue Ogwang introduced Covidex™ herbal remedy at Mbarara University of Science and Technology. The Uganda National Drug Authority has approved the herbal formulation Covidex™ to be used as a complementary therapy to cure viral infections. It contains berberine as an active ingredient to treat SARS-COV-2 (Cacciabue et al. 2022). Covidex™ formulation incorporated two more antiviral agents, *Zanthoxylum gilleti* and *Warbugia ugandensis,* along with some alkaloids, saponins, and sugars. Covidex™ available in nasal or oral drops and manufactured by Jena Herbals Ltd. An experiment was conducted to investigate the safety with efficacy profile of Covidex™ in managing viral infection in Uganda.

In March 2022, a random, placebo-controlled, double-blinded, and interventional assignment was carried out in collaboration with Makerere University and Mbarara University. Recruited 584 COVID-19 positive individuals above 18 years and grouped them into two equal categories. Group one, assigned as control, received placebo drops as an experiment for seven days. Group two was given to individuals treated with Covidex™ three times a day, either given two drops through nasally with the addition of six drops orally or six drops orally, continued for seven days. Evaluation of investigation was done by adverse events score, viral load, and presence of antigen in the nasopharynx. Unfortunately, Covidex™ safety and efficacy outcome is still unknown on viral infection.

9.4.5 Kan-Jang® (Registered Number: NCT04847518)

Despite the complexity of viral respiratory infection in COVID-19, research must develop various therapeutics. Traditional herbal preparations have proven their potential efficacy in curing viral infections in ancient literature. Therefore, the blend of conventional herbal extracts, *Eleutherococcus senticosus (Rupr. & Maxim.) and Andrographis paniculata (Burm. F.) Wall. ex. Nees* incorporated into a capsule in a fixed amount called Kan-Jang® (Mkrtchyan et al. 2005). Both herbs exhibit antiviral activity, immunostimulant, and anti-inflammatory activity in respiratory tract viral infections (Spasov et al. 2004).

Owing to EMEA guidelines and good manufacturing practices, Kan-Jang® was formulated into an appropriate dosage form, optimized, evaluated, and assessed for clinical application at the Swedish Herbal Institute. Each capsule of Kan-Jang® contains 195 mg of *Andrographis paniculate,* 11.4 mg of *Eleutherococcus senticosus,* 10 mg of 14-deoxy-11,12-didehydroandrograholide, andrographolide, and diterpene lactones along with inactive ingredients. Phase II was carried out to produce scientific proof of the effectiveness and tolerability of the Kan-Jang® capsule against respiratory viral infection.

A randomized, quadrupole-blinded, and placebo-controlled phase II study was carried out by enrolling 140 COVID-19 diagnosed participants with URTI (Ratiani et al. 2022). A qualified pharmacist (QP) randomly divided the 140 participants into groups A and B and kept confidential details about such groups. To group A, 70 participants, QP distributed a treatment code of two capsules of Kan-Jang® supplements thrice a day for up to 2 weeks. Group B remaining 70 participants received two capsules of inactive excepts with the same strength for 2 weeks. The main effectiveness endpoints were lowering of the acute stage with the average URTI syndromes scoring scales, such as throat infection, watery nose, difficulty breathing, hoarseness, cough, dizziness, pain, and fatigue. Kan-Jang® capsules noticeably improved the recovery rate by significantly lowering the rate of sick leaves by more than 21% compared to placebo groups. Also, Kan-Jang® capsules significantly reduced the severe UTRI symptoms.

9.4.6 Chinese Herbal Medicine (Registered Number: NCT05269511)

In Hong Kong, randomized interventional, open-labeled control analysis was designed to investigate the safety and effectiveness of Chinese herbal medicine (CHM) and vitamin C combination to avoid the proliferation of COVID-19 among many hospital healthcare practitioners who directly come into contact with COVID-19-infected patients. CHM is a traditional medicine of China that mostly contains granules blend (9 gms/ sachet) of 10 different herbs *Astragali Radix, Glycyrrhizae, Schizonepetae Herba, Codonopsis Radix, Saposhnikovia Radix, Mori Folium, Poria, Atractylodis Rhizoma, Atractylodis Macrocephala Rhizoma, Radix et Rhizoma, Lonicerae Japonicae Flos.* The investigation aimed to study the potential therapeutic activity and tolerability of CHM alone or combined with vitamin C. Around 652 COVID-19-negative individuals were randomly selected. The investigation was done by excluding participants with a history of another severe disease, nephropathy, or hypersensitivity. Experimental participants were assigned to consume CHM (one sachet) in combination with Vit C (1000 mg per day) mixed in 150 mL of water after meals for 14 days. The remaining placebo participants were assigned to consume only vitamin C 1000 mg for 14 days. The study did not post the study outcomes.

9.4.7 Xagrotin (Registered Number: NCT05017493)

Xagrotin is an herbal compound that claimed its potential activity in COVID-19. Phase I and phase II trial studies demonstrated an understanding of the impact of Xagrotin herbal compound activity on different features, including gender, age, disease severity, lifestyle, and concurrent illness, on the results. Six hundred sixty-one subjects were enrolled for the interventional, multi-center, randomized controlled study. All the subjects were adults and diagnosed with COVID-19. Three hundred sixty-one subjects were recruited to receive the standard Xagrotin herbal compound, and 178 participants were chosen randomly to take a placebo for 1 month. Mortality after 30 days of treatment was the primary outcome measurement, and the time for the presence and duration of the symptoms of hospitalization was measured as a secondary outcome. The study was completed in December 2021, and the results were not disclosed.

9.4.8 Curcumin C3 Complex (Registered Number: NCT04266275)

Clinical evidence demonstrated that a rise in cervical cancer must be correlated with HPV. Women specifically diagnosed with HPV 16/ HPV 18 must develop metastatic cervical carcinoma. Immunosuppression increases the rates of cervical cancer and HPV infection, especially in HIV-positive women. There are limited or no proper therapies clinically available. Curcumin is a traditional medicine extracted from *Curcuma longa* and is widely used for its anti-inflammatory and anti-infectious properties. Emerging research indicated that curcumin could potentially decrease the tumor's early signs and act as a chemo-preventive by targeting various cellular pathways in human cancer cells.

Curcumin C3 complex capsules were formulated by Sabinsa cooperation to treat cervical intraepithelial cancer and HPV infection. Women diagnosed with symptoms of low-grade squamous intraepithelial lesions (LSIL) and high-grade squamous intraepithelial lesions (HSIL). It also indicated the potential effect of curcumin capsules on HIV-infected with mild symptoms of HPV infection and uninfected women. An interventional, random, triple-blind placebo-control phase II study was carried out to assess the potency of the curcumin C3 complex. Two hundred female participants were recruited, among which the experimental study assigned participants given 2000 mg of curcumin capsule once a week to administer intravaginally for 20 weeks by excluding the menstrual period. Remain participants were assigned to take a 2000 mg placebo capsule for 20 weeks. Intravaginal administration was instructed to the subject to reduce errors in the study. The study was evaluated for 6 months by scoring cervical cytology and HPV clearance test. The outcome of this study was not disclosed.

9.4.9 Septilin® (Registered Number: NCT02013323)

A polyherbal preparation named Septilin® tablets constitutes powders of *Shankha bhasma and Balsamodendron mukul* along with hydroalcoholic extracts of *Emblica officianalis, Maharasnadi quath, Moringa pterygosperma, Tinospora cordifolia, Glycyrrhiza glabra,* and *Rubia cordifolia.* Septilin® 500 mg tablets as oral dietary health supplements are manufactured by Himalaya Drug Company Private Ltd., India (Daswani and Yegnanarayan 2002). It is claimed to be effective in chronic URTI, cutaneous infection, dental infection, and as an immunomodulator (Khanna and Sharma 2001). Septilin® phase II trial was aimed to examine the systemin immunomodulator activity and microbiological efficacy in chronic periodontitis. In 2013, a quadruple-blinded, randomized, placebo-controlled interventional study was performed by recruiting 60 subjects. Septilin® group was assigned to receive two tablets thrice a week, whereas the placebo group received control tablets thrice a day. Post-treatment was monitored for clinical and microbiological activity at baseline for 10 days, 1 month, and 6 months. Evaluation of the study was characterized by scoring Plaque Index (PI), percent bleeding, Gingival Index (GI), and Probing depth (PD). Compared to the placebo group, positive outcomes from clinical trials of Septilin® proven dual effect on the immune system and suppress periodontitis.

Phase-I

Apart from antiviral herbal formulations mentioned in phase III and phase II trial studies, a few formulations like modified Qing Fei Pai Du Tang (mQFPD), Hesperidin & diosmin mixture, and Sho-Saiko-to are now being examined in phase I clinical trial. Data collected from *https://clinicaltrials.gov.*

9.4.10 Modified Qing Fei Pai Du Tang (mQFPD) (Registered Number: NCT04939415)

Modified Qing Fei Pai Du Tang (mQFPD) is a Chinese herbal médicine composed of 21 different herbs. It includes a blend of *Rhizoma Zingiberis Recens, Flos Farfarae, Herba Eohedre, Poria, Radix Bupleuri, Raw Gypsum, Semen Armeniacae Amarum, Ramulus CinnamomiFructus Auranii Immaturus, Polyporus Umbellatus, Rhizoma Aslimatis, Rhizoma Belamcandae, Rhizoma Atractylodis, Radix Glycyrrhizae, Macrocephalae, Pericarpium Citri Reticulatae, Radix Scutellaria, Rhizome Pinelliae Preparata, Radix Asteris, Herba Asari, Rhizoma Dioscoreae, and Herba Pogostemonis.* mQFPD capsules are used to treat a viral infection like COVID-19 with mild to severe symptoms. This triple-blinded randomized analysis aims to characterize the preclinical tolerability and potency of mQFPD compared to placebo in 66 participants. The experimental group received modified QFPD eight capsules thrice a day for up to 14 days. A placebo-controlled group received organic brown rice as a control capsule for 14 days. mQFPD capsule improved antiviral activity through different signaling pathways and promoted immunostimulant activity vial T and B lymphocyte cell function, indicating positive outcomes with respect to efficacy and safety.

9.4.11 Hesperidin and Diosmin Mixture (Registered Number: NCT04452799)

Hesperidin is a traditional herbal blend used for its anti-inflammatory and antioxidant properties. Also reported, Hesperidin potently acts to cure COVID-19 by inhibiting virus entry through ACE2 blockage (Haggag et al. 2020). Hesperidin, in combination with diosmin coadministration synergistically, reduces the complications in SARS-CoV-2 (Utomo 2020). An early clinical investigation was conducted to characterize the role of combination in treating infection. A randomized, double-blinded, and placebo-controlled parallel study was carried out. Thrice a day 1000 mg of the dose was given to group one for one week. Outcomes were measured with negative PCR, respiratory rate, serum IL-1 β, and TNF-α level. Herbal combination therapy concluded speedy recovery and lowered symptoms.

9.4.12 Sho-Saiko-to (SST) (Registered Number: NCT00633230)

Sho-Saiko-to (SST) is a Japanese traditional medicine made of seven different herbs (Ohtake et al. 2004). Sho-Saiko-to (SST) is widely used to suppress chronic viral liver diseases and liver cancers (Chang et al. 2007). It is formulated into an oral capsule by incorporating 700 mg of SST extract and 28 mg of inactive ingredients. A randomized quadruple study was conducted to evaluate three major objectives (a) cytokine production through various parameters, viral load, and liver enzyme levels, (b) In Vitro response of SST on target, and (c) to analyze the mechanism of action SST on disease target. Subjects were categorized as the experimental group given three capsules twice a day and compared with a placebo group. SST slightly and effectively reduced IL-10 levels and increased IL-4 and IL-5 levels over average limits. SST was also reported to reduce viral load, promote lymphocytes and T-helper cell count, and maintain liver enzymes within normal limits. Thus, SST can lower the risk of liver damage and promote potential benefits for hepatitis C viral infection.

9.5 Patented Antiviral Herbal Formulations

In recent years, several patent applications have been filed on herbal antiviral compositions through various institutes and corporations. Despite there being very few applications available to reach the world market due to poor novelty in herbal preparations and limited scientific evidence on pharmacological activity. However, few manufacturers and institutes generate scientific evidence through In Vitro and In Vivo studies. Data were collected from google patent (https://patents.google.com), and comprehensive patents on antiviral herbal preparations are described in detail below

9.5.1 Antiviral CHM Composition, Its Preparation Method, and Use

Patent No. JP2020519701A, Status- Active

The patent application was granted on February 10, 2022. The invention relates to antiviral Chinese herb composition prepared using Chinese medicinal materials, namely Zikoppi and Sakuhaku, with the primary drug substance being sardine, juniper bud, sanjiko, cuckoo, and juniper. This invention discloses the formulation of herbal compounds to many pharmaceutically acceptable dosage forms, such as powders, granules, pills, dragees, tablets, enteric-coated tablets, film-coated tablets, capsules, oral solutions, and suspension. As per the results of In Vitro pharmacodynamic tests and In Vivo experiments, it is proven that CHM has a broad-spectrum antiviral effect against influenza A, H1N1, H7N7, and H9N2 viruses, Zika virus, dengue type I and dengue type II viruses, chikungunya virus, etc.(ホンチャン 2020).

9.5.2 *Treatment of Hepatitis C*

Patent No. WO2019204471A1; Status - Published

The patent was published on October 24, 2019. This invention is used to treat hepatitis C virus infection. It comprises herbal extracts of herbs such as Areca catechu, Albizia julibrissin, Ficus pumila Linn, and Vaccaria segetalis formulates as tablets, caplets, capsules, gel caps, pills, troches, suspensions, dispersions, syrups, solutions, granules, transdermal patches, beads, gels, pellets, powders, magmas, creams, lozenges, plasters, pastes, suppositories, discs, lotions, liquid sprays for nasal or oral administration, aerosolized powder formulation for inhalation in intravesical administration. The invention also mentions that developed formulations can be administered through various routes, including oral, parenteral, transmucosal, transdermal, intrapulmonary, intravesical, intragastrical, intraduodenal, intrathecal, intramuscular, intraperitoneal, subcutaneous, intradermal, intra-arterial, intrabronchial, intravenous, inhalation, and topical administration (Cheng 2019).

9.5.3 *Antivirus Chinese Herbal Medicine Composition, Traditional Chinese Medicine Composition, and Preparation Method Thereof*

Patent No. CN102068492B, Status- Active

The patent application was granted on July 4, 2012. This invention comprises of herbal composition, mainly flos lonicerae flower (4–6 parts), milkvetch root (4–6 parts), the root of baical skullcap (3–5 parts), perilla leaf (2–4 parts), and indigowoad root (3–5 parts). The invention mentions that the said composition can be used clinically in clearing lung obstruction, treating acute URTI and influenza with significant efficacy due to its benefits and strengthening vital qi, clearing heat toxins, and eliminating pathogens properties (Cai et al. 2012).

9.5.4 *Composition for Prevention and Treatment of Influenza Virus and Composition for Inhibiting the Activity of Neuraminidase Comprising Extracts of Turmeric*

Patent No. KR101566441B1

The patent was published on November 5, 2011. This invention claims about a pharmaceutical composition containing neuraminidase activity inhibitor as an active ingredient used for the prevention and treatment of influenza infection. The active component, according to the claims, was either a urogram or a turmeric extract, or a fraction thereof. The fractions may be a water fraction, ethyl acetate fraction, or hexane fraction. The composition was formulated to oral and parenteral formulations such as powders, granules, tablets, capsules, pills, solutions, syrups, suspensions, emulsions, sterilized aqueous solutions, nonaqueous solutions, lyophilized powders, and suppositories (이우송 et al. 2015).

9.5.5 *Medicine Composition for Preventing and Treating AIDS and its Preparation Process and Use*

Patent No. CN101020027A, Status- Expired

The invention was published on April 26, 2010. This patent discloses a composition for preventing AIDS, its preparation process, and its use. This composition includes the active and auxiliary components. Active ingredients include Radix Ilicis Pubescentis (5–60%), Rhizoma Polygoni Cuspidati (5–60%), Radix Glycyrrhizae (5–60%), Herba Epimedii (5–60%), Radix Astragali (0–50%), Fructus Ligustri Lucidi (5–60%), Fructus Arctic (0–50%), Rhizoma Curcumae (0–50%), Radix Arnebiae (0–50%), Rhizoma Drynariae (0–50%). The mass percent for complementary elements is 0–50%. The complementary components include Carapax Trionycis, Radix Scutellariae, Herba Violae, Herba Taraxaci, Herba Cistanches, Herba ecliptae, Flos Chrysanthemi Indici, Fructus rhois chinensis, Fructus Lycii, Spica Prunellae, Radix Salviae Miltiorrhizae, Radix Angelicae Sinensis, and Radix Ginseng. This invented formulation may be prepared into tablets, granules, capsules, oral liquid, or injections. This formulation helps to inhibit HIV replication by delaying disease, promoting immunologic reconstruction, and attenuating resisting therapy (Hu 2007).

9.5.6 *Herbal Compositions for the Inhibition of Human Immunodeficiency Virus (HIV)*

Patent No. US20090053335A1, Status- Abandoned

This patent was published on February 26, 2009. The invention discloses an herbal composition comprised of an organic extract of Hedyotis diffusa, Scutellaria barbata, Lonicera japonica, Prunella Vulgaris, and Solanum nigrum for the inhibition of HIV. It can be prepared into oral liquid, granules, and tablets. The invention also claims a significant reduction in the virus load in HIV patients after 2 months of treatment with this herbal composition in a clinical trial study (Chih-hui 2009).

9.5.7 Polyherbal Compositions and Methods for Treating Viral Infections

Patent No. WO2006078424A2

This patent was published on July 27, 2006. This invention is regarding the herbal extracts of Withania somnifera, Mangifera indica, and purified Shilajit (rock exudated extract). It is beneficial as an immune-supporting and antiviral agent. Further studies proved that this polyherbal composition decreases the mean HIV viral load, which can be used as an adjuvant to conventional antiviral drug therapy (Goshal 2006).

9.5.8 Antiretroviral Herbal Formulation

Patent No. WO2006008761A2

The patent was published on January 26, 2006. This antiretroviral synergistic herbal combination is formulated by extracting and pulverizing Glycyrrhiza glabra (root), Allium sativum (bulb), and Azadirachta indica (seed kernel). The pulverized fine powder is further extracted with cold water, and the filtrate is collected and air-dried. The acetic acid solution is prepared from the extract of Annona Squamosa, Pichcorhiza kurroa, Piper betel, Tinospora Cordifolia, and Withania Somnifera. The synergistic herbal extract is prepared by soaking above obtained air-dried filtrate and acetic acid solution of extract at room temperature. This invention boosts the immune response and inactivates the growth and replication of human retroviruses by increasing CD 4 and CD 8 cells count. This non-toxic invented formulation can be taken orally as a drug or vaccine by children and adults. However, bolus administration will be the effective route of administration with a dosage range of 4 to 100 g per day, more specifically, 1.0 to 50 g per day. Preferably, the composition is administered at least three times per day (Raja 2006)

9.6 Clinically Available Antiviral Herbal Formulations

This segment comprises extensive details regarding commercially approved herbal formulations. Currently, a few number of herbal marketed formulations are available to treat SARS-Cov-2, Hepatitis, HSV, Measles, respiratory tract viral infections, and immunomodulators. These products are available in oral solid dosage forms (granules, tablets, capsules, and powder) and decoction. Commercially, these herbal formulations are available as dietary supplements and health promoters and are enlisted in (Table 9.3).

Table 9.3 Commercially available various herbal products to treat a viral infection

Sr. N	Type of formulation	Natural product evaluated	Action studied for	Proposed mechanism	References
1	Tablet	Ayurvedic polyherbal formulation-NOQ19	SARS-Cov-2	Reduce and inhibit viral load	(Kanchibhotla et al. 2022)
2	Capsules	Xiao chai hu tang	Coxsackie B1 virus	Inhibit viral replication early stage by inducing the interferon expression	(Cheng et al. 2006c)
3	Granules	Sho-saiko-to (Xiao-Chai-Hu-Tang)	Hepatitis B	Inhibition of the replication of hepatitis B	(Chang et al. 2007)
4	Capsules	Yin Chen Hao Tang	Herpes simplex viruses 1 and 2	Direct inactivation of the virus infectivity	(Cheng et al. 2008)
5	Radix gentianae decoction	Long Dan Xie Gan Tan	Herpes simplex virus 2	Inactivation of the virus infectivity	(Cheng et al. 2008)
6	Tablet	Shengma-gegen-tang	Measles virus	Inhibit measles virus by secreting cytokine TNF-alpha	(Huang et al. 1997)
7	Powder	Sheng-Ma-Ge-gen-Tang (Shoma-kakkon-to)	Respiratory syncytial virus	Stimulate secretion of cytokines like IFN-β and TNF-α which inhibit viral attachment	(Wang et al. 2011)

9.7 Need for Regulations in Herbal Medicine

Medicinal herbs have a wide diversity of bioactive compounds which may alter within the same species of plants (Sahoo et al. 2010) based on the factors like physical, chemical, and environmental factors (including soil composition and fertilizer type, geographical diversity, plant maturity, harvesting techniques, and handling and preservation) (Pant et al. 2021). Most people assume that herbal medicine is natural and has no adverse effects; however, this is not always correct (Giveon et al. 2004). Literature scientifically demonstrated that plant-based medicines and related clinical applications could produce harmful and undesirable consequences if the plant material is poor (Niggemann and Gruber 2003). Thus, regulating the quality and efficacy of crude plant material concerning bioactive compounds is challenging and tedious (Bagade and Patil 2021). Most herbal medicines are not subjected to preclinical and clinical analysis, and the lack of information on bioavailability, toxicity, efficacy, and safety profile. As a result, many herbs fail to reach the global market (Zhou et al. 2019). The quality of crude plant material or herbal medicine directly affects the safety and efficacy (Street et al. 2008). Quality is crucial for production and sale and is the primary concern to consumers and the public health sector. It is controlled by regulatory bodies of respective countries and globally by various organizations. Regulatory bodies can implement laws, rules, and harmonization to improve herbal medicine's quality, safety, and efficacy to bring greater quality of living by promoting the economy in developing countries and health care (Sen et al. 2011).

Besides these quality control and safety requirements on herbal drugs, another major limitation in the Indian herbal drug industry is the lack of production and poor commercialization of herbal medicine across the country (Sahoo and Manchikanti 2013). Inadequate research and development, poor focus on brand promotion, and minimal investments are severe concerns for the poor progress of the Indian herbal sector. Global harmonization in the development and implementation of herbal medicines minimizes delays and promises an inevitable approach and desirable benefits from a clinical and commercial perspective (Vaidya and Devasagayam 2007; Wei et al. 2020). Thus, it is essential to be aware and understand various aspects of the regulatory guidelines and the requirements of India and the world.

9.7.1 Herbal Drugs Regulation and Guidelines in India

9.7.1.1 Drug and Cosmetic Act 1940 and its Rules 1945

In India, plant-based therapeutics and their formulations are governed by the Drug and Cosmetic Act (D and C) 1940 and its Rules 1945. According to the D and C act of 1940, amendments were made in 1964 in chapter four to regulate, import, manufacture, distribute, and commercialize Ayurveda, Siddha, and Unani (ASU) medicines (Singh and Shukla 2021). D and C Acts described each regulation for (ASU) medicines categorized into various sections, rules, and schedules, as mentioned in Table 9.4 (Mukherjee 2019).

D and C Act 1940 established different administrative bodies for the smooth and transparent implementation of the rules in the country. Chapter IVA of the Drugs and Cosmetic Act 1940 describes provisions relating to ASU drugs. This chapter describes 18 sections to explain the various aspects of ASU medicine. According to sect. 33-C, the Central Government of India established Ayurvedic, Siddha, and Unani Drug Technical Advisory Board (ASUDTAB) to guide and assist the central and state governments on technical errors (Mukherjee 2003). Also, as mentioned in sect. 33-D, Ayurveda, Siddha, and Unani Drugs Consultative Committee (ASUDCC) were formed by the central government as an advisory committee to advise and maintain uniformity throughout India (Sharma and Pundarikakshudu 2019). Also, to conduct testing of raw materials and herbal products, the Central Drugs Laboratory (CDL) at Kasauli, State Drugs Control Laboratories (SDCL), and Government Analysts was established.

Table 9.4 Different schedules and the role of India herbal regulatory guidelines

Category	Guideline	Function
Drug and Cosmetics Rules 1945 (sections for license)	Sect. XVI rules 151 to 160	Manufacture for commercialization, approval/ renewal for a license on ASU drugs
	Sect. XVI (A) rule 160 A to 160 J	Official approval of organizations to conduct analytical studies on ASU raw materials and primary materials used in their manufacturing
	Sect. XVII rule 161 A	Packaging and labeling of ASU medicines for export
	Sect. XVII rule 161 B	Provision for storage and expiry conditions for ASU medicines
	Sect. XVIII rules 162 to 167	Government analyst and inspector's duties, Pharmacopeial lab to study or testing protocols for ASU validation
	Sect. XIX rules 168	Requirements and standards fulfilled by manufacturing and distribution of ASU drugs.
Drug and Cosmetics Rules 1945 (Schedules)	Schedule-A	Various types of proforma forms (24D, 24E, 25D, and 25E), applications, and licenses for the manufacture and sale of ASU drugs
	Schedule-E1	List of poisonous materials under the ASU drugs
	Schedule-T	Provisions for good manufacturing practices for ASU medicines

9.7.2 *Ministry of AYUSH*

To create awareness of education and research and to improve healthcare availability in rural areas, the Indian government formed the Department of Indian System of Medicine and Homeopathy (ISM and H) within the Ministry of Health and Family Welfare in 1995 (Singh and Shukla 2021). To focus on overall development independently, the ISM&H is currently renamed as AYUSH in 2003. Further, it is elevated permanently into the Ministry of Department of Ayurvedic, Yoga, Unani Siddha, Naturopathy, Sowa-RIGPA, and Homeopathy (AYUSH). The aim of this department is to address various aspects of educational organizations, primary health care, research and development, and investment in human source development (Singhal and Roy 2018; Mukherjee 2019). Also, to maintain good agriculture practices (GAP) to cultivate medicinal herbs, collection and storage, implementation of good manufacturing protocols (GMP), and (GCP) (Fig. 9.3).

9.7.3 *Regulatory across Countries*

Various countries developed regulatory bodies to control the manufacture and sale of herbal preparations and their formulations intended to protect public health (Mukherjee 2019). Regulatory standards across countries provide several

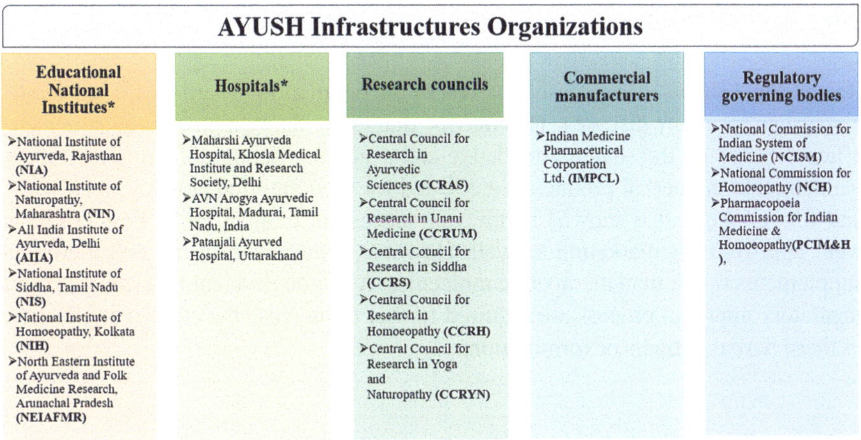

Fig. 9.3 Various organizations' aspects and their role under AYUSH in India

Table 9.5 Regulatory standards for herbal medicine in various countries

Sr. No.	Country	Regulatory bodies
1.	Australia	Australian regulatory guidelines for complementary medicines (ARGCM)
2.	Brazil	National health surveillance agency (ANVISA) and the Ministry of Agriculture
3.	Canada	Natural and non-prescription health products directorate (NNHPD)
4.	China	National medical products administration of China
5.	Europe	Committee on herbal medicinal products
6.	Egypt	Head of herbal medicines registration department- central Administration of Pharmaceutical Products – EDA
7.	India	Ministry of AYUSH, Govt of India
8.	Kuwait	Kuwait drug and food control administration
9.	Sri Lanka	National medicines regulatory authority
10.	Mexico	Comision federal Para la protection contra Reisgos Sanitarios
11.	Myanmar	Department of traditional medicine, ministry of health
12.	Nepal	Dept of Ayurveda and alternative medicine
13.	Philippines	Philippine Institute of Traditional and Alternative Health Care (PITAHC)
14.	Thailand	Department for development of Thai traditional and alternative medicine
15.	Ukraine	State Service of Ukraine on medicines and drugs control (SMDC)

advantages by assuring favorable commercialization circumstances to enable early access to pharmaceuticals, encouraging competitiveness and efficiency, and minimizing unnecessary clinical repetition. Several regulatory bodies worldwide are mentioned in Table 9.5. The data were collected from the respective country's regulatory authority website.

9.8 Conclusion

Recent trend shows the increased interest of the population toward traditional medicine usage for many diseases. This may be due to the thought process of safety and efficacy based on the shared knowledge and available literature. During the recent pandemic, many herbal products were tested for their antiviral potency worldwide, and few of them also showed significant success in clinical trials. Few products were approved for marketing as well. However, many of them are considered as supplements rather than therapeutic molecules. A thorough scientific evaluation and regulatory approval process are required further to increase the clinical application of these herbal extracts or formulations.

References

Adianti M, Aoki C, Komoto M, Deng L, Shoji I, Wahyuni TS, Lus MI, Soetjipto, Fuchino H, Kawahara N, Hotta H (2014) Anti-hepatitis C virus compounds obtained from Glycyrrhiza uralensi s and other Glycyrrhiza species. Microbiol Immunol 58(3):180–187. https://doi.org/10.1111/1348-0421.12127

Aggarwal R (2011) Clinical presentation of hepatitis E. Virus Res 161(1):15–22. https://doi.org/10.1016/j.virusres.2011.03.017

Akturk H, Sık G, Salman N, Sutcu M, Tatli B, Akcay Ciblak M, Bulent Erol O, Hancerli Torun S, Citak A, Somer A (2015) Atypical presentation of human bocavirus: severe respiratory tract infection complicated with encephalopathy. J Med Virol 87(11):1831–1838. https://doi.org/10.1002/jmv.24263

Al-Ansari MM, Ranjit Singh AJA, Al-Khattaf FS, Michael JS (2021) Nano-formulation of herbo-mineral alternative medicine from linga chenduram and evaluation of antiviral efficacy. Saudi J Biol Sci 28(3):1596–1606. https://doi.org/10.1016/j.sjbs.2020.12.005

Alfajaro MM, Kim H-J, Park J-G, Ryu E-H, Kim J-Y, Jeong Y-J, Kim D-S, Hosmillo M, Son K-Y, Lee J-H, Kwon H-J, Ryu YB, Park S-J, Park S-I, Lee WS, Cho K-O (2012) Anti-rotaviral effects of Glycyrrhiza uralensis extract in piglets with rotavirus diarrhea. Virol J 9(1):310. https://doi.org/10.1186/1743-422X-9-310

Ansari SH, Islam F, Sameem M (2012) Influence of nanotechnology on herbal drugs: a review. J Adv Pharm Technol Res 3(3):1–5. https://doi.org/10.4103/2231-4040.101006

Feehan J, Apostolopoulos V (2020) Is COVID-19 the worst pandemic? Maturitas 149:56–58. https://doi.org/10.1016/j.maturitas.2021.02.001

Arjanova OV, Prihoda ND, Yurchenko LV, Sokolenko NI, Vihrova LA, Pylypchuk VS, Frolov VM, Kutsyna GA (2009) Enhancement of efficacy of tuberculosis drugs with Immunoxel (Dzherelo™) in HIV-infected patients with active pulmonary tuberculosis. Immunotherapy 1(4):549–556. https://doi.org/10.2217/imt.09.25

Baba M, Shigeta S (1987) Antiviral activity of glycyrrhizin against varicella-zoster virus in vitro. Antivir Res 7(2):99–107. https://doi.org/10.1016/0166-3542(87)90025-8

Badam L (1997) In vitro antiviral activity of indigenous glycyrrhizin, licorice and glycyrrhizic acid (sigma) on Japanese encephalitis virus. J Commun Dis 29(2):91–99

Bagade SB, Patil M (2021) Recent advances in microwave assisted extraction of bioactive compounds from complex herbal samples: a review. Crit Rev Anal Chem 51(2):138–149. https://doi.org/10.1080/10408347.2019.1686966

Baltina LA, Tasi Y-T, Huang S-H, Lai H-C, Baltina LA, Petrova SF, Yunusov MS, Lin C-W (2019) Glycyrrhizic acid derivatives as dengue virus inhibitors. Bioorg Med Chem Lett 29(20):126645. https://doi.org/10.1016/j.bmcl.2019.126645

Banerjee S, Baidya SK, Adhikari N, Ghosh B, Jha T (2023) Glycyrrhizin as a promising kryptonite against SARS-CoV-2: clinical, experimental, and theoretical evidences. J Mol Struct 1275:134642. https://doi.org/10.1016/j.molstruc.2022.134642

Batbold U, Butov DO, Kutsyna GA, Damdinpurev N, Grinishina EA, Mijiddorj O, Kovolev ME, Baasanjav K, Butova TS, Sandagdorj M, Batbold O, Tseveendorj A, Chunt E, Zaitzeva SI, Stepanenko HL, Makeeva NI, Mospan IV, Pylypchuk VS, Rowe JL, Nyasulu P, Jirathitikal V, Bain AI, Tarakanovskaya MG, Bourinbaiar AS (2017) Double-blind, placebo-controlled, 1:1 randomized phase III clinical trial of Immunoxel honey lozenges as an adjunct immunotherapy in 269 patients with pulmonary tuberculosis. Immunotherapy 9(1):13–24. https://doi.org/10.2217/imt-2016-0079

Blacklow NR, Greenberg HB (1991) Viral gastroenteritis. N Engl J Med 325(4):252–264. https://doi.org/10.1056/NEJM199107253250406

Borrows CL, Turner PC (2014) Seasonal screening for viral gastroenteritis in young children and elderly hospitalized patients: is it worthwhile? J Hosp Infect 87(2):98–102. https://doi.org/10.1016/j.jhin.2014.02.011

Brook K, Bennett J, Desai SP (2017) The chemical history of morphine: an 8000-year journey, from resin to de-novo synthesis. Journal of Anesthesia History 3(2):50–55. https://doi.org/10.1016/j.janh.2017.02.001

Byeon JC, Ahn JB, Jang WS, Lee S-E, Choi J-S, Park J-S (2019) Recent formulation approaches to oral delivery of herbal medicines. J Pharm Investig 49(1):17–26. https://doi.org/10.1007/s40005-018-0394-4

Cacciabue M, Aguilera P, Gismondi MI, Taboga O (2022) Covidex: an ultrafast and accurate tool for SARS-CoV-2 subtyping. Infect Genet Evol 99:1–3. https://doi.org/10.1016/j.meegid.2022.105261

Cai, G., Zhu, G., Zhang, S., Tang, C., Huang, J., Yang, Y., Wang, Y., Du, H.E., and Yue, W.E.I. 2012. Anti-virus chinese herbal medicine composition, traditional chinese medicine composition and preparation method thereof

Calixto JB (2000) Efficacy, safety, quality control, marketing and regulatory guidelines for herbal medicines (phytotherapeutic agents). Braz J Med Biol Res 33(2):179–189. https://doi.org/10.1590/S0100-879X2000000200004

Calland N, Albecka A, Belouzard S, Wychowski C, Duverlie G, Descamps V, Hober D, Dubuisson J, Rouillé Y, Séron K (2012) (−)-Epigallocatechin-3-gallate is a new inhibitor of hepatitis C virus entry. Hepatology 55(3):720–729. https://doi.org/10.1002/hep.24803

Carmona F, Soares Pereira AM (2013) Herbal medicines: old and new concepts, truths and misunderstandings. Rev Bras 23(2):379–385. https://doi.org/10.1590/S0102-695X2013005000018

Chang J, Liu H, Wang K, Chen M, Chiang L, Hua Y, Lin C (2005) Ethanol extract of Polygonum cuspidatum inhibits hepatitis B virus in a stable HBV-producing cell line. Antivir Res 66(1):29–34. https://doi.org/10.1016/j.antiviral.2004.12.006

Chang JS, Wang KC, Liu HW, Chen MC, Chiang LC, Lin CC (2007) Sho-Saiko-to (Xiao-Chai-Hu-Tang) and crude Saikosaponins inhibit hepatitis B virus in a stable HBV-producing cell line. Am J Chin Med 35(02):341–351. https://doi.org/10.1142/S0192415X07004862

Chen SD, Gao H, Zhu QC, Wang YQ, Li T, Mu ZQ, Wu HL, Peng T, Yao XS (2012) Houttuynoids A–E, anti-herpes simplex virus active flavonoids with novel skeletons from Houttuynia cordata. Org Lett 14(7):1772–1775. https://doi.org/10.1021/ol300017m

Cheng HY, Yang CM, Lin TC, Shieh DE, Lin CC (2006a) Ent-Epiafzelechin-(4α→8)-epiafzelechin extracted from Cassia javanica inhibits herpes simplex virus type 2 replication. J Med Microbiol 55(2):201–206. https://doi.org/10.1099/jmm.0.46110-0

Cheng PW, Ng LT, Chiang LC, Lin CC (2006b) Antiviral effects of saikosaponins on human coronavirus 229E in vitro. Clin Exp Pharmacol Physiol 33(7):612–616. https://doi.org/10.1111/j.1440-1681.2006.04415.x

Cheng PW, Ng LT, Lin CC (2006c) Xiao Chai Hu Tang inhibits CVB1 virus infection of CCFS-1 cells through the induction of type I interferon expression. Int Immunopharmacol 6(6):1003–1012. https://doi.org/10.1016/j.intimp.2006.01.011
Cheng HY, Huang HH, Yang CM, Lin LT, Lin CC (2008) The in vitro anti-herpes simplex virus type-1 and type-2 activity of Long Dan Xie Gan Tan, a prescription of traditional Chinese medicine. Chemotherapy 54(2):77–83. https://doi.org/10.1159/000119705
Cheng HY, Yang CM, Lin TC, Lin LT, Chiang LC, Lin CC (2011) Excoecarianin, isolated from Phyllanthus urinaria Linnea, inhibits herpes simplex virus type 2 infection through inactivation of viral particles. Evid Based Complement Alternat Med 2011:1–10. https://doi.org/10.1093/ecam/nep157
Cheng Y (2019) Treatment of hepatitis C. Retrieved from https://patents.google.com/patent/WO2019204471A1/en?oq=WO2019204471A1
Chiang LC, Ng LT, Cheng PW, Chiang W, Lin CC (2005) Antiviral activities of extracts and selected pure constituents of Ocimum basilicum. Clin Exp Pharmacol Physiol 32(10):811–816. https://doi.org/10.1111/j.1440-1681.2005.04270.x
Chiang et al (2003) Cytotoxicity and anti-hepatitis B virus activities of Saikosaponins from Bupleurum species. Planta Med 69(8):705–709. https://doi.org/10.1055/s-2003-42797
Chih-hui LIN (2009) Herbal compositions for the inhibition of human immunodeficiency virus (HIV)
Choi HJ, Lim CH, Song JH, Baek SH, Kwon DH (2009) Antiviral activity of raoulic acid from Raoulia australis against picornaviruses. Phytomedicine 16(1):35–39. https://doi.org/10.1016/j.phymed.2008.10.012
Choi HJ, Song JH, Park KS, Baek SH (2010) In vitro anti-enterovirus 71 activity of gallic acid from Woodfordia fruticosa flowers. Lett Appl Microbiol 50(4):438–440. https://doi.org/10.1111/j.1472-765X.2010.02805.x
Chowdhury P, Bora T, Khan SA, Chakraborty B, Senapati K, Sengupta M, Borchetia S, Bandyopadhyay T (2016) Inhibition of Japanese encephalitis virus infection by biogenic catechin silver nanoparticles: an in-vitro study. Int J Infect Dis 45:276. https://doi.org/10.1016/j.ijid.2016.02.611
Danaher RJ, Wang C, Dai J, Mumper RJ, Miller CS (2011) Antiviral effects of blackberry extract against herpes simplex virus type 1. Oral Surg Oral Med Oral Pathol Oral Radiol Endod 112(3):e31–e35. https://doi.org/10.1016/j.tripleo.2011.04.007
Dao TT, Nguyen PH, Lee HS, Kim E, Park J, Lim SI, Oh WK (2011) Chalcones as novel influenza A (H1N1) neuraminidase inhibitors from Glycyrrhiza inflata. Bioorg Med Chem Lett 21(1):294–298. https://doi.org/10.1016/j.bmcl.2010.11.016
Dao TT, Dang TT, Nguyen PH, Kim E, Thuong PT, Oh WK (2012) Xanthones from polygala karensium inhibit neuraminidases from influenza A viruses. Bioorg Med Chem Lett 22(11):3688–3692. https://doi.org/10.1016/j.bmcl.2012.04.028
Daswani BR, Yegnanarayan R (2002) Immunomodulatory activity of septilin, a polyherbal preparation. Phytother Res 16(2):162–165. https://doi.org/10.1002/ptr.996
Doris Kumadoh KO (2017) Dosage forms of herbal medicinal products and their stability considerations-an overview. J Crit Rev 4(4):1–8. https://doi.org/10.22159/jcr.2017v4i4.16077
Duan E, Wang D, Fang L, Ma J, Luo J, Chen H, Li K, Xiao S (2015) Suppression of porcine reproductive and respiratory syndrome virus proliferation by glycyrrhizin. Antivir Res 120:122–125. https://doi.org/10.1016/j.antiviral.2015.06.001
Eccles R (2002) An explanation for the seasonality of acute upper respiratory tract viral infections. Acta Otolaryngol 122(2):183–191. https://doi.org/10.1080/00016480252814207
Efremenko YV, Arjanova OV, Prihoda ND, Yurchenko LV, Sokolenko NI, Mospan IV, Pylypchuk VS, Rowe J, Jirathitikal V, Bourinbaiar AS, Kutsyna GA (2012) Clinical validation of sublingual formulations of Immunoxel (Dzherelo) as an adjuvant immunotherapy in treatment of TB patients. Immunotherapy 4(3):273–282. https://doi.org/10.2217/imt.11.176
Ekor M (2014) The growing use of herbal medicines: issues relating to adverse reactions and challenges in monitoring safety. Front Pharmacol 4:1–10. https://doi.org/10.3389/fphar.2013.00177

Fatima M, Zaidi NSS, Amraiz D, Afzal F (2016) In vitro antiviral activity of Cinnamomum cassia and its nanoparticles against H7N3 influenza A virus. J Microbiol Biotechnol 26(1):151–159. https://doi.org/10.4014/jmb.1508.08024

Fazil M, Nikhat S (2022) Therapeutic and palliative role of an Unani herbal decoction in COVID-19 and similar respiratory viral illnesses: phytochemical & pharmacological perspective. J Ethnopharmacol 297:1–15. https://doi.org/10.1016/j.jep.2022.115526

Feng Yeh C, Chih Wang K, Chai Chiang L, Shieh DE, Hong Yen M, San Chang J (2013) Water extract of licorice had anti-viral activity against human respiratory syncytial virus in human respiratory tract cell lines. J Ethnopharmacol 148(2):466–473. https://doi.org/10.1016/j.jep.2013.04.040

Ferreira CE, Pinto-Neto AM, Conde DM, Costa-Paiva L, Morais SS, Magalhães J (2007) Menopause symptoms in women infected with HIV: prevalence and associated factors. Gynecol Endocrinol 23(4):198–205. https://doi.org/10.1080/09513590701253743

Field H (2001) Herpes simplex virus antiviral drug resistance – current trends and future prospects. J Clin Virol 21(3):261–269. https://doi.org/10.1016/S1386-6532(00)00169-4

Fraser C, Riley S, Anderson RM, Ferguson NM (2004) Factors that make an infectious disease outbreak controllable. Proc Natl Acad Sci 101(16):6146–6151. https://doi.org/10.1073/pnas.0307506101

Funk S, Gilad E, Watkins C, Jansen VAA (2009) The spread of awareness and its impact on epidemic outbreaks. Proc Natl Acad Sci 106(16):6872–6877. https://doi.org/10.1073/pnas.0810762106

Garcia-Zalisnak D, Rapuano C, Sheppard JD, Davis AR (2018) Adenovirus ocular infections: prevalence, pathology, pitfalls, and practical pointers. Eye Contact Lens Sci Clin Pract 44(1):1–7. https://doi.org/10.1097/ICL.0000000000000226

Gescher K, Kühn J, Hafezi W, Louis A, Derksen A, Deters A, Lorentzen E, Hensel A (2011a) Inhibition of viral adsorption and penetration by an aqueous extract from Rhododendron ferrugineum L. as antiviral principle against herpes simplex virus type-1. Fitoterapia 82(3):408–413. https://doi.org/10.1016/j.fitote.2010.11.022

Gescher K, Kühn J, Lorentzen E, Hafezi W, Derksen A, Deters A, Hensel A (2011b) Proanthocyanidin-enriched extract from Myrothamnus flabellifolia Welw. Exerts antiviral activity against herpes simplex virus type 1 by inhibition of viral adsorption and penetration. J Ethnopharmacol 134(2):468–474. https://doi.org/10.1016/j.jep.2010.12.038

Ghosh A, Nundy S, Mallick TK (2020) How India is dealing with COVID-19 pandemic. Sensors International 1:1–8. https://doi.org/10.1016/j.sintl.2020.100021

Gibbs A (2020) Binomial nomenclature for virus species: a long view. Arch Virol 165(12):3079–3083. https://doi.org/10.1007/s00705-020-04828-8

Giveon SM, Liberman N, Klang S, Kahan E (2004) Are people who use "natural drugs" aware of their potentially harmful side effects and reporting to family physician? Patient Educ Couns 53(1):5–11. https://doi.org/10.1016/S0738-3991(03)00241-6

Goshal S (2006) Polyherbal compositions and methods for treating viral infections. Retrieved from https://patents.google.com/patent/US7250181B2/en

Gralinski LE, Baric RS (2015) Molecular pathology of emerging coronavirus infections. J Pathol 235(2):185–195. https://doi.org/10.1002/path.4454

Granados A, Peci A, McGeer A, Gubbay JB (2017) Influenza and rhinovirus viral load and disease severity in upper respiratory tract infections. J Clin Virol 86:14–19. https://doi.org/10.1016/j.jcv.2016.11.008

Haggag E, Elshamy A, Rabeh M, Gabr N, Salem M, Youssif K, Samir A, Bin Muhsinah A, Alsayari A, Abdelmohsen UR (2019) Antiviral potential of green synthesized silver nanoparticles of Lampranthus coccineus and Malephora lutea. Int J Nanomedicine 14:6217–6229. https://doi.org/10.2147/IJN.S214171

Haggag YA, El-Ashmawy NE, Okasha KM (2020) Is hesperidin essential for prophylaxis and treatment of COVID-19 infection? Med Hypotheses 144(30):1–3. https://doi.org/10.1016/j.mehy.2020.109957

Haid S, Novodomská A, Gentzsch J, Grethe C, Geuenich S, Bankwitz D, Chhatwal P, Jannack B, Hennebelle T, Bailleul F, Keppler OT, Poenisch M, Bartenschlager R, Hernandez C, Lemasson M, Rosenberg AR, Wong-Staal F, Davioud-Charvet E, Pietschmann T (2012) A plant-derived flavonoid inhibits entry of all HCV genotypes into human hepatocytes. Gastroenterology 143(1):213–222.e5. https://doi.org/10.1053/j.gastro.2012.03.036

Halberstein RA (2005) Medicinal plants: historical and cross-cultural usage patterns. Ann Epidemiol 15(9):686–699. https://doi.org/10.1016/j.annepidem.2005.02.004

Hao B-J, Wu Y-H, Wang J-G, Hu S-Q, Keil DJ, Hu H-J, Lou J-D, Zhao Y (2012) Hepatoprotective and antiviral properties of isochlorogenic acid A from Laggera alata against hepatitis B virus infection. J Ethnopharmacol 144(1):190–194. https://doi.org/10.1016/j.jep.2012.09.003

Hasan A, Devi MS, S., Sharma, G., Narayanan, V., Sathiyarajeswaran, P., Vinayak, S., and Sunil, S. (2023) Vathasura Kudineer, an Andrographis based polyherbal formulation exhibits immunomodulation and inhibits chikungunya virus (CHIKV) under in vitro conditions. J Ethnopharmacol 302:115762. https://doi.org/10.1016/j.jep.2022.115762

Hassan MM, El Zowalaty ME, Khan SA, Islam A, Nayem MRK, Järhult JD (2020) Role of environmental temperature on the attack rate and case fatality rate of coronavirus disease 2019 (COVID-19) pandemic. Infection Ecol Epidemiol 10(1):1–6. https://doi.org/10.1080/20008686.2020.1792620

He W, Han H, Wang W, Gao B (2011) Anti-influenza virus effect of aqueous extracts from dandelion. Virol J 8(1):538. https://doi.org/10.1186/1743-422X-8-538

Ho HY, Cheng ML, Weng SF, Leu YL, Chiu DTY (2009) Antiviral effect of epigallocatechin gallate on enterovirus 71. J Agric Food Chem 57(14):6140–6147. https://doi.org/10.1021/jf901128u

Hu F (2007) Medicine composition for preventing and treating AIDS and its prepn process and use. Retrieved form https://patents.google.com/patent/CN101020027A/en?oq=CN101020027A

Huang S, Chang W (2009) Advantages of nanotechnology- based chinese herb drugs on biological activities. Curr Drug Metab 10(8):905–913. https://doi.org/10.2174/138920009790274603

Huang SP, Shieh GJ, Lee L, Teng HJ, Kao ST, Lin JG (1997) Inhibition effect of Shengma-Gegen-Tang on measles virus in vero cells and human peripheral blood mononuclear cells. Am J Chin Med 25(01):89–96. https://doi.org/10.1142/S0192415X97000123

Huang W, Zhang X, Wang Y, Ye W, Ooi VE, Chung HY, Li Y (2010) Antiviral biflavonoids from Radix Wikstroemiae (Liaogewanggen). Chin Med 5(1):23. https://doi.org/10.1186/1749-8546-5-23

Hull R, Rima B (2020) Virus taxonomy and classification: naming of virus species. Arch Virol 165(11):2733–2736. https://doi.org/10.1007/s00705-020-04748-7

Imamura T, Oshitani H (2015) Global reemergence of enterovirus D68 as an important pathogen for acute respiratory infections. Rev Med Virol 25(2):102–114. https://doi.org/10.1002/rmv.1820

Ito M, Sato A, Hirabayashi K, Tanabe F, Shigeta S, Baba M, De Clercq E, Nakashima H, Yamamoto N (1988) Mechanism of inhibitory effect of glycyrrhizin on replication of human immunodeficiency virus (HIV). Antivir Res 10(6):289–298. https://doi.org/10.1016/0166-3542(88)90047-2

Jacobson IM, Cacoub P, Dal Maso L, Harrison SA, Younossi ZM (2010) Manifestations of chronic hepatitis C virus infection beyond the liver. Clin Gastroenterol Hepatol 8(12):1017–1029. https://doi.org/10.1016/j.cgh.2010.08.026

Jain J, Kumar A, Narayanan V, Ramaswamy RS, Sathiyarajeswaran P, Shree Devi MS, Kannan M, Sunil S (2020) Antiviral activity of ethanolic extract of Nilavembu Kudineer against dengue and chikungunya virus through in vitro evaluation. J Ayurveda Integrat Med 11(3):329–335. https://doi.org/10.1016/j.jaim.2018.05.006

Jaiswal YS, Williams LL (2017) A glimpse of Ayurveda – the forgotten history and principles of Indian traditional medicine. J Tradit Complement Med 7(1):50–53. https://doi.org/10.1016/j.jtcme.2016.02.002

Jeong HJ, Kim YM, Kim JH, Kim JY, Park J-Y, Park S-J, Ryu YB, Lee WS (2012) Homoisoflavonoids from Caesalpinia sappan displaying viral neuraminidases inhibition. Biol Pharm Bull 35(5):786–790. https://doi.org/10.1248/bpb.35.786

Jiang ZY, Liu WF, Zhang XM, Luo J, Ma YB, Chen JJ (2013) Anti-HBV active constituents from Piper longum. Bioorg Med Chem Lett 23(7):2123–2127. https://doi.org/10.1016/j.bmcl.2013.01.118

Jones AW (2011) Early drug discovery and the rise of pharmaceutical chemistry. Drug Test Anal 3(6):337–344. https://doi.org/10.1002/dta.301

Joye A, Gonzales JA (2018) Ocular manifestations of cytomegalovirus in immunocompetent hosts. Curr Opin Ophthalmol 29(6):535–542. https://doi.org/10.1097/ICU.0000000000000521

Kanchibhotla D, Subramanian S, Ravi Kumar RM, Venkatesh Hari KR, Pathania M (2022) An in-vitro evaluation of a polyherbal formulation, against SARS-Cov-2. J Ayurveda Integrat Med 13(3):100581. https://doi.org/10.1016/j.jaim.2022.100581

Khanna N, Sharma SB (2001) Anti-inflammatory and analgesic effect of herbal preparation: septilin. Indian J Med Sci 55(4):195–202

Kim K, Kim KH, Kim HY, Cho HK, Sakamoto N, Cheong J (2010) Curcumin inhibits hepatitis C virus replication via suppressing the Akt-SREBP-1 pathway. FEBS Lett 584(4):707–712. https://doi.org/10.1016/j.febslet.2009.12.019

Kim DW, Seo KH, Curtis-Long MJ, Oh KY, Oh J-W, Cho JK, Lee KH, Park KH (2014) Phenolic phytochemical displaying SARS-CoV papain-like protease inhibition from the seeds of Psoralea corylifolia. J Enzyme Inhib Med Chem 29(1):59–63. https://doi.org/10.3109/14756366.2012.753591

Koishi AC, Zanello PR, Bianco ÉM, Bordignon J, Duarte N, dos Santos C (2012) Screening of dengue virus antiviral activity of marine seaweeds by an in-situ enzyme-linked immunosorbent assay. PLoS One 7(12):e51089. https://doi.org/10.1371/journal.pone.0051089

Koonin EV, Krupovic M, Agol VI (2021) The Baltimore classification of viruses 50 years later: how does it stand in the light of virus evolution? Microbiol Mol Biol Rev 85(3):1–19. https://doi.org/10.1128/MMBR.00053-21

Kramer-Hämmerle S, Rothenaigner I, Wolff H, Bell JE, Brack-Werner R (2005) Cells of the central nervous system as targets and reservoirs of the human immunodeficiency virus. Virus Res 111(2):194–213. https://doi.org/10.1016/j.virusres.2005.04.009

Krawitz C, Mraheil MA, Stein M, Imirzalioglu C, Domann E, Pleschka S, Hain T (2011) Inhibitory activity of a standardized elderberry liquid extract against clinically-relevant human respiratory bacterial pathogens and influenza A and B viruses. BMC Complement Altern Med 11(1):16. https://doi.org/10.1186/1472-6882-11-16

Kudo E, Taura M, Matsuda K, Shimamoto M, Kariya R, Goto H, Hattori S, Kimura S, Okada S (2013) Inhibition of HIV-1 replication by a tricyclic coumarin GUT-70 in acutely and chronically infected cells. Bioorg Med Chem Lett 23(3):606–609. https://doi.org/10.1016/j.bmcl.2012.12.034

Kuhn JH (2012) Virus taxonomy. Encyclopedia of Virology, 28-37. https://doi.org/10.1016/B978-0-12-809633-8.21231-4

Lau SKP, To, W, Tse PWT, Chan AKH, Woo PCY, Tsoi H, Leung AFY, Li KSM, Chan PKS, Lim WWL, Yung RWH, Chan K, Yuen K (2005) Human parainfluenza virus 4 outbreak and the role of diagnostic tests. J Clin Microbiol 43(9):4515–4521. https://doi.org/10.1128/JCM.43.9.4515-4521.2005

Lau KM, Lee KM, Koon CM, Cheung CSF, Lau CP, Ho HM, Lee MYH, Au SWN, Cheng CHK, Lau CBS, Tsui SKW, Wan DCC, Waye MMY, Wong KB, Wong CK, Lam CWK, Leung PC, Fung KP (2008) Immunomodulatory and anti-SARS activities of Houttuynia cordata. J Ethnopharmacol 118(1):79–85. https://doi.org/10.1016/j.jep.2008.03.018

Lefkowitz EJ, Dempsey DM, Hendrickson RC, Orton RJ, Siddell SG, Smith DB (2018) Virus taxonomy: the database of the international committee on taxonomy of viruses (ICTV). Nucleic Acids Res 46(1):708–717. https://doi.org/10.1093/nar/gkx932

Li S, Chen C, Zhang H, Guo H, Wang H, Wang L, Zhang X, Hua S, Yu J, Xiao P (2005) Identification of natural compounds with antiviral activities against SARS-associated coronavirus. Antivir Res 67(1):18–23. https://doi.org/10.1016/j.antiviral.2005.02.007

Li S, Li SK, Gan RY, Song FL, Kuang L, Li HB (2013) Antioxidant capacities and total phenolic contents of infusions from 223 medicinal plants. Ind Crop Prod 51:289–298. https://doi.org/10.1016/j.indcrop.2013.09.017

Liang S, Li M, Yu X, Jin H, Zhang Y, Zhang L, Zhou D, Xiao S (2019) Synthesis and structure-activity relationship studies of water-soluble β-cyclodextrin-glycyrrhetinic acid conjugates as potential anti-influenza virus agents. Eur J Med Chem 166:328–338. https://doi.org/10.1016/j.ejmech.2019.01.074

Lin JC (2003) Mechanism of action of glycyrrhizic acid in inhibition of Epstein-Barr virus replication in vitro. Antivir Res 59(1):41–47. https://doi.org/10.1016/S0166-3542(03)00030-5

Lin Y-M, Flavin MT, Schure R, Chen FC, Sidwell R, Barnard DI, Huffmann JH, Kern ER (1999) Antiviral activities of Biflavonoids. Planta Med 65(2):120–125. https://doi.org/10.1055/s-1999-13971

Lin CW, Tsai FJ, Tsai CH, Lai CC, Wan L, Ho TY, Hsieh CC, Chao PDL (2005) Anti-SARS coronavirus 3C-like protease effects of Isatis indigotica root and plant-derived phenolic compounds. Antivir Res 68(1):36–42. https://doi.org/10.1016/j.antiviral.2005.07.002

Lin LT, Chen TY, Chung CY, Noyce RS, Grindley TB, McCormick C, Lin TC, Wang GH, Lin CC, Richardson CD (2011) Hydrolyzable tannins (chebulagic acid and punicalagin) target viral glycoprotein-glycosaminoglycan interactions to inhibit herpes simplex virus 1 entry and cell-to-cell spread. J Virol 85(9):4386–4398. https://doi.org/10.1128/JVI.01492-10

Lin LT, Chen TY, Lin SC, Chung CY, Lin TC, Wang GH, Anderson R, Lin CC, Richardson CD (2013) Broad-spectrum antiviral activity of chebulagic acid and punicalagin against viruses that use glycosaminoglycans for entry. BMC Microbiol 13(1):187. https://doi.org/10.1186/1471-2180-13-187

Lin GL, McGinley JP, Drysdale SB, Pollard AJ (2018) Epidemiology and immune pathogenesis of viral sepsis. Front Immunol 9:1–21. https://doi.org/10.3389/fimmu.2018.02147

Liu L, Hu L, Yao Z, Qin Z, Idehara M, Dai Y, Kiyohara H, Yamada H, Yao X (2019) Mucosal immunomodulatory evaluation and chemical profile elucidation of a classical traditional Chinese formula, Bu-Zhong-Yi-qi-Tang. J Ethnopharmacol 228:188–199. https://doi.org/10.1016/j.jep.2018.08.003

Loc, HN, Lan TTN, Huong DTL, Tuyen NT, Quang TM, Dao LM, & Vuong NL (2022a) Effectiveness of traditional Vietnamese medicine Kovir capsule in adults with mild COVID-19: A propensity score-matched study. Phytotherapy research : PTR, 37(3), 777–780. https://doi.org/10.1002/ptr.7668

Loc HN, Lan TTN, Huong DTL, Tuyen NT, Quang TM, Dao LM, Van PH, Vuong NL (2022b) Traditional Vietnamese medicine Kovir capsule in patients with mild COVID -19: aa double-blind randomized controlled trial. Phytother Res 36(7):2878–2888. https://doi.org/10.1002/ptr.7455

Loutfy SA, Elberry MH, Farroh KY, Mohamed HT, Mohamed AA, Mohamed EB, Faraag AHI, Mousa SA (2020) Antiviral activity of chitosan nanoparticles encapsulating curcumin against hepatitis C virus genotype 4a in human hepatoma cell lines. Int J Nanomedicine 15:2699–2715. https://doi.org/10.2147/IJN.S241702

Low JSY, Wu KX, Chen KC, Ng MML, Chu JJH (2011) Narasin, a novel antiviral compound that blocks dengue virus protein expression. Antivir Ther 16(8):1203–1218. https://doi.org/10.3851/IMP1884

Lu DY, Lu TR (2019) Herbal medicine in new era. Hospice Palliative Med Int J 3(4):125–130. https://doi.org/10.15406/hpmij.2019.03.00165

Lubbe A, Seibert I, Klimkait T, van der Kooy F (2012) Ethnopharmacology in overdrive: the remarkable anti-HIV activity of Artemisia annua. J Ethnopharmacol 141(3):854–859. https://doi.org/10.1016/j.jep.2012.03.024

Ma LY, Ma SC, Wei F, Lin RC, But PPH, Lee SHS, Lee SF (2003) Uncinoside A and B, two new antiviral chromone glycosides from Selaginella uncinata. Chem Pharm Bull 51(11):1264–1267. https://doi.org/10.1248/cpb.51.1264

Ma SG, Gao RM, Li YH, Jiang JD, Gong NB, Li L, Lü Y, Tang WZ, Liu YB, Qu J, Lü HN, Li Y, Yu SS (2013) Antiviral Spirooliganones A and B with unprecedented skeletons from the roots of Illicium oligandrum. Org Lett 15(17):4450–4453. https://doi.org/10.1021/ol401992s

Mahdi JG (2010) Medicinal potential of willow: A chemical perspective of aspirin discovery. J Saudi Chem Soc 14(3):317–322. https://doi.org/10.1016/j.jscs.2010.04.010

Mahmoud DB, Ismail WM, Moatasim Y, Kutkat O, ElMeshad AN, Ezzat SM, El Deeb KS, El-Fishawy AM, Gomaa MR, Kandeil A, Al-karmalawy AA, Ali MA, Mostafa A (2021) Delineating a potent antiviral activity of Cuphea ignea extract loaded nano-formulation against SARS-CoV-2: in silico and in vitro studies. J Drug Deliv Sci Technol 66:102845

Masarani M, Dinneen M, Coyne K, Hawkins D (2008) The genitourinary complications of HIV infection in men. Br J Hosp Med 69(3):141–146. https://doi.org/10.12968/hmed.2008.69.3.28750

Maver T, Maver U, Stana Kleinschek K, Smrke DM, Kreft S (2015) A review of herbal medicines in wound healing. Int J Dermatol 54(7):740–751. https://doi.org/10.1111/ijd.12766

McWhorter JH (1996) Spicebush. A Cherokee remedy for the measles. N C Med J 57(5):306

Meléndez-Villanueva MA, Morán-Santibañez K, Martínez-Sanmiguel JJ, Rangel-López R, Garza-Navarro MA, Rodríguez-Padilla C, Zarate-Triviño DG, Trejo-Ávila LM (2019) Virucidal activity of gold nanoparticles synthesized by green chemistry using garlic extract. Viruses 11(12):1111. https://doi.org/10.3390/v11121111

Mkrtchyan A, Panosyan V, Panossian A, Wikman G, Wagner H (2005) A phase I clinical study of Andrographis paniculata fixed combination Kan Jang™ versus ginseng and valerian on the semen quality of healthy male subjects. Phytomedicine 12(6–7):403–409. https://doi.org/10.1016/j.phymed.2004.10.004

Mohamed HEA, Afridi S, Khalil AT, Ali M, Zohra T, Akhtar R, Ikram A, Shinwari ZK, Maaza M (2020) Promising antiviral, antimicrobial and therapeutic properties of green nanoceria. Nanomedicine 15(5):467–488. https://doi.org/10.2217/nnm-2019-0368

Montgomery JR, Carroll RB, McCollum AM (2011) Ocular vaccinia: a consequence of unrecognized contact transmission. Mil Med 176(6):699–701. https://doi.org/10.7205/MILMED-D-10-00473

Monto AS (2002) Epidemiology of viral respiratory infections. Am J Med 112(6):4–12. https://doi.org/10.1007/978-3-030-26961-6_28

Mouler Rechtman M, Har-Noy O, Bar-Yishay I, Fishman S, Adamovich Y, Shaul Y, Halpern Z, Shlomai A (2010) Curcumin inhibits hepatitis B virus via down-regulation of the metabolic coactivator PGC-1α. FEBS Lett 584(11):2485–2490. https://doi.org/10.1016/j.febslet.2010.04.067

Mourya DT, Yadav PD, Ullas PT, Bhardwaj SD, Sahay RR, Chadha MS, Shete AM, Jadhav S, Gupta N, Gangakhedkar RR, Khasnobis P, Singh SK (2019) Emerging/re-emerging viral diseases & new viruses on the Indian horizon. Indian J Med Res 149(4):447–467. https://doi.org/10.7205/MILMED-D-10-00473

Mukherjee PK (2003) Exploring botanicals in Indian system of medicine – regulatory perspectives. Clin Res Regul Aff 20(3):249–264. https://doi.org/10.1081/CRP-120023840

Mukherjee PK (2019) Regulatory harmonization and good quality practices for the development of herbal medicine. In: Morrissey K (ed) Quality control and evaluation of herbal drugs evaluating natural products and traditional medicine. Elsevier, Kolkata, India, pp 723–739. https://doi.org/10.1016/B978-0-12-813374-3.00021-1

Nekoua MP, Hober D (2022) Coxsackievirus infection and associated diseases. Microorganisms 10(8):1–2. https://doi.org/10.3390/microorganisms10081566

Newman H, Gooding C (2013) Viral ocular manifestations: a broad overview. Rev Med Virol 23(5):281–294. https://doi.org/10.1002/rmv.1749

Ng OT, Koh V, Chiew CJ, Marimuthu K, Thevasagayam NM, Mak TM, Chua JK, Ong SSH, Lim YK, Ferdous Z, Johari AK b, Chen MIC, Maurer-Stroh S, Cui L, Lin RTP, Tan KB, Cook AR,

Leo YS, Lee VJ (2021) Impact of delta variant and vaccination on SARS-CoV-2 secondary attack rate among household close contacts. The Lancet Regional Health – Western Pacific 17:1–8. https://doi.org/10.1016/j.lanwpc.2021.100299

Niggemann B, Gruber C (2003) Side-effects of complementary and alternative medicine. Allergy 58(8):707–716. https://doi.org/10.1034/j.1398-9995.2003.00219.x

Nishimura N, Nishio H, Lee MJ, Uemura K (2005) The clinical features of respiratory syncytial virus: lower respiratory tract infection after upper respiratory tract infection due to influenza virus. Pediatr Int 47(4):412–416. https://doi.org/10.1111/j.1442-200x.2005.02099.x

Nooreen Z, Rai VK, Yadav NP (2018) Phytopharmaceuticals: A new class of drug in India. Annals of phytomedicine: an international journal 7(1):27–37. https://doi.org/10.21276/ap.2018.7.1.4

Nwodo UU, Ngene AA, Iroegbu CU, Onyedikachi OAL, Chigor VN, Okoh AI (2011) In vivo evaluation of the antiviral activity of Cajanus cajan on measles virus. Arch Virol 156(9):1551–1557. https://doi.org/10.1007/s00705-011-1032-x

Ohtake N, Nakai Y, Yamamoto M, Sakakibara I, Takeda S, Amagaya S, Aburada M (2004) Separation and isolation methods for analysis of the active principles of Sho-saiko-to (SST) oriental medicine. J Chromatogr B 812(1–2):135–148. https://doi.org/10.1016/j.jchromb.2004.06.051

Olilaa D, Olwa-Odyek, and Opuda-Asibo, J (2002) Screening extracts of Zanthoxylum chalybeum and Warburgia ugandensis for activity against measles virus (Swartz and Edmonston strains) in vitro. Afr Health Sci 2(1):2–10

Olszanecka-Glinianowicz M, Doniec Z, Schönknecht K, Almgren-Rachtan A (2020) The herbal medicine containing of ivy leaf dry extract in the treatment of productive cough in children. Wiad Lek 73(4):668–673

Orvedahl A, MacPherson S, Sumpter R, Tallóczy Z, Zou Z, Levine B (2010) Autophagy protects against Sindbis virus infection of the central nervous system. Cell Host Microbe 7(2):115–127. https://doi.org/10.1016/j.chom.2010.01.007

Pant P, Pandey S, Dall'Acqua, S. (2021) The influence of environmental conditions on secondary metabolites in medicinal plants: a literature review. Chem Biodivers 18(11):1–42. https://doi.org/10.1002/cbdv.202100345

Patil P, Alagarasu K, Chowdhury D, Kakade M, Cherian S, Kaushik S, Yadav JP, Kaushik S, Parashar D (2022) In-vitro antiviral activity of Carica papaya formulations against dengue virus type 2 and chikungunya virus. Heliyon 8(12):e11879. https://doi.org/10.1016/j.heliyon

Pekar JE, Magee A, Parker E, Moshiri N, Izhikevich K, Havens JL, Gangavarapu K, Malpica Serrano LM, Crits-Christoph A, Matteson NL, Zeller M, Levy JI, Wang JC, Hughes S, Lee J, Park H, Park M-S, Ching Zi Yan K, Lin RTP, Mat Isa MN, Noor YM, Vasylyeva TI, Garry RF, Holmes EC, Rambaut A, Suchard MA, Andersen KG, Worobey M, Wertheim JO (2022) The molecular epidemiology of multiple zoonotic origins of SARS-CoV-2. Science 377(6609):960–966. https://doi.org/10.1126/science.abp8337

Perry RT, Halsey NA (2004) The clinical significance of measles: a review. J Infect Dis 189(1):4–16. https://doi.org/10.1086/377712

Petrera E, Coto CE (2009) Therapeutic effect of meliacine, an antiviral derived from Melia azedarach L., in mice genital herpetic infection. Phytother Res 23(12):1771–1777. https://doi.org/10.1002/ptr.2850

Polyak SJ, Morishima C, Shuhart MC, Wang CC, Liu Y, Lee DYW (2007) Inhibition of T-cell inflammatory cytokines, hepatocyte NF-κB signaling, and HCV infection by standardized silymarin. Gastroenterology 132(5):1925–1936. https://doi.org/10.1053/j.gastro.2007.02.038

Raja, A.K. 2006. Anti-retroviral herbal formulation

Ranjitsingh AJA, Devanesan S, AlSalhi MS, Paul P, Padmalatha C (2022) Antiviral and cytotoxic effects of a traditional drug KanthaRasaVillai with a cocktail of metallic nanoparticles. J King Saud Univ Science 34(1):101693. https://doi.org/10.1016/j.jksus.2021.101693

Ratiani L, Pachkoria E, Mamageishvili N, Shengelia R, Hovhannisyan A, Panossian A (2022) Efficacy of Kan Jang® in patients with mild COVID-19: interim analysis of a randomized, quadruple-blind, placebo-controlled trial. Pharmaceuticals 15(8):77–93. https://doi.org/10.3390/ph15081013

Ryu YB, Jeong HJ, Kim JH, Kim YM, Park J-Y, Kim D, Naguyen TTH, Park SJ, Chang JS, Park KH (2010) Biflavonoids from Torreya nucifera displaying SARS-CoV 3CLpro inhibition. Bioorg Med Chem 18(22):7940–7947. https://doi.org/10.1016/j.bmc.2010.09.035

Sahoo N, Manchikanti P (2013) Herbal drug regulation and commercialization: an Indian industry perspective. J Altern Complement Med 19(12):957–963

Sahoo N, Manchikanti P, Dey S (2010) Herbal drugs: standards and regulation. Fitoterapia 81(6):462–471. https://doi.org/10.1089/acm.2012.0275

Sasidharan S, Chen Y, Saravanan D, Sundram K, Latha L (2010) Extraction, isolation and characterization of bioactive compounds from plants' extracts. Afr J Tradit Complement Altern Med 8(1):1–10

Scarmozzino F, Visioli F (2020) Covid-19 and the subsequent lockdown modified dietary habits of almost half the population in an Italian sample. Foods 9(5):1–8. https://doi.org/10.3390/foods9050675

Sen S, Chakraborty R (2017) Revival, modernization and integration of Indian traditional herbal medicine in clinical practice: importance, challenges and future. J Tradit Complement Med 7(2):234–244. https://doi.org/10.1016/j.jtcme.2016.05.006

Sen S, Chakraborty R, De B (2011) Challenges and opportunities in the advancement of herbal medicine: India's position and role in a global context. J Herbal Med 1(3–4):67–75. https://doi.org/10.1016/j.hermed.2011.11.001

Sharma AK, Pundarikakshudu K (2019) Regulatory aspects of traditional Indian medicines (TIM) in India and in international purview. J AOAC Int 102(4):993–1002. https://doi.org/10.5740/jaoacint.18-0379

Sharma V, Kaushik S, Pandit P, Dhull D, Yadav JP, Kaushik S (2019) Green synthesis of silver nanoparticles from medicinal plants and evaluation of their antiviral potential against chikungunya virus. Appl Microbiol Biotechnol 103(2):881–891. https://doi.org/10.1007/s00253-018-9488-1

Singh S, Shukla VK (2021) Current regulations for herbal medicines in India. Int J Drug Regulat Affair 9(2):30–34. https://doi.org/10.22270/ijdra.v9i2.466

Singhal S, Roy V (2018) Awareness, practice and views about integrating AYUSH in allopathic curriculum of allopathic doctors and interns in a tertiary care teaching hospital in New Delhi, India. J Integrat Med 16(2):113–119. https://doi.org/10.1016/j.joim.2018.02.001

Spasov AA, Ostrovskij OV, Chernikov MV, Wikman G (2004) Comparative controlled study of Andrographis paniculata fixed combination, Kan Jang® and an Echinacea preparation as adjuvant, in the treatment of uncomplicated respiratory disease in children. Phytother Res 18(1):47–53. https://doi.org/10.1002/ptr.1359

Sreekanth TVM, Nagajyothi PC, Muthuraman P, Enkhtaivan G, Vattikuti SVP, Tettey CO, Kim DH, Shim J, Yoo K (2018) Ultra-sonication-assisted silver nanoparticles using Panax ginseng root extract and their anti-cancer and antiviral activities. J Photochem Photobiol B Biol 188:6–11. https://doi.org/10.1016/j.jphotobiol.2018.08.013

Stewart AD, Logsdon JM, Kelley SE (2005) An empirical study of the evolution of virulence under both horizontal and vertical transmission. Int J Organic Evaluat 59(4):730–739. https://doi.org/10.1554/03-330

Street RA, Stirk WA, Van Staden J (2008) South African traditional medicinal plant trade—challenges in regulating quality, safety and efficacy. J Ethnopharmacol 119(3):705–710. https://doi.org/10.1016/j.jep.2008.06.019

Tamura S, Yang G-M, Yasueda N, Matsuura Y, Komoda Y, Murakami N (2010) Tellimagrandin I, HCV invasion inhibitor from Rosae Rugosae Flos. Bioorg Med Chem Lett 20(5):1598–1600. https://doi.org/10.1016/j.bmcl.2010.01.084

Teshome H (2019) Review on principles of zoonoses prevention, control and eradication. Am J Biomed Sci Res 3(2):188–197. https://doi.org/10.34297/AJBSR.2019.03.000660

Theisen LL, Muller CP (2012) EPs® 7630 (Umckaloabo®), an extract from pelargonium sidoides roots, exerts anti-influenza virus activity in vitro and in vivo. Antivir Res 94(2):147–156. https://doi.org/10.1016/j.antiviral.2012.03.006

Ting D, Dong N, Fang L, Lu J, Bi J, Xiao S, Han H (2018) Multisite inhibitors for enteric coronavirus: antiviral cationic carbon dots based on curcumin. ACS Appl Nano Mater 1(10):5451–5459. https://doi.org/10.1021/acsanm.8b00779

Utomo RY (2020) The chemopreventive potential of diosmin and hesperidin for COVID-19 and its comorbid diseases. Indonesian J Cancer Chemoprevent 11(3):154–167. https://doi.org/10.14499/indonesianjcanchemoprev

Vaidya ADB, Devasagayam TPA (2007) Current status of herbal drugs in India: an overview. J Clin Biochem Nutr 41(1):1–11. https://doi.org/10.3164/jcbn.2007001

Van Regenmortel MHV (2007) Virus species and virus identification: past and current controversies. Infect Genet Evol 7(1):133–144. https://doi.org/10.1016/j.meegid.2006.04.002

Wang KC, Chang JS, Chiang LC, Lin CC (2011) Sheng-Ma-Ge-gen-Tang (Shoma-kakkon-to) inhibited cytopathic effect of human respiratory syncytial virus in cell lines of human respiratory tract. J Ethnopharmacol 135(2):538–544. https://doi.org/10.1016/j.jep.2011.03.058

Wang KC, Chang JS, Lin LT, Chiang LC, Lin CC (2012a) Antiviral effect of cimicifugin from Cimicifuga foetida against human respiratory syncytial virus. Am J Chin Med 40(05):1033–1045. https://doi.org/10.1142/S0192415X12500760

Wang Y, Chen M, Zhang J, Zhang XL, Huang XJ, Wu X, Zhang QW, Li YL, Ye WC (2012b) Flavone C -glycosides from the leaves of Lophatherum gracile and their in vitro antiviral activity. Planta Med 78(01):46–51. https://doi.org/10.1055/s-0031-1280128

Wang J, Chen X, Wang W, Zhang Y, Yang Z, Jin Y, Ge HM, Li E, Yang G (2013) Glycyrrhizic acid as the antiviral component of Glycyrrhiza uralensis Fisch. Against coxsackievirus A16 and enterovirus 71 of hand foot and mouth disease. J Ethnopharmacol 147(1):114–121. https://doi.org/10.1016/j.jep.2013.02.017

Wang CC, Prather KA, Sznitman J, Jimenez JL, Lakdawala SS, Tufekci Z, Marr LC (2021) Airborne transmission of respiratory viruses. Science 373(6558). https://doi.org/10.3164/jcbn.2007001

Wei XC, Cao B, Luo CH, Huang HZ, Tan P, Xu XR, Xu RC, Yang M, Zhang Y, Han L, Zhang DK (2020) Recent advances of novel technologies for quality consistency assessment of natural herbal medicines and preparations. Chin Med 15(1):1–24. https://doi.org/10.1186/s13020-020-00335-9

Welliver RC (2003) Review of epidemiology and clinical risk factors for severe respiratory syncytial virus (RSV) infection. J Pediatr 143(5):112–117. https://doi.org/10.1067/S0022-3476(03)00508-0

Wolkerstorfer A, Kurz H, Bachhofner N, Szolar OHJ (2009) Glycyrrhizin inhibits influenza A virus uptake into the cell. Antivir Res 83(2):171–178. https://doi.org/10.1016/j.antiviral.2009.04.012

Woodhead M (2005) Guidelines for the management of adult lower respiratory tract infections. Eur Respir J 26(6):1138–1180. https://doi.org/10.1183/09031936.05.00055705

Xu F, Zhao C, Li Y, Li J, Deng Y, Shi T (2011) Exploring virus relationships based on virus-host protein-protein interaction network. BMC Syst Biol 5(3):2–7. https://doi.org/10.1186/1752-0509-5-S3-S11

Yang CM, Cheng HY, Lin TC, Chiang LC, Lin CC (2007a) Hippomanin a from acetone extract of Phyllanthus urinaria inhibited HSV-2 but not HSV-1 infectionin vitro. Phytother Res 21(12):1182–1186. https://doi.org/10.1002/ptr.2232

Yang CM, Cheng HY, Lin TC, Chiang LC, Lin CC (2007b) The in vitro activity of geraniin and 1,3,4,6-tetra-O-galloyl-β-d-glucose isolated from Phyllanthus urinaria against herpes simplex virus type 1 and type 2 infection. J Ethnopharmacol 110(3):555–558. https://doi.org/10.1016/j.jep.2006.09.039

Yang XX, Li CM, Huang CZ (2016) Curcumin modified silver nanoparticles for highly efficient inhibition of respiratory syncytial virus infection. Nanoscale 8(5):3040–3048. https://doi.org/10.1039/C5NR07918G

You H, Wu X, Guo X (2020) Distribution of COVID-19 morbidity rate in association with social and economic factors in Wuhan, China: implications for urban development. Int J Environ Res Public Health 17(10):1–14. https://doi.org/10.3390/ijerph17103417

Yu MS, Lee J, Lee JM, Kim Y, Chin YW, Jee JG, Keum YS, Jeong YJ (2012) Identification of myricetin and scutellarein as novel chemical inhibitors of the SARS coronavirus helicase, nsP13. Bioorg Med Chem Lett 22(12):4049–4054. https://doi.org/10.1016/j.bmcl.2012.04.081

Zaal MJW, Völker-dieben HJ, Wienesen M, D'Amaro J, Kijlstra A (2001) Longitudinal analysis of varicella-zoster virus DNA on the ocular surface associated with herpes zoster ophthalmicus. Am J Ophthalmol 131(1):25–29. https://doi.org/10.1016/S0002-9394(00)00652-8

Zandi K, Teoh BT, Sam SS, Wong PF, Mustafa MR, AbuBakar S (2011) Antiviral activity of four types of bioflavonoid against dengue virus type-2. Virol J 8(1):560. https://doi.org/10.1186/1743-422X-8-560

Zandi K, Teoh BT, Sam SS, Wong PF, Mustafa MR, AbuBakar S (2012) Novel antiviral activity of baicalein against dengue virus. BMC Complement Altern Med 12(1):214. https://doi.org/10.1186/1472-6882-12-214

Zang N, Xie X, Deng Y, Wu S, Wang L, Peng C, Li S, Ni K, Luo Y, Liu E (2011) Resveratrol-mediated gamma interferon reduction prevents airway inflammation and airway hyper-responsiveness in respiratory syncytial virus-infected immunocompromised mice. J Virol 85(24):13061–13068. https://doi.org/10.1128/JVI.05869-11

Zardini A, Galli M, Tirani M, Cereda D, Manica M, Trentini F, Guzzetta G, Marziano V, Piccarreta R, Melegaro A, Ajelli M, Poletti P, Merler S (2021) A quantitative assessment of epidemiological parameters required to investigate COVID-19 burden. Epidemics 37:100530. https://doi.org/10.1016/j.epidem.2021.100530

Zeng FL, Xiang YF, Liang ZR, Wang X, Huang D, Zhu SN, Li MM, Yang DP, Wang DM, Wang YF (2013) Anti-hepatitis B virus effects of Dehydrocheilanthifoline from Corydalis saxicola. Am J Chin Med 41(01):119–130. https://doi.org/10.1142/S0192415X13500092

Zhang Y, Li Y (2022) Exploring the clinical significance of Yinqiao powder based on the compatibility of Chinese herbal formulas. Med Theory Hypothesis 5(1):1–4. https://doi.org/10.53388/TMRTH202203001

Zheng X, Li X, Liu J, Shi L, Wang H, Tian K, Pan X (2022) Horizontal transmission might be a common route of hepatitis B virus exposure in highly endemic areas. J Med Virol 94(10):4983–4992. https://doi.org/10.1002/jmv.27905

Zhou X, Li CG, Chang D, Bensoussan A (2019) Current status and major challenges to the safety and efficacy presented by chinese herbal medicine. Medicines 6(1):1–10. https://doi.org/10.3390/medicines6010014

ホンチャンジア (2020) 抗ウィルス漢方薬組成物及びその調製方法並びに使用

이우송, 노문철, 박수진, 류영배, 장종선, 정형재, and 권형준 (2015) Composition for prevention and treatment of influenza virus and composition for inhibiting the activity of neuraminidase comprising extracts of Turmeric

Chapter 10
The Use of Animal Models for Antiviral Therapeutics Development: Opportunities and Challenges

Islam Abbadi, Mustapha Lkhider, Ahd Ouladlahsen, Haya Altawalah, Ali A. Rabaan, Fadila Guessous, and Sayeh Ezzikouri

Abstract Significant scientific advances have been made through the use of In Vitro techniques (such as cell culture). However, the entire organism is needed to study physiological processes and relationships between different organs. Animal models of infectious diseases have a significant advantage over In Vitro studies because they help researchers comprehend the complex interplay between pathogens and the immune system.

Viral infections are among the most common causes of morbidity and mortality worldwide. Moreover, drug resistance is a key challenge that necessitates vigilant management of treatment regimens and the ongoing discovery of novel antiretroviral medications. It is crucial to understand the fundamentals of using model systems because evaluating new antiviral drugs and vaccines in animal models is necessary before conducting clinical trials in human patients. An animal model of human viral infection should ideally mirror the host-pathogen interactions and disease progression found in the natural disease course. A suitable animal model of viral infection should enable the analysis of numerous infection parameters, such as clinical symptoms, viral growth, clinicopathological parameters, cellular and humoral immune responses, and virus-host interactions.

In this chapter, we summarize the findings on the use of animal models for the development and evaluation of therapeutic approaches and vaccines to combat some of the most dangerous viral infections that still pose a burden to public health.

I. Abbadi
Virology Unit, Viral Hepatitis Laboratory, Institut Pasteur du Maroc, Casablanca, Morocco

Laboratory of virology, microbiology, quality and biotechnology/Eco-toxicology and biodiversity, University Hassan II Casablanca Faculty des Sciences Techniques, Mohammedia, Morocco

M. Lkhider
Laboratory of virology, microbiology, quality and biotechnology/Eco-toxicology and biodiversity, University Hassan II Casablanca Faculty des Sciences Techniques, Mohammedia, Morocco

N. Kumar et al. (eds.), *Advances in Antiviral Research*, Livestock Diseases and Management, https://doi.org/10.1007/978-981-99-9195-2_10

Keywords Animal models · Antivirals · Vaccines · In Vivo

Abbreviations

AAV	Adeno-associated viruses
Ad5-Hdpp4	Adenovirus 5 encoding human DPP4
AIDS	Acquired immunodeficiency syndrome
ART	Antiviral
BLT	Bone-liver-thymus
bNAbs	Broadly neutralizing antibodies
cccDNA	Covalently closed circular DNA
CCR5	CC chemokine receptor 5
CD4	Cluster of differentiation 4
CoVs	Coronaviruses
CRISPR/Cas9	Clustered regularly interspaced short palindromic repeats
CTLA	Cytotoxic T lymphocyte-associated
CYP3A4	Cytochrome P450 3A4
DHBV	Duck hepatitis B virus
FAH	Fumaryl acetoacetate hydrolase
FDA	Food and Drug Administration
GBV-B	GB virus B
gRNA	guide RNA
HAP	Hetero-aryldihydropyramidines
HAV	Hepatitis A virus
HBeAg	HBV e antigen
HBsAg	HBV surface antigen
HBV	Hepatitis B virus
HBx	Hepatitis B protein X
HCC	Hepatocellular carcinoma

A. Ouladlahsen
Faculté de Médecine et de Pharmacie, Université Hassan II, Casablanca, Morocco

H. Altawalah
Department of Microbiology, Faculty of Medicine, Kuwait University, Kuwait City, Kuwait

Virology Unit, Yacoub Behbehani center, Sabah Hospital, Ministry of Health, Kuwait City, Kuwait

A. A. Rabaan
Molecular Diagnostic Laboratory, Johns Hopkins Aramco Healthcare, Dhahran, Saudi Arabia

F. Guessous
Mohammed VI University of Health Sciences (UM6SS), Casablanca, Morocco

Department of Microbiology, Immunology and Cancer Biology, School of Medicine, University of Virginia, Charlottesville, VA, USA

S. Ezzikouri (✉)
Virology Unit, Viral Hepatitis Laboratory, Institut Pasteur du Maroc, Casablanca, Morocco
e-mail: sayeh.ezzikouri@pasteur.ma

HCoV	Human coronavirus
HCV	Hepatitis C virus
HIV	Human immunodeficiency virus
Hu-HSC	Human hematopoietic stem cells
I-BET151	Bromodomain inhibitor
IFN	Interferon
IL-7	Interleukin 7
JAK1/JAK2	Janus kinase 1/Janus Kinase 2
MERS	Middle east respiratory syndrome coronavirus
miR-122	Micro RNA
mRNA	messenger RNA
MVA	Modified vaccinia virus Ankara
NAs	Nucleos(t)ide analogs
NHPs	Non-human primates
NKT	Naturel killer T-cell
NNRTIs	Non-nucleoside reverse transcriptase inhibitors
NPHV	Non-primate hepacivirus
NS5B	Non-structural
NTCP	Sodium taurocholate cotransporting polypeptide
OCLN	Occludin
RBD	Receptor binding domain
RHV	Rodent hepacivirus
RNA	Ribonucleic acid
RNAi	RNA interference SARS-CoV: Severe acute respiratory syndrome coronavirus
RT	Reverse transcriptase
SCID-hu	Severe combined immune deficiency human
SHIV	Simian human immunodeficiency virus
siRNA	Small interfering RNA
SIV	Simian immunodeficiency virus
SOCS3	Suppressor of cytokine signaling 3
SSK1	Senescence-specific killing
TLR 7	Toll-like receptor 7
TLR 9	Toll-like receptor 9
TMPRSS2	Transmembrane serine protease 2
uPA	Urokinase-type plasminogen activator
UTR	Untranslated region
WHO	World health Organization
WHV	Woodchuck hepatitis virus

10.1 Introduction

Viral diseases are among the main causes of morbidity and mortality worldwide (Howard et al. 2012). Virus-specific vaccines and antiviral drugs are the most effective defense mechanisms against viral diseases (Andersen et al. 2020; De Clercq and Li 2016).

Animals have been used for scientific purposes for a long time in biological studies and medicine. Since humans and animals, especially mammals, share many anatomical and physiological similarities, researchers have studied various mechanisms and evaluated innovative therapies in animal models before applying their findings to humans (Barré-Sinoussi and Montagutelli 2015).

In Vitro techniques (e.g., cell culture) constitute significant scientific advancements and have replaced the use of animals to some extent. However, the investigation of physiological activities and systemic connections between organs requires the entire organism. Thus, the use of animal models of infectious diseases rather than In Vitro studies has the major advantage of helping researchers better understand the complex interactions between the immune system and pathogens (Barré-Sinoussi and Montagutelli 2015).

A variety of scientific issues have been investigated using animal models, from basic research to the development and evaluation of the efficacy of new vaccination techniques and treatments. This chapter presents the results of the use of animal models in the development of antiviral drugs against fatal viral infections caused by various viruses such as HIV, hepatitis C and B viruses (HCV and HBV), severe acute respiratory syndrome coronavirus 1 (SARS-CoV-1), middle east respiratory syndrome coronaviruse (MERS-CoV), and severe acute respiratory syndrome coronavirus 2 (SARS-CoV-2).

10.2 Human Immunodeficiency Virus Type 1 (HIV-1)

Human Immunodeficiency Virus (HIV) infection in humans induces immunodepression, leading to the development of acquired immunodeficiency syndrome (AIDS) and mortality in HIV-infected individuals in the absence of antiretroviral therapy (ART) (Policicchio et al. 2016). Furthermore, considerable resources and effort have been invested in the development of drugs and vaccines for the treatment and prevention of this infection. Similarly, significant progress has been made in understanding the pathogenesis, prevention, and therapeutic management of AIDS over the last few decades. In particular, the availability of effective antiretroviral therapies has significantly reduced HIV-related morbidity and mortality (Evans et al. 2013).

Animal models play an essential role in assessing the efficacy of antiretroviral therapies used in the treatment of HIV infection in humans, as well as in acquiring the scientific knowledge needed to develop a reliable and safe HIV-1 vaccine (Garcia 2016). In curative AIDS research, non-human primates (NHPs), humanized mice, and to a lesser extent felines infected with their respective immunodeficiency viruses are the most widely used animal models (Policicchio et al. 2016).

10.2.1 Non-Human Primate Models (NHPs)

Only humans and chimpanzees are susceptible to HIV-1 infections. Therefore, chimpanzees were the first subjects of research on the development of animal models for use in the HIV/AIDS fight, a model that has since been suspended for ethical considerations (Hatziioannou and Evans 2012; Nath et al. 2000). However, studies on other NHPs have contributed to the identification of key models of how HIV infections spread, have been used to test different treatments, and are crucial for developing AIDS vaccines (Sui et al. 2013; Van Rompay 2011; Van Rompay 2010).

In AIDS research, Macaques are commonly used as animal models, because of the different disease outcomes that a variety of simian immunodeficiency virus (SIV) strains may induce in diverse NHP species (Terrade et al. 2021). The basic features of HIV-1 infection in humans are mirrored in SIV and SHIV infections in macaques. These include progressive deterioration in health status leading to significant immunodeficiency and eventual mortality from opportunistic infections or tumors. These findings persist despite the development of antiretroviral immune responses, including the presence of virus-targeting neutralizing antibodies and cytotoxic T cells (Kline et al. 2013).

The three macaque species commonly used in HIV research models are rhesus macaques (*Macaca mulatta*) of Chinese and Indian origin, cynomolgus macaques (*Macaca fascicularis*), and pig-tailed macaques (*Macaca nemestrina*) (Huot et al. 2019; Pandrea 2016). These macaque species exhibit variable susceptibility to different strains of SIV, leading to a spectrum of disease outcomes. Some macaques show elite-controlled infection with minimal disease progression, while others show AIDS-defining symptoms or rapid disease progression, culminating in death within months of infection (Policicchio et al. 2016).

Although SIV and HIV-1 share structural and sequence similarities, there are significant differences that hinder the application of certain treatments in SIV-infected macaque models. For example, non-nucleoside reverse transcriptase inhibitors (NNRTIs) are ineffective against the reverse transcriptases (RTs) present in HIV-2 and SIV (Balzarini et al. 1995). To address this limitation, the researchers developed RT-SHIVs by replacing the RT-encoding region of specific SIV clones with that of HIV-1. This modification enables RT-SHIVs to be targeted with RT inhibitors, including NNRTIs, in studies involving macaques (Ambrose et al. 2004). In a recent study, (Gelé et al. 2022)sought to validate a non-human primate (NHP) model using Cynomolgus macaques administered antiretroviral drugs (tenofovir disoproxil fumarate, emtricitabine, and dolutegravir) for a prolonged period after infection with the pathogenic SIVmac251 strain. The researchers felt that this model of SIV+ NHP treated with a combination of antiretroviral therapies for two years would provide valuable insights into the immunological, virological, and pharmacological aspects essential to the control and cure of HIV infection (Gelé et al. 2022).

NHP models have played an essential role in scientific research aimed at identifying the specific immune responses required for effective protection against HIV-1. These models have been recognized as essential preclinical tools for evaluating the efficacy of various vaccination strategies (Evans et al. 2013). The SIV animal model has been extensively used since its inception to evaluate various vaccination strategies and identify the most protective ones (Van Rompay 2011). In NHP models, a range of immunotherapies, including antibody-based treatments and therapeutic vaccinations, have been evaluated to enhance immune-mediated eradication of infected cells (Terrade et al. 2021).

Antibodies have been extensively studied in the search for an HIV vaccine, and now play an essential role in the search for a cure. Previous studies have investigated whether neutralizing and non-neutralizing antibodies can destroy infected cells that generate viral proteins (Eroshkin et al. 2014; Scheid et al. 2009). In the SHIV-infected macaque model, early injection of two broadly neutralizing antibodies (bNAbs) reduces persistent viremia, promotes T-cell immunity, and controls infection (Nishimura et al. 2017). Moreover, the ability of monoclonal antibody-mediated suppression of host-expressed integrin $\alpha4\beta7$ to improve virological regulation in NHP models has attracted the attention of many researchers (Abravanel et al. 2013; Byrareddy et al. 2016; Calenda et al. 2018).

models have been used to test a number of vaccines. Live attenuated vaccines have been widely recognized as highly effective due to their ability to elicit sustained and widespread antiviral immune responses (Abel et al. 2003; Daniel et al. 1992; Marthas et al. 1990). However, this technique is not currently being investigated in humans owing to safety concerns. The use of NHP models has demonstrated the efficacy of T-cell vaccines that express SIV proteins in eliciting cellular immune responses, reducing levels of viral particles in the bloodstream and preserving memory CD4 T cells (Letvin et al. 2006; Mattapallil et al. 2006; Wilson et al. 2006).

10.2.2 Small HIV Animal Models

Previous attempts to infect small animal models, such as mice, rats, and rabbits, with HIV-1 have been unsuccessful (Garcia 2016; Morrow et al. 1987). In particular, mice are a good example of small animals that can be genetically manipulated (Hatziioannou and Evans 2012). Consequently, the small animal models that have proved most effective for studying HIV/AIDS are known as "humanized mice." These mice are genetically engineered to have a weakened immune system and are then grafted with human tissue to mimic the human immune system (Agarwal et al. 2020; Marsden 2020; Weichseldorfer et al. 2020). Various mouse models have been developed to mimic human biology. Among these models, the most frequently used in HIV-1 research are SCID-hu (severe combined immune deficiency human) mice, hu-HSC (humane hematopoietic stem cells) mice, and BLT (bone-liver-thymus) mice (Akkina 2013; Policicchio et al. 2016).

In the SCID-hu model, it has been demonstrated that prostratin and IL-7 can both activate HIV-1 in latently infected cells (Korin et al. 2002; Scripture-Adams et al. 2002), with these cells becoming vulnerable to an anti-HIV immunotoxin as a result of the reactivation of viral expression (Brooks et al. 2003). The efficacy of zinc-finger nucleases targeting CCR5 and small interfering RNA (siRNA) constructs targeting human T cells in suppressing HIV-1 infection was evaluated using hu-HSC models (Holt et al. 2010; Kumar et al. 2008). Finally, in hu-HSC model, combinations of inducers (Vorinostat, I-BET151, and CTLA) and broadly neutralizing antibodies bNAbs form a treatment approach that affects the formation and maintenance of the HIV-1 reservoir (Halper-Stromberg et al. 2014). The protective effect of CCR5-targeted RNA interference (RNAi) on HIV-1 replication has been demonstrated in BLT mouse models (Shimizu et al. 2015). The combination of ART and an HIV-specific cytotoxic immunotoxin demonstrated a significant reduction in the number of actively infected cells in various organs of BLT mice (Denton et al. 2014).

10.3 Hepatitis Viruses

Hepatitis viruses are recognized as important human pathogens that invade hepatocytes and cause liver damage. Chimpanzees have been used extensively as a research model for hepatitis viruses. Today, thanks to chimpanzee-derived vaccines that are widely used worldwide, entire generations are immune to hepatitis A virus (HAV) and hepatitis B virus (HBV) (Lanford et al. 2017).

The inaccessibility to chimps' models due to ethical, economical, and technical issues had hampered the development of curative treatments for HBV chronic infection with liver cancer being a prevalent outcome. The use of chimpanzee models has played an important role in the understanding and treatment of hepatitis C virus (HCV). These models have contributed to the development of antiviral therapies that require short treatment times and have minimal side effects. What's more, these therapies have achieved remarkable cure rates, in excess of 95% (Bourlière and Pietri 2019; Carrat et al. 2019). However, these antiviral therapies are insufficient and prophylactic vaccines are needed for HCV eradication.

Because of this situation, scientists all over the world are looking for other animal models to study HBV and HCV infections so that they can develop new antiviral drugs and vaccines. Here, we provide a summary of what has been learned about HCV and HBV infections in different animal models so far.

10.3.1 Hepatitis C Virus (HCV)

The HCV, responsible for hepatitis C in humans, was identified in 1989. It has been classified as a member of the *Hepacivirus* genus within the *Flaviviridae* family (Scheel et al. 2015b). This virus can spread through the blood and cause acute or chronic hepatitis (Hoofnagle 2002). The use of robust in vitro cell culture methodologies has made a significant contribution to the study of the viral life cycle, leading to the identification of direct-acting antivirals that have brought about significant changes in standard treatment protocols (Lindenbach and Rice 2013; Scheel and Rice 2013). Although it is possible to conduct functional and inhibition studies of HCV using cell culture systems, a comprehensive examination of host-virus interactions in the liver, particularly during the early stages of infection, when the outcome of infection resolution or persistence is determined, cannot be performed (Raghuraman et al. 2012; Rehermann 2009), posing challenges in these systems. Consequently, exploring these interactions using In Vivo models remains a crucial objective. The lack of a suitable animal model is a major obstacle to the development of cost-effective pharmaceuticals and vaccines against HCV (Callaway 2014).

Currently, chimpanzees are the only animal species recognized as susceptible to HCV infection (Farci et al. 1992; Houghton and Abrignani 2005). The future prospects for the NHP model appear dim, due to the high financial burden, ethical considerations, and regulatory limitations (Hoofnagle 2002; Lanford et al. 2017). Hepatitis caused by HCV in chimpanzees is usually less serious than that caused by HCV in humans; however, severe hepatitis has been reported (Farci et al. 1992). In addition, chronic hepatitis and liver cirrhosis are rarely seen in chimpanzees, compared to humans (Major et al. 2004). The reason for this difference is still unknown, but it is plausible that it is associated with the absence of cofactors, such as alcohol consumption and obesity, which are known to induce liver disease in humans (Lanford et al. 2017).

Hence, the main objective of researchers studying HCV is to discover a new animal model that is both manageable and receptive to experimentation (Barth et al. 2008; Billerbeck et al. 2013; Drexler et al. 2013; Hartlage et al. 2016; Kapoor et al. 2013; Pybus and Thézé 2016; Scheel et al. 2015b; Scull et al. 2015; Verstrepen et al. 2015; Von Schaewen and Ploss 2014). A number of immune-compromised mouse models have been developed, in which human liver cells have been successfully transplanted (Bukh 2012; Vercauteren et al. 2014; Von Schaewen and Ploss 2014). In addition, previous studies have revealed the persistence of HCV replication in mice genetically engineered to express human HCV entry factors, in particular CD81 and OCLN (Dorner et al. 2013; Von Schaewen and Ploss 2014). In another study, transgenic mice with an ICR genetic background and the inclusion of the human CD81 and occludin genes showed susceptibility to HCV infection (Chen et al. 2014). HCV replication was completed in this animal model, resulting in persistent viremia for more than 12 months after infection, with anticipated progression to fibrosis and cirrhosis. For the first time, persistent HCV infection with a full replication cycle and hepatopathological symptoms was recapitulated in immune-competent mice. The aforementioned model provides a new context for examining the etiology of chronic hepatitis C and designing more effective therapeutic interventions (Chen et al. 2014). Together, these models have improved our ability to study HCV infection In Vivo. However, it is crucial to recognize and address notable constraints such as suboptimal infection levels and suppression of innate and adaptive immune responses in order to achieve a desirable HCV replication rate (Billerbeck et al. 2013; Von Schaewen and Ploss 2014).

Due to the strict species tropism of HCV, limited to humans and chimpanzees, and the lack of suitable immunocompetent animal models that accurately reproduce robust HCV infection, it is imperative to study other animal hepaciviruses as potential models for the study of HCV (Hartlage et al. 2016; Scheel et al. 2015b). GB virus B (GBV-B) is widely recognized as a homologue of HCV and is known to induce only transient, self-limiting infections in tamarins, in contrast to the persistent nature of HCV infection in humans (Beames et al. 2001).

In recent years, a number of new hepacivirus species have been identified in various animal species, including dogs, horses, rodents, bats, colobus monkeys, cattle and, more recently, the graceful cat sharks. These discoveries have broadened our understanding of the genetic variability and host range of hepaciviruses (Baechlein et al. 2015; Drexler et al. 2013; Hartlage et al. 2016; Kapoor et al. 2011; Quan et al. 2013; Scheel et al. 2015b; Thézé et al. 2015). Nevertheless, until now, only the genomic characteristics of these viruses have been recognized. For a surrogate model to be considered ideal, it must faithfully reproduce the characteristics of HCV, such as its ability to target the liver, its susceptibility to long-term infection, the resulting immune response and disease progression (Bukh 2012; Hartlage et al. 2016). Non-primate hepacivirus (NPHV), which shares homology with HCV, was first discovered in dogs, then in horses. Like HCV, NPHV is hepatotropic and capable of inducing persistent infections in horses (Scheel et al. 2015a). However, horses are not genetically manageable or amenable. Similar constraints exist for bovine, *Colobus* sp., and bats because of laboratory inaccessibility and lack of host-specific reagents (Hartlage et al. 2016).

Prior to 2012, GBV-B was the only animal hepacivirus used as a surrogate for HCV. However, between 2012 and 2017, researchers shifted their focus to NPHV (Pfaender et al. 2017; Scheel et al. 2015a; Tanaka et al. 2014). However, a rat-specific hepacivirus (RHV) has recently become the focus of research for developing a substitute model for HCV (Billerbeck et al. 2017; Ploss and Kapoor 2020; Trivedi et al. 2017). RHV and HCV share several genetic features, including the arrangement of their UTRs, the structure of their NS5B, and the presence of polyprotein cleavage sites and miR-122 seeding sites (Hartlage et al. 2016).

Additionally, a deeper molecular understanding of the involvement of innate, humoral and cellular immunity in the regulation of RHV infection and its persistent presence in immunocompetent rat models could potentially serve as a crucial basis for the development of new vaccination strategies in humans (Hartlage et al. 2016).

10.3.2 Hepatitis B Virus

The human hepatitis B virus (HBV) belongs to the *Hepadnaviridae* family. The HBV is a small DNA virus responsible for both acute and chronic forms of hepatitis. Infection leads to a series of liver complications, such as cirrhosis and hepatocellular carcinoma (Dandri et al. 2017). HBV infection is still a big health problem worldwide, and breakthroughs in basic and clinical research on HBV are needed to find a cure for HBV infection.

Treatments for hepatitis B can be classified into two distinct groups according to their mechanisms of action. The first category includes nucleos(t)ide analogues (NAs), which are the most widely used treatments. NAs act directly by inhibiting the reverse transcription process. The second category of treatments consists of enhancing the host immune system's response to the virus (Li et al. 2020). Although current medications inhibit HBV replication, infection clearance is rare. The main contributing factor is the enduring presence of the viral genome, known as covalently closed circular DNA (ccDNA), in the nuclei of infected hepatocytes. In addition, the immune system's inability to generate robust antibody responses specific to HBV exacerbates the situation (Nassal 2015). Due to the unique replication mechanism of HBV, new antiviral approaches are needed to effectively resolve HBV infection. These strategies aim to specifically target the cccDNA located in the nucleus of infected hepatocytes, while facilitating restoration of the immune system (Petersen et al. 2016). Challenges to HBV research and preclinical testing of antiviral drugs include the particular species and tissue tropism of the virus, the limited availability of infection models, and the constraints of using chimpanzees, which are the only animals likely to be infected with HBV (Wieland 2015).

HBV has a limited host range due to its specific affinity for hepatocytes. Chimpanzees are currently the only known animal species with an inherent susceptibility to HBV infection (Guidotti et al. 1999), Mauritian cynomolgus monkeys (Dupinay et al. 2013), tree shrews (Köck et al. 2001; Yan et al. 1996), and woodchucks (Kulkarni et al. 2007).

These animals have been used to study virus-host interactions and pathogenesis, including but not limited to acute and chronic hepatitis, fibrosis, cirrhosis, and hepatocellular carcinoma (HCC). In addition, HBV-related hepadnaviruses, such as duck hepatitis B virus (DHBV) and woodchuck hepatitis virus (WHV), have been used as animal models for HBV. These models have contributed to our understanding of the mechanisms of HBV replication and disease development in their respective host species (D'Ugo et al. 2010; Reaiche et al. 2010; Sprengel et al. 1985; Yang et al. 2012).

10.3.3 HBV-Related Viruses

Although DHBV has only 40% similarity to human HBV (Sprengel et al. 1985), the availability of duck and duckling hepatocyte cultures makes them readily accessible for research purposes. Various antiviral drugs have been evaluated in a duck model to address this problem (Colledge et al. 2000; Foster et al. 2005; Wang et al. 2012). Nevertheless, it seems that these animals, including the woodchuck, are relatively less sensitive to the toxicological effects of certain substances, such as the PAH (heteroaryldihydropyrimidine) family. Interestingly, this family effectively inhibits HBV (hepatitis B virus) replication, although it is not effective against DHBV and WHV (Campagna et al. 2013; Deres et al. 2003). The importance of the duck model in understanding the different facets of the replication mechanism used by hepadnaviruses, as a virus family, is paramount. An additional constraint of this model concerns the use of carboxypeptidase D by DHBV, as opposed to sodium taurocholate co-transporting polypeptide (NTCP), as a receptor for viral entry. Consequently, DHBV infection is not generally associated with liver pathology or HCC progression.

On the other hand, the genetic structure of WHV and HBV is comparable. In addition, the woodchuck has been used as an animal model to study the development and treatment of chronic hepatitis B infection in humans (Menne and Cote 2007). The preclinical evaluation of most antiviral drugs currently used to treat HBV infection has benefited from studies on naturally and artificially infected woodchucks, such as the American marmot (*Marmota monax*) (Dandri et al. 2000; Mason et al. 1998; Menne et al. 2005). The study by Menne et al. (2015) highlighted the ability of the oral TLR7 agonist GS-9620 to induce sustained antiviral responses and seroconversion in woodchucks chronically infected with WHV. These results have excited scientists, who are now considering using GS-9620 as a potential new therapeutic strategy to achieve a functional cure in patients with HBV infection (Menne et al. 2015). The outcome of the disease in woodchucks depends on age; indeed experimental infection of newborns often leads to chronic infection, in contrast to that in adults, where infected animals typically develop an acute and resolved infection (Cote et al. 2000). Transcriptomic analysis of woodchucks revealed that chronic infection with WHV, which shares similarities with HBV, is associated with a restricted intrahepatic type I interferon response, activation of markers associated with T-cell depletion in the liver, increased levels of suppressor of cytokine signaling 3 (SOCS3) in the liver, and accumulation of neutrophils in the liver (Fletcher et al. 2012). However, it is important to highlight that the carcinogenic process differs between WHV and HBV (Tennant et al. 2004).

Therefore, when working with HBV-related systems such as WHV and DHBV, it is essential to take into account both virus- and host-related variations.

10.3.4 HBV Infection Animal Models

The chimpanzee is the only immunocompetent species with complete susceptibility to HBV infection. Chimpanzees have the capacity to develop chronic and acute HBV infections, with pathogenesis and immunological profiles similar to those observed in HBV-infected human patients (Wieland 2015). Consequently, this animal model has played a pivotal role in safety studies and the development of the HBV vaccine (McAuliffe et al. 1980), as well as, the evaluation of the effectiveness of therapeutic antibodies (Eren et al. 2000) and research aiming to improve antiviral immunity (TLR 7 activation) (Lanford et al. 2013). The mechanisms of HBV clearance have also been studied in the chimpanzee model. The above studies demonstrated that successful eradication of infection requires a strong and diverse immune response from polyclonal CD8+ T cells during the peak of infection (Thimme et al. 2003). In addition, research on experimentally infected chimpanzees has demonstrated the presence of residual amounts of ccDNA in the liver, even after the infection has been resolved (Wieland et al. 2004). Although the chimpanzee model is highly predictive of a human HBV infection, it has been largely disregarded due to ethical concerns. After experimental intrahepatic injection of HBV DNA, macaques were shown to promote HBV replication (Gheit et al. 2002).

Tupaia or tree shrews is another animal model suitable for experimental HBV infection. In Vitro *and* In Vivo infection with HBV causes mild infection characterized by minimal viral replication (Walter et al. 1996). Nevertheless, infection of neonatal Tupaia leads to persistent infection characterized by minimal viremia and certain pathological alterations, including the development of fibrosis and HCC (Yang et al. 2015). Notably, the use of primary Tupaia hepatocytes has been crucial in determining the cellular receptor and other factors involved in HBV attachment (Schulze et al. 2007; Yan et al. 2012).

10.3.5 HBV Transfection Mouse Models

Mice are widely recognized as the most studied and practical of small laboratory animals. Nevertheless, it is important to note that mice are not susceptible to HBV infection.

To solve this problem, transgenic mice have been created to express proteins associated with HBV, such as HBsAg, HBeAg, and HBx (Chisari et al. 1985; Kim et al. 1991; Milich et al. 1990). Murine hepatocytes have been shown to generate infectious HBV virions using transgenic mice endowed with the complete HBV genome (Guidotti et al. 1995). The efficacy of polymerase inhibitors, specifically lamivudine and entecavir, has been assessed in HBV transgenic mice (Weber et al. 2002). These animal models were also used to test the efficacy of cytokines (Wieland et al. 2000) and HBV-specific siRNAs in inhibiting viral replication (McCaffrey et al. 2003). Genetically engineered mice are extensively employed as a model for studying chronic HBV infection (Dembek et al. 2015). Therapeutic vaccination techniques have been also evaluated in transgenic models in different studies (Buchmann et al. 2013; Farag et al. 2012; Li et al. 2015a). The use of HBV transgenic mice has proven to be beneficial in preclinical evaluations of antiviral agents and vaccination therapeutic strategies. However, it is important to note that this model has a significant drawback, as it is not suitable for investigating the mechanisms and substances that interfere with the progression of HBV infection, the production of cccDNA, and viral clearance (Dandri et al. 2017).

An alternative model for assessing HBV replication involved the use of hydrodynamic injection or virus vector-mediated transduction techniques for inserting HBV genome into murine subjects.

The hydrodynamic injection is a reliable method for delivering DNA into hepatocytes (Liu et al. 1999). However, this technique causes severe liver damage and ALT elevation directly after injection (Huang et al. 2012). This approach enabled researchers to explore the immunological responses that emerge during acute infections. This system also has the advantage of an In Vivo analysis of different HBV genotypes and variations in a short time (Qin et al. 2013). In addition, animal models based on hydrodynamics have been employed to assess the efficacy of antiviral medications, such as the antiviral activity of hairpin RNA against HBV mRNA sequence (McCaffrey et al. 2003). The CRISPR/Cas9 system is a novel technique used for precise cleavage of DNA at specific sites, employing a synthetic guide RNA (gRNA) that corresponds to the target DNA sequence (Gasiunas et al. 2012; Jinek et al. 2012). In hydrodynamics HBV persistence mouse model, viral replication markers and HBV-expressing vector levels were reduced without any signs of toxicity by using HBV-specific gRNAs (Lin et al. 2014; Ramanan et al. 2015). Indicating that by preventing HBV replication and expression, the CRISPR/Cas9 system can be a therapeutic approach that helps treat HBV infection (Dong et al. 2015; Zhen et al. 2015).

A less harmful method for animals has been developed to facilitate HBV genome delivery into hepatocytes using viral vectors. Adenoviral vectors provide reliable and efficient delivery of foreign DNA into the liver of immunologically competent animals, and recombinant DNA techniques make it relatively simple to manipulate these vectors (Bramson et al. 1995). For a couple of months following intravenous administration of viral vectors derived from adenoviruses containing HBV genomes (Ad-HBV) in mice with intact immune systems, the synthesis of HBV proteins occurred prior to the elimination of the virus through immune responses specifically targeting HBV (John von Freyend et al. 2011; Sprinzl et al. 2001). The adoption of this particular acute HBV infection model facilitates the study of immune-mediated pathways involved in viral clearance, specifically focusing on the role of NKT cells (Zeissig et al. 2012). The administration of small doses of Ad-HBV can further extend the infection phase, creating conditions that are comparable to chronic infection (Huang et al. 2012).

In addition, recombinant adeno-associated viruses (AAV) have been employed for the delivery of replication-competent HBV genomes. In the AAV-HBV mouse model, HBV particles and HBsAg can be produced in a sustained manner for more than a year (Dion et al. 2013; Yang et al. 2014). Therefore, this approach enables the establishment of chronic HBV infection in mouse models (Dion et al. 2013). Therapeutic vaccination with the TLR9 agonist CpG-B was evaluated in AAV-HBV-infected mice, resulting in not only the elimination of viremia but also robust antibody and T-cell responses (Yang et al. 2014). The use of mice infected with AAV/HBV as an experimental model holds promise in enhancing the comprehension of the underlying mechanism(s) responsible for HBV immunotolerance, as well as facilitating the development of novel immunotherapeutic approaches to combat HBV infections (Yang et al. 2014). Furthermore, the presence of HBV cccDNA was observed in the hepatic tissues of mice that were transduced with AAV carrying the HBV genome. In light of these findings, it is plausible that the AAV-HBV/mouse model could serve as a valuable tool for assessing the effectiveness of drugs aimed at altering the regulation and persistence of HBV cccDNA (Lucifora et al. 2017).

10.3.6 Liver Humanized Mouse Models for HBV Infection

Extensive research and development efforts have been dedicated to the investigation of human hepatocyte transplantation in mice in recent years. Mice whose livers have been effectively repopulated with human hepatocytes demonstrate permissiveness to HBV infection and the subsequent generation of HBV cccDNA (Bissig et al. 2010).

Alb-urokinase-type plasminogen activator (uPA) transgenic mice were initially used to describe the transplanted hepatocytes' remarkable regeneration capacity (Rhim et al. 1995). The use of uPA-based humanized mice with chimeric livers has spread quickly throughout the hepatitis research community, and numerous uPA animals have been produced using the same techniques in mice with different genetic origins (Lütgehetmann et al. 2012; Petersen et al. 2008). Fumaryl acetoacetate hydrolase (FAH) defective mice provide an alternative humanized model in which hepatocyte failure can be produced as desired (Azuma et al. 2007). The double immunological deletion of Rag2 and the gamma chain of the IL-2 receptor (Il2rg) were paired with the Fah deficient chimeric mice to create "the FRG mouse model," which increased levels of human chimerism (Bissig et al. 2007; Bissig et al. 2010).

To determine whether HBV can evade antiviral defense induction and alter the expression profile of genes implicated in bile acid metabolism, chimeric mice with human livers were used (Lütgehetmann et al. 2011; Oehler et al. 2014). Furthermore, these chimeric mice serve as a valuable model for conducting In Vivo experiments to evaluate the efficacy of direct antiviral agents and to gain a better understanding of the mechanism by which authorized treatment regimens, such as IFN, exert their effects (Allweiss et al. 2014; Lütgehetmann et al. 2011; Nakagawa et al. 2013). Also, humanized mice infected with HBV were used to evaluate the potential of RNA interference (RNAi) therapy for preventing the spread of drug-resistant virus genotypes (Shih et al. 2015). An additional investigation revealed that the application of a novel monoclonal antibody, which exhibits specificity toward a particular epitope on the HBsAg, facilitated sustained inhibition of HBV in various persistent HBV mouse models, including chimeric mice infected with HBV (Zhang et al. 2016).

However, the examination of adaptive immune responses and immunotherapeutic techniques is not feasible in human liver chimera mice due to their inherent genetic immune deficiency. (Dandri et al. 2017). To overcome these constraints, mice with both human hepatocytes and an immune system were developed: AFC8 and A2/NSG models (Bility et al. 2014; Washburn et al. 2011).

In conclusion, the human hepatocyte chimeric mouse, characterized by the repopulation of mouse livers with human hepatocytes, emerges as the most favorable mouse model for investigating chronic HBV infection and its subsequent establishment of cccDNA. However, these mice will require human immune system reconstitution in order to be used for vaccine research.

10.4 Coronaviruses

Coronaviruses (CoVs) are a group of enveloped, positive-sense RNA viruses that are part of the Nidovirales order. Viruses within this taxonomic family have the capacity to infect both humans and a wide range of other animal species (Weiss et al. 2005). *Alphacoronaviruses*, *Betacoronaviruses*, *Gammacoronaviruses*, and *Deltacoronaviruses* are the four genera of coronaviruses (Adams et al. 2017). Only the first two genera can infect humans. Human coronaviruses were divided into seven species. Four prevalent species consistently circulate within the human population and are responsible for inducing respiratory infections of varying severity. These species are identified as HCoV-OC43, HCoV-HKU1, HCoV-229E, and HCoV-NL63 (Vabret et al. 2003; Van Der Hoek et al. 2006; Walsh et al. 2013; Woo et al. 2005). The three remaining species, severe acute respiratory syndrome coronavirus (SARS-CoV), Middle East respiratory syndrome coronavirus (MERS-CoV), and the newly discovered SARS-CoV-2, all cause severe acute respiratory infections that threaten public health (Huang et al. 2020; Liang et al. 2020; Peiris et al. 2004).

A host reservoir for the three fatal CoVs, SARS-CoV, MERS-CoV, and SARSCoV-2, has been hypothesized to be wild bats (Cui et al. 2019; Wu et al. 2021; Zhou et al. 2020). The potential intermediate hosts of SARS-CoV and MERS-CoV have been identified as civets and dromedary camels, respectively (Raj et al. 2014; Shi and Hu 2008). Pangolins were proposed as a possible intermediate host for the recently discovered SARS-CoV-2 (Zhang et al. 2020). To date, scientific evidence has shown that hamsters, ferrets, and NHPs exhibit susceptibility to SARS-CoV-2 infection (Chan et al. 2020; Kim et al. 2020; Lu et al. 2020).

Animal models are essential for studying the cause and pathogenesis of such infectious diseases. Furthermore, their contributions could extend to the realm of scientific inquiry and advancement, specifically in the domain of vaccine and antiviral research and development, aimed at mitigating the impact of human coronavirus infections through treatment and prevention strategies. In this section, we provide a concise overview of the outcomes derived from animal models employed in the advancement of antiviral agents and vaccines targeting the three lethal CoVs infections.

10.4.1 Severe Acute Respiratory Syndrome Coronavirus (SARS-CoV)

The advancement of antiviral therapies and vaccines targeting SARS-CoV infection has been facilitated by the research conducted in animal models. Several animal models, such as rhesus macaques, mice, hamsters, and ferrets, have been employed to assess various compounds' efficacy in inhibiting SARS-CoV infection. These compounds include siRNA inhibitors (Li et al. 2005), interferon-α (Barnard et al. 2006), stinging nettle lectin (Kumaki et al. 2011) as well as two monoclonal antibodies, namely MAB201 and CR3014 (Greenough et al. 2005; Roberts et al. 2006; Ter Meulen et al. 2004).

Moreover, various animal models, including Rhesus macaques, cynomolgus macaques, mice, ferrets, and hamsters, have been used as experimental models in the research and development of diverse vaccines targeting SARS-CoV infection. The vaccines under consideration encompass a range of types, including adenovirus vector vaccines (Gao et al. 2003; Kobinger et al. 2007), inactivated vaccines (Qin et al. 2006), subunit vaccine (Bisht et al. 2005), recombinant modified vaccinia virus Ankara (MVA) vector vaccine (Bisht et al. 2005; Zhiwei Chen et al. 2005; Weingartl et al. 2004), recombinant BHPIV3/SARS vector vaccine (Buchholz et al. 2004) and DNA vaccine (Yang et al. 2004).

10.4.2 Middle East Respiratory Syndrome Coronavirus (MERS-CoV)

Scientists have shown that a MERS treatment regimen, including interferon (IFN-α2b) and ribavirin, improves clinical outcomes and suppresses viral replication in rhesus macaque models (Falzarano et al. 2013). Furthermore, the efficacy of monoclonal antibodies (MCA1, hMS-1, and 4C2h) and a polyclonal antibody (SAB-301) was tested and validated using the common marmoset, CAG-hDPP4 transgenic mouse model, and Ad5-hDPP4-transduced mouse model (Chen et al. 2017; Li et al. 2015b; Luke et al. 2016; Qiu et al. 2016). Also, Remdesivir was tested on transgenic mice and was found to significantly reduce MERS titers (Sheahan et al. 2020). The viral titer and infection-related pathology in marmosets can be efficiently reduced by lopinavir/ritonavir or interferon-1b, either alone or in combination (Chan et al. 2015).

Regarding MERS vaccine development only transgenic mice and NHPs have been used as animal model. Consequently, vaccines such as the recombinant subunit vaccine (Du et al. 2013), Measles Virus vector vaccine (Malczyk et al. 2015), recombinant MVA-MERS-S vaccine (Volz et al. 2015), and recombinant bivalent vaccine (MERS-CoV and Rabies Virus) (Kato et al. 2019) have all been evaluated using transgenic mouse models. In contrast, rhesus macaque models have been employed in the advancement of various vaccines, such as recombinant VSV vector vaccines (Liu et al. 2018), recombinant S protein receptor binding domain (RBD) vaccines (Lan et al. 2015), and ChAdOx1 MERS vaccines (Doremalen et al. 2020a). A DNA-based vaccine targeting the Spike protein of Middle East Respiratory Syndrome Coronavirus (MERS-CoV) was assessed in animal models, including mice and NHPs. The findings demonstrated its efficacy in providing protection against MERS-CoV in rhesus macaques (Muthumani et al. 2015).

10.4.3 SARS-CoV-2

Various animal models, such as NHPs (specifically rhesus macaques and cynomolgus macaques), mice, hamsters, and ferrets, have been extensively employed in the assessment and advancement of diverse therapeutic strategies and vaccines aimed at mitigating the SARS-CoV pandemic (Lin et al. 2022).

Currently, the World Health Organization (WHO) has included a total of eleven COVID-19 vaccines on its emergency use list (WHO 2023). The Food and Drug Administration (FDA) has licensed four medications and six monoclonal antibodies for emergency use (FDA 2023). Although not all vaccines and therapies have been tested in animals, here, our primary focus will be on the use of animal models to test and develop therapeutic and preventive interventions to combat SARS-CoV-2 infection.

10.4.3.1 Animal Models and SARS-CoV-2 Drugs Development

The efficacy of Molnupivarir in reducing viral loads in the upper respiratory system and preventing viral transmission has been demonstrated in ferrets infected with SARS-CoV-2 (Cox et al. 2021b). The administration of Molnupivarir resulted in a significant reduction in viral RNA copies in the lungs of hamsters infected with various strains of SARS-CoV-2 (Abdelnabi et al. 2021).

Paxlovid is a pharmaceutical formulation that combines Nirmatrelvir (PF-07321332) and ritonavir. Nirmatrelvir is an orally administered protease inhibitor specifically designed to target the SARS-CoV-2 virus. On the other hand, Ritonavir acts as an inhibitor of the CYP3A4 isoenzyme (Heskin et al. 2022). In experimental murine models infected with the SARS-CoV-2 MA10 strain,

PF-07321332 efficiently minimized lung lesions and viral load (Owen et al. 2021). Additionally, the compound PF-07321332 has been observed to provide protection to Syrian hamsters against various strains of SARS-CoV-2. Moreover, it effectively inhibits the transmission of the viral delta variant to untreated animals (Abdelnabi et al. 2022). Rhesus macaque models have been used to assess the Remdesivir efficacy, it was demonstrated that early injection of the Remdesivir into SARS-CoV-2 infected animals results in reducing viral replication and immune system invasion in the lungs (Williamson et al. 2020). Remdesivir family prodrug (GS-441524) was shown to suppress SARS-CoV-2 viral replication and attenuate pulmonary inflammation in ferret and mouse models when administered orally or intraperitoneally (Cox et al. 2021a; Li et al. 2022).

Developing drugs for SARS-CoV-2 should also involve targeting host factors (Kausar et al. 2021). The administration of the TMPRSS2 protease inhibitor (nafomastat) as a prophylactic measure resulted in a significant reduction in pulmonary SARS-CoV-2 viral loads in mouse models (Li et al. 2021b). The infection of SARS-CoV-2 leads to an exaggerated inflammatory response characterized by the release of numerous cytokines, known as a cytokine storm. This dysregulated immune response has been identified as a significant contributor to mortality in individuals with severe cases of COVID-19 (Ye et al. 2020). Both the prodrug SSK1, which is activated by β-galactosidase, and the JAK1/JAK2 inhibitor Baricitinib have demonstrated efficacy in mitigating the inflammatory response and providing protection against SARS-CoV-2 infection in rhesus macaques (Hoang et al. 2021; Lu et al. 2021).

10.4.3.2 Animal Models and Monoclonal Antibodies

Monoclonal antibodies have rapidly emerged as an indispensable instrument in combating the COVID-19 pandemic due to their high degree of specificity (Hwang et al. 2022). Notably, a research team has created a delivery method for CB6 monoclonal antibody to the lungs of mouse models infected with the MP7 strain of SARS-CoV-2. This innovative vehicle, known as VEEV-VRP-CB6, was administered intranasally. The study demonstrated that this approach successfully averted body weight loss and reduced the viral titer in the lungs of the infected mice (Li et al. 2021a). Additionally, in mouse models pretreatment with the anti-SARS-CoV-2 monoclonal antibody (1B07), which targets SARS-CoV-2 S protein, was demonstrated to successfully block SARS-CoV-2 infection, prevent body weight loss, and reduce the production of many pro-inflammatory cytokines and chemokines (Hassan et al. 2020a). Furthermore, the efficacy of the monoclonal antibody COVA1–18 in limiting SARS-CoV-2 infections has been demonstrated across three distinct animal models, namely hamsters, mice, and NHPs (Maisonnasse et al. 2021).

Therapeutics based on a mixture of several monoclonal antibodies demonstrated enhanced efficacy against SARS-CoV-2 variants in contrast to single monoclonal antibody therapy (Ning et al. 2021). For example, a research investigation demonstrated that rhesus macaques, upon receiving intravenous injections of LY-CoV555 monoclonal antibodies, exhibited a notable reduction in viral replication and loads within their lower and upper respiratory tracts when infected with SARS-CoV-2 after a 24-hour interval (Jones et al. 2021). Similarly, the administration of REGEN-COV therapy, both prior and subsequent to infection, demonstrated protective effects against weight loss in rhesus macaques and hamsters (Baum et al. 2020).

10.4.3.3 Animal Models and SARS-CoV-2 Vaccines

The evaluation of potential vaccines against SARS-CoV-2 heavily relies on animal models, with a particular emphasis on NHPs, transgenic mice, and receptor-transduced mice. Among these models, rhesus macaques have played a vital role in assessing various SARS-CoV-2 vaccines (Lu et al. 2020). These vaccines include DNA vaccines (Li et al. 2021c; Yu et al. 2020), inactivated virus vaccines (PiCoVacc, BBIBF-corV, BBV152) (Gao et al. 2020; Wang et al. 2020; Yadav et al. 2021), mRNA vaccines (BNT162b1, BNT162b2, mRNA-1273) Corbett et al. 2021; Vogel et al. 2021), subunit vaccines (S-Trimer) (Liang et al. 2021), receptor binding domain (RBD)-based nanoparticle vaccines (Ma et al. 2020) and different adenovirus vector-based vaccines (Feng et al. 2020; Mercado et al. 2020; Doremalen et al. 2020b).

The ChAd-SARS-CoV-2-S vaccine uses chimpanzee adenoviruses to encode the entire SARS-CoV-2 S protein and has been proven effective in protecting transgenic mice against respiratory tract infections caused by SARS-CoV-2 with just a single dose (Hassan et al. 2020b). Moreover, PiCoVacc, a vaccine made from purified and inactivated SARS-CoV-2, has been effective in producing neutralizing antibodies specific to SARS-CoV-2 in mice, rats, and rhesus macaques. (Gao et al. 2020). Various animals, including mice, rats, guinea pigs, rabbits, and NHPs, have shown that BBIBP-CorV, an inactivated SARS-CoV-2 vaccine, was effective in producing neutralizing antibodies (Wang et al. 2020). In hamsters, Ad26 vaccines provided protection against pneumonia and death caused by a high-dose SARS-CoV-2 infection (Tostanoski et al. 2020). Another study found that Ad5-nCoV, which encodes the full-length spike protein of SARS-CoV-2, was effective in protecting mice challenged with mouse-adapted SARS-CoV-2 and ferrets vaccinated with the same (Wu et al. 2020).

In the ongoing battle against COVID-19, messenger RNA (mRNA) vaccines have proven to be crucial. NPHs vaccinated with mRNA-1273 showed strong neutralizing activity against SARS-CoV-2 and experienced rapid protection in the upper and lower respiratory tract without any lung damage (Corbett et al. 2020). Furthermore, mRNA-1273 vaccine has been shown to be effective in protecting NPHs from the SARS-CoV-2 β variant (Corbett et al. 2021). Additionally, BNT162b2 protects the lower respiratory tract from SARS-CoV-2 viral RNA and does not appear to worsen the disease in macaque models (Vogel et al. 2021).

10.5 Conclusion

This research delves into the global impact of viral infections on public health, with a particular emphasis on viruses like HIV-1, HCV, HBV, and CoVs like SARS-CoV, MERS-CoV, and SARS-CoV-2. The following section provides an outline of the role animal models have played in assessing therapeutic treatments and creating vaccines to combat viral infections.

References

Abdelnabi R, Foo CS, De Jonghe S, Maes P, Weynand B, Neyts J (2021) Molnupiravir inhibits replication of the emerging SARS-CoV-2 variants of concern in a hamster infection model. J Infect Dis 224(5):749–753

Abdelnabi R, Foo CS, Jochmans D, Vangeel L, De Jonghe S, Augustijns P et al (2022) The oral protease inhibitor (PF-07321332) protects Syrian hamsters against infection with SARS-CoV-2 variants of concern. Nat Commun 13(1):71

Abel K, Compton L, Rourke T, Montefiori D, Lu D, Rothaeusler K et al (2003) Simian-human immunodeficiency virus SHIV89 6-induced protection against intravaginal challenge with pathogenic SIVmac239 is independent of the route of immunization and is associated with a combination of cytotoxic T-lymphocyte and alpha interferon responses. Virology 77(5):3099–3118

Abravanel F, Lhomme S, Dubois M, Peron JM, Alric L, Kamar N, Izopet J (2013) Hepatitis E virus. Med Mal Infect 43(7):263–270

Adams MJ, Lefkowitz EJ, King AM, Harrach B, Harrison RL, Knowles NJ et al (2017) 50 years of the international committee on taxonomy of viruses: progress and prospects. Arch Virol 162(5):1441–1446

Agarwal Y, Beatty C, Biradar S, Castronova I, Ho S, Melody K, Bility MT (2020) Moving beyond the mousetrap: current and emerging humanized mouse and rat models for investigating prevention and cure strategies against HIV infection and associated pathologies. Retrovirology 17(1):1–11

Akkina RJ (2013) New generation humanized mice for virus research: comparative aspects and future prospects. Virology 435(1):14–28

Allweiss L, Volz T, Lütgehetmann M, Giersch K, Bornscheuer T, Lohse AW et al (2014) Immune cell responses are not required to induce substantial hepatitis B virus antigen decline during pegylated interferon-alpha administration. Hepatology 60(3):500–507

Ambrose Z, Boltz V, Palmer S, Coffin JM, Hughes SH, KewalRamani VN (2004) In vitro characterization of a simian immunodeficiency virus-human immunodeficiency virus (HIV) chimera expressing HIV type 1 reverse transcriptase to study antiviral resistance in pigtail macaques. Virology 78(24):13553–13561

Andersen PI, Ianevski A, Lysvand H, Vitkauskiene A, Oksenych V, Bjørås M et al (2020) Discovery and development of safe-inman broad-spectrum antiviral agents. Int J Infect Dis 93:268–276

Azuma H, Paulk N, Ranade A, Dorrell C, Al-Dhalimy M, Ellis E et al (2007) Robust expansion of human hepatocytes in Fah−/−/Rag2−/−/Il2rg−/− mice. Nat Biotechnol 25(8):903–910

Baechlein C, Fischer N, Grundhoff A, Alawi M, Indenbirken D, Postel A et al (2015) Identification of a novel Hepacivirus in domestic cattle from Germany. J Virol 89(14):7007–7015

Balzarini J, Weeger M, Camarasa M-J, Declercq E, Uberla KJB, Communications BR (1995) Sensitivity/resistance profile of a simian immunodeficiency virus containing the reverse transcriptase gene of human immunodeficiency virus type 1 (HIV-1) toward the HIV-1-specific non-nucleoside reverse transcriptase inhibitors. Biochem Biophys Res Commun 211(3):850–856

Barnard DL, Day CW, Bailey K, Heiner M, Montgomery R, Lauridsen L et al (2006) Evaluation of immunomodulators, interferons and known in vitro SARS-coV inhibitors for inhibition of SARS-coV replication in BALB/c mice. Antiviral Chem Chemother 17(5):275–284

Barré-Sinoussi F, Montagutelli X (2015) Animal models are essential to biological research: issues and perspectives. Future Sci 1(4):FSO63

Barth H, Robinet E, Liang TJ, Baumert TF (2008) Mouse models for the study of HCV infection and virus–host interactions. Hepatology 49(1):134–142

Baum A, Ajithdoss D, Copin R, Zhou A, Lanza K, Negron N et al (2020) REGN-COV2 antibodies prevent and treat SARS-CoV-2 infection in rhesus macaques and hamsters. Science 370(6520):1110–1115

Beames B, Chavez D, Lanford RE (2001) GB virus B as a model for hepatitis C virus. ILAR J 42(2):152–160

Bility MT, Cheng L, Zhang Z, Luan Y, Li F, Chi L et al (2014) Hepatitis B virus infection and immunopathogenesis in a humanized mouse model: induction of human-specific liver fibrosis and M2-like macrophages. PLoS Pathog 10(3):e1004032

Billerbeck E, De Jong Y, Dorner M, De La Fuente C, Ploss A (2013) Animal models for hepatitis C. Springer, Berlin, Heidelberg, pp 49–86

Billerbeck E, Wolfisberg R, Fahnøe U, Xiao JW, Quirk C, Luna JM et al (2017) Mouse models of acute and chronic hepacivirus infection. Science 357(6347):204–208

Bisht H, Roberts A, Vogel L, Subbarao K, Moss B (2005) Neutralizing antibody and protective immunity to SARS coronavirus infection of mice induced by a soluble recombinant polypeptide containing an N-terminal segment of the spike glycoprotein. Virology 334(2):160–165

Bissig K-D, Le TT, Woods N-B, Verma IM (2007) Repopulation of adult and neonatal mice with human hepatocytes: a chimeric animal model. PNAS 104(51):20507–20511

Bissig K-D, Wieland SF, Tran P, Isogawa M, Le TT, Chisari FV, Verma IM (2010) Human liver chimeric mice provide a model for hepatitis B and C virus infection and treatment. Clin Investig 120(3):924–930

Bourlière M, Pietri O (2019) Hepatitis C virus therapy: no one will be left behind. Int J Antimicrob Agents 53(6):755–760

Bramson JL, Graham FL, Gauldie J (1995) The use of adenoviral vectors for gene therapy and gene transfer in vivo. Curr Opin Biotechnol 6(5):590–595

Brooks DG, Hamer DH, Arlen PA, Gao L, Bristol G, Kitchen CM et al (2003) Molecular characterization, reactivation, and depletion of latent HIV. Immunity 19(3):413–423

Buchholz UJ, Bukreyev A, Yang L, Lamirande EW, Murphy BR, Subbarao K, Collins PL (2004) Contributions of the structural proteins of severe acute respiratory syndrome coronavirus to protective immunity. PNAS 101(26):9804–9809

Buchmann P, Dembek C, Kuklick L, Jäger C, Tedjokusumo R, von Freyend MJ et al (2013) A novel therapeutic hepatitis B vaccine induces cellular and humoral immune responses and breaks tolerance in hepatitis B virus (HBV) transgenic mice. Vaccine 31(8):1197–1203

Bukh J (2012) Animal models for the study of hepatitis C virus infection and related liver disease. Gastroenterology 142(6):1279–1287. e1273

Byrareddy SN, Arthos J, Cicala C, Villinger F, Ortiz KT, Little D et al (2016) Sustained virologic control in SIV+ macaques after antiretroviral and α4β7 antibody therapy. Science 354(6309):197–202

Calenda G, Keawvichit R, Arrode-Brusés G, Pattanapanyasat K, Frank I, Byrareddy SN et al (2018) Integrin α4β7 blockade preferentially impacts CCR6+ lymphocyte subsets in blood and mucosal tissues of naive rhesus macaques. Immunology 200(2):810–820

Callaway EJN (2014) Hepatitis C drugs not reaching poor: treatment guidelines for virus highlight challenge of paying for expensive drugs in low-income countries. Nature 508(7496):295–297

Campagna MR, Liu F, Mao R, Mills C, Cai D, Guo F et al (2013) Sulfamoylbenzamide derivatives inhibit the assembly of hepatitis B virus nucleocapsids. Virology 87(12):6931–6942

Carrat F, Fontaine H, Dorival C, Simony M, Diallo A, Hezode C et al (2019) Clinical outcomes in patients with chronic hepatitis C after direct-acting antiviral treatment: a prospective cohort study. Lancet 393(10179):1453–1464

Chan JF-W, Yao Y, Yeung M-L, Deng W, Bao L, Jia L et al (2015) Treatment with lopinavir/ritonavir or interferon-β1b improves outcome of MERS-CoV infection in a nonhuman primate model of common marmoset. Infect Dis 212(12):1904–1913

Chan JF-W, Zhang AJ, Yuan S, Poon VK-M, Chan CC-S, Lee AC-Y et al (2020) Simulation of the clinical and pathological manifestations of coronavirus disease 2019 (COVID-19) in a golden Syrian hamster model: implications for disease pathogenesis and transmissibility. Clin Infect Dis 71(9):2428–2446

Chen Z, Zhang L, Qin C, Ba L, Yi CE, Zhang F et al (2005) Recombinant modified vaccinia virus Ankara expressing the spike glycoprotein of severe acute respiratory syndrome coronavirus induces protective neutralizing antibodies primarily targeting the receptor binding region. Cell Res 79(5):2678–2688

Chen J, Zhao Y, Zhang C, Chen H, Feng J, Chi X et al (2014) Persistent hepatitis C virus infections and hepatopathological manifestations in immune-competent humanized mice. Infect Dis 24(9):1050–1066

Chen Z, Bao L, Chen C, Zou T, Xue Y, Li F et al (2017) Human neutralizing monoclonal antibody inhibition of Middle East respiratory syndrome coronavirus replication in the common marmoset. Virology 215(12):1807–1815

Chisari F, Pinkert C, Milich D, Filippi P, McLachlan A, Palmiter R, Brinster RJS (1985) A transgenic mouse model of the chronic hepatitis B surface antigen carrier state. Science 230(4730):1157–1160

Colledge D, Civitico G, Locarnini S, Shaw T (2000) In vitro antihepadnaviral activities of combinations of penciclovir, lamivudine, and adefovir. Antimicrob Agents Chemother 44(3):551–560

Corbett KS, Flynn B, Foulds KE, Francica JR, Boyoglu-Barnum S, Werner AP, Flach B, O'Connell S, Bock KW, Minai M, Nagata BM, Andersen H, Martinez DR, Noe AT, Douek N, Donaldson MM, Nji NN, Alvarado GS, Edwards DK, Flebbe DR, Lamb E, Doria-Rose NA, Lin BC, Louder MK, O'Dell S, Schmidt SD, Phung E, Chang LA, Yap C, Todd J-PM, Pessaint L, Van Ry A, Browne S, Greenhouse J, Putman-Taylor T, Strasbaugh A, Campbell T-A, Cook A, Dodson A, Steingrebe K, Shi W, Zhang Y, Abiona OM, Wang L, Pegu A, Yang ES, Leung K, Zhou T, Teng I-T, Widge A, Gordon I, Novik L, Gillespie RA, Loomis RJ, Moliva JI, Stewart-Jones G, Himansu S, Kong W-P, Nason MC, Morabito KM, Ruckwardt TJ, Ledgerwood JE, Gaudinski MR, Kwong PD, Mascola JR, Carfi A, Lewis MG, Baric RS, McDermott A, Moore IN, Sullivan NJ, Roederer M, Seder RA, Graham BS (2020) Evaluation of the mRNA-1273 vaccine against SARS-CoV-2 in nonhuman primates. N Engl J Med 383(16):1544–1555. https://doi.org/10.1056/NEJMoa2024671

Corbett KS, Werner AP, Connell SO, Gagne M, Lai L, Moliva JI, Flynn B, Choi A, Koch M, Foulds KE (2021) mRNA-1273 protects against SARS-CoV-2 beta infection in nonhuman primates. Nat Immunol 22(10):1306–1315. https://doi.org/10.1038/s41590-021-01021-0

Cote PJ, Korba BE, Miller RH, Jacob JR, Baldwin BH, Hornbuckle WE et al (2000) Effects of age and viral determinants on chronicity as an outcome of experimental woodchuck hepatitis virus infection. Hepatology 31(1):190–200

Cox RM, Wolf JD, Lieber CM, Sourimant J, Lin MJ, Babusis D et al (2021a) Oral prodrug of remdesivir parent GS-441524 is efficacious against SARS-CoV-2 in ferrets. Nat Commun 12(1):6415

Cox RM, Wolf JD, Plemper RK (2021b) Therapeutically administered ribonucleoside analogue MK-4482/EIDD-2801 blocks SARS-CoV-2 transmission in ferrets. Nat Commun 6(1):11–18

Cui J, Li F, Shi ZL (2019) Origin and evolution of pathogenic coronaviruses. Nat Rev Microbiol 17(3):181–192

D'Ugo E, Argentini C, Giuseppetti R, Canitano A, Catone S, Rapicetta M (2010) The woodchuck hepatitis B virus infection model for the evaluation of HBV therapies and vaccine therapies. Expert Opin Drug Discovery 5(12):1153–1162

Dandri M, Burda MR, Will H, Petersen J (2000) Increased hepatocyte turnover and inhibition of woodchuck hepatitis B virus replication by adefovir in vitro do not lead to reduction of the closed circular DNA. Hepatology 32(1):139–146
Dandri M, Petersen J (2017) Animal models of HBV infection. Best Pract Res Clin Gastroenterol 31(3):273–279
Daniel MD, Kirchhoff F, Czajak SC, Sehgal PK, Desrosiers RC (1992) Protective effects of a live attenuated SIV vaccine with a deletion in the nef gene. Science 258(5090):1938–1941
De Clercq E, Li G (2016) Approved antiviral drugs over the past 50 years. Clin Microbiol Rev 29(3):695–747
Dembek C, Protzer U (2015) Mouse models for therapeutic vaccination against hepatitis B virus. Med Microbiol Immunol 204(1):95–102
Denton PW, Long JM, Wietgrefe SW, Sykes C, Spagnuolo RA, Snyder OD et al (2014) Targeted cytotoxic therapy kills persisting HIV infected cells during ART. PLoS Pathog 10(1):e1003872
Deres K, Schroder CH, Paessens A, Goldmann S, Hacker HJ, Weber O et al (2003) Inhibition of hepatitis B virus replication by drug-induced depletion of nucleocapsids. Science 299(5608):893–896
Dion S, Bourgine M, Godon O, Levillayer F, Michel M-L (2013) Adeno-associated virus-mediated gene transfer leads to persistent hepatitis B virus replication in mice expressing HLA-A2 and HLA-DR1 molecules. Virology 87(10):5554–5563
Dong C, Qu L, Wang H, Wei L, Dong Y, Xiong S (2015) Targeting hepatitis B virus cccDNA by CRISPR/Cas9 nuclease efficiently inhibits viral replication. Antivir Res 118:110–117
Dorner M, Rice CM, Ploss A (2013) Study of hepatitis C virus entry in genetically humanized mice. Methods 59(2):249–257
Drexler JF, Corman VM, Müller MA, Lukashev AN, Gmyl A, Coutard B et al (2013) Evidence for novel hepaciviruses in rodents. PLoS Pathog 9(6):e1003438
Du L, Kou Z, Ma C, Tao X, Wang L, Zhao G et al (2013) A truncated receptor-binding domain of MERS-CoV spike protein potently inhibits MERS-CoV infection and induces strong neutralizing antibody responses: implication for developing therapeutics and vaccines. PLoS One 8(12):e81587
Dupinay T, Gheit T, Roques P, Cova L, Chevallier-Queyron P, Tasahsu S-i et al (2013) Discovery of naturally occurring transmissible chronic hepatitis B virus infection among Macaca fascicularis from Mauritius Island. Hepatology 58(5):1610–1620
Eren R, Ilan E, Nussbaum O, Lubin I, Terkieltaub D, Arazi Y et al (2000) Preclinical evaluation of two human anti–hepatitis B virus (HBV) monoclonal antibodies in the HBV-trimera mouse model and in HBV chronic carrier chimpanzees. Hepatology 32(3):588–596
Eroshkin AM, LeBlanc A, Weekes D, Post K, Li Z, Rajput A et al (2014) bNAber: database of broadly neutralizing HIV antibodies. Nucleic Acids Res 42(D1):D1133–D1139
Evans DT, Silvestri G (2013) Non-human primate models in AIDS research. Curr Opin HIV AIDS 8(4):255
Falzarano D, De Wit E, Rasmussen AL, Feldmann F, Okumura A, Scott DP et al (2013) Treatment with interferon-α2b and ribavirin improves outcome in MERS-CoV–infected rhesus macaques. Nat Med 19(10):1313–1317
Farag MM, Tedjokusumo R, Flechtenmacher C, Asen T, Stremmel W, Müller M et al (2012) Immune tolerance against HBV can be overcome in HBV transgenic mice by immunization with dendritic cells pulsed by HBVsvp. Vaccine 30(42):6034–6039
Farci P, London WT, Wong DC, Dawson GJ, Vallari DS, Engle R, Purcell RH (1992) The natural history of infection with hepatitis C virus (HCV) in chimpanzees: comparison of serologic responses measured with first-and second-generation assays and relationship to HCV viremia. Infect Dis 165(6):1006–1011
FDA (2023) Emergency use authorization. Retrieved from https://www.fda.gov/emergency-preparedness-and-response/mcmlegal-regulatory-and-policy-framework/emergency-use-authorization#coviddrugs

Feng L, Wang Q, Shan C, Yang C, Feng Y, Wu J et al (2020) An adenovirus-vectored COVID-19 vaccine confers protection from SARS-COV-2 challenge in rhesus macaques. Nat Commun 11(1):4207

Fletcher SP, Chin DJ, Ji Y, Iniguez AL, Taillon B, Swinney DC et al (2012) Transcriptomic analysis of the woodchuck model of chronic hepatitis B. Hepatology 56(3):820–830

Foster WK, Miller DS, Scougall CA, Kotlarski I, Colonno RJ, Jilbert AR (2005) Effect of antiviral treatment with entecavir on age-and dose-related outcomes of duck hepatitis B virus infection. Virology 79(9):5819–5832

Gao W, Tamin A, Soloff A, D'Aiuto L, Nwanegbo E, Robbins PD et al (2003) Effects of a SARS-associated coronavirus vaccine in monkeys. Science 362(9399):1895–1896

Gao Q, Bao L, Mao H, Wang L, Xu K, Yang M et al (2020) Development of an inactivated vaccine candidate for SARS-CoV-2. Lancet 369(6499):77–81

Garcia JV (2016) In vivo platforms for analysis of HIV persistence and eradication. Clin Invest 126(2):424–431

Gasiunas G, Barrangou R, Horvath P, Siksnys V (2012) Cas9–crRNA ribonucleoprotein complex mediates specific DNA cleavage for adaptive immunity in bacteria. PNAS 109(39):E2579–E2586

Gelé T, Gouget H, Dereuddre-Bosquet N, Furlan V, Le Grand R, Lambotte O et al (2022) Pharmacological validation of longterm treatment with antiretroviral drugs in a model of siv-infected non-human primates. J Pharm 14(11):2282

Gheit T, Sekkat S, Cova L, Chevallier M, Petit MA, Hantz O et al (2002) Experimental transfection of Macaca sylvanus with cloned human hepatitis B virus. Gen Virol 83(7):1645–1649

Greenough TC, Babcock GJ, Roberts A, Hernandez HJ, Thomas WD Jr, Coccia JA et al (2005) Development and characterization of a severe acute respiratory syndrome—associated coronavirus—neutralizing human monoclonal antibody that provides effective immunoprophylaxis in mice. Infect Dis 191(4):507–514

Guidotti LG, Matzke B, Schaller H, Chisari FV (1995) High-level hepatitis B virus replication in transgenic mice. Virology 69(10):6158–6169

Guidotti LG, Rochford R, Chung J, Shapiro M, Purcell R, Chisari FV (1999) Viral clearance without destruction of infected cells during acute HBV infection. Science 284(5415):825–829

Halper-Stromberg A, Lu C-L, Klein F, Horwitz J, Bournazos S, Nogueira L et al (2014) Broadly neutralizing antibodies and viral inducers decrease rebound from HIV-1 latent reservoirs in humanized mice. Cell 158(5):989–999

Hartlage AS, Cullen JM, Kapoor A (2016) The strange, expanding world of animal Hepaciviruses. Ann Rev Virol 3(1):53–75. https://doi.org/10.1146/annurev-virology-100114-055104

Hassan AO, Case JB, Winkler ES, Thackray LB, Kafai NM, Bailey AL et al (2020a) A SARS-CoV-2 infection model in mice demonstrates protection by neutralizing antibodies. Cell 182(3):744–753.e744. https://doi.org/10.1016/j.cell.2020.06.011

Hassan AO, Kafai NM, Dmitriev IP, Fox JM, Smith BK, Harvey IB et al (2020b) A single-dose intranasal Chad vaccine protects upper and lower respiratory tracts against SARS-CoV-2. Cell 183(1):169–184.e113. https://doi.org/10.1016/j.cell.2020.08.026

Hatziioannou T, Evans DT (2012) Animal models for HIV/AIDS research. Nat Rev Microbiol 10(12):852–867

Heskin J, Pallett SJC, Mughal N, Davies GW, Moore LSP, Rayment M, Jones R (2022) Caution required with use of ritonavirboosted PF-07321332 in COVID-19 management. Lancet 399(10319):21–22

Hoang TN, Pino M, Boddapati AK, Viox EG, Starke CE, Upadhyay AA et al (2021) Baricitinib treatment resolves lower-airway macrophage inflammation and neutrophil recruitment in SARS-CoV-2-infected rhesus macaques. Cell 184(2):460–475.e421. https://doi.org/10.1016/j.cell.2020.11.007

Holt N, Wang J, Kim K, Friedman G, Wang X, Taupin V et al (2010) Zinc finger nuclease-mediated CCR5 knockout hematopoietic stem cell transplantation controls HIV-1 in vivo. Nat Biotechnol 28(8):839

Hoofnagle J (2002) Course and outcome of hepatitis C. Hepatology 36(S1):S21–S29

Houghton M, Abrignani S (2005) Prospects for a vaccine against the hepatitis C virus. Nature 436(7053):961–966

Howard CR, Fletcher NF (2012) Emerging virus diseases: can we ever expect the unexpected? J Emerging Microbes Infect 1(1):1–9

Huang LR, Gäbel YA, Graf S, Arzberger S, Kurts C, Heikenwalder M et al (2012) Transfer of HBV genomes using low doses of adenovirus vectors leads to persistent infection in immune competent mice. Lancet 142(7):1447–1450. e1443

Huang C, Wang Y, Li X, Ren L, Zhao J, Hu Y et al (2020) Clinical features of patients infected with 2019 novel coronavirus in Wuhan, China. Gastroenterology 395(10223):497–506

Huot N, Rascle P, Müller-Trutwin M (2019) Apport des modèles animaux dans la recherche sur le VIH. Virologie 23(4):229–240

Hwang Y-C, Lu R-M, Su S-C, Chiang P-Y, Ko S-H, Ke F-Y et al (2022) Monoclonal antibodies for COVID-19 therapy and SARS-CoV-2 detection. J Biomed Sci 29(1):1. https://doi.org/10.1186/s12929-021-00784-w

Jinek M, Chylinski K, Fonfara I, Hauer M, Doudna JA, Charpentier E (2012) A programmable dual-RNA–guided DNA endonuclease in adaptive bacterial immunity. Science 337(6096):816–821

John von Freyend M, Untergasser A, Arzberger S, Oberwinkler H, Drebber U, Schirmacher P, Protzer U (2011) Sequential control of hepatitis B virus in a mouse model of acute, self-resolving hepatitis B. J Viral Hepat 18(3):216–226

Jones BE, Brown-Augsburger PL, Corbett KS, Westendorf K, Davies J, Cujec TP et al (2021) The neutralizing antibody, LYCoV555,protects against SARS-CoV-2 infection in nonhuman primates. Science 13(593):eabf1906. https://doi.org/10.1126/scitranslmed.abf1906

Kapoor A, Simmonds P, Gerold G, Qaisar N, Jain K, Henriquez JA et al (2011) Characterization of a canine homolog of hepatitis C virus. PNAS 108(28):11608–11613

Kapoor A, Simmonds P, Scheel TKH, Hjelle B, Cullen JM, Burbelo PD et al (2013) Identification of rodent homologs of hepatitis C virus and pegiviruses. *MBio* 4(2):e00216–e00213. https://doi.org/10.1128/mBio.00216-13

Kato H, Takayama-Ito M, Iizuka-Shiota I, Fukushi S, Posadas-Herrera G, Horiya M et al (2019) Development of a recombinant replication-deficient rabies virus-based bivalent-vaccine against MERS-CoV and rabies virus and its humoral immunogenicity in mice. PLoS One 14(10):e0223684

Kausar S, Said Khan F, Ishaq Mujeeb Ur Rehman M, Akram M, Riaz M, Rasool G et al (2021) A review: mechanism of action of antiviral drugs. Int J Immunopathol Pharmacol 35:20587384211002621

Kim C-M, Koike K, Saito I, Miyamura T, Jay G (1991) HBx gene of hepatitis B virus induces liver cancer in transgenic mice. Nature 351(6324):317–320

Kim Y-I, Kim S-G, Kim S-M, Kim E-H, Park S-J, Yu K-M et al (2020) Infection and rapid transmission of SARS-CoV-2 in ferrets. Cell Host Microbe 27(5):704–709. e702

Kline C, Ndjomou J, Franks T, Kiser R, Coalter V, Smedley J et al (2013) Persistence of viral reservoirs in multiple tissues after antiretroviral therapy suppression in a macaque RT-SHIV model. PLoS One 8(12):e84275

Kobinger GP, Figueredo JM, Rowe T, Zhi Y, Gao G, Sanmiguel JC et al (2007) Adenovirus-based vaccine prevents pneumonia in ferrets challenged with the SARS coronavirus and stimulates robust immune responses in macaques. Vaccine 25(28):5220–5231

Köck J, Nassal M, MacNelly S, Baumert TF, Blum HE, von Weizsäcker F (2001) Efficient infection of primary tupaia hepatocytes with purified human and woolly monkey hepatitis B virus. Virology 75(11):5084–5089

Korin YD, Brooks DG, Brown S, Korotzer A, Zack JA (2002) Effects of prostratin on T-cell activation and human immunodeficiency virus latency. Virology 76(16):8118–8123

Kulkarni K, Jacobson IM, Tennant BC (2007) The role of the woodchuck model in the treatment of hepatitis B virus infection. Clin Liver Dis 11(4):707–725

Kumaki Y, Wandersee MK, Smith AJ, Zhou Y, Simmons G, Nelson NM et al (2011) Inhibition of severe acute respiratory syndrome coronavirus replication in a lethal SARS-CoV BALB/c mouse model by stinging nettle lectin, Urtica dioica agglutinin. Antivir Res 90(1):22–32
Kumar P, Ban H-S, Kim S-S, Wu H, Pearson T, Greiner DL et al (2008) T cell-specific siRNA delivery suppresses HIV-1 infection in humanized mice. Cell 134(4):577–586
Lan J, Yao Y, Deng Y, Chen H, Lu G, Wang W et al (2015) Recombinant receptor binding domain protein induces partial protective immunity in rhesus macaques against Middle East respiratory syndrome coronavirus challenge. EBioMedicine 2(10):1438–1446
Lanford RE, Guerra B, Chavez D, Giavedoni L, Hodara VL, Brasky KM et al (2013) GS-9620, an oral agonist of Toll-like receptor-7, induces prolonged suppression of hepatitis B virus in chronically infected chimpanzees. Gastroenterology 144(7):1508–1517. e1510
Lanford RE, Walker CM, Lemon SM (2017) The chimpanzee model of viral hepatitis: advances in understanding the immune response and Treatment of viral hepatitis. ILAR J 58(2):172–189. https://doi.org/10.1093/ilar/ilx028
Letvin NL, Mascola JR, Sun Y, Gorgone DA, Buzby AP, Xu L et al (2006) Preserved CD4+ central memory T cells and survival in vaccinated SIV-challenged monkeys. Science 312(5779):1530–1533
Li B-j, Tang Q, Cheng D, Qin C, Xie FY, Wei Q et al (2005) Using siRNA in prophylactic and therapeutic regimens against SARS coronavirus in Rhesus macaque. Nat Med 11(9):944–951
Li J, Ge J, Ren S, Zhou T, Sun Y, Sun H et al (2015a) Hepatitis B surface antigen (HBsAg) and core antigen (HBcAg) combine CpG oligodeoxynucletides as a novel therapeutic vaccine for chronic hepatitis B infection. Vaccine 33(35):4247–4254
Li Y, Wan Y, Liu P, Zhao J, Lu G, Qi J et al (2015b) A humanized neutralizing antibody against MERS-CoV targeting the receptor-binding domain of the spike protein. Cell Res 25(11):1237–1249
Li F, Wang Z, Hu F, Su L (2020) Cell culture models and animal models for HBV study. Adv Exp Med Biol 1179:109–135
Li J-Q, Zhang Z-R, Zhang H-Q, Zhang Y-N, Zeng X-Y, Zhang Q-Y et al (2021a) Intranasal delivery of replicating mRNA encoding neutralizing antibody against SARS-CoV-2 infection in mice. Signal Transduct Target Ther 6(1):369. https://doi.org/10.1038/s41392-021-00783-1
Li K, Meyerholz DK, Bartlett JA, McCray PB (2021b) The TMPRSS2 inhibitor nafamostat reduces SARS-CoV-2 pulmonary infection in mouse models of COVID-19. MBio 12(4):e00970–e00921
Li Y, Bi Y, Xiao H, Yao Y, Liu X, Hu Z et al (2021c) A novel DNA and protein combination COVID-19 vaccine formulation provides full protection against SARS-CoV-2 in rhesus macaques. Emerg Microb Infect 10(1):342–355. https://doi.org/10.1080/22221751.2021.1887767
Li Y, Cao L, Li G, Cong F, Li Y, Sun J et al (2022) Remdesivir metabolite GS-441524 effectively inhibits SARS-CoV-2 infection in mouse models. J Med Chem 65(4):2785–2793. https://doi.org/10.1021/acs.jmedchem.0c01929
Liang Y, Wang M-L, Chien C-S, Yarmishyn AA, Yang Y-P, Lai W-Y et al (2020) Highlight of immune pathogenic response and hematopathologic effect in SARS-CoV, MERS-CoV, and SARS-Cov-2 infection. Front Immunol 11:1022
Liang JG, Su D, Song T-Z, Zeng Y, Huang W, Wu J et al (2021) S-trimer, a COVID-19 subunit vaccine candidate, induces protective immunity in nonhuman primates. Nat Commun 12(1):1346. https://doi.org/10.1038/s41467-021-21634-1
Lin S-R, Yang H-C, Kuo Y-T, Liu C-J, Yang T-Y, Sung K-C et al (2014) The CRISPR/Cas9 system facilitates clearance of the intrahepatic HBV templates in vivo. Mol Ther-Nucleic Acids 3:e186
Lin Q, Lu C, Hong Y, Li R, Chen J, Chen W, Chen J (2022) Animal models for studying coronavirus infections and developing antiviral agents and vaccines. Antivir Res 105345:203
Lindenbach BD, Rice CM (2013) The ins and outs of hepatitis C virus entry and assembly. Nat Rev Microbiol 11(10):688–700
Liu F, Song Y, Liu D (1999) Hydrodynamics-based transfection in animals by systemic administration of plasmid DNA. Gene Ther 6(7):1258–1266

Liu R, Wang J, Shao Y, Wang X, Zhang H, Shuai L et al (2018) A recombinant VSV-vectored MERS-CoV vaccine induces neutralizing antibody and T cell responses in rhesus monkeys after single dose immunization. Antivir Res 150:30–38

Lu S, Zhao Y, Yu W, Yang Y, Gao J, Wang J et al (2020) Comparison of nonhuman primates identified the suitable model for COVID-19. Signal Transduct Target Ther 5(1):157. https://doi.org/10.1038/s41392-020-00269-6

Lu S, Zhao J, Dong J, Liu H, Zhu Y, Li H et al (2021) Effective treatment of SARS-CoV-2-infected rhesus macaques by attenuating inflammation. Cell Res 31(2):229–232. https://doi.org/10.1038/s41422-020-00414-4

Lucifora J, Salvetti A, Marniquet X, Mailly L, Testoni B, Fusil F et al (2017) Detection of the hepatitis B virus (HBV) covalently-closed-circular DNA (cccDNA) in mice transduced with a recombinant AAV-HBV vector. Antivir Res 145:14–19

Luke T, Wu H, Zhao J, Channappanavar R, Coleman CM, Jiao J-A et al (2016) Human polyclonal immunoglobulin G from transchromosomic bovines inhibits MERS-CoV in vivo. Sci Transl Med 8(326):326ra321-326ra321

Lütgehetmann M, Bornscheuer T, Volz T, Allweiss L, Bockmann JH, Pollok JM et al (2011) Hepatitis B virus limits response of human hepatocytes to interferon-α in chimeric mice. Gastroenterology 140(7):2074–2083. e2072

Lütgehetmann M, Mancke LV, Volz T, Helbig M, Allweiss L, Bornscheuer T et al (2012) Humanized chimeric uPA mouse model for the study of hepatitis B and D virus interactions and preclinical drug evaluation. Hepatology 55(3):685–694

Ma X, Zou F, Yu F, Li R, Yuan Y, Zhang Y et al (2020) Nanoparticle vaccines based on the receptor binding domain (RBD) and heptad repeat (HR) of SARS-CoV-2 elicit robust protective immune responses. Immunity 53(6):1315–1330.e1319. https://doi.org/10.1016/j.immuni.2020.11.015

Maisonnasse P, Aldon Y, Marc A, Marlin R, Dereuddre-Bosquet N, Kuzmina NA et al (2021) COVA1-18 neutralizing antibody protects against SARS-CoV-2 in three preclinical models. Nat Commun 12(1):6097. https://doi.org/10.1038/s41467-021-26354-0

Major ME, Dahari H, Mihalik K, Puig M, Rice CM, Neumann AU, Feinstone SM (2004) Hepatitis C virus kinetics and host responses associated with disease and outcome of infection in chimpanzees. Hepatology 39(6):1709–1720

Malczyk AH, Kupke A, Prüfer S, Scheuplein VA, Hutzler S, Kreuz D et al (2015) A highly immunogenic and protective Middle East respiratory syndrome coronavirus vaccine based on a recombinant measles virus vaccine platform. Virology 89(22):11654–11667

Marsden MD (2020) Benefits and limitations of humanized mice in HIV persistence studies. Retrovirology 17(1):1–6

Marthas M, Sutjipto S, Higgins J, Lohman B, Torten J, Luciw PA et al (1990) Immunization with a live, attenuated simian immunodeficiency virus (SIV) prevents early disease but not infection in rhesus macaques challenged with pathogenic SIV. Virology 64(8):3694–3700

Mason WS, Cullen J, Moraleda G, Saputelli J, Aldrich CE, Miller DS et al (1998) Lamivudine therapy of WHV-infected woodchucks. Virology 245(1):18–32

Mattapallil JJ, Douek DC, Buckler-White A, Montefiori D, Letvin NL, Nabel GJ, Roederer M (2006) Vaccination preserves CD4 memory T cells during acute simian immunodeficiency virus challenge. J Exp Med 203(6):1533–1541

McAuliffe VJ, Purcell RH, Gerin JL (1980) Type B hepatitis: a review of current prospects for a safe and effective vaccine. Rev Infect Dis 2(3):470–492

McCaffrey AP, Nakai H, Pandey K, Huang Z, Salazar FH, Xu H et al (2003) Inhibition of hepatitis B virus in mice by RNA interference. Nat Biotechnol 21(6):639–644

Menne S, Cote PJ (2007) The woodchuck as an animal model for pathogenesis and therapy of chronic hepatitis B virus infection. World J Gastroenterol 13(1):104

Menne S, Cote PJ, Korba BE, Butler SD, George AL, Tochkov IA et al (2005) Antiviral effect of oral administration of tenofovir disoproxil fumarate in woodchucks with chronic woodchuck hepatitis virus infection. Antimicrob Agents Chemother 49(7):2720–2728

Menne S, Tumas DB, Liu KH, Thampi L, AlDeghaither D, Baldwin BH et al (2015) Sustained efficacy and seroconversion with the Toll-like receptor 7 agonist GS-9620 in the Woodchuck model of chronic hepatitis B. Hepatology 62(6):1237–1245

Mercado NB, Zahn R, Wegmann F, Loos C, Chandrashekar A, Yu J et al (2020) Single-shot Ad26 vaccine protects against SARS-CoV-2 in rhesus macaques. Nature 586(7830):583–588. https://doi.org/10.1038/s41586-020-2607-z

Milich DR, Jones JE, Hughes JL, Price J, Raney AK, McLachlan A (1990) Is a function of the secreted hepatitis B e antigen to induce immunologic tolerance in utero? PNAS 87(17):6599–6603

Morrow W, Wharton M, Lau D, Levy J (1987) Small animals are not susceptible to human immunodeficiency virus infection. J Gen Virol 68(8):2253–2257

Muthumani K, Falzarano D, Reuschel EL, Tingey C, Flingai S, Villarreal DO et al (2015) A synthetic consensus anti–spike protein DNA vaccine induces protective immunity against Middle East respiratory syndrome coronavirus in nonhuman primates. Sci Transl Med 7(301):301ra132–301ra132

Nakagawa S-i, Hirata Y, Kameyama T, Tokunaga Y, Nishito Y, Hirabayashi K et al (2013) Targeted induction of interferon-λ in humanized chimeric mouse liver abrogates hepatotropic virus infection. PLoS One 8(3):e59611

Nassal M (2015) HBV cccDNA: viral persistence reservoir and key obstacle for a cure of chronic hepatitis B. Gut 64(12):1972–1984

Nath BM, Schumann KE, Boyer JD (2000) The chimpanzee and other non-human-primate models in HIV-1 vaccine research. Trends Microbiol 8(9):426–431

Ning L, Abagna HB, Jiang Q, Liu S, Huang J (2021) Development and application of therapeutic antibodies against COVID-19. Int J Biol Sci 17(6):1486

Nishimura Y, Gautam R, Chun T-W, Sadjadpour R, Foulds KE, Shingai M et al (2017) Early antibody therapy can induce longlasting immunity to SHIV. Nature 543(7646):559–563

Oehler N, Volz T, Bhadra OD, Kah J, Allweiss L, Giersch K et al (2014) Binding of hepatitis B virus to its cellular receptor alters the expression profile of genes of bile acid metabolism. Hepatology 60(5):1483–1493

Owen DR, Allerton CMN, Anderson AS, Aschenbrenner L, Avery M, Berritt S et al (2021) An oral SARS-CoV-2 Mpro inhibitor clinical candidate for the treatment of COVID-19. Science 374(6575):1586–1593

Peiris JS, Guan Y, Yuen K (2004) Severe acute respiratory syndrome. Nat Med 10(12):S88–S97

Petersen J, Dandri M, Mier W, Lütgehetmann M, Volz T, Von Weizsäcker F et al (2008) Prevention of hepatitis B virus infection in vivo by entry inhibitors derived from the large envelope protein. Nat Biotechnol 26(3):335–341

Petersen J, Thompson AJ, Levrero M (2016) Aiming for cure in HBV and HDV infection. Hepatology 65(4):835–848

Pfaender S, Walter S, Grabski E, Todt D, Bruening J, Romero-Brey I et al (2017) Immune protection against reinfection with nonprimate hepacivirus. Proc Natl Acad Sci U S A 114(12):E2430–E2439. https://doi.org/10.1073/pnas.1619380114

Ploss A, Kapoor A (2020) Animal models of hepatitis C virus infection. Cold Spring Harb Perspect Med 10(5):a036970

Policicchio BB, Pandrea I, Apetrei C (2016) Animal models for HIV cure research. Front Immunol 7:170939

Pybus OG, Thézé J (2016) Hepacivirus cross-species transmission and the origins of the hepatitis C virus. Curr Opin Virol 16:1–7. https://doi.org/10.1016/j.coviro.2015.10.002

Qin E, Shi H, Tang L, Wang C, Chang G, Ding Z et al (2006) Immunogenicity and protective efficacy in monkeys of purified inactivated Vero-cell SARS vaccine. Vaccine 24(7):1028–1034

Qin B, Budeus B, Cao L, Wu C, Wang Y, Zhang X et al (2013) The amino acid substitutions rtP177G and rtF249A in the reverse transcriptase domain of hepatitis B virus polymerase reduce the susceptibility to tenofovir. Antivir Res 97(2):93–100

Qiu H, Sun S, Xiao H, Feng J, Guo Y, Tai W et al (2016) Single-dose treatment with a humanized neutralizing antibody affords full protection of a human transgenic mouse model from lethal Middle East respiratory syndrome (MERS)-coronavirus infection. Antivir Res 132:141–148

Quan P-L, Firth C, Conte JM, Williams SH, Zambrana-Torrelio CM, Anthony SJ et al (2013) Bats are a major natural reservoir for hepaciviruses and pegiviruses. PNAS 110(20):8194–8199

Raghuraman S, Park H, Osburn WO, Winkelstein E, Edlin BR, Rehermann B (2012) Spontaneous clearance of chronic hepatitis C virus infection is associated with appearance of neutralizing antibodies and reversal of T-cell exhaustion. Infect Dis 205(5):763–771

Raj VS, Farag EA, Reusken CB, Lamers MM, Pas SD, Voermans J et al (2014) Isolation of MERS coronavirus from a dromedary camel, Qatar. Emerg Infect Dis 20(8):1339

Ramanan V, Shlomai A, Cox DB, Schwartz RE, Michailidis E, Bhatta A et al (2015) CRISPR/Cas9 cleavage of viral DNA efficiently suppresses hepatitis B virus. Sci Rep 5(1):1–9

Reaiche GY, Le Mire MF, Mason WS, Jilbert ARJV (2010) The persistence in the liver of residual duck hepatitis B virus covalently closed circular DNA is not dependent upon new viral DNA synthesis. Virology 406(2):286–292

Rehermann B (2009) Hepatitis C virus versus innate and adaptive immune responses: a tale of coevolution and coexistence. J Clin Invest 119(7):1745–1754

Rhim JA, Sandgren EP, Palmiter RD, Brinster RL (1995) Complete reconstitution of mouse liver with xenogeneic hepatocytes. PNAS 92(11):4942–4946

Roberts A, Thomas WD, Guarner J, Lamirande EW, Babcock GJ, Greenough TC et al (2006) Therapy with a severe acute respiratory syndrome associated coronavirus neutralizing human monoclonal antibody reduces disease severity and viral burden in golden Syrian Hamsters. Infect Dis 193(5):685–692

Scheel TK, Rice CM (2013) Understanding the hepatitis C virus life cycle paves the way for highly effective therapies. Nat Med 19(7):837–849

Scheel TKH, Kapoor A, Nishiuchi E, Brock KV, Yu Y, Andrus L et al (2015a) Characterization of nonprimate hepacivirus and construction of a functional molecular clone. PNAS 112(7):2192–2197

Scheel TKH, Simmonds P, Kapoor A (2015b) Surveying the global virome: identification and characterization of HCV-related animal hepaciviruses. Antivir Res 115:83–93. https://doi.org/10.1016/j.antiviral.2014.12.014

Scheid JF, Mouquet H, Feldhahn N, Seaman MS, Velinzon K, Pietzsch J et al (2009) Broad diversity of neutralizing antibodies isolated from memory B cells in HIV-infected individuals. Nat Rev Gastroenterol Hepatol 458(7238):636–640

Schulze A, Gripon P, Urban S (2007) Hepatitis B virus infection initiates with a large surface protein–dependent binding to heparan sulfate proteoglycans. Hepatology 46(6):1759–1768

Scripture-Adams DD, Brooks DG, Korin YD, Zack JA (2002) Interleukin-7 induces expression of latent human immunodeficiency virus type 1 with minimal effects on T-cell phenotype. Virology 76(24):13077–13082

Scull MA, Shi C, de Jong YP, Gerold G, Ries M, von Schaewen M et al (2015) Hepatitis C virus infects rhesus macaque hepatocytes and simianized mice. Hepatology 62(1):57–67

Sheahan TP, Sims AC, Leist SR, Schäfer A, Won J, Brown AJ et al (2020) Comparative therapeutic efficacy of remdesivir and combination lopinavir, ritonavir, and interferon beta against MERS-CoV. Nat Commun 11(1):1–14

Shi Z, Hu Z (2008) A review of studies on animal reservoirs of the SARS coronavirus. Virus Res 133(1):74–87

Shih Y-M, Sun C-P, Chou H-H, Wu T-H, Chen C-C, Wu P-Y et al (2015) Combinatorial RNA interference therapy prevents selection of pre-existing HBV variants in human liver chimeric mice. Sci Rep 5(1):1–15

Shimizu S, Ringpis G-E, Marsden MD, Cortado RV, Wilhalme HM, Elashoff D et al (2015) RNAi-mediated CCR5 knockdown provides HIV-1 resistance to memory T cells in humanized BLT mice. Mol Ther-Nucleic Acids 4:e227

Sprengel R, Kuhn C, Will H, Schaller H (1985) Comparative sequence analysis of duck and human hepatitis B virus genomes. Med Virol 15(4):323–333

Sprinzl MF, Oberwinkler H, Schaller H, Protzer U (2001) Transfer of hepatitis B virus genome by adenovirus vectors into cultured cells and mice: crossing the species barrier. Virology 75(11):5108–5118

Sui Y, Gordon S, Franchini G, Berzofsky JA (2013) Nonhuman primate models for HIV/AIDS vaccine development. Curr Protoc Immunol 102(1):12.14.11–12.14.30

Tanaka T, Kasai H, Yamashita A, Okuyama-Dobashi K, Yasumoto J, Maekawa S et al (2014) Hallmarks of hepatitis C virus in equine hepacivirus. J Virol 88(22):13352–13366. https://doi.org/10.1128/JVI.02280-14

Tennant BC, Toshkov IA, Peek SF, Jacob JR, Menne S, Hornbuckle WE et al (2004) Hepatocellular carcinoma in the woodchuck model of hepatitis B virus infection. Gastroenterology 127(5):S283–S293

Ter Meulen J, Bakker AB, Van Den Brink EN, Weverling GJ, Martina BE, Haagmans BL et al (2004) Human monoclonal antibody as prophylaxis for SARS coronavirus infection in ferrets. Lancet 363(9427):2139–2141

Terrade G, Huot N, Petitdemange C, Lazzerini M, Orta Resendiz A, Jacquelin B, Müller-Trutwin MJV (2021) Interests of the non-human primate models for HIV cure research. Vaccine 9(9):958

Thézé J, Lowes S, Parker J, Pybus OG (2015) Evolutionary and phylogenetic analysis of the Hepaciviruses and Pegiviruses. Genome Biol Evol 7(11):2996–3008. https://doi.org/10.1093/gbe/evv202

Thimme R, Wieland S, Steiger C, Ghrayeb J, Reimann KA, Purcell RH, Chisari FV (2003) CD8+ T cells mediate viral clearance and disease pathogenesis during acute hepatitis B virus infection. Virology 77(1):68–76

Tostanoski LH, Wegmann F, Martinot AJ, Loos C, McMahan K, Mercado NB et al (2020) Ad26 vaccine protects against SARSCoV-2 severe clinical disease in hamsters. Nat Med 26(11):1694–1700

Trivedi S, Murthy S, Sharma H, Hartlage Alex S, Kumar A, Gadi S et al (2017) Viral persistence, liver disease and host response in hepatitis C-like virus rat model. Hepatology 0(ja):435. https://doi.org/10.1002/hep.29494

Vabret A, Mourez T, Gouarin S, Petitjean J, Freymuth F (2003) An outbreak of coronavirus OC43 respiratory infection in Normandy, France. Clin Infect Dis 36(8):985–989

Van Der Hoek L, Pyrc K, Berkhout B (2006) Human coronavirus NL63, a new respiratory virus. FEMS Microbiol Rev 30(5):760–773

van Doremalen N, Haddock E, Feldmann F, Meade-White K, Bushmaker T, Fischer RJ et al (2020a) A single dose of ChAdOx1 MERS provides protective immunity in rhesus macaques. Sci Adv 6(24):eaba8399

van Doremalen N, Lambe T, Spencer A, Belij-Rammerstorfer S, Purushotham JN, Port JR et al (2020b) ChAdOx1 nCoV-19 vaccine prevents SARS-CoV-2 pneumonia in rhesus macaques. Nature 586(7830):578–582. https://doi.org/10.1038/s41586-020-2608-y

Van Rompay KK (2010) Evaluation of antiretrovirals in animal models of HIV infection. Antivir Res 85(1):159–175

Van Rompay KK (2011) The use of nonhuman primate models of HIV infection for the evaluation of antiviral strategies. AIDS Res Hum Retrovir 28(1):16–35. https://doi.org/10.1089/aid.2011.0234

Vercauteren K, de Jong YP, Meuleman P (2014) HCV animal models and liver disease. Hepatology 61(1):S26–S33

Verstrepen BE, Boonstra A, Koopman G (2015) Immune mechanisms of vaccine induced protection against chronic hepatitis C virus infection in chimpanzees. World J Hepatol 7(1):53

Vogel AB, Kanevsky I, Che Y, Swanson KA, Muik A, Vormehr M et al (2021) BNT162b vaccines protect rhesus macaques from SARS-CoV-2. Nature 592(7853):283–289. https://doi.org/10.1038/s41586-021-03275-y

Volz A, Kupke A, Song F, Jany S, Fux R, Shams-Eldin H et al (2015) Protective efficacy of recombinant modified vaccinia virus Ankara delivering Middle East respiratory syndrome coronavirus spike glycoprotein. Virology 89(16):8651–8656

Von Schaewen M, Ploss A (2014) Murine models of hepatitis C: what can we look forward to? Antivir Res 104:15–22

Walsh EE, Shin JH, Falsey AR (2013) Clinical impact of human coronaviruses 229E and OC43 infection in diverse adult populations. Infect Dis 208(10):1634–1642

Walter E, Keist R, Niederöst B, Pult I, Blum HE (1996) Hepatitis B virus infection of tupaia hepatocytes in vitro and in vivo. Hepatology 24(1):1–5

Wang Y-X, Wen Y-M, Nassal M (2012) Carbonyl J acid derivatives block protein priming of hepadnaviral P protein and DNAdependent DNA synthesis activity of hepadnaviral nucleocapsids. Virology 86(18):10079–10092

Wang H, Zhang Y, Huang B, Deng W, Quan Y, Wang W et al (2020) Development of an inactivated vaccine candidate, BBIBP-CorV, with potent protection against SARS-CoV-2. Cell 182(3):713–721.e719. https://doi.org/10.1016/j.cell.2020.06.008

Washburn ML, Bility MT, Zhang L, Kovalev GI, Buntzman A, Frelinger JA et al (2011) A humanized mouse model to study hepatitis C virus infection, immune response, and liver disease. Gastroenterology 140(4):1334–1344

Weber O, Schlemmer K-H, Hartmann E, Hagelschuer I, Paessens A, Graef E et al (2002) Inhibition of human hepatitis B virus (HBV) by a novel non-nucleosidic compound in a transgenic mouse model. Antivir Res 54(2):69–78

Weichseldorfer M, Heredia A, Reitz M, Bryant JL, Latinovic OS (2020) Use of humanized mouse models for studying HIV-1 infection, pathogenesis and persistence. J AIDS HIV Treat 2(1):23

Weingartl H, Czub M, Czub S, Neufeld J, Marszal P, Gren J et al (2004) Immunization with modified vaccinia virus Ankarabased recombinant vaccine against severe acute respiratory syndrome is associated with enhanced hepatitis in ferrets. Virology 78(22):12672–12676

Weiss SR, Navas-Martin S (2005) Coronavirus pathogenesis and the emerging pathogen severe acute respiratory syndrome coronavirus. Microbiol Mol Biol Rev 69(4):635–664

WHO (2023) COVID-19 vaccines with WHO emergency use list

Wieland SF (2015) The chimpanzee model for hepatitis B virus infection. Cold Spring Harb Perspect Med 5(6):a021469

Wieland SF, Guidotti LG, Chisari FV (2000) Intrahepatic induction of alpha/beta interferon eliminates viral RNA-containing capsids in hepatitis B virus transgenic mice. Virology 74(9):4165–4173

Wieland SF, Spangenberg HC, Thimme R, Purcell RH, Chisari FV (2004) Expansion and contraction of the hepatitis B virus transcriptional template in infected chimpanzees. PNAS 101(7):2129–2134

Williamson BN, Feldmann F, Schwarz B, Meade-White K, Porter DP, Schulz J et al (2020) Clinical benefit of remdesivir in rhesus macaques infected with SARS-CoV-2. Nature 585(7824):273–276. https://doi.org/10.1038/s41586-020-2423-5

Wilson NA, Reed J, Napoe GS, Piaskowski S, Szymanski A, Furlott J et al (2006) Vaccine-induced cellular immune responses reduce plasma viral concentrations after repeated low-dose challenge with pathogenic simian immunodeficiency virus SIVmac239. J Virol 80(12):5875–5885

Woo PC, Lau SK, Tsoi H-w, Huang Y, Poon RW, Chu C-m et al (2005) Clinical and molecular epidemiological features of coronavirus HKU1–associated community-acquired pneumonia. J Infect Dis 192(11):1898–1907

Wu S, Zhong G, Zhang J, Shuai L, Zhang Z, Wen Z et al (2020) A single dose of an adenovirus-vectored vaccine provides protection against SARS-CoV-2 challenge. Nature 11(1):1–7

Wu T, Kang S, Peng W, Zuo C, Zhu Y, Pan L et al (2021) Original hosts, clinical features, transmission routes, and vaccine development for coronavirus disease (COVID-19). Front Med (Lausanne) *8*:702066

Yadav PD, Ella R, Kumar S, Patil DR, Mohandas S, Shete AM et al (2021) Immunogenicity and protective efficacy of inactivated SARS-CoV-2 vaccine candidate, BBV152 in rhesus macaques. Nat Commun 12(1):1386. https://doi.org/10.1038/s41467-021-21639-w
Yan RQ, Su JJ, Huang DR, Gan YC, Yang C, Huang GH (1996) Human hepatitis B virus and hepatocellular carcinoma I. Experimental infection of tree shrews with hepatitis B virus. J Cancer Res Clin Oncol 122(5):283–288
Yan H, Zhong G, Xu G, He W, Jing Z, Gao Z et al (2012) Sodium taurocholate cotransporting polypeptide is a functional receptor for human hepatitis B and D virus. elife 1:e00049
Yang Z-y, Kong W-p, Huang Y, Roberts A, Murphy BR, Subbarao K, Nabel GJ (2004) A DNA vaccine induces SARS coronavirus neutralization and protective immunity in mice. Nature 428(6982):561–564
Yang Q, Zhao X, Zang L, Fang X, Zhao J, Yang X et al (2012) Anti-hepatitis B virus activities of α-DDB–FNC, a novel nucleoside–biphenyldicarboxylate compound in cells and ducks, and its anti-immunological liver injury effect in mice. Antivir Res 96(3):333–339
Yang D, Liu L, Zhu D, Peng H, Su L, Fu Y-X et al (2014) A mouse model for HBV immunotolerance and immunotherapy. Cell Mol Immunol 11(1):71–78
Yang C, Ruan P, Ou C, Su J, Cao J, Luo C et al (2015) Chronic hepatitis B virus infection and occurrence of hepatocellular carcinoma in tree shrews (Tupaia belangeri chinensis). Virol J 12(1):1–10
Ye Q, Wang B, Mao J (2020) The pathogenesis and treatment of the 'cytokine Storm' in COVID-19. J Inf Secur 80(6):607–613. https://doi.org/10.1016/j.jinf.2020.03.037
Yu J, Tostanoski LH, Peter L, Mercado NB, McMahan K, Mahrokhian SH et al (2020) DNA vaccine protection against SARSCoV-2 in rhesus macaques. Science 369(6505):806–811
Zeissig S, Murata K, Sweet L, Publicover J, Hu Z, Kaser A et al (2012) Hepatitis B virus–induced lipid alterations contribute to natural killer T cell–dependent protective immunity. Nat Med 18(7):1060–1068
Zhang T-Y, Yuan Q, Zhao J-H, Zhang Y-L, Yuan L-Z, Lan Y et al (2016) Prolonged suppression of HBV in mice by a novel antibody that targets a unique epitope on hepatitis B surface antigen. Gut 65(4):658–671
Zhang T, Wu Q, Zhang Z (2020) Probable pangolin origin of SARS-CoV-2 associated with the COVID-19 outbreak. Curr Biol 30(7):1346–1351. e1342
Zhen S, Hua L, Liu Y, Gao L, Fu J, Wan D et al (2015) Harnessing the clustered regularly interspaced short palindromic repeat (CRISPR)/CRISPR-associated Cas9 system to disrupt the hepatitis B virus. Gene Ther 22(5):404–412
Zhou P, Yang X-L, Wang X-G, Hu B, Zhang L, Zhang W et al (2020) A pneumonia outbreak associated with a new coronavirus of probable bat origin. Nature *579*(7798):270–273

Chapter 11
In Vitro and in Vivo Evaluation Tools of SARS-CoV-2 Antiviral Drugs

Gaurav Kumar Sharma, Sanketkumar Nehul, Shweta Choudhary, and Sonalika Mahajan

Abstract We have outlined the in vitro and preclinical methods for the assessment of the therapeutic properties of compounds against SARS-CoV2 in this chapter. Various criteria, which should be considered before selecting appropriate models, have been underlined. Recent updates and utilities of biophysical assays for the preliminary evaluation of the antiviral effects of any compounds have been discussed. Several cell culture systems, both primary and cell lines, have been studied for SARS-CoV2 studies, hence parameters required for the selection of suitable cell culture systems for antiviral studies are discussed. Similarly, several animal models have been studied and are available for preclinical trials, hence parameters required for the selection of the in vivo models have been discussed in this chapter. Precise quantification of the virus is critical to estimate the impact of antiviral drugs, hence methods for quantifications have also been described here.

Keywords SARS-CoV2 · Antiviral · Drug repurposing · Animal models

11.1 Introduction

The tools for in vitro and preclinical trials of substances for repurposing as SARS-CoV-2 antivirals are described in the chapter. Before submitting to clinical trials, it is crucial to evaluate the antiviral potential by utilizing a variety of techniques. Viruses hijack the metabolic pathways of the host cell to complete their replication cycle. It is challenging to specifically target viral proteins and enzymes because, during the infection and replication process, viruses utilize a variety of host machinery. The main components of the antiviral mechanism are their defense against host

G. K. Sharma (✉) · S. Mahajan
ICAR-Indian Veterinary Research Institute Izatnagar Bareilly, Bareilly, India
e-mail: Gaurav.sharma1@icar.gov.in

S. Nehul · S. Choudhary
Indian Institute of Technology, Roorkee, India

N. Kumar et al. (eds.), *Advances in Antiviral Research*, Livestock Diseases and Management, https://doi.org/10.1007/978-981-99-9195-2_11

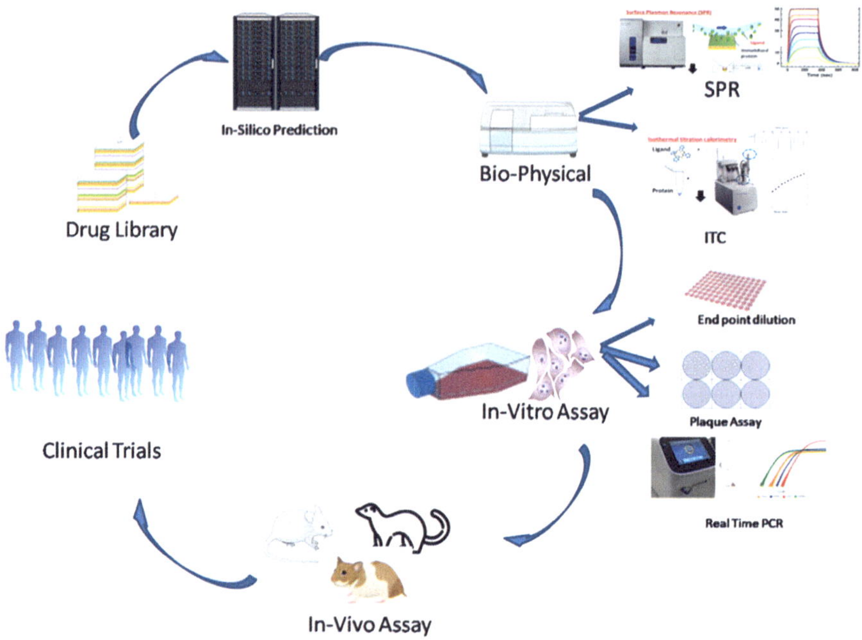

Fig. 11.1 Overview of in silico, in vitro, and in vivo evaluation tools for discovering antiviral drugs against SARS-CoV-2

entry, their ability to prevent viral reproduction by reducing DNA/RNA synthesis, and their antioxidant and scavenging properties. Since SARS-CoV-2 is a risk group 3 pathogen and must be handled in a biosafety level 3 (BSL-3) laboratory, biosafety issues are one of the obstacles to the antiviral evaluation of SARS-CoV-2.

The spike protein, proteases, helicase, and RNA-dependent RNA polymerase are the key proteins of SARS-CoV-2 that are primarily targeted for identifying direct-acting antiviral drugs. Screening for antiviral activity is a crucial step in the creation of new drugs. The next stage is to do cytotoxicity studies on the compounds to ensure your effective findings are reliable and fall within an acceptable therapeutic window. This is important after choosing the appropriate compounds and mechanisms. A brief description of in vitro methods and preclinical trials in laboratory models for the estimation of the antiviral potential of the compounds has been discussed. Figure 11.1 outlines the general procedure adopted for the evaluation of drugs' antiviral potential against SARS-CoV-2.

11.2 In Vitro Methods

A reasonably cheap and straightforward in vitro technique for high-throughput compound screening is the cell culture system. In vitro techniques reduce the need for experimental animals while saving time and money. This method is based on the virus' capacity to multiply and infect cells at high titers, producing certain cytopathic effects (CPE). There is a decrease in virus titer and discernible CPE when virus replication is inhibited in the presence of particular chemicals. Various methods of virus quantification are used to quantify the drugs' dose-dependent suppression of viruses. However, choosing appropriate cellular models is crucial for screening and assessing the antiviral efficacy of compounds, particularly in studies that are aimed at assessing the broad-spectrum antiviral effects of compounds beyond inhibition of the structural and nonstructural viral proteins.

11.2.1 Biophysical Characterization of Binding Affinities of Antivirals against Specific Proteins of SARS-CoV-2

Biophysical methods, which can involve a wide range of techniques such as surface plasmon resonance (SPR), isothermal titration calorimetry (ITC), nuclear magnetic resonance, differential scanning calorimetry, fluorescence resonance energy transfer, circular dichroism spectroscopy, etc., have been increasingly employed in the early drug discovery process to quantify interactions between protein and ligands (Biswas 2018). In the process of drug discovery, a thorough understanding of the protein–ligand interactions is essential to understand biology at the molecular level and to determine the thermodynamic parameters such as binding kinetics, enthalpy, entropy, structural changes in proteins upon ligand interactions, stoichiometry, and binding affinity of compounds toward a specific protein (Patching 2014). These techniques have upgraded with time and are now widely used for primary or secondary screening of molecules as well as in lead optimization for improving their affinity toward a target.

ITC has become the gold standard method for precisely determining a ligand's binding affinity and thermodynamics toward a certain protein over the past decade. Additionally, the technique determines the enthalpy, affinity, and stoichiometry of a protein–ligand interaction in a single reaction in a very short period (Liu et al. 2022). ITC monitors the heat changes produced (for an exothermic binding event) or absorbed (for an endothermic binding event) during a protein–ligand interaction with the ligand solution usually titrated from the injection syringe into the protein solution contained within the calorimeter cell. To determine the binding parameters, the compound (ligand) is dissolved in assay buffer and 2–3 μL (to be optimized) of it is titrated from an injection syringe into a calorimeter cell containing a specific concentration of recombinant purified protein at an interval of 3 min (to be optimized), and the number of injections is set as per requirement. The measurements are to be performed at 25 °C on a MicroCal PEAQ isothermal titration calorimeter instrument. The data obtained are fitted using the one-site or two-site binding model, and the parameters obtained will help to understand the binding kinetics of protein–ligands.

Another powerful tool to characterize small-molecule binding to proteins is SPR. This technique measures the binding kinetics of protein–ligand complexes in a real-time environment and requires only very small volumes of proteins and ligands to conduct the experiment (Patching 2014) (Gupta et al. 2020). The method relies on the immobilization of one of the binding partners (called a ligand) on the surface of a sensor chip (Ni-NTA or CM5) in an assay buffer. This can be a recombinant purified targeted protein solution. Association time and dissociation time were set. After immobilization, the second molecule (called an analyte) is allowed to flow over the surface of the sensor chip. The binding parameters or the kinetics of the reaction are monitored using the Biacore T200 instrument (or its latest models) by measuring the changes in the refractive index of the medium close to the sensor surface upon injection of the analyte. The sensorgrams obtained by subtracting the reference flow cell (without protein) will help to quantify the affinity values by fitting the data to a one-site (1:1 molecular ratio) Langmuir binding model.

11.2.2 Selection of Cell Lines

Not all human cell culture lines support the development or infection of SARS-CoV-2 in laboratory settings; SARS-CoV-2 preferentially targets lung epithelium and causes respiratory infection in infected patients (Cagno 2020). SARS-CoV-2 utilizes ACE2 and transmembrane serine proteinase 2 (TMPRSS2) through viral S protein (the SARS-CoV-2 spike protein) for entry inside the cells. Hence, the proper selection of cell lines for in vitro evaluation of drugs that can mimic human cells is highly essential. The selection of cell lines has extensively been discussed by Pires De Souza et al. (Pires De Souza et al. 2022).

However, because primary human epithelial cells are highly expensive and cannot grow continuously, established laboratory cell lines are preferred over them (Takayama 2020). In SARS-CoV-2 research, several indefinitely proliferating cell lines, including Vero, Vero E6, Calu-3, Caco-2, HEK293T-ACE2, and Huh7 are often used (Takayama 2020). One drawback of employing immortalized cell lines is that they cannot faithfully reproduce human physiological circumstances, thus it is important to confirm any biological findings using primary human cells or animal models. Table 11.1 describes the cell lines with applications in in vitro antiviral assays against SARS-CoV-2.

11.2.3 Assessment of Cellular Cytotoxicity of Molecules

Assessment of the cellular cytotoxicity of compounds is the first and most important step before moving further with the compounds for antiviral tests. When assessing possible pharmaceutical therapies, a substance that is causing a cytotoxic response may be excluded from further screening rounds. Numerous processes, including

Table 11.1 List of SARS-CoV-2 cell lines currently used in research

Cell line	Source	Application for antiviral studies	References
HEK293T ACE2 cell line (human)	Human kidney-derived epithelial cells	• Easy to grow and maintain, with high reproducibility. • For isolation of SARS-CoV-2, evaluating the efficacy of antiviral drugs and vaccines, and studying syncytia formation. • Supports modest virus replication.	(Chu et al. 2020; Wang et al. 2020; Xia et al. 2020; Buchrieser et al. 2020)
Vero and VeroE6 cells	African green monkey kidney epithelial cells	• Express ACE2 and low levels of TMPRSS2 support high virus replication for isolation. • Drug screening. • Superior to other cell lines for isolation of variants. • The low ACE2 polymorphism relative to human ACE2 may be a drawback for their use in research on human viruses like SARS-CoV-2,	(Ogando et al. 2020; Takayama 2020).
Huh-7 cell line	Human	• For studying virus tropism, replication kinetics, cell damage profiling, and cellular damages in response to infections by SARS-CoV-2 and other coronaviruses. • Supports modest virus replication.	(Mautner et al. 2022; Pires De Souza et al. 2022; Pandamooz et al. 2022).
A549 cells	Human lung Adenocarcinoma	• Studying replication of SARS-CoV-2, anti-SARS-Cov-2 drug screening (Chang et al. 2022). • Permissive to SARS-CoV-2 replication (Pandamooz et al. 2022).	(Mautner et al. 2022)
Caco-2 cells	Human epithelial Colorectal adenocarcinoma	• Screening of antiviral compounds and studying pathways that are triggered in response to virus infection. • Supports modest virus replication but cytopathic effect is not observed.	(Pires De Souza et al. 2022; Pandamooz et al. 2022).
Calu-3 cells	Human lung adenocarcinoma	• Studying virus replication kinetics and antiviral effect of compounds. • Robust virus replication, cytopathic effect is not observed. • Furin cleavage site of SARS-CoV-2 appears to be preserved in virus isolated from calu-3 cells.	(Pires De Souza et al. 2022; Pandamooz et al. 2022)
LLC-MK2 cell line	(rhesus monkey)	• Permissive for SARS-CoV-2. • Supports high virus replication, cytopathic effect is not observed.	(Pires De Souza et al. 2022).

(continued)

Table 11.1 (continued)

Cell line	Source	Application for antiviral studies	References
U251 cells	(human brain glioblastoma)	• Studying virus tropism and replication kinetics of SARS-CoV-2 (Pires De Souza et al. 2022). • The cytopathic effect is absent.	(Mautner et al. 2022; Pires De Souza et al. 2022).

membrane stability, enzyme activity, or metabolic activity, can be used to study cell viability (Johnson et al. 2013). Each viability reagent serves as the cornerstone of a viability kit, which is an assay to assess cellular health utilizing a single parameter readout or a number of detection methods. The most popular technique for evaluating the cytotoxic effects of substances or molecules is the MTT (3-(4,5-dimethylthi azol-2-yl)-2,5-diphenyltetrazolium bromide) assay. MTT dye is a yellow water-soluble dye that is converted by mitochondrial enzymes in living cells into soluble, purple-colored insoluble formazan (Ghasemi et al. 2021). The cytotoxic effect of compounds in cells will be indicated by a decrease in the amount of purple-colored product in compound-treated cells compared to mere cell controls (cells incubated with only maintenance media).

The brief procedure to carry out the assay is as follows: protocol adopted from Garcia et al. (Garcia et al. 2021):

(a) In a standard BSL-2 facility, count and seed cells in a 96-well plate inside a laminar hood to get 70–80% confluency.
(b) A twofold serial dilution of the test compound was prepared in a cell culture medium devoid of serum, after 24 h.
(c) Postincubation, the media were discarded from the plate, and the diluted compounds were added in wells of a 96-well plate. Add only serum-free media in control wells (untreated control).
(d) The plate is sealed and then kept in a humidified CO_2 incubator at 37 °C for an incubation period of 24–48 h.
(e) After the drug treatment is over, add 20 μL MTT solution (5 mg/mL in PBS) to all wells. The plate is covered in aluminum foil and is incubated for 3–4 h at 37 °C.
(f) Postincubation, the media and MTT solutions were decanted from all wells, and 100 μL/well of DMSO was added to each well to dissolve the formazan crystals.
(g) Keep the plate in the incubator and read the absorbance at 570 nm in a plate reader.
(h) Finally, the percentage cytotoxicity of the test compound can be determined using the formula $[(A - B)/A \times 100]$.

Here, A and B denote the absorbance signals of control and compound-treated cells, respectively. The concentration of the test compound that was able to reduce the absorbance by 50% in comparison to cell control will be determined as CC_{50}.

11.2.4 General Procedure for Analysis of the Antiviral Efficacy of Test Compounds

All steps for handling live viruses should be conducted in the BSL-3 facility following all institute biosafety guidelines. The detailed stepwise protocol is as follows (Fu et al. 2021; Qiao et al. 2021):

(a) To study the anti-SARS-CoV-2 potential of compounds, a confluent monolayer of Vero E6 cells pre-seeded in wells of a 24-well plate was treated with different concentrations of compounds for 1–3 h (pretreatment) and was infected with SARS-CoV-2 (MOI 0.1) (in BSL-3 facility).
(b) Postincubation, the virus–drug mixture is subsequently removed from the plate (discarded in an appropriate hazardous waste container), and the cells are again treated with fresh medium containing dilutions of the compound (posttreatment).

The plate is incubated at 37 °C in a 5% CO_2 incubator for 48 h in the BSL-3 facility. After completion of the incubation period, the antiviral effect of compounds is quantified by visual CPE-based assay, end-point dilution method ($TCID_{50}$), qRT-PCR method, and/or plaque reduction assay which are described briefly in the following sections.

11.2.5 General Procedure of Visual CPE-Based Antiviral Assay

Observing virus infection-induced visible CPE on a confluent monolayer of cells under an inverted microscope is the most crucial step in CPE-based antiviral evaluation (Dilnessa and Zeleke 2017). CPE refers to morphological changes in cells brought on by cytopathogenic virus infection. The CPE inhibition assay is the most cost- and time-effective method for high-throughput screening of drugs' antiviral activity.

A series of cell lines (Table 11.1) were published to be highly susceptible to SARS-CoV-2 infection. However, CPE was reported to occur in cell lines such as VeroE6 showing cellular damages, such as cell rounding, degeneration, syncytium formation, and detachment. CPE was not reported in Calu-3, LLCMK2, Caco 2, RK-13, and PK-15 and cell lines even after a week of virus infection (Murgolo et al. 2021). Figure 11.2 depicts the visual CPE of SARS-CoV-2 in Vero cells infected with different concentrations of the virus.

The standard test procedure is as follows which may require minor modifications under different conditions (Case et al. 2020):

(a) Plate cells for infection into the plate with media compatible with the chosen cell type, in a standard BSL-2 laboratory. Cells were allowed to grow in plates

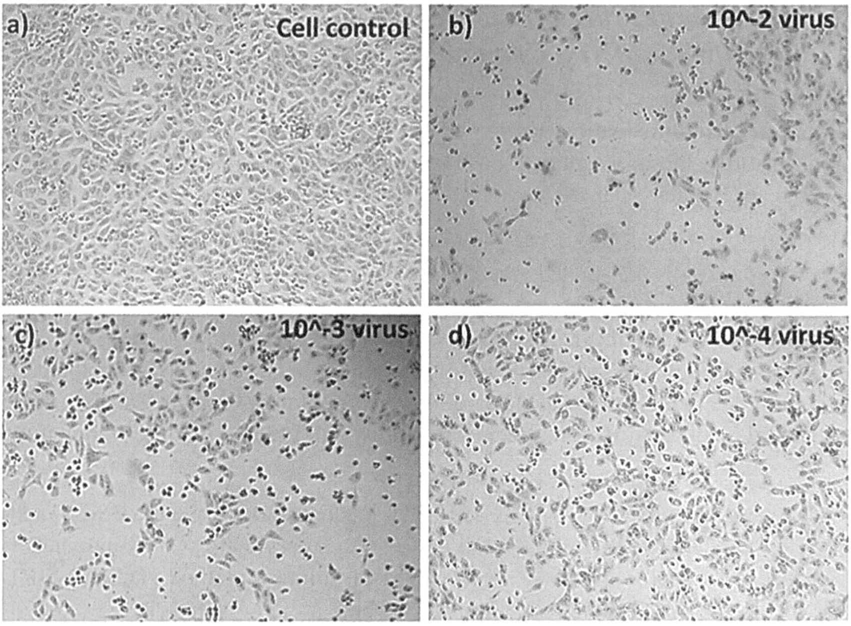

Fig. 11.2 Representative images for CPE produced in Vero cells on infection with SARS-CoV-2. (**a**) Cell control (uninfected) (**b–d**) depict Vero cells infected with increasing dilutions of SARS-CoV-2

to attain a monolayer of 80–90% confluency. Carefully remove growth medium (10% serum) from cells and add maintenance medium (fresh medium with 2% serum) to the plate.

(b) On the following day, transfer the plate to the BSL-3 facility. Thaw a vial of SARS-CoV-2 virus stock and infect the cells at the desired multiplicity of infection (MOI).

(c) Inoculate SARS-CoV-2 virus dilutions into the well of cells, leaving one uninoculated control well (cell control). For instance, 200 μL of virus dilution/well will be used to infect cells in a 24-well plate. Cells and viruses should be incubated for 1.5–2 h in a humidified environment at 37 °C in an incubator in the BSL-3 facility itself.

(d) Following virus infection, the viral inoculum will be emptied, and each concentration of the test substance diluted in maintenance medium (2% serum) will be added to the selected wells (posttreatment). The plate will be incubated for 48–72 h at 37 °C and 5% CO_2 till the appearance of a prominent CPE. If the effect of the compound needs to be evaluated as a pretreatment, the cells can be treated with the test compounds for 15 min to 6 h before virus infection.

(e) The plate is to be examined under an inverted light microscope for any differences in cell appearance compared to normal control cells every 24 h until the experiment is complete under BSL-3 conditions.

(f) To quantify the inhibition of the virus in the presence of compounds, $TCID_{50}$ or dye-based cytotoxicity assays can be used to quantify the CPE inhibition on cellular monolayers infected with the virus.

11.2.6 General Procedure for Endpoint Dilution-Based Method

Endpoint dilution method or $TCID_{50}$ assays are frequently used to measure virus titer and are regarded as a key tool for the precise assessment of viral replication and infectivity, both of which are essential for the assessment and investigation of novel therapeutics and treatments (Jo et al. 2021; Lei et al. 2021). The assay looks at the CPE of a virus on a host cell culture that has been infected. The $TCID_{50}$ tests provide an extra benefit over the conventional and popular plaque assays for quantifying viruses that are unable to generate plaques or infect cell monolayers. The amount of virus dilution that is required to induce CPE in 50% of wells containing the inoculated cell culture is termed $TCID_{50}$.

The steps to carry out the experiment in a 96-well plate are as follows (Driouich et al. 2021; Touret et al. 2021):

(a) Vero E6 cells were seeded at a cell density of 2.5×10^4 cells/well in a total volume of 100 μL/well, in a 96-well plate. Incubate the plate in a humidified 5% CO_2 incubator at 37 °C overnight.
(b) In a separate 96-well plate, perform twofold serial dilutions of all samples harvested from the antiviral assay.
(c) Remove medium from 96-well microplates having a confluent monolayer of cells and transfer 100 μL of each dilution of each sample to this plate. Incubate the plate for 3–4 days at 37 °C and 5% CO_2 and check for the presence of CPE.
(d) After 48–72 h, the CPE is visualized under an inverted microscope, and then the cells are fixed with formaldehyde for 1 h at room temperature. Postincubation, the formaldehyde solution was discarded, and the cells were stained with 50 μL/well volume of crystal violet stain for 1 h. The stain is then decanted followed by washing with distilled water till the dye is completely removed.
(e) $TCID_{50}$ value can be calculated by following Reed and Muench method (Reed and Muench 1938).

11.2.7 General Procedure for Plaque Reduction Assay

By measuring the plaques that develop in cell culture after infection with repeated dilutions of a viral specimen, the plaque assay is a common quantitative way of assessing infectious SARS-CoV-2 (Mendoza et al. 2020; Amarilla et al. 2021). This test counts "plaques" which are the areas of cellular death that develop after an infectious unit enters a cell and spreads to other cells during the incubation period. This assay has several drawbacks, including being labor-intensive, low-throughput, and unreliable when the drugs being tested are cytotoxic or the virus cannot produce CPE in a certain cell type.

A detailed stepwise protocol for the assay is explained further (Case et al. 2020; Mendoza et al. 2020; Amarilla et al. 2021):

(a) Plate Vero E6 cells into wells of a 24-well plate. Incubate the cells overnight (12–18 h) at 37 °C till a confluent monolayer cell is obtained.
(b) In a BSL-3 facility, prepare serial dilution of compound-treated virus and virus control samples harvested from antiviral assay (starting from 10–1 to 10–6; ten-fold dilutions) in a plate. Remove the media from wells and add 200 μL of each dilution to wells of a 24-well plate.
(c) Incubate the plate for 1–2 h in a 5% CO_2 incubator, gently rocking the plates after every 15–20 min.
(d) Postincubation, empty content from all wells in appropriate hazardous waste containers and add 500 μL of MEM: methylcellulose mixture to each well of the 24-well plate.
(e) Incubate the plates at 37 °C in 5% CO_2 for 3–4 days. Postincubation, gently remove the methylcellulose overlays and fix the cells by adding 4% paraformaldehyde (PFA) to each well. Incubate the plate for 20–40 min at room temperature and then discard the PFA in an appropriate hazardous waste container.
(f) Add 0.05% (w/v) crystal violet prepared in 20% methanol to each well and incubate the plate at room temperature for 20–30 min. Remove the crystal violet from each well and wash the plate with distilled water until the plaques are clearly visualized.
(g) The plaques are enumerated in dilutions in which there are 10–100 plaques. The titer is calculated as PFU/mL using the following formula:

$$\text{Virus Titer}\left(\frac{\text{PFU}}{\text{ml}}\right) = \frac{\text{AverageNumber of plaques counted}}{\text{Dilution factor} \times \text{Inoculum volume of virus}}$$

(h) To calculate the percentage of inhibition of the virus, the following formula [1 – (VD/VC)] × 100% can be used, where VD and VC refer to the virus titer in the presence (VD) and absence (VC) of the test compound, respectively.

The percentage reduction in virus titer in the presence of different concentrations of the compounds can be calculated to enumerate the 50% inhibitory concentration or IC_{50} values.

11.2.8 General Procedure for Quantitative Real-Time PCR Assay

A qRT-PCR assay can quickly identify the presence of viral RNA in patient samples or clinical samples and is the most extensively used method for the detection, clinical diagnosis, and quantification of SARS-CoV-2 (Case et al. 2020; Perchetti et al.

2020). Due to this assay's increased speed, sensitivity, reproducibility, low intra-assay variability, ease of use, and decreased danger of carryover contamination, it has gained greater acceptability on a global scale. The most popular technique for quickly and easily detecting SARS-CoV-2 in clinical samples is one-step qRT-PCR. Reverse transcription and quantitative PCRS (qPCRs) can be carried out in the same reaction mixture when using the one-step qRT-PCR assay (Perchetti et al. 2020; Abasiyanik et al. 2021). As a result, the operator can mix all the required ingredients at the start of the reaction, establish the proper thermal cycle settings in the qPCR machine, and then analyze the data. This is ideally suited for a high-throughput screening environment of antiviral drugs or clinical diagnostics within a completely automated process due to the minimum experimental handling mistakes. To accelerate the deployment of new therapies for COVID-19, one-step RT-PCR has been widely used for quick high-throughput screening of antivirals against SARS-CoV-2 as well as for quantification of antiviral efficacy of compounds both in vitro as well as in vivo models, as evident from published studies (Touret et al. 2020, 2021). One-step qRT-PCR kits that can be used with the TaqMan probe hydrolysis technology have recently been created by a number of businesses and offer speedy results.

For the calculation of the antiviral potency of compounds, the currently widely used procedure for one-step qRT-PCR involves the following steps (Case et al. 2020; Toptan et al. 2020; Fu et al. 2021):

(a) **RNA extraction:** RNA from supernatant samples harvested from antiviral assay is to be extracted either manually using the TRIzol method or the kit-based method in the BSL-3 facility. For the kit-based method, RNA is to be purified from cell-culture supernatants using kits following the manufacturer's instructions. Currently, both magnetic bead-based and column-based kits are commercially available and highly effective in the purification of SARS-CoV-2 viral RNA from cell culture supernatants or animal tissues.

(b) **Setting up qRT-PCR reaction:** The final qRT-PCR reaction of 20 μL is to be assembled using a commercially available kit or in-house designed assay (COVISure-COVID-19 Real-Time PCR kit (Genetix)). Generally, 5 μL RNA is mixed with the primers-probe mix, and RNase-free water to fill the reaction as per the manufacturer's instructions. The primers-probe mixture targeted the envelope (E-gene) and the RdRP gene of the SARS-CoV-2, whereas internal control targeted the host RNP gene. All the probes are labeled with different dyes to enable multiplexing in real-time PCR. All the reaction tubes are accompanied by a pair of positive and negative controls provided in the kit. The thermal cycling steps are to be followed as per kit instructions. RT-PCR results are analyzed by calculating Delta delta ($\Delta\Delta$) Ct as the difference between Ct values obtained for compound-treated samples and infection control to quantify the viral load in the presence of compounds. For all PCR amplifications, positive and negative controls (water or RNA isolated from only cell control) are included.

(c) To quantify the percentage inhibition, the following formula [1 − (VD/VC)] × 100% can be used, where VC will refer to the virus titer in the absence of the compound and VD will indicate the virus titer in the presence of the test compound. The percentage reduction in virus titer in the presence of different concentrations of the compounds can be calculated to enumerate the 50% inhibitory concentration or IC_{50} values.

11.2.9 General Procedure for Immunofluorescence Assay

Immunofluorescence (IF) assay in virus-infected cells is a classical approach in virus diagnostics as well as for the assessment of the antiviral effect of compounds. The assay can be carried out only for viruses for which antibodies are commercially available. With a fewer number of steps, the assay can be performed faster than $TCID_{50}$ and has lower variability. The whole process is under a physiological condition; therefore, the assay can provide information related to morphological changes within the cell when infected with a virus and the antiviral effect of the compound, if any, for its capability to clear virus infection. This assay can provide qualitative and quantitative estimates of viruses but it cannot discriminate between noninfectious and infectious viral particles present in the sample. This assay has wide-scale applications in biological and medicinal research to analyze the distribution of glycans, proteins, and small biological molecules. In a recent study, the IF assay technique has been used for in vitro screening of antivirals against SARS-CoV-2 (Sangeun et al. 2020).

The steps to conduct the assay are as follows (Riva et al. 2020; Cao et al. 2020):

(a) To carry out this experiment, VeroE6 cells are seeded in wells of 24-well/12-well/6-well as per the seeding density and are allowed to grow overnight at 37 °C in 5% CO_2.
(b) Next day, the plate is transferred to the BSL-3 facility and is treated with different increasing concentrations of a compound in a manner similar to an antiviral study. Postincubation period of 2–3 h, the compound containing medium is decanted, and the cells are infected with SARS-CoV-2 at the desired MOI. After 1.5–2 h of virus adsorption, the virus inoculum is discarded followed by washing with media.
(c) Fresh media containing dilutions of compounds are added to the cells, and the plate is kept for 24–36 h (tentatively 50% of the total time required to produce CPE).
(d) Postincubation, remove the media and fix the cells with 4% PFA at room temperature for 30–60 min. After washing with PBS twice or thrice, cells are permeabilized with 0.5% triton X-100 for 8–10 min at room temperature.
(e) Then cells are blocked with 5% bovine serum albumin at room temperature for 30 min and are then incubated with a primary antibody targeting a specific viral

protein for 1–2 h. After several washing steps with 1X-PBS, the cells are incubated with a fluorescent-tagged secondary antibody for 1 h.

(f) Postincubation, the cells are washed with PBS twice or thrice, and cellular nuclei can be stained with DAPI or other nucleus-staining dyes. After incubation, wash the cells with PBS, and now the plate can be visualized under a fluorescent microscope.

(g) The fluorescence signal can be compared between mock- and compound-treated cells using a fluorescent-activated cell counter or confocal microscopy.

11.3 In Vivo Methods

Robust SARS-CoV-2 animal models are important to gain insight into viral transmission and pathogenesis, to evaluate the activity of antivirals and monoclonal antibodies, and to test the efficacy of a vaccine against SARS-CoV-2. Currently, there are different models of SARS-CoV-2 such as transgenic and nontransgenic mice, ferrets, hamsters, and nonhuman primates (Parolin et al. 2021).

11.3.1 Mouse Model

Mice have been the first choice of researchers to study the pathogenesis of different viruses due to several advantages such as low cost of maintenance, rapid reproduction rate, small size, detailed understanding of their physiology, and availability of research tools. The spike protein of SARS-CoV-2 does not have a natural affinity toward the mice ACE2 protein, thus naturally, SARS-CoV-2 does not infect the nontransgenic bread mice. To overcome this limitation, the virus needs to be adapted by several passages in wild-type (WT) mice to increase its pathogenicity (Dinnon et al. 2020; Gu et al. 2020). Published studies have reported the establishment of strain-based mouse-adapted models for conducting research on SARS-CoV and MERS (Roberts et al. 2007; Matthew et al. 2012). Intriguingly, it is also postulated that some variants of SARS-CoV-2, such as the Beta B.1.351 variant, have better binding affinity to the mouse receptor and hence display better replication in different WT mouse models (Abdelnabi et al. 2022). Beta B.1.351 variant infection model of SARS-CoV-2 has been developed by Abdelnabie et al. in SCID mice as a method to evaluate the antiviral potency of possible SARS-CoV-2 small-molecule inhibitors (Abdelnabi et al. 2021). For this purpose, male SCID mice were intranasally infected with 10^5 $TCID_{50}$ of the Beta B.1.351 variant which resulted in high viral loads in mice (Abdelnabi et al. 2021). On treatment of infected mice with the antiviral drugs, the symptoms were reduced and lung pathology was improved. Together, the data demonstrated the applicability of this SCID mouse-based infection model of Beta B.1.351 as a reliable preclinical model for evaluating the efficacies of antivirals against SARS-CoV-2. Another way is to adapt the mice by replacing the mice ACE2 gene with the human ACE2 (hACE2) gene (Jiang et al. 2020). When transgenic hACE2 mice were infected with SARS-CoV-2, symptoms of disease developed with active replication in the lungs, and mortality due to lethal encephalitis was also observed (Jiang et al. 2020). Furthermore, the study suggested infiltration of immune cells into the lungs after histopathological examination that resulted in interstitial pneumonia (Bao et al. 2020). Moreover, hACE2 mice have now been widely used for in vivo validation of antivirals against SARS-CoV-2 (Moreau et al. 2020; Winkler et al. 2020; Rathnasinghe et al. 2020; Qiao et al. 2021). Despite significant observable immune-mediated lung damages upon SARS-CoV-2 infection, a drawback of using the hACE2 model for studying viral pathogenesis is the lethal effects caused by neuroinvasion, relatively slow growth rate of mice, and limited availability. K-18hACE2 mice have been used to detect the neutralizing human monoclonal antibody efficacy as well as to detect the efficacy of the vaccine against SARS-CoV-2 (Rosenfeld et al. 2021; An et al. 2023).

There are some limitations of the mice model such as low transmission rate. Even though hACE2-transgenic mice can tolerate only modest amounts (about 50%) of close contact transmission, respiratory droplet transmission proved to be low (Bao et al. 2020).

11.3.2 Ferrets

As the lung physiology and response to infection are similar in ferrets (*Mustela putorius furo*) as in humans, ferrets have been vital models for the study of respiratory diseases (Sun et al. 2010). Coughing is one of the most important symptoms of COVID-19 exhibited by ferrets (Guan et al. 2020). Another symptom is the rise in body temperature of the ferret on infection with SARS-CoV-2. The blood, nasal passages, trachea, lungs, and intestines are among the body parts where the virus can be detected after replication (Kim et al. 2020). In addition, the ferret model has also been used for studying the pathogenesis of SARS-CoV (Chu et al. 2008) and influenza (Maher and DeStefano 2004; Belser et al. 2011). In the case of SARS-CoV-2, the ferret model has been used to detect the therapeutic application of ribonucleoside analog MK-4482/EIDD-2801 in ferrets, to evaluate intranasal SARS-CoV-2 fusion inhibitory activity of lipopeptide (Cox et al. 2021; de Vries et al. 2021). It was discovered that infected ferrets could transmit the infection to animals they came in contact with, and these animals would then begin to show symptoms such as a reduction in body weight (Kim et al. 2020).

11.3.3 Golden Hamster Model

SARS-CoV-2 binds to the ACE2 receptor of Golden or Syrian hamsters (*Mesocricetus auratus*) as it is similar to the hACE2. Thus, the golden hamster is naturally susceptible to SARS-CoV-2 infection. Hamsters are widely used in other viral infections also (Miao et al. 2019). Recent research has shown that the ACE2–RBD interface between Syrian hamster ACE2 and hACE2 is highly similar, and Syrian hamster ACE2 binds effectively to the SARS-CoV-2 RBD (Rosenke et al. 2020). Following intranasal infection with SARS-CoV-2, these animals displayed significant viral replication in the lungs. According to a histopathological study, the lungs developed significant pulmonary edema, inflammation, and cell death. Hamsters with the infection were reported to lose weight and display increased respiratory rate (Chan et al. 2020). The hamster model has been used to detect the anti-SARS-CoV-2 activity of Molnupiravir (Abdelnabi et al. 2021), TEMPOL (Maio et al. 2022), Favipiravir (Driouich et al. 2021), and Galidesivir (Taylor et al. 2022).

11.3.4 Nonhuman Primates

Because of higher genetic similarity with humans, NHPs are the best animal models to study human diseases. The most frequently utilized species in this category of preclinical animal models are the rhesus macaque (*Macaca mulatta*) and *Macaca fascicularis*, the macaque species of cynomolgus monkey (Cauvin et al. 2015). Remdesivir's treatment effectiveness has also been examined using the rhesus monkey model (Williamson et al. 2020), which has been proven to be efficacious in the treatment of infections in humans (Beigel et al. 2020). Interestingly, NHPs were also used to study the antiviral activity of hydroxychloroquine (Maisonnasse et al. 2020), molnupiravir (Johnson et al. 2023), and remdesivir against SARS-CoV-2 (Vermillion et al. 2023). Despite the wide applicability of these models for assessing the pathogenesis of SARS-CoV and SARS-CoV-2 and drug discovery, the limitations of these models cannot be ignored. In comparison to the extent of disease severity in humans, NHP models are reported to exhibit less severe disease pathology after viral infection (Cockrell et al. 2018; Chandrashekar et al. 2020). In contrast to the acute lung injury observed in mouse models for other related highly pathogenic CoVs, NHPs fail to develop severe lung damage (Cockrell et al. 2016).

11.4 Conclusions

To conclude, it is essential to discover compounds with effective antiviral activities against SARS-CoV2. Drug repurposing is a promising method for drug discovery. Currently, several in vitro and in vivo methods are available for this purpose.

However, choosing a suitable system and protocol is highly essential to estimate the antiviral potential of the compounds. In addition, high-throughput methods are required to rapidly screen the effective compounds among the potential compounds. Several of the available methods for in vitro and preclinical evaluation have been discussed along with their advantage and disadvantages which will help the researchers to select appropriate model(s).

Acknowledgments The authors are thankful to ICAR, NASF, and DST-SERB for providing funds and infrastructure to take up the work. The authors are also thankful to Joint Director CADRAD, ICAR IVRI, and Director ICAR-IVRI for providing the necessary facility to compile the book chapter.

References

Abasiyanik MF, Flood B, Lin J et al (2021) Sensitive detection and quantification of SARS-CoV-2 in saliva. Sci Rep 11:12425. https://doi.org/10.1038/s41598-021-91835-7

Abdelnabi R, Foo CS, De Jonghe S et al (2021) Molnupiravir inhibits replication of the emerging SARS-CoV-2 variants of concern in a hamster infection model. The Journal of Infectious Diseases 224:749–753. https://doi.org/10.1093/infdis/jiab361

Abdelnabi R, Foo CS, Kaptein SJF et al (2022) A SCID mouse model to evaluate the efficacy of antivirals against SARS-CoV-2 infection. J Virol 96:e00758–e00722. https://doi.org/10.1128/jvi.00758-22

Amarilla AA, Modhiran N, Setoh YX et al (2021) An optimized high-throughput Immuno-plaque assay for SARS-CoV-2. Front Microbiol 12:1–17. https://doi.org/10.3389/fmicb.2021.625136

An D, Li K, Rowe DK et al (2023) Protection of K18-hACE2 mice and ferrets against SARS-CoV-2 challenge by a single-dose mucosal immunization with a parainfluenza virus 5–based COVID-19 vaccine. Sci Adv 7:eabi5246. https://doi.org/10.1126/sciadv.abi5246

Bao L, Deng W, Huang B et al (2020) The pathogenicity of SARS-CoV-2 in hACE2 transgenic mice. Nature 583:830–833. https://doi.org/10.1038/s41586-020-2312-y

Beigel JH, Tomashek KM, Dodd LE et al (2020) Remdesivir for the treatment of Covid-19 — final report. N Engl J Med 383:1813–1826. https://doi.org/10.1056/NEJMoa2007764

Belser JA, Katz JM, Tumpey TM (2011) The ferret as a model organism to study influenza a virus infection. Disease Model Mech 4:575–579. https://doi.org/10.1242/dmm.007823

Biswas P (2018) Modern biophysical approaches to study protein–ligand interactions. Biophys Rev Lett 13:133–155. https://doi.org/10.1142/S1793048018300013

Buchrieser J, Dufloo J, Hubert M et al (2020) Syncytia formation by SARS-CoV-2-infected cells. The EMBO Journal 39:e106267. https://doi.org/10.15252/embj.2020106267

Cagno V (2020) SARS-CoV-2 cellular tropism. Lancet Microbe 1:e2–e3. https://doi.org/10.1016/S2666-5247(20)30008-2

Cao R, Hu H, Li Y et al (2020) Anti-SARS-CoV-2 potential of Artemisinins in-vitro. ACS Infectious Disease 6:2524–2531. https://doi.org/10.1021/acsinfecdis.0c00522

Case JB, Bailey AL, Kim AS et al (2020) Growth, detection, quantification, and inactivation of SARS-CoV-2. Virology 548:39–48. https://doi.org/10.1016/j.virol.2020.05.015

Cauvin AJ, Peters C, Brennan F (2015) Chapter 19 - advantages and limitations of commonly used nonhuman primate species in Research and Development of biopharmaceuticals. In: Bluemel J, Korte S, Schenck E, Weinbauer GF (eds) The nonhuman primate in nonclinical drug development and safety assessment. Academic Press, San Diego, pp 379–395

Chan JF-W, Zhang AJ, Yuan S et al (2020) Simulation of the clinical and pathological manifestations of coronavirus disease 2019 (COVID-19) in a Golden Syrian hamster model: implications for disease pathogenesis and transmissibility. Clin Infect Dis 71:2428–2446. https://doi.org/10.1093/cid/ciaa325

Chandrashekar A, Liu J, Martinot AJ et al (2020) SARS-CoV-2 infection protects against rechallenge in rhesus macaques. Science 369:812–817. https://doi.org/10.1126/science.abc4776

Chang C-W, Parsi KM, Somasundaran M et al (2022) A newly engineered A549 cell line expressing ACE2 and TMPRSS2 is highly permissive to SARS-CoV-2, including the Delta and omicron variants. Viruses 14(7):1369. https://doi.org/10.3390/v14071369

Chu Y-K, Ali GD, Jia F et al (2008) The SARS-CoV ferret model in an infection–challenge study. Virology 374:151–163. https://doi.org/10.1016/j.virol.2007.12.032

Chu H, Chan JF-W, Yuen TT-T et al (2020) Comparative tropism, replication kinetics, and cell damage profiling of SARS-CoV-2 and SARS-CoV with implications for clinical manifestations, transmissibility, and laboratory studies of COVID-19: an observational study. Lancet Microbe 1:e14–e23. https://doi.org/10.1016/S2666-5247(20)30004-5

Cockrell AS, Yount BL, Scobey T et al (2016) A mouse model for MERS coronavirus-induced acute respiratory distress syndrome. Nat Microbiol 2:16226. https://doi.org/10.1038/nmicrobiol.2016.226

Cockrell AS, Johnson JC, Moore IN et al (2018) A spike-modified Middle East respiratory syndrome coronavirus (MERS-CoV) infectious clone elicits mild respiratory disease in infected rhesus macaques. Sci Rep 8:10727. https://doi.org/10.1038/s41598-018-28900-1

Cox RM, Wolf JD, Plemper RK (2021) Therapeutically administered ribonucleoside analogue MK-4482/EIDD-2801 blocks SARS-CoV-2 transmission in ferrets. Nature Microbiology 6:11–18. https://doi.org/10.1038/s41564-020-00835-2

de Vries RD, Schmitz KS, Bovier FT et al (2021) Intranasal fusion inhibitory lipopeptide prevents direct-contact SARS-CoV-2 transmission in ferrets. Science 371:1379–1382. https://doi.org/10.1126/science.abf4896

Dilnessa T, Zeleke H (2017) Cell culture, cytopathic effect and immunofluorescence diagnosis of viral infection. J Microbiol Modern Techniq 2:102–110

Dinnon KH, Leist SR, Schäfer A et al (2020) A mouse-adapted model of SARS-CoV-2 to test COVID-19 countermeasures. Nature 586:560–566. https://doi.org/10.1038/s41586-020-2708-8

Driouich J-S, Cochin M, Lingas G et al (2021) Favipiravir antiviral efficacy against SARS-CoV-2 in a hamster model. Nat Commun 12:1735. https://doi.org/10.1038/s41467-021-21992-w

Fu Z, Huang B, Tang J et al (2021) The complex structure of GRL0617 and SARS-CoV-2 PLpro reveals a hot spot for antiviral drug discovery. Nat Commun 12:488. https://doi.org/10.1038/s41467-020-20718-8

Garcia G, Sharma A, Ramaiah A et al (2021) Antiviral drug screen identifies DNA-damage response inhibitor as potent blocker of SARS-CoV-2 replication. Cell Rep 35:108940. https://doi.org/10.1016/j.celrep.2021.108940

Ghasemi M, Turnbull T, Sebastian S, Kempson I (2021) The MTT assay: utility, limitations, pitfalls, and interpretation in bulk and single-cell analysis. Int J Mol Sci 22(23):12827. https://doi.org/10.3390/ijms222312827

Gu H, Chen Q, Yang G et al (2020) Adaptation of SARS-CoV-2 in BALB/c mice for testing vaccine efficacy. Science 369:1603–1607. https://doi.org/10.1126/science.abc4730

Guan W, Ni Z, Hu Y et al (2020) Clinical characteristics of coronavirus disease 2019 in China. N Engl J Med 382:1708–1720. https://doi.org/10.1056/NEJMoa2002032

Gupta A, Rani C, Pant P et al (2020) Structure-based virtual screening and biochemical validation to discover a potential inhibitor of the SARS-CoV-2 Main protease. ACS Omega 5:33151–33161. https://doi.org/10.1021/acsomega.0c04808

Jiang R-D, Liu M-Q, Chen Y et al (2020) Pathogenesis of SARS-CoV-2 in transgenic mice expressing human angiotensin-converting enzyme 2. Cell 182:50–58.e8. https://doi.org/10.1016/j.cell.2020.05.027

Jo E, Kim H, König A et al (2021) Determination of infectious hepatitis B virus particles by an end-point dilution assay identifies a novel class of inhibitors. Antivir Res 196:105195. https://doi.org/10.1016/j.antiviral.2021.105195

Johnson S, Nguyen V, Coder D (2013) Assessment of cell viability. Current Protocols 64:9.2.1–9.2.26. https://doi.org/10.1002/0471142956.cy0902s64

Johnson DM, Brasel T, Massey S et al (2023) Evaluation of molnupiravir (EIDD-2801) efficacy against SARS-CoV-2 in the rhesus macaque model. Antivir Res 209:105492. https://doi.org/10.1016/j.antiviral.2022.105492

Kim Y-I, Kim S-G, Kim S-M et al (2020) Infection and rapid transmission of SARS-CoV-2 in ferrets. Cell Host and Microbe 27:704–709.e2. https://doi.org/10.1016/j.chom.2020.03.023

Lei C, Yang J, Hu J, Sun X (2021) On the calculation of $TCID_{50}$ for quantitation of virus infectivity. Virol Sin 36:141–144. https://doi.org/10.1007/s12250-020-00230-5

Liu H, Iketani S, Zask A et al (2022) Development of optimized drug-like small molecule inhibitors of the SARS-CoV-2 3CL protease for treatment of COVID-19. Nat Commun 13:1891. https://doi.org/10.1038/s41467-022-29413-2

Maher JA, DeStefano J (2004) The ferret: An animal model to study influenza virus. Lab Animal (NY) 33:50–53. https://doi.org/10.1038/laban1004-50

Maio N, Cherry S, Schultz DC et al (2022) TEMPOL inhibits SARS-CoV-2 replication and development of lung disease in the Syrian hamster model. iScience 25:105074. https://doi.org/10.1016/j.isci.2022.105074

Maisonnasse P, Guedj J, Contreras V et al (2020) Hydroxychloroquine use against SARS-CoV-2 infection in non-human primates. Nature 585:584–587. https://doi.org/10.1038/s41586-020-2558-4

Matthew F, Boyd Y, Sudhakar A et al (2012) Molecular determinants of severe acute respiratory syndrome coronavirus pathogenesis and virulence in young and aged mouse models of human disease. J Virol 86:884–897. https://doi.org/10.1128/JVI.05957-11

Mautner L, Hoyos M, Dangel A et al (2022) Replication kinetics and infectivity of SARS-CoV-2 variants of concern in common cell culture models. Virol J 19:76. https://doi.org/10.1186/s12985-022-01802-5

Mendoza EJ, Manguiat K, Wood H, Drebot M (2020) Two detailed plaque assay protocols for the quantification of infectious SARS-CoV-2. Curr Protoc Microbiol 57:cpmc105. https://doi.org/10.1002/cpmc.105

Miao J, Chard LS, Wang Z, Wang Y (2019) Syrian hamster as an animal model for the study on infectious diseases. Front Immunol 10:2329. https://doi.org/10.3389/fimmu.2019.02329

Moreau GB, Burgess SL, Sturek JM et al (2020) Evaluation of K18-hACE2 mice as a model of SARS-CoV-2 infection. The American Journal of Tropical Medicine and Hygiene 103:1215–1219. https://doi.org/10.4269/ajtmh.20-0762

Murgolo N, Therien AG, Howell B et al (2021) SARS-CoV-2 tropism, entry, replication, and propagation: considerations for drug discovery and development. PLoS Pathogens 17:e1009225. https://doi.org/10.1371/journal.ppat.1009225

Ogando NS, Dalebout TJ, Zevenhoven-Dobbe JC et al (2020) SARS-coronavirus-2 replication in Vero E6 cells: replication kinetics, rapid adaptation and cytopathology. J Gen Virol 101:925–940. https://doi.org/10.1099/jgv.0.001453

Pandamooz S, Jurek B, Meinung C-P et al (2022) Experimental models of SARS-CoV-2 infection: possible platforms to study COVID-19 pathogenesis and potential treatments. Annu Rev Pharmacol Toxicol 62:25–53. https://doi.org/10.1146/annurev-pharmtox-121120-012309

Parolin C, Virtuoso S, Giovanetti M et al (2021) Animal hosts and experimental models of SARS-CoV-2 infection. Chemotherapy 66:8–16. https://doi.org/10.1159/000515341

Patching SG (2014) Surface plasmon resonance spectroscopy for characterisation of membrane protein–ligand interactions and its potential for drug discovery. Biochimica et Biophysica Acta(BBA) – Biomembranes 1838:43–55. https://doi.org/10.1016/j.bbamem.2013.04.028

Perchetti GA, Nalla AK, Huang M-L et al (2020) Multiplexing primer/probe sets for detection of SARS-CoV-2 by qRT-PCR. J Clin Virol 129:104499. https://doi.org/10.1016/j.jcv.2020.104499

Pires De Souza GA, Le Bideau M, Boschi C et al (2022) Choosing a cellular model to study SARS-CoV-2. Front Cell Infect Microbiol 12:1003608. https://doi.org/10.3389/fcimb.2022.1003608

Qiao J, Li Y-S, Zeng R et al (2021) SARS-CoV-2 Mpro inhibitors with antiviral activity in a transgenic mouse model. Science 371:1374–1378. https://doi.org/10.1126/science.abf1611
Rathnasinghe R, Strohmeier S, Amanat F et al (2020) Comparison of transgenic and adenovirus hACE2 mouse models for SARS-CoV-2 infection. Emerging Microbes and Infections 9:2433–2445. https://doi.org/10.1080/22221751.2020.1838955
Reed LJ, Muench H (1938) A simple method of estimating fifty per cent endpoint. Am J Plant Sci 27:493–497. https://doi.org/10.1093/oxfordjournals.aje.a118408
Riva L, Yuan S, Yin X et al (2020) Discovery of SARS-CoV-2 antiviral drugs through large-scale compound repurposing. Nature 586:113–119. https://doi.org/10.1038/s41586-020-2577-1
Roberts A, Deming D, Paddock CD et al (2007) A mouse-adapted SARS-coronavirus causes disease and mortality in BALB/c mice. PLoS Pathog 3:e5. https://doi.org/10.1371/journal.ppat.0030005
Rosenfeld R, Noy-Porat T, Mechaly A et al (2021) Post-exposure protection of SARS-CoV-2 lethal infected K18-hACE2 transgenic mice by neutralizing human monoclonal antibody. Nat Commun 12:944. https://doi.org/10.1038/s41467-021-21239-8
Rosenke K, Meade-White K, Letko M et al (2020) Defining the Syrian hamster as a highly susceptible preclinical model for SARS-CoV-2 infection. Emerging Microbes Infections 9:2673–2684. https://doi.org/10.1080/22221751.2020.1858177
Sangeun J, Meehyun K, Jihye L et al (2020) Identification of antiviral drug candidates against SARS-CoV-2 from FDA-approved drugs. Antimicrob Agents Chemother 64:e00819–e00820. https://doi.org/10.1128/AAC.00819-20
Sun X, Sui H, Fisher JT et al (2010) Disease phenotype of a ferret CFTR-knockout model of cystic fibrosis. J Clin Invest 120:3149–3160. https://doi.org/10.1172/JCI43052
Takayama K (2020) In-vitro and animal models for SARS-CoV-2 research. Trends Pharmacol Sci 41:513–517. https://doi.org/10.1016/j.tips.2020.05.005
Taylor R, Bowen R, Demarest JF et al (2022) Activity of Galidesivir in a hamster model of SARS-CoV-2. Viruses 14(1):8. https://doi.org/10.3390/v14010008
Toptan T, Hoehl S, Westhaus S et al (2020) Optimized qRT-PCR approach for the detection of intra- and extra-cellular SARSCoV-2 RNAs. Int J Mol Sci 21(12):4396. https://doi.org/10.3390/ijms21124396
Touret F, Gilles M, Barral K et al (2020) In vitro screening of a FDA approved chemical library reveals potential inhibitors of SARS-CoV-2 replication. Sci Rep 10:13093. https://doi.org/10.1038/s41598-020-70143-6
Touret F, Driouich J-S, Cochin M et al (2021) Preclinical evaluation of Imatinib does not support its use as an antiviral drug against SARS-CoV-2. Antivir Res 193:105137. https://doi.org/10.1016/j.antiviral.2021.105137
Vermillion MS, Murakami E, Ma B et al (2023) Inhaled remdesivir reduces viral burden in a nonhuman primate model of SARS-CoV-2 infection. Sci Transl Med 14:eabl8282. https://doi.org/10.1126/scitranslmed.abl8282
Wang C, Li W, Drabek D et al (2020) A human monoclonal antibody blocking SARS-CoV-2 infection. Nat Commun 11:2251. https://doi.org/10.1038/s41467-020-16256-y
Williamson BN, Feldmann F, Schwarz B et al (2020) Clinical benefit of remdesivir in rhesus macaques infected with SARS-CoV-2. Nature 585:273–276. https://doi.org/10.1038/s41586-020-2423-5
Winkler ES, Bailey AL, Kafai NM et al (2020) SARS-CoV-2 infection of human ACE2-transgenic mice causes severe lung inflammation and impaired function. Nat Immunol 21:1327–1335. https://doi.org/10.1038/s41590-020-0778-2
Xia S, Liu M, Wang C et al (2020) Inhibition of SARS-CoV-2 (previously 2019-nCoV) infection by a highly potent pan-coronavirus fusion inhibitor targeting its spike protein that harbors a high capacity to mediate membrane fusion. Cell Res 30:343–355. https://doi.org/10.1038/s41422-020-0305-x

Chapter 12
Discovery of SARS-CoV-2 Antiviral Drugs Through Large-Scale Virtual Screening of FDA-Approved Drugs

Wahiba Ezzemani, Ahd Ouladlahsen, Haya Altawalah, Rachid Saile, M'hammed Sarih, Anass Kettani, and Sayeh Ezzikouri

Abstract Severe acute respiratory syndrome coronavirus 2 (SARS-CoV-2) is the etiological agent responsible for the viral pneumonia outbreak of coronavirus disease 2019 (COVID-19). Because of the time-consuming process of new drug development, drug repositioning may be the only solution to this epidemic. By performing structure-based virtual screening, two targets M^{pro} and RNA-dependent RNA polymerase were screened against 2659 FDA-approved drugs. As a result, three drugs can block both targets based on their good binding affinity and lower binding. This study provides new lead compounds for further in vitro and in vivo studies and new insights for those drugs currently ongoing clinical studies to treat this infection.

W. Ezzemani
Virology Unit, Viral Hepatitis Laboratory, Institut Pasteur du Maroc, Casablanca, Morocco

Laboratoire de Biologie et Santé (URAC34), Départment de Biologie, Faculté des Sciences Ben Msik, Hassan II University of Casablanca, Casablanca, Morocco

A. Ouladlahsen
Faculté de médecine et de pharmacie, Université Hassan II, Casablanca, Morocco

Service des maladies infectieuses, CHU Ibn Rochd, Casablanca, Morocco

H. Altawalah
Department of Microbiology, Faculty of Medicine, Kuwait University, Kuwait, Kuwait

Virology Unit, Yacoub Behbehani Center, Sabah Hospital, Ministry of Health, Kuwait, Kuwait

R. Saile · A. Kettani
Laboratoire de Biologie et Santé (URAC34), Départment de Biologie, Faculté des Sciences Ben Msik, Hassan II University of Casablanca, Casablanca, Morocco

M. Sarih
Service de Parasitologie et des Maladies Vectorielles, Institut Pasteur du Maroc, Casablanca, Morocco

S. Ezzikouri (✉)
Virology Unit, Viral Hepatitis Laboratory, Institut Pasteur du Maroc, Casablanca, Morocco
e-mail: sayeh.ezzikouri@pasteur.ma

N. Kumar et al. (eds.), *Advances in Antiviral Research*, Livestock Diseases and Management, https://doi.org/10.1007/978-981-99-9195-2_12

Keywords Drug repurposing · SARS-CoV-2 · Antiviral drugs · COVID-19 · Drug candidates · M^{pro} · RdRp · Virtual screening

12.1 Introduction

The novel severe acute respiratory syndrome coronavirus 2 (SARS-CoV-2) is a single-stranded enveloped RNA virus with a symmetrical nucleocapsid. It targets cells via the viral structural spike (S) protein that binds to the angiotensin-converting enzyme 2 receptor to enter the host cell. Host type 2 transmembrane serine protease (TMPRSS2) facilitates cell entry via the S protein (Hoffmann et al. 2020). Mechanistically, the viral polyproteins that encode the replicase–transcriptase complex are synthesized, and the RNA-dependent RNA polymerase (RdRp) replicates viral RNA inside the cell. Then, the structural proteins are synthesized leading to the completion of assembly and release of viral particles (Y. Chen et al. 2020; Fehr and Perlman 2015; Fung and Liu 2014). The outbreak of the novel emerging infection coronavirus disease 2019 (COVID-19) was forcing the scientific community to select appropriate therapeutics with very little time available for new drug discovery because no vaccine was available yet (Asselah et al. 2021; Lotfi et al. 2020; Tsang et al. 2021). General strategies were used to search for inhibitors against SARS-CoV-2 currently including viral protein-based screening and viral replication-based phenotypic screening (Alibakhshi et al. 2021, p. 2; Farouk et al. 2021; Mahmoudinobar et al. 2021; Murgolo et al. 2021). Therefore, drug repurposing is a rapid strategy for identifying new uses for approved drugs (Ashburn and Thor 2004). Using this approach, significant achievements, such as SARS-CoV-2 inhibitors, were quickly determined (Chan et al. 2022; Fan et al. 2020; Jourdan et al. 2020; Li et al. 2020). In this chapter, we focus on repurposed drugs that act through SARS-CoV-2-related targets Mpro and RdRp (Al-Bari 2017; Sanders et al. 2020; Savarino et al. 2003) as positive controls by performing a structure-based virtual screening (SBVS) approach of 2659 FDA-approved drugs of the DrugBank Database to identify inhibitors efficiently targeting the SARS-CoV-2.

12.2 Materials and Methods

12.2.1 Database Preparation

The library 2659 of FDA-approved drugs from the DrugBank database was downloaded as an SD file with simplified molecular-input line-entry system representations (Wishart et al. 2006, 2008).

12.2.2 Protein Preparation

The sequences of the target proteins (M^{pro} and RdRp) of SARS-CoV-2 were retrieved from the Protein Data Bank using the accession ID: 6 LU7–7BV2 (Al-Bari 2017; Ionescu 2020, p. 2; Jin et al. 2020; Yin et al. 2020). Thereafter, an energy minimization program (Swiss-PdbViewer) was run with the GROMOS96 force field, to determine the proper molecular arrangement in space. Hydrogens were added in protein residues as well as Gasteiger partial charges. To enhance the analysis, the target proteins were first standardized and converted into .pdbqt format using AutoDock in the PyRx virtual screening tool (Python Prescription 0.8) (Dallakyan and Olson 2015).

12.2.3 Active Site Identification

SARS-CoV-2 M^{pro} shares 96% sequence homology with SARS-CoV and has similar catalytic efficiency. Its tertiary structure consists of the catalytic domains I and II and the helical domain III. The noncanonical catalytic dyad with residues His41 and Cys145 is buried in the active-site cavity that is located on the surface of the protein flanked by residues from both domains I and II (Dai et al. 2020; Huang et al. 2004; Muramatsu et al. 2016; Xu et al. 2020; Xue et al. 2007). Indeed, the RdRp residues: Val557, Asp623, Thr680, and Asn691 are known to play an essential role in catalytic activity and are conserved in the CoV family as per the previous literature (Barage et al. 2022; Jia and Gong 2019; Kirchdoerfer and Ward 2019).

12.2.4 Ligand Preparation

The 2659 small-molecule drugs were imported to virtual screening software, PyRx, for computational drug discovery (Dallakyan and Olson 2015). All the ligands were optimized with the universal force field and converted into the .pdbqt format compatible file for AutoDock, using OpenBabel (Geldenhuys et al. 2006). The nonpolar hydrogen is merged with carbons and polar hydrogen; Gasteiger partial charges were used for geometry optimization. Then, the torsional degrees of freedom were set to zero. In this study, we considered the lopinavir–ritonavir and remdesivir–favipiravir–galidesivir as positive controls for the M^{pro} and RdRp proteins, respectively. Remdesivir, favipiravir, and galidesivir were identified to act on viral replication. Additionally, lopinavir and ritonavir were found to act on viral entry (Singh et al. 2020, p. 19).

12.2.5 Virtual Screening Based on Molecular Docking

The active site residues of M^{pro} and RdRp were treated as flexible, and the grid box was set to 16.4127 × 17.3200 × 65.7850 and 97.9508 × 97.5025 × 103.6861, respectively. The grid box was centered on selected flexible residues and left adequate space with dimensions X (31.5128 Å), Y (23.5079 Å), and Z (25.2003 Å)/M^{pro}, and X (54.6628 Å), Y (50.3403 Å), and Z (44.3573 Å)/RdRp. The eight docking runs were performed for each ligand. The generated docked conformation was ranked by predicted binding affinity and root mean square deviation (RMSD). Furthermore, to represent the receptor–ligand binding energy, the best complexes were evaluated by redocking using SwissDock (Grosdidier et al. 2011). Finally, the molecular interactions of the docking site were visualized using Discovery Studio 2020.

12.3 Results

The top ligand molecules from the virtual screening based on molecular docking were filtered and selected with the lowest binding affinity and a value of RMSD equal to 0. Subsequently, the obtained docked complexes were analyzed for their binding energy, molecular interactions with the active site residues and hydrogen, and hydrophobic interaction. It was observed that, among all compounds, we picked the top 9/10 compounds that showed promising interaction for each active site of M^{pro} and RdRp proteins respectively, as summarized in Tables 12.1 and 12.2, and molecularly represented in Figs. 12.1, 12.2, 12.3, 12.4 and 12.5.

12.4 Discussion

In the last 20 years, new infectious agents such as MERS (2012) and SARS (2002–2003) have emerged and caused epidemics (de Wit et al. 2016). Since COVID-19 is spreading uncontrollably across the globe, it is essential to discover new drugs. The time-consuming effective antiviral drug development is extremely challenging because conventional drug development approaches normally are costly and take years to implement. However, the repurposing of approved pharmaceutical drugs (also called drug repositioning, retasking, or reprofiling) may be the fastest option for emerging infectious diseases (Parvathaneni et al. 2019). Among its advantages, the time frame for drug development can be reduced as this strategy uses de-risked compounds with known details on potential toxicity, formulation, and pharmacology and drug-like molecules with potential activity (Barrows et al. 2016; Botta et al. 2018; Lv et al. 2018; Xu et al. 2016). It focuses on US Food and Drug Administration FDA-approved drugs (Konreddy et al. 2019), which can

Table 12.1 Molecular docking analysis of FDA-approved drugs against SARS-CoV-2 M^{pro}

Compounds	Binding affinity Kcal/mol	Hydrogen bonds	Binding energy kJ/mol
Daunorubicin	-8.0	4H: TYR A: 54 – ASN A: 142 – CYS A: 145 – HIS A: 164	-38.032
Sitagliptin	-8.1	5H: TYR A: 54 – PHE A: 140 – GLY A: 143 – GLU A: 166	-41.129
Nelfinavir	-8.3	2H: HIS A: 164 – GLU A: 166	-41.798
Lapatinib	-8.3	5H: HIS A: 41 – PHE A: 140 – 2GLY A: 143 – CYS A: 145	-49.580
Imatinib	-8.4	1H: PHE A: 140	-41.505
Irbesartan	-8.6	4H: GLU A: 166 – MET A: 165 – ARG A: 188 – THR A: 190	-37.907
Dihydroergotamine	-9.3	3H: SER A: 46 – ASN A: 142 – GLU A: 166-	-44.769
Tadalafil	-9.7	1H: GLY A: 143	-35.815
Retapamulin	-16.9	0H	-36.317
Lopinavir	-8.3	4H: PHE A: 140 – GLY A: 143 – HIS A: 164 – GLU A: 166	-42.383
Ritonavir	-8.0	5H: 2 ASN A: 142 – 2 GLU A: 166 – GLN A:	-45.563

directly enter phase III or IV clinical trials making the drug development process potential (Breckenridge and Jacob 2019; Pushpakom et al. 2019). In the current study, to search for potential SARS-CoV-2 therapeutic drugs, we first screened 2659 FDA-approved drugs from the DrugBank database. Furthermore, known positive control drugs $(\text{lopinavir–ritonavir})_{M}{}^{pro}$ and $(\text{favipiravir–galidesivir–remdesivir})_{RdRp}$ (Singh et al. 2020, p. 19), which are highly potent for COVID-19 infection, were evaluated in parallel to validate the results. Based on current knowledge about the use of these positive controls, remdesivir/NCT04292730, favipiravir/NCT04529499, and galidesivir/NCT03891420 are considered as drugs acting on viral replication. Furthermore, lopinavir and ritonavir (NCT04455958) were acting on viral entry (GlobalData Healthcare 2020, p. 19; Singh et al. 2020, p. 19). Based on the SBVS, we picked the top-9/10 hits for M^{pro}/RdRp protein depending on the binding affinity,

Table 12.2 Molecular docking analysis of FDA-approved drugs against SARS-CoV-2 RdRp

Compounds	Binding affinity Kcal/mol	Hydrogen bonds	Binding energy kJ/mol
Retapamulin	-15.2	0H	-36.819
Dihydroergotamine	-9.0	0H	-37.697
Argatroban	-8.9	8H: PHE A: 442 – LYS A: 545 – 2LYS A: 551 – ARG A: 553 – ARG A: 555 – LYS A: 621 – SER A: 795	-37.948
Sorafenib	-8.6	5H: PHE A: 442 – GLN A: 444 – SER A: 549 – 2ASP A: 623	-35.396
Imatinib	-8.5	3H: TYR A: 619 – 2ASP A: 760	-34.936
Methadone	-8.4	1H: ARG A: 553	-34.309
Liothyronine	-8.3	6H: PHE A: 442 – LYS A: 545 – ALA A: 547 – 2ARG A: 553 – TYR A: 619	-41.965
Idarubicin	-8.1	4H: ARG A: 555 – LYS A: 621 – 2ASP A: 760	-36.986
Raltitrexed	-8.0	4H: PHE A: 442 – ALA A: 547 – ARG A: 553 – ASP A: 623	-48.325
Rifampicin	-13,1	0H	-36,718
Favipiravir	-5.9	5H: LYS A: 545 – SER A: 549 – LYS A: 551 – ARG A: 553 – ARG A: 555	-28.409
Galidesivir	-7.0	3H: 2PHE A: 442 – ARG A: 553	-35.103
Remdesivir	-7.3	7H: LYS A: 551 – TYR A: 619 – LYS A: 621 – CYS A: 622 – ASP A: 760 - PHE A: 793 – SER A: 795	-36.104

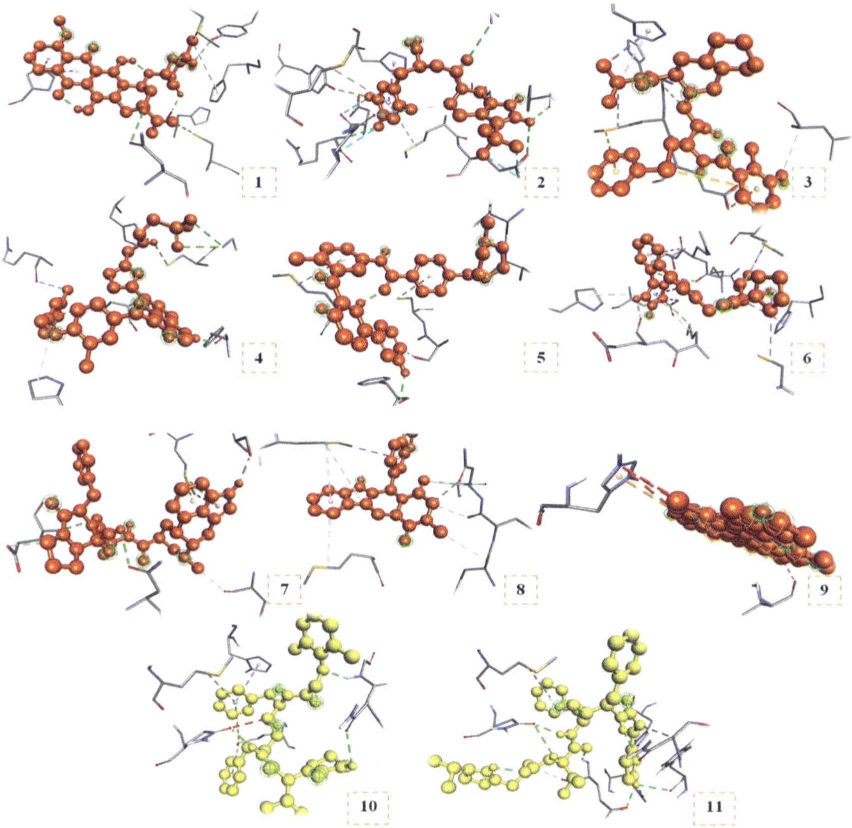

Fig. 12.1 Docking interaction structure of the SARS-CoV-2 M^{pro} and nine favorable inhibitors (1–daunorubicin, 2–sitagliptin, 3–nelfinavir, 4–lapatinib, 5–imatinib, 6–irbesartan, 7–dihydroergotamine, 8–tadalafil, and 9–rétapamulin) with hydrogen bonds/10–lopinavir and 11–ritonavir: positive controls

hydrogen bonds, and binding energy. Hydrogen bonds play a critical role in protein–ligand interaction and lower the binding energy to stabilize the docked complex (D. Chen et al. 2016). All the compounds showed an RMSD value less than 2.0 Å from the experimentally observed conformation and we considered it as a successful prediction (Xiao et al. 2018). The 9 top compounds of M^{pro} showed the docking score in a range of −8.0 to −16.9 Kcal/mol, and the 10 top compounds of RdRp showed the docking score in a range of −8.0 to −15.2 Kcal/mol. They displayed a higher binding affinity (most negative) and low binding energies ΔG with a significant number of hydrogen bonds compared to the positive controls. Knowing that the binding affinity and binding energy values of interaction of the positive controls referred (lopinavir and ritonavir) M^{pro}–(favipiravir, galidesivir, and remdesivir) RdRp is (−8.3 and − 8.0 Kcal/mol) M^{pro}–(−5.9, −7.0, and − 7.3 Kcal/mol) RdRp/(−42.383 and − 45.563 kJ/mol) M^{pro}–(−28.409, −35.103, and − 36.104 kJ/mol) RdRp, respectively. Daunorubicin is an anthracycline antibiotic used to induce

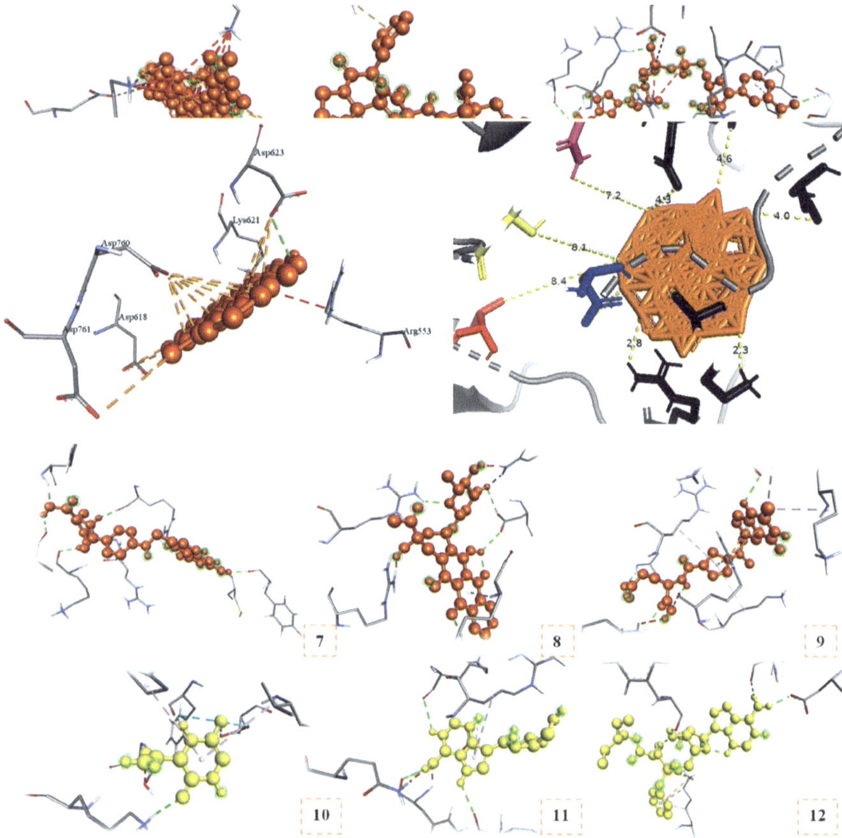

Fig. 12.2 Docking interaction structure of the SARS-CoV-2 RdRp and nine favorable inhibitors (1–retapamulin, 2–dihydroergotamine, 3–argatroban, 4–sorafenib, 5–imatinib, 6–methadone, 7–liothyronine, 8–idarubicin, and 9–raltitrexed) with hydrogen bonds/10–favipiravir, 11–galidesivir, and 12–remdesivir: positive controls

remission of nonlymphocytic leukemia in adults and acute lymphocytic leukemia in children and adults (Riggs 1984). Our docking results showed five of the hydrogen bonds involving TYR54: 3.3 Å with a distance of 3.6 Å, ASN142: 2.9 Å, CYS145: 3.6 Å, and HIS164: 3.0 Å maintained upon the binding of daunorubicin with the active site of SARS-CoV-2 M^{pro}. Two hydrogen bonds (ASN142–HIS164) were in common with the positive controls ritonavir and lopinavir, respectively. The other amino acid residues participating in the interaction at the active site were HIS41: 3.2 Å, MET49: 3.9 Å, and PRO168: 4.6 Å. Furthermore, daunorubicin was found to exhibit a binding affinity of −8.0 Kcal/mol and a lower binding energy of −41.798 kJ/mol. Sitagliptin is an oral dipeptidyl peptidase-4 inhibitor that is used in combination with diet and exercise in the therapy of type 2 diabetes (Scott 2017). Our docking results showed that the hydrogen bonds involving TYR54: 3.6 Å, PHE140: 4.0 Å, GLY143: 2.4 Å, and GLU166: 3.3 Å maintained upon the binding of

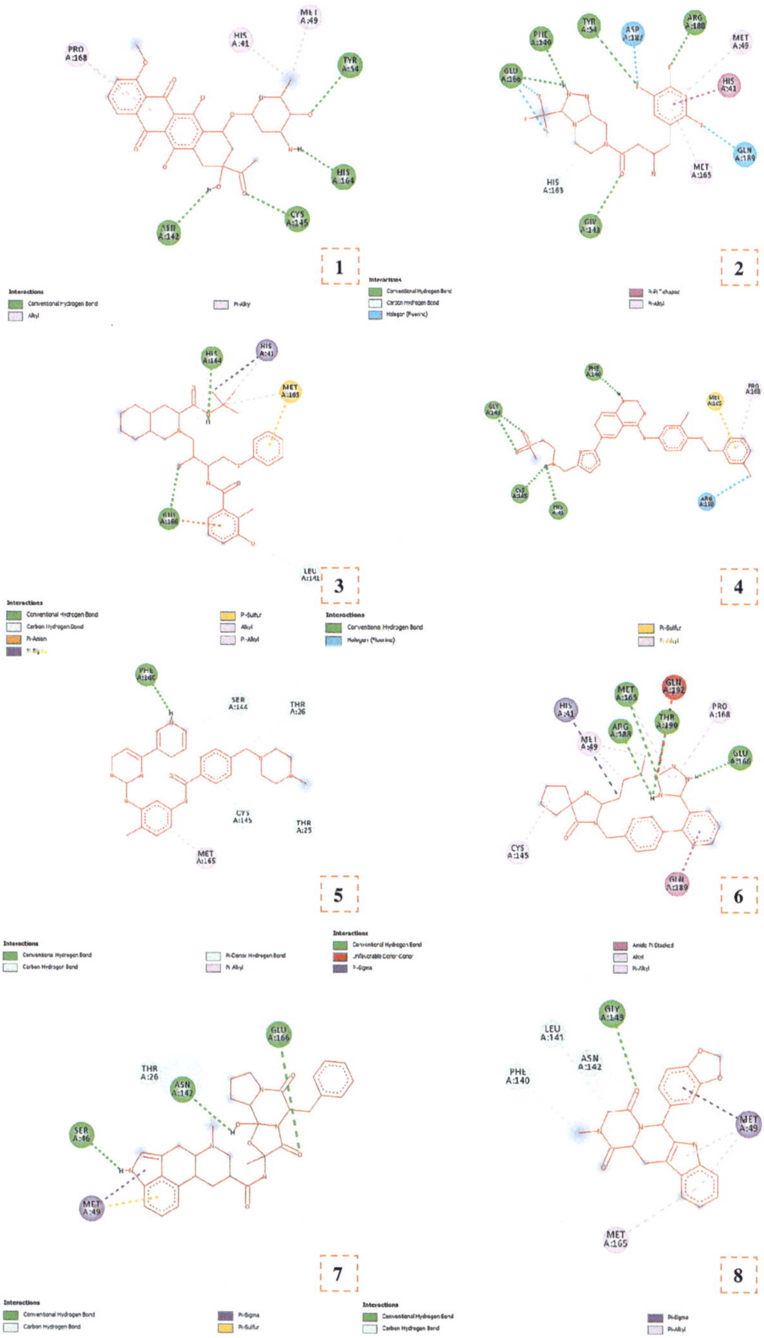

Fig. 12.3 Docking interaction structure of the SARS-CoV-2 RdRp and rifampicin

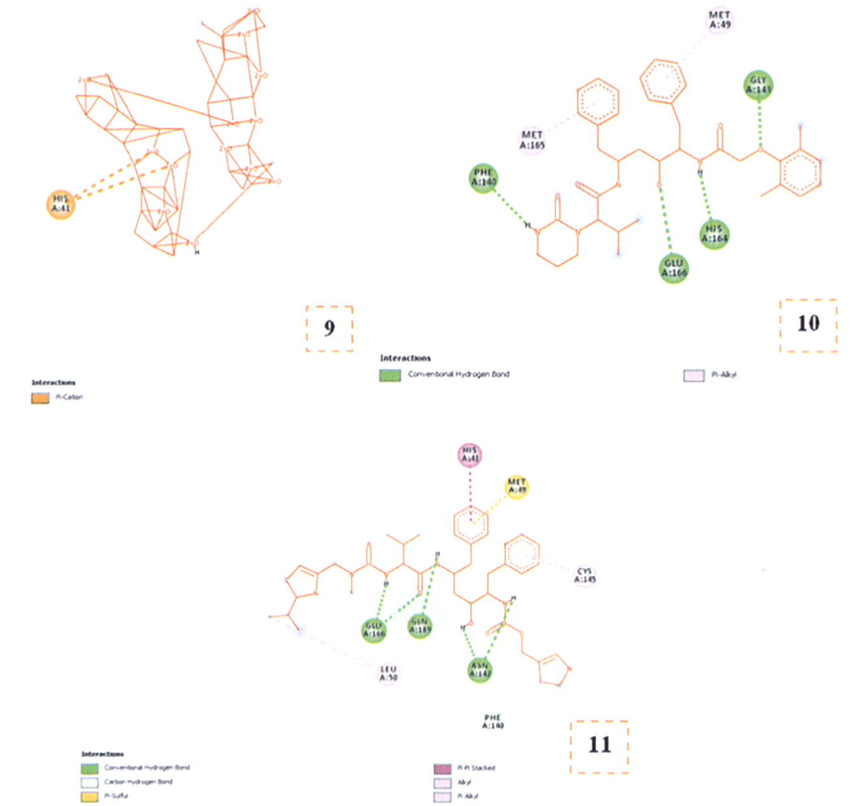

Fig. 12.4 The 2D protein–ligand interactions map between SARS-CoV-2 M^{pro}. 1–daunorubicin, 2–sitagliptin, 3–nelfinavir, 4–lapatinib, 5–imatinib, 6–irbesartan, 7–dihydroergotamine, 8–tadalafil, 9–retapamulin/10–lopinavir and 11–ritonavir: positive controls, obtained by Discovery Studio Visualizer

sitagliptin and the active site of M^{pro}, with a binding affinity of −8.1 Kcal/mol. PHE140 and GLU166 hydrogen bonds were in common with lopinavir. Sitagliptin was also combined with HIS41: 3.7 Å, MET49: 4.3 Å, HIS163: 1.3 Å, MET165:4.9 Å, and GLN189: 2.9 Å amino acid residues. We note that sitagliptin exhibits a binding energy value of −41.129 kJ/mol. Our docking results showed that both nelfinavir (potent HIV-1 protease inhibitor) (Wang et al. 2020) and lapatinib (anti-cancer drug) (Jia et al. 2020) had a binding affinity of −8.3 Kcal/mol with binding energies of −41.798 and – 49.580 kJ/mol, respectively. Furthermore, Ohashi et al. identified that nelfinavir has a higher antiviral potential than the drug candidates (remdesivir and chloroquine) in VeroE6/TMPRSS2 cells. The nelfinavir suppressed viral replication partly by protease inhibition (Ohashi et al. 2021). This is consistent with our results based on the binding affinity. Nelfinavir could bind to the interface active site of the SARS-CoV-2 M^{pro} ($Nelfinavir_{M^{pro}}$). The amino acid residues participating in the interaction at the active site were HIS41: 3.9 Å, LEU141: 3.8 Å, and MET165: 4.2 Å, with two amino acids that were found to be involved in forming hydrogen

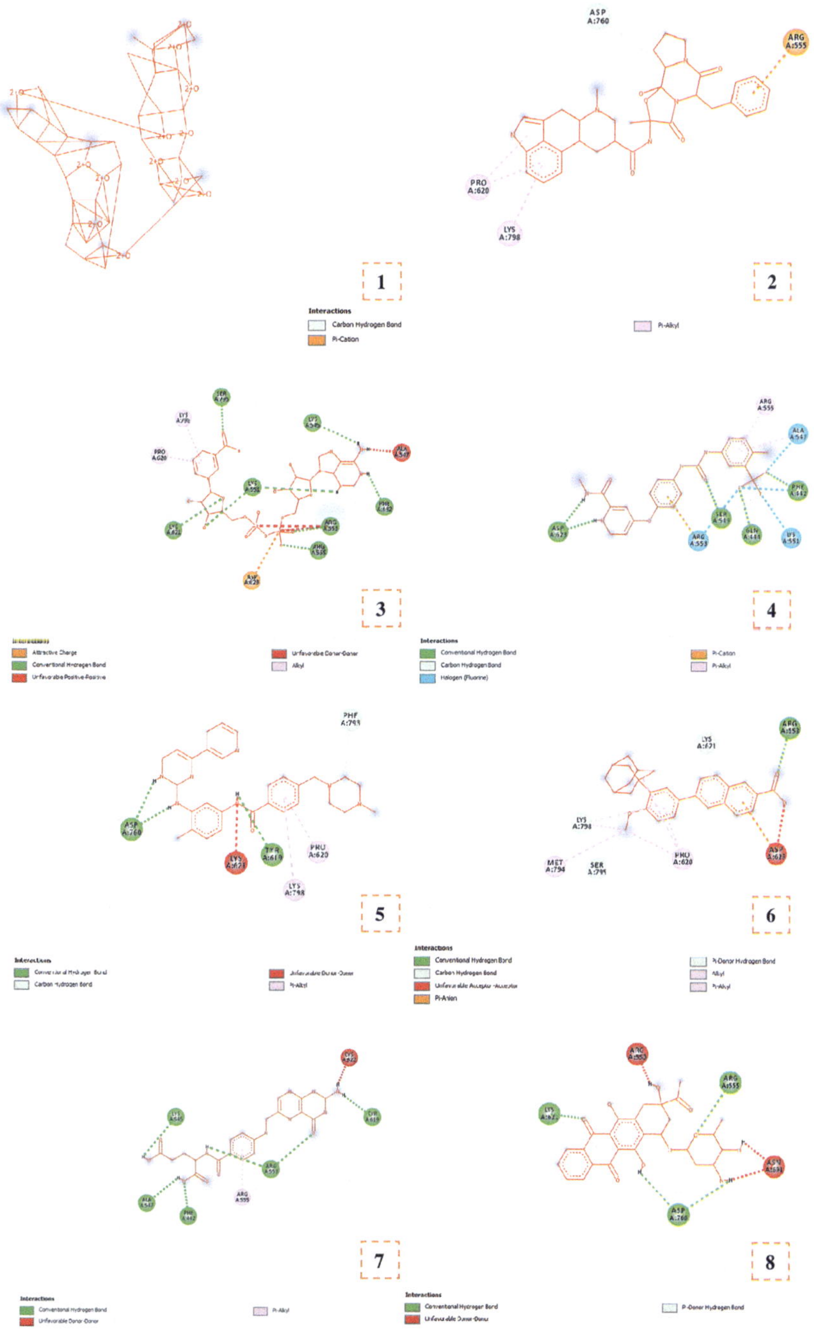

Fig. 12.5 The 2D protein–ligand interactions map between SARS-CoV-2 RdRp. 1–retapamulin, 2–dihydroergotamine, 3–argatroban, 4–sorafenib, 5–imatinib, 6–methadone, 7–liothyronine, 8–idarubicin, 9–raltitrexed/10–favipiravir, 11–galidesivir, and 12–remdesivir: positive controls, obtained by Discovery Studio Visualizer

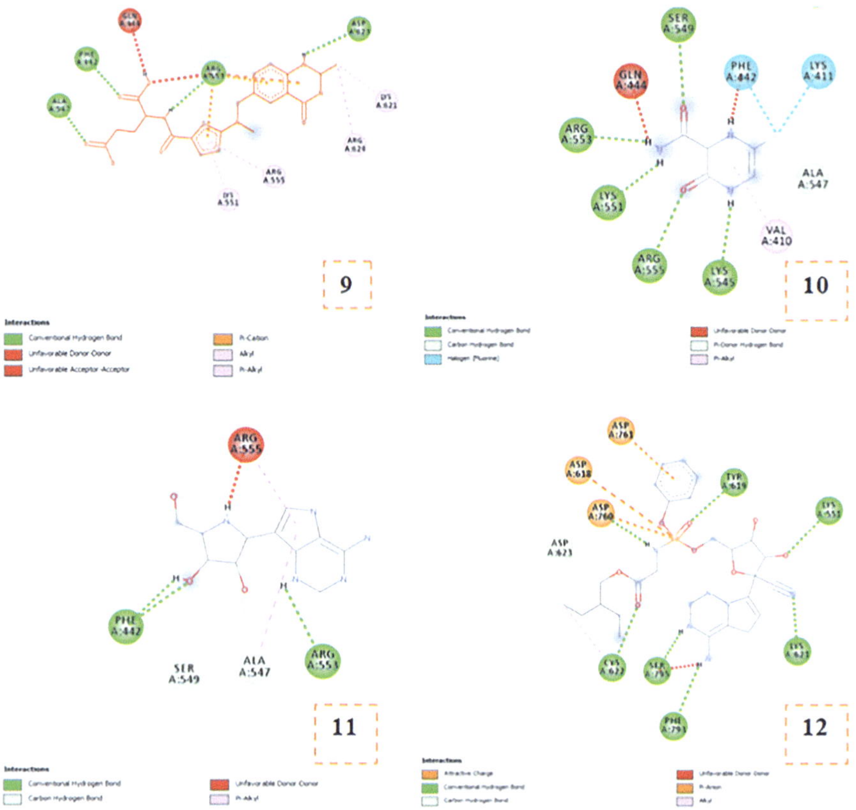

Fig. 12.5 (continued)

bonds with this compound (HIS164: 2.2 Å and GLU166: 4.7 Å). These conventional hydrogen bonds were in common with the lopinavir. Then, lapatinib could bind to the interface active site of the SARS-CoV-2 M^{pro} (Lapatinib$_{M}{}^{pro}$). The amino acid residues participating in the interaction at the active site were MET165: 4.0 Å, PRO168: 3.6 Å, and ARG188:3.3 Å, with five amino acids that were found to be involved in forming hydrogen bonds with this compound (HIS41: 3.5 Å, PHE140: 2.8 Å, 2GLY143: 2.4 Å, and CYS145: 3.5 Å). PHE140 and 2GLY143 conventional hydrogen bonds were also in common with the lopinavir. Imatinib is a small-molecule kinase inhibitor used to treat certain types of cancer (Suttorp et al. 2018). Our docking results showed that imatinib exhibits a binding affinity of −8.4 Kcal/mol with a binding energy of −41.505 kJ/mol. Furthermore, Aman et al. identified that imatinib might confer clinical benefit in hospitalized patients with COVID-19. However, further studies are required (Aman et al. 2021). Then, another study evaluated the efficacy and safety of imatinib combined with the best conventional care (BCC) versus placebo plus BCC in hospitalized patients with COVID-19 (Emadi et al. 2020). This is consistent with our results based on the binding affinity. Imatinib could bind to the interface active site of the SARS-CoV-2 M^{pro} (Imatinib$_{M}{}^{pro}$). The amino acid residues participating in the interaction at the active site were THR25: 3.7 Å, THR26: 2.9 Å, SER144: 3.3 Å, CYS145: 3.4 Å, and MET165: 3.4 Å, with one amino acid that was found to be involved in forming hydrogen bond with this compound (PHE140: 3.2 Å). This conventional hydrogen bond was in common with the lopinavir. Irbesartan is an angiotensin II receptor blocker used in the therapy of hypertension and diabetic nephropathy (Hachiya et al. 2021). Our docking results showed four of the hydrogen bonds involving MET165: 4.1 Å, GLU166: 4.2 Å, ARG188: 3.3 Å, and THR190: 4.4 Å maintained on the binding of irbesartan with the active site of M^{pro}. Then, the amino acid residues participating in the interaction molecular were HIS41: 3.5 Å, MET49: 5.4 Å, CYS145: 5.9 Å, PRO168: 6.7 Å, and GLN189: 4.6 Å. Irbesartan exhibited a binding energy of −37.907 kJ/mol and a binding affinity of −8.6 Kcal/mol. Dihydroergotamine is an antimigraine drug (Dahlöf and Maassen Van Den Brink 2012). Tadalafil is a selective phosphodiesterase-5 inhibitor used to treat erectile dysfunction, benign prostatic, hyperplasia, and pulmonary arterial hypertension (Mónica and De Nucci 2019). Gurung et al. found that dihydroergotamine binds to the active site with a lower binding energy (Gurung et al. 2020). This is consistent with our results based on a lower binding energy of −44.769 kJ/mol. Dihydroergotamine could bind to the interface active site of the SARS-CoV-2 M^{pro} (His41 and Cys145) with a binding affinity of −9.3 Kcal/mol. The amino acid residues participating in the interaction at the active site were THR26 (3.1 Å) and MET49 (3.6 Å), with three amino acids that were found to be involved in forming hydrogen bonds with this compound (SER46: 2.8 Å, ASN142: 3.5 Å, and GLU166: 4.1 Å). These conventional hydrogen bonds were in common with the ritonavir. Then, tadalafil could bind to the interface active site of the SARS-CoV-2 M^{pro} (tadalafil M^{pro}). The amino acid residues participating in the interaction at the active site were MET49: 4.1 Å, PHE140: 4.7 Å, LEU141: 3.7 Å, ASN142: 3.1 Å, and MET165: 3.6 Å with one amino acid that was found to be involved in forming a hydrogen bond with this compound (GLY143: 2.4 Å) and which is in common with the lopinavir. Retapamulin is a topical antibiotic agent used for the treatment of impetigo caused by methicillin-susceptible *Staphylococcus aureus* and *Streptococcus pyogenes* ("Retapamulin for impetigo and other infections" 2008; Yang and Keam 2008). Our docking results showed that the retapamulin exhibits the best binding affinity of −16.9 Kcal/mol with a binding energy of −36.317 kJ/mol. It is rare to exceed −15 Kcal/mol. Kuntz and coworkers suggested that other biological factors may be the cause of the limit, molecules with too high binding affinity can exhibit clearance problems in the body (Kuntz et al. 1999). The amino acid residues participating in the interaction at the active site were His41: 3.2 Å and Thr190: 3.4 Å. Retapamulin forms a cation–π interaction with a positively charged histidine residue (41) in the active site. Cation–π interactions play an important role in determining protein structure and function. They influence biological structures, catalysis, and molecular recognition (Dougherty 2007; Johnston et al. 2014; Mantaj et al. 2017). Computational studies show that cation–π interactions are prevalent in protein–ligand interactions (Ma and Dougherty 1997; Zacharias and Dougherty 2002). Associated with the RdRp, retapamulin $(\text{retapamulin})_{\text{RdRp}}$ and $(\text{dihydroergotamine})_{\text{RdRp}}$ have been found to have the best binding affinity of −15.2

and − 9.0 Kcal/mol, respectively. The amino acid residues participating in the interaction at the active site with retapamulin were ASP618: 3.8 Å, LYS621: 3.7 Å, and SER795: 3.6 Å and then, ARG555: 3.5 Å, PRO620: 3.7 Å, ASP760: 4.5 Å, and LYS798: 4.3 Å with dihydroergotamine. Argatroban$_{RdRp}$ is a direct, selective thrombin inhibitor used for the prevention and treatment of thrombosis related to heparin use (Aliter and Al-Horani 2021, p. 19). Thus, argatroban presents a substantial promise in treating severe cases of COVID-19 (Aliter and Al-Horani 2021, p. 19; Sagardia and Daniels 2020, p. 19). Argatroban could bind to the interface active site of the SARS-CoV-2 RdRp (VAL557, ASP623, THR680, and ASN691) with a binding affinity of −8.9 Kcal/mol and a binding energy of −37.948 kJ/mol**.** The amino acid residues participating in the interaction at the active site were PRO620: 4.1 Å, ASP623: 5.7 Å, and LYS798: 4.0 Å, with eight amino acids that were found to be involved in forming hydrogen bonds with this compound (PHE442: 3.1 Å, LYS545: 2.9 Å, LYS551: 4.6 Å, ARG553: 3.5 Å, ARG555: 3.9 Å, LYS621: 2.3 Å, and SER795: 3.3 Å). These conventional hydrogen bonds were in common with the three positive controls (remdesivir, favipiravir, and galidesivir). Sorafenib$_{RdRp}$ is a kinase inhibitor used in the treatment of unresectable liver carcinoma and advanced renal carcinoma (Tang et al. 2020). Our docking results showed the hydrogen bonds involving PHE442: 2.9 Å, GLN444: 3.9 Å, SER549: 5.3 Å, and 2 ASP623: 3.9 Å maintained on the binding of sorafenib and the active site of RdRp, with a binding affinity of −8.6 Kcal/mol. PHE442 and SER549 hydrogen bonds were in common with galidesivir and favipiravir, respectively. Sorafenib was also combined with ALA547: 4.3 Å, LYS551: 7.6 Å, ARG553: 3.0 Å, and ARG555: 3.6 Å amino acid residues. We note that sorafenib exhibits a binding energy value of −35.396 kJ/mol. Associated with the RdRp, imatinib (Imatinib)$_{RdRp}$ has been found to have a binding affinity of −8.5 Kcal/mol and a binding energy of −34.936 kJ/mol. The amino acid residues participating in the interaction at the active site with imatinib were PRO620: 3.7 Å, PHE793: 5.5 Å, LYS798: 5.5 Å with TYR619: 4.6 Å, and 2 ASP760: 3.2 Å conventional hydrogen bonds. Methadone$_{RdRp}$ is an opioid analgesic indicated for the management of severe pain that is not responsive to alternative treatments (Kreutzwiser and Tawfic 2020). Our docking results showed that methadone exhibits a binding affinity of −8.4 Kcal/mol with a binding energy of −34.309 kJ/mol. Methadone could bind to the interface active site of SARS-CoV-2 RdRp. The amino acid residues participating in the interaction at the active site were PRO620: 3.7 Å, LYS621: 2.6 Å, MET794: 3.8 Å, SER795: 3.6 Å, and LYS798: 3.9 Å with one amino acid that was found to be involved in forming hydrogen bond with this compound (ARG553: 4.1 Å). This conventional hydrogen bond was in common with favipiravir and galidesivir. (Liothyronine)^{RdRp} is a thyroid hormone replacement therapy used to treat hypothyroidism (Jonklaas et al. 2021). Our docking results showed six of the hydrogen bonds involving PHE442: 2.3 Å, LYS545: 4.8 Å, ALA547: 4.9 Å, 2ARG553: 3.5 Å, and TYR619: 5.2 Å maintained on the binding of liothyronine with the active site of SARS-CoV-2 RdRp. Five hydrogen bonds (PHE442, LYS545, 2 ARG553, and TYR619) were in common with the positive controls galidesivir, favipiravir, and remdesivir, respectively. The other amino acid residue participating in the interaction at the active site was ARG555: 5.0 Å. Furthermore, liothyronine was found to exhibit a binding affinity of −8.3 Kcal/mol and a lower binding energy of −41.965 kJ/mol. $Idarubicin_{RdRp}$ is an anthracycline antineoplastic agent used to treat acute myeloid leukemia (Shi et al. 2021). Our docking results showed the hydrogen bonds involving ARG555: 3.0 Å, LYS621: 4.2 Å, and 2 ASP760: 2.8 Å maintained on the binding of idarubicin and the active site of RdRp, with a binding affinity of −8.1 Kcal/mol. The hydrogen bonds were in common with favipiravir and remdesivir, respectively. We note that idarubicin exhibits a binding energy value of −36.986 kJ/mol. $Raltitrexed_{RdRp}$ is a folate analog thymidylate synthase inhibitor used in the treatment of advanced colorectal cancer (Liu et al. 2014). Our docking results showed that raltitrexed exhibits a binding affinity of −8.0 Kcal/mol with a binding energy of −48.325 kJ/mol. Raltitrexed could bind to the interface active site of SARS-CoV-2 RdRp. The amino acid residues participating in the interaction at the active site were LYS551: 4.7 Å, ARG555: 3.4 Å, LYS621: 4.2 Å, and ARG624: 4.5 Å, with four amino acids that were found to be involved in forming hydrogen bonds with this compound (PHE442: 2.7 Å, ALA547: 2.2 Å, ARG553: 3.7 Å, and ASP623: 2.5 Å). PHE442 and ARG553 conventional hydrogen bonds were in common with the galidesivir. $Rifampicin_{RdRp}$ is an antibiotic used to treat several types of mycobacterial infections, including *Mycobacterium avium* complex, leprosy, and in combination with other antibacterials to treat latent or active tuberculosis ("Rifampicin" 1971). Our docking results showed the amino acid residues involved in the interaction at the active site were LYS621: 2.3 Å, ASP760: 4.3 Å, ASP618: 4.0 Å, ASP761: 4.6 Å, and ARG553: 2.8 Å. Thus, the distances of residues VAL557, ASP623, THR680, and ASN691 from the active site were 8.4 Å, 4.4 Å, 8.1 Å, and 7.2 Å, respectively. In addition, we found that rifampicin exhibits a binding affinity of −13.1 Kcal/mol and a lower binding energy of −36.718 kJ/mol.

Among all these results mentioned above, retapamulin, dihydroergotamine, and imatinib can block both M_{pro} and RdRp based on their good binding affinity and the lower binding energy using re-docking. Therefore, this study suggests that the top compounds can be tested in vitro for their anti-SARS-CoV-2 effect. These compounds offer a new direction to the researcher to design target-specific inhibitors, as drug repurposing has not been clinically effective in COVID-19 patients.

12.5 Conclusion

The indiscriminate spread of the coronavirus poses a threat to human health around the world. M^{pro} and RdRp SARS-CoV-2 proteins are potential targets to fight against the virus. Thus, this study carried out high-throughput FDA-approved drugs. Subsequently, 19 compounds were selected for further evaluation based on virtual screening, docking scores, and binding energy. Among them, retapamulin, dihydroergotamine, and imatinib can simultaneously be used as candidate drugs for the treatment of SARS-CoV-2 infection, making them possible for broad-spectrum antiviral drugs. Therefore, this study suggests that these compounds can be tested in vitro for their anti-SARS-CoV-2 effect, providing more choices for clinical treatment and prevention of SARS-CoV-2 infection.

References

Al-Bari MAA (2017) Targeting endosomal acidification by chloroquine analogs as a promising strategy for the treatment of emerging viral diseases. Pharmacol Res Perspect 5(1):e00293. https://doi.org/10.1002/prp2.293

Alibakhshi A, Ranjbar MM, Javanmard SH, Yarian F, Ahangarzadeh S (2021) Virtual screening for the identification of potential candidate molecules against envelope (E) and membrane (M) proteins of SARS-CoV-2. J Comput Biophys Chem 20(03):209–224. https://doi.org/10.1142/S2737416521500083

Aliter KF, Al-Horani RA (2021) Thrombin inhibition by Argatroban: potential therapeutic benefits in COVID-19. Cardiovasc Drugs Ther 35(2):195–203. https://doi.org/10.1007/s10557-020-07066-x

Aman J, Duijvelaar E, Botros L, Kianzad A, Schippers JR, Smeele PJ et al (2021) Imatinib in patients with severe COVID-19: a randomised, double-blind, placebo-controlled, clinical trial. Lancet Respir Med 9(9):957–968. https://doi.org/10.1016/S2213-2600(21)00237-X

Ashburn TT, Thor KB (2004) Drug repositioning: identifying and developing new uses for existing drugs. Nat Rev Drug Discov 3(8):673–683. https://doi.org/10.1038/nrd1468

Asselah T, Durantel D, Pasmant E, Lau G, Schinazi RF (2021) COVID-19: discovery, diagnostics and drug development. J Hepatol 74(1):168–184. https://doi.org/10.1016/j.jhep.2020.09.031

Barage S, Karthic A, Bavi R, Desai N, Kumar R, Kumar V, Lee KW (2022) Identification and characterization of novel RdRp and Nsp15 inhibitors for SARS-COV2 using computational approach. J Biomol Struct Dyn 40(6):2557–2574. https://doi.org/10.1080/07391102.2020.1841026

Barrows NJ, Campos RK, Powell ST, Prasanth KR, Schott-Lerner G, Soto-Acosta R et al (2016) A screen of FDA-approved drugs for inhibitors of Zika virus infection. Cell Host Microbe 20(2):259–270. https://doi.org/10.1016/j.chom.2016.07.004

Botta L, Rivara M, Zuliani V, Radi M (2018) Drug repurposing approaches to fight dengue virus infection and related diseases. Front Biosci (Landmark Edition) 23(6):997–1019. https://doi.org/10.2741/4630

Breckenridge A, Jacob R (2019) Overcoming the legal and regulatory barriers to drug repurposing. Nat Rev Drug Discov 18(1):1–2. https://doi.org/10.1038/nrd.2018.92

Chan WKB, Olson KM, Wotring JW, Sexton JZ, Carlson HA, Traynor JR (2022) In silico analysis of SARS-CoV-2 proteins as targets for clinically available drugs. Sci Rep 12(1):5320. https://doi.org/10.1038/s41598-022-08320-y

Chen D, Oezguen N, Urvil P, Ferguson C, Dann SM, Savidge TC (2016) Regulation of protein-ligand binding affinity by hydrogen bond pairing. Sci Adv 2(3):e1501240. https://doi.org/10.1126/sciadv.1501240

Chen Y, Liu Q, Guo D (2020) Emerging coronaviruses: genome structure, replication, and pathogenesis. J Med Virol 92(4):418–423. https://doi.org/10.1002/jmv.25681

Dahlöf C, Maassen Van Den Brink A (2012) Dihydroergotamine, ergotamine, methysergide and sumatriptan – basic science in relation to migraine treatment. Headache 52(4):707–714. https://doi.org/10.1111/j.1526-4610.2012.02124.x

Dai W, Zhang B, Jiang X-M, Su H, Li J, Zhao Y et al (2020) Structure-based design of antiviral drug candidates targeting the SARS-CoV-2 main protease. Science (New York, N.Y.) 368(6497):1331–1335. https://doi.org/10.1126/science.abb4489

Dallakyan S, Olson AJ (2015) Small-molecule library screening by docking with PyRx. Methods Mol Biol (Clifton, N.J.) 1263:243–250. https://doi.org/10.1007/978-1-4939-2269-7_19

de Wit E, van Doremalen N, Falzarano D, Munster VJ (2016) SARS and MERS: recent insights into emerging coronaviruses. Nat Rev Microbiol 14(8):523–534. https://doi.org/10.1038/nrmicro.2016.81

Dougherty DA (2007) Cation-pi interactions involving aromatic amino acids. J Nutr 137(6 Suppl 1):1504S–1508S.; discussion 1516S-1517S. https://doi.org/10.1093/jn/137.6.1504S

Emadi A, Chua JV, Talwani R, Bentzen SM, Baddley J (2020) Safety and efficacy of Imatinib for hospitalized adults with COVID-19: a structured summary of a study protocol for a randomised controlled trial. Trials 21(1):897. https://doi.org/10.1186/s13063-020-04819-9

Fan H-H, Wang L-Q, Liu W-L, An X-P, Liu Z-D, He X-Q et al (2020) Repurposing of clinically approved drugs for treatment of coronavirus disease 2019 in a 2019-novel coronavirus-related coronavirus model. Chin Med J 133(9):1051–1056. https://doi.org/10.1097/CM9.0000000000000797

Farouk A-E, Baig MH, Khan MI, Park T, Alotaibi SS, Dong J-J (2021) Screening of inhibitors against SARS-CoV-2 spike protein and their capability to block the viral entry mechanism: a viroinformatics study. Saudi J Biol Sci 28(6):3262–3269. https://doi.org/10.1016/j.sjbs.2021.02.066

Fehr AR, Perlman S (2015) Coronaviruses: an overview of their replication and pathogenesis. Methods Mol Biol (Clifton, N.J.) 1282:1–23. https://doi.org/10.1007/978-1-4939-2438-7_1

Fung TS, Liu DX (2014) Coronavirus infection, ER stress, apoptosis and innate immunity. Front Microbiol 5:296. https://doi.org/10.3389/fmicb.2014.00296

Geldenhuys WJ, Gaasch KE, Watson M, Allen DD, Van der Schyf CJ (2006) Optimizing the use of open-source software applications in drug discovery. Drug Discov Today 11(3–4):127–132. https://doi.org/10.1016/S1359-6446(05)03692-5

GlobalData Healthcare. (2020). Influenza drug favipiravir is being tested to treat Covid-19, Mars 30. https://www.clinicaltrialsarena.com/comment/influenza-favipiravir-covid-19/

Grosdidier A, Zoete V, Michielin O (2011) SwissDock, a protein-small molecule docking web service based on EADock DSS. Nucleic Acids Res 39((Web Server issue)):W270–W277. https://doi.org/10.1093/nar/gkr366

Gurung AB, Ali MA, Lee J, Abul Farah M, Al-Anazi KM (2020) In silico screening of FDA approved drugs reveals ergotamine and dihydroergotamine as potential coronavirus main protease enzyme inhibitors. Saudi J Biol Sci 27(10):2674–2682. https://doi.org/10.1016/j.sjbs.2020.06.005

Hachiya K, Masuya M, Kuroda N, Yoneda M, Tsuboi J, Nagaharu K et al (2021) Irbesartan, an angiotensin II type 1 receptor blocker, inhibits colitis-associated tumourigenesis by blocking the MCP-1/CCR2 pathway. Sci Rep 11(1):19943. https://doi.org/10.1038/s41598-021-99412-8

Hoffmann M, Kleine-Weber H, Schroeder S, Krüger N, Herrler T, Erichsen S et al (2020) SARS-CoV-2 cell entry depends on ACE2 and TMPRSS2 and is blocked by a clinically proven protease inhibitor. Cell 181(2):271–280.e8. https://doi.org/10.1016/j.cell.2020.02.052

Huang C, Wei P, Fan K, Liu Y, Lai L (2004) 3C-like proteinase from SARS coronavirus catalyzes substrate hydrolysis by a general base mechanism. Biochemistry 43(15):4568–4574. https://doi.org/10.1021/bi036022q

Ionescu MI (2020) An overview of the crystallized structures of the SARS-CoV-2. Protein J 39(6):600–618. https://doi.org/10.1007/s10930-020-09933-w

Jia H, Gong P (2019) A structure-function diversity survey of the RNA-dependent RNA polymerases from the positive-Strand RNA viruses. Front Microbiol 10:1945. https://doi.org/10.3389/fmicb.2019.01945

Jia J-N, Yin X-X, Li Q, Guan Q-W, Yang N, Chen K-N et al (2020) Neuroprotective effects of the anti-cancer drug Lapatinib against epileptic seizures via suppressing glutathione peroxidase 4-dependent Ferroptosis. Front Pharmacol 11:601572. https://doi.org/10.3389/fphar.2020.601572

Jin Z, Du X, Xu Y, Deng Y, Liu M, Zhao Y et al (2020) Structure of Mpro from SARS-CoV-2 and discovery of its inhibitors. Nature 582(7811):289–293. https://doi.org/10.1038/s41586-020-2223-y

Johnston CP, Kothari A, Sergeieva T, Okovytyy SI, Jackson KE, Paton RS, Smith MD (2014) Catalytic enantioselective synthesis of indanes by a cation-directed 5-endo-trig cyclization. Nat Chem 7(2):171–177. https://doi.org/10.1038/nchem.2150

Jonklaas J, Bianco AC, Cappola AR, Celi FS, Fliers E, Heuer H et al (2021) Evidence-based use of Levothyroxine/Liothyronine combinations in treating hypothyroidism: a consensus document. Thyroid Offic J Am Thyroid Assoc 31(2):156–182. https://doi.org/10.1089/thy.2020.0720

Jourdan J-P, Bureau R, Rochais C, Dallemagne P (2020) Drug repositioning: a brief overview. J Pharm Pharmacol 72(9):1145–1151. https://doi.org/10.1111/jphp.13273

Kirchdoerfer RN, Ward AB (2019) Structure of the SARS-CoV nsp12 polymerase bound to nsp7 and nsp8 co-factors. Nat Commun 10(1):2342. https://doi.org/10.1038/s41467-019-10280-3

Konreddy AK, Rani GU, Lee K, Choi Y (2019) Recent drug-repurposing-driven advances in the discovery of novel antibiotics. Curr Med Chem 26(28):5363–5388. https://doi.org/10.2174/0929867325666180706101404

Kreutzwiser D, Tawfic QA (2020) Methadone for pain management: a pharmacotherapeutic review. CNS Drugs 34(8):827–839. https://doi.org/10.1007/s40263-020-00743-3

Kuntz ID, Chen K, Sharp KA, Kollman PA (1999) The maximal affinity of ligands. Proc Natl Acad Sci U S A 96(18):9997–10002. https://doi.org/10.1073/pnas.96.18.9997

Li Z, Li X, Huang Y-Y, Wu Y, Liu R, Zhou L et al (2020) Identify potent SARS-CoV-2 main protease inhibitors via accelerated free energy perturbation-based virtual screening of existing drugs. Proc Natl Acad Sci U S A 117(44):27381–27387. https://doi.org/10.1073/pnas.2010470117

Liu Y, Wu W, Hong W, Sun X, Wu J, Huang Q (2014) Raltitrexed-based chemotherapy for advanced colorectal cancer. Clin Res Hepatol Gastroenterol 38(2):219–225. https://doi.org/10.1016/j.clinre.2013.11.006

Lotfi M, Hamblin MR, Rezaei N (2020) COVID-19: transmission, prevention, and potential therapeutic opportunities. Clin Chim Acta Int J Clin Chem 508:254–266. https://doi.org/10.1016/j.cca.2020.05.044

Lv B-M, Tong X-Y, Quan Y, Liu M-Y, Zhang Q-Y, Song Y-F, Zhang H-Y (2018) Drug repurposing for Japanese encephalitis virus infection by systems biology methods. Molecules (Basel, Switzerland) 23(12):E3346. https://doi.org/10.3390/molecules23123346

Ma JC, Dougherty DA (1997) The Cationminus signpi interaction. Chem Rev 97(5):1303–1324. https://doi.org/10.1021/cr9603744

Mahmoudinobar F, Britton D, Montclare JK (2021) Protein-based lateral flow assays for COVID-19 detection. Protein Eng Des Sel PEDS 34:gzab010. https://doi.org/10.1093/protein/gzab010

Mantaj J, Jackson PJM, Rahman KM, Thurston DE (2017) From Anthramycin to Pyrrolobenzodiazepine (PBD)-containing Antibody-Drug Conjugates (ADCs). Angew Chem Int Ed Engl 56(2):462–488. https://doi.org/10.1002/anie.201510610

Mónica FZ, De Nucci G (2019) Tadalafil for the treatment of benign prostatic hyperplasia. Expert Opin Pharmacother 20(8):929–937. https://doi.org/10.1080/14656566.2019.1589452

Muramatsu T, Takemoto C, Kim Y-T, Wang H, Nishii W, Terada T et al (2016) SARS-CoV 3CL protease cleaves its C-terminal autoprocessing site by novel subsite cooperativity. Proc Natl Acad Sci U S A 113(46):12997–13002. https://doi.org/10.1073/pnas.1601327113

Murgolo N, Therien AG, Howell B, Klein D, Koeplinger K, Lieberman LA et al (2021) SARS-CoV-2 tropism, entry, replication, and propagation: considerations for drug discovery and development. PLoS Pathog 17(2):e1009225. https://doi.org/10.1371/journal.ppat.1009225

Ohashi H, Watashi K, Saso W, Shionoya K, Iwanami S, Hirokawa T et al (2021) Potential anti-COVID-19 agents, cepharanthine and nelfinavir, and their usage for combination treatment. iScience 24(4):102367. https://doi.org/10.1016/j.isci.2021.102367

Parvathaneni V, Kulkarni NS, Muth A, Gupta V (2019) Drug repurposing: a promising tool to accelerate the drug discovery process. Drug Discov Today 24(10):2076–2085. https://doi.org/10.1016/j.drudis.2019.06.014

Pushpakom S, Iorio F, Eyers PA, Escott KJ, Hopper S, Wells A et al (2019) Drug repurposing: progress, challenges and recommendations. Nat Rev Drug Discov 18(1):41–58. https://doi.org/10.1038/nrd.2018.168

Retapamulin for impetigo and other infections (2008) Drug Ther Bull 46(10):76–79. https://doi.org/10.1136/dtb.2008.09.0023

Rifampicin: a review (1971) Drugs 1(5):354–398. https://doi.org/10.2165/00003495-197101050-00002

Riggs CE (1984) Clinical pharmacology of daunorubicin in patients with acute leukemia. Semin Oncol 11(4 Suppl 3):2–11

Sagardia LM, Daniels LM (2020) Thrombolysis and use of argatroban for the treatment of massive pulmonary embolism following anticoagulation failure in a patient with COVID-19. Am J Health Syst Pharm AJHP Offic J Am Soc Health Syst Phar 77(23):1961–1964. https://doi.org/10.1093/ajhp/zxaa287

Sanders JM, Monogue ML, Jodlowski TZ, Cutrell JB (2020) Pharmacologic treatments for coronavirus disease 2019 (COVID-19): a review. JAMA 323(18):1824–1836. https://doi.org/10.1001/jama.2020.6019

Savarino A, Boelaert JR, Cassone A, Majori G, Cauda R (2003) Effects of chloroquine on viral infections: an old drug against today's diseases? Lancet Infect Dis 3(11):722–727. https://doi.org/10.1016/s1473-3099(03)00806-5

Scott LJ (2017) Sitagliptin: a review in type 2 diabetes. Drugs 77(2):209–224. https://doi.org/10.1007/s40265-016-0686-9

Shi P, Zha J, Feng J, Jiang Z, Zhao H, Deng M et al (2021) Low-dose Triptolide enhanced activity of idarubicin against acute myeloid Leukemia stem-like cells via inhibiting DNA damage repair response. Stem Cell Rev Rep 17(2):616–627. https://doi.org/10.1007/s12015-020-10054-1

Singh TU, Parida S, Lingaraju MC, Kesavan M, Kumar D, Singh RK (2020) Drug repurposing approach to fight COVID-19. Pharmacol Rep PR 72(6):1479–1508. https://doi.org/10.1007/s43440-020-00155-6

Suttorp M, Bornhäuser M, Metzler M, Millot F, Schleyer E (2018) Pharmacology and pharmacokinetics of imatinib in pediatric patients. Expert Rev Clin Pharmacol 11(3):219–231. https://doi.org/10.1080/17512433.2018.1398644

Tang W, Chen Z, Zhang W, Cheng Y, Zhang B, Wu F et al (2020) The mechanisms of sorafenib resistance in hepatocellular carcinoma: theoretical basis and therapeutic aspects. Signal Transduct Target Ther 5(1):87. https://doi.org/10.1038/s41392-020-0187-x

Tsang HF, Chan LWC, Cho WCS, Yu ACS, Yim AKY, Chan AKC et al (2021) An update on COVID-19 pandemic: the epidemiology, pathogenesis, prevention and treatment strategies. Expert Rev Anti-Infect Ther 19(7):877–888. https://doi.org/10.1080/14787210.2021.1863146

Wang R-G, Zhang H-X, Zheng Q-C (2020) Revealing the binding and drug resistance mechanism of amprenavir, indinavir, ritonavir, and nelfinavir complexed with HIV-1 protease due to double mutations G48T/L89M by molecular dynamics simulations and free energy analyses. Phys Chem Chem Phys PCCP 22(8):4464–4480. https://doi.org/10.1039/c9cp06657h

Wishart DS, Knox C, Guo AC, Shrivastava S, Hassanali M, Stothard P et al (2006) DrugBank: a comprehensive resource for in silico drug discovery and exploration. Nucleic Acids Res 34(Database issue):D668–D672. https://doi.org/10.1093/nar/gkj067

Wishart DS, Knox C, Guo AC, Cheng D, Shrivastava S, Tzur D et al (2008) DrugBank: a knowledgebase for drugs, drug actions and drug targets. Nucleic Acids Res 36(Database issue):D901–D906. https://doi.org/10.1093/nar/gkm958

Xiao W, Wang D, Shen Z, Li S, Li H (2018) Multi-body interactions in molecular docking program devised with key water molecules in protein binding sites. Molecules (Basel, Switzerland) 23(9):E2321. https://doi.org/10.3390/molecules23092321

Xu M, Lee EM, Wen Z, Cheng Y, Huang W-K, Qian X et al (2016) Identification of small-molecule inhibitors of Zika virus infection and induced neural cell death via a drug repurposing screen. Nat Med 22(10):1101–1107. https://doi.org/10.1038/nm.4184

Xu J, Zhao S, Teng T, Abdalla AE, Zhu W, Xie L et al (2020) Systematic comparison of two animal-to-human transmitted human coronaviruses: SARS-CoV-2 and SARS-CoV. Viruses 12(2):E244. https://doi.org/10.3390/v12020244

Xue X, Yang H, Shen W, Zhao Q, Li J, Yang K et al (2007) Production of authentic SARS-CoV M(pro) with enhanced activity: application as a novel tag-cleavage endopeptidase for protein overproduction. J Mol Biol 366(3):965–975. https://doi.org/10.1016/j.jmb.2006.11.073

Yang LPH, Keam SJ (2008) Retapamulin: a review of its use in the management of impetigo and other uncomplicated superficial skin infections. Drugs 68(6):855–873. https://doi.org/10.2165/00003495-200868060-00008

Yin W, Mao C, Luan X, Shen D-D, Shen Q, Su H et al (2020) Structural basis for inhibition of the RNA-dependent RNA polymerase from SARS-CoV-2 by remdesivir. Science (New York, N.Y.) 368(6498):1499–1504. https://doi.org/10.1126/science.abc1560

Zacharias N, Dougherty DA (2002) Cation-pi interactions in ligand recognition and catalysis. Trends Pharmacol Sci 23(6):281–287. https://doi.org/10.1016/s0165-6147(02)02027-8

Chapter 13
A High-Throughput Computational Pipeline for Selection of Effective Antibody Therapeutics Against Viruses

Rahul Kaushik, Naveen Kumar, and Thomas Launey

Abstract Continuous evolution of viruses such as severe acute respiratory syndrome-coronavirus-2 (SARS-CoV-2) generates multiple variants that can evade approved neutralizing antibodies (NAbs) or other antibody therapeutics. Frequently retesting whether a panel of NAbs has retained their neutralization potential against such emerging variants is costly and time-consuming. In such a case, computational tools that mimic the experiment binding affinity of the target protein of the virus and NAbs could play a pivotal role in providing researchers with quick understanding and guidance for using a set of NAbs effective against new virus variants. Hence, we developed and validated a computational pathway for binding affinity prediction in protein complexes. We demonstrated the application of this computational pathway in the rapid screening of a panel of NAbs for their efficacy in targeting the SARS-CoV-2 Omicron variant using benchmarking with the experimental data. We believe that the proposed computational pathway could help antiviral researchers in the initial screening of multiple NAbs against viruses of interest.

Keywords Binding strength · Protein-protein interactions · Predicted dissociation constant · Neutralizing antibodies · SARS-CoV-2 · Variants of concern

13.1 Introduction

The wild-type severe acute respiratory syndrome-coronavirus-2 (SARS-CoV-2) has assimilated several mutations in the course of the progression of the coronavirus disease 2019 (COVID-19) pandemic, leading to the evolution of multiple variants

R. Kaushik (✉) · T. Launey
Biotechnology Research Center, Technology Innovation Institute, Masdar City, Abu Dhabi, UAE
e-mail: rahul.kaushik@tii.ae

N. Kumar
ICAR-National Institute of High Security Animal Diseases, Bhopal, India

N. Kumar et al. (eds.), *Advances in Antiviral Research*, Livestock Diseases and Management, https://doi.org/10.1007/978-981-99-9195-2_13

(Rambaut et al. 2020; Alai et al. 2021; Faria et al. 2021; Tegally et al. 2021). These multiple variants are responsible for a change in the increased virulence and transmissibility, clinical illness presentation, and decreased effectiveness of treatment or vaccination. Therefore, these variants are characterized as variants of concern (VOCs) by the World Health Organization (WHO). Attributing to the resistance of the VOCs against several neutralizing antibodies (Nabs), including certain clinical antibodies approved by the United States Food and Drug Administration (FDA) as therapeutics, the situation has been further aggravated (Mlcochova et al. 2021; Wei et al. 2021; Planas et al. 2022). The NAbs are known to be very crucial in protecting to combat SARS-CoV-2. However, the rate at which new variants are emerging, the effect of these mutations on the receptor-binding domain (RBD) of Spike proteins, and the resulting ability to escape the defense imparted by NAbs have raised concerns about the creation of new potent Nabs, keeping pace with SARS-CoV-2 evolution (Augusto et al. 2022; Cao et al. 2022; Cele et al. 2022). In early 2023, Omicron VOC and its descendent lineages, such as BA.1, BA.2, BA.3, BA.4, BA.5, and recombinant forms (XBB and BQ.1), are predominantly circulating across the world while other variants that were once major VOCs (Alpha, Beta, Gamma, and Delta) now only circulate at residual level (Trevor et al. 2021; Who 2022a, b). The Omicron VOC emerged in South Africa by the accumulation of 30 amino acid residues substitutions in its Spike protein (32 mutations at the nucleotide level), in comparison to wild-type SARS-CoV-2 (Zahradník et al. 2021; Kannan et al. 2022). These amino acid residues substitutions in the receptor-binding domain (RBD) of Spike protein of Omicron VOC strongly influence the interactions of the SARS-CoV-2 to the host cells in terms of binding affinity, via their angiotensin-converting enzyme 2 (ACE2) receptor and thus an augmented transmissibility (Han et al. 2021; Augusto et al. 2022) or curtails the neutralizing ability of several NAbs, particularly for Omicron VOC (Khoury et al. 2021; Cele et al. 2022; Dejnirattisai et al. 2022). Furthermore, a high rate of substitutions among circulating VOCs and their presence in overlapping geographical areas, new variants with better binding affinity to human ACE2 are likely to emerge in the near future (Kumar et al. 2021; Kaushik et al. 2022) and may compromise the efficacy of NAbs.

The identification of the most effective NAbs among all existing NAbs necessitates a labor-intensive, time-consuming, and costly extensive experimental screening (Greaney et al. 2021a, b; Augusto et al. 2022; Cao et al. 2022; Cele et al. 2022; Dejnirattisai et al. 2022; Kannan et al. 2022; Liu et al. 2022; Mohapatra et al. 2022; Planas et al. 2022). Considering the primary use of NAbs in hospital settings with patients already harboring respiratory symptoms, the importance of identifying effective NAbs in a timely and cost-efficient manner cannot be understated. Our goal is to present an approach combining state-of-the-art computational methods into a pipeline to deliver reliable solutions for narrowing down NAbs candidates, as well as for preemptively exploring the consequences of potential mutations. We show that this strategy allows for fast and targeted experimental confirmation.

Here, we describe a consensus-based computational pipeline for reliable binding affinity prediction between the RBD of Spike protein and a panel of NAbs. The presented computational pipeline's robustness and reliability are demonstrated by its high correlation with the experimental data of available NAbs for effectiveness against SARS-CoV-2. The developed pipeline has the potential to furnish crucial insights for predicting the neutralization potential of existing NAbs upon the emergence of new variants in the future and could be implemented to forecast the neutralization potential for the most probable mutations.

In this chapter, we provide detailed insight into the development of a computational pipeline for making reliable predictions about the neutralization potential of a panel of NAbs against SARS-CoV-2 VOCs. The chapter is divided into six sections for clarity: (i) Introduction—describes the current situation with SARS-CoV-2 variants and the importance of having a computational pipeline for screening NAbs for efficacy in combating SARS-CoV-2, (ii) Dataset Compilation—presents the publicly available sequences and structure repositories used in the proposed pipeline to compile initial dataset for SARS-CoV-2 variants and NAbs, (iii) Protein Structure Modeling—describes the modeling of protein sequence of interest using the state-of-the-art approaches and selecting the best model structure, (iv) Modeling the Protein-Protein Complex—covers the modeling of RBD-NAb complex using homology-based protein structural modeling approaches and selecting the optimal model for the protein-protein complex among the multiple model structures, (v) Characterization of Binding Affinities and Benchmarking—focuses on the prediction and evaluation of binding affinities of RBD-NAbs complexes, as well as their comparison with experimental datasets, and (vi) Application/Case Study—demonstrates a use case of screening the NAbs against SARS-CoV-2 VOCs using the proposed computational pipeline, in order to find the NAbs that maintained their neutralizing capacity to counter the diverse SARS-CoV-2 variants.

13.2 Sequence and Structural Dataset Compilation for SARS-CoV-2 and NAbs

13.2.1 Genomic Sequences of SARS-CoV-2

Global Initiative on Sharing All Influenza Data (GISAID) is a continuously maintained and updated comprehensive repository of genetic sequences, associated clinical, geographical, and epidemiological data, along with species-specific data for diverse viruses including human, avian, and animal viruses. It fosters the data sharing of influenza viruses and the SARS-CoV-2 (https://gisaid.org) (Elbe and Buckland-Merrett 2017; Shu and Mccauley 2017). The database has more than 14.5 million genomes from 215 countries and territories, allowing researchers to investigate and comprehend the evolution and transmission of viruses during pandemics and epidemics. The available data at GISAID can be accessed through a simple procedure of registration and acceptance of the database access agreement, which includes perpetuating scientific ethics by properly recognizing the source laboratories for the data and supporting collaborative study. Researchers can consult the EpiCoV™ database maintained by the GISAID for the most recent genomic sequences of previously and currently circulating SARS-CoV-2 variants (Elbe and Buckland-Merrett 2017; Shu and Mccauley 2017; Gisiad 2021). The corresponding RBD protein of SARS-CoV-2 VOCs serves as the first step for protein structural modeling in this study. The database provides an easy-to-use interactive platform equipped with several bioinformatics tools to perform preliminary analysis to explore the gene(s)/genome(s) of interest.

13.2.2 Sequence and Structural Information of NAbs

The structural characterization of NAbs against SARS-CoV-2 is very critical to exploring the vaccine efficacy, therapeutics development, and designing of broadly acting NAbs. The Coronavirus Antibody Database (CoV-AbDab, http://opig.stats.ox.ac.uk/webapps/covabdab/), which is maintained by the University of Oxford, provides updated and exhaustive details for many NAbs, compiled from more than 1500 published/patented nanobodies and antibodies, in terms of their potential of cross-neutralization for beta-coronaviruses, origin, germline assignments, corresponding protein databank (PDB) entries (subject to availability)/homolog structural models, and source literature (Raybould et al. 2021). The CoV-AbDab rationales to deliver consolidated knowledge through coordinated efforts for identifying and characterizing novel NAbs against SARS-CoV-2. The researcher can access the entire CoV-AbDab and filter the desired results in a CSV file format along with their structural models and information (Raybould et al. 2021).

13.3 Structural Modeling of Spike Protein's RBD and Its Refinement

13.3.1 Protein Structure Prediction of Spike Protein's RBD

Recent breakthroughs in protein structure prediction approaches have armored researchers with highly reliable protein structural models in time- and cost-efficient manners. The biennial Critical Assessment of Protein Structure Prediction (CASP) experiments document the success of protein structure prediction approaches that use the intrinsic structural features from experimental protein structures (Haas et al. 2018; Kaushik et al. 2018; Baek et al. 2021; Kryshtafovych et al. 2021; Kwon et al. 2021). The methods such as AlphaFold2 (Jumper et al. 2021), RoseTTAFold (Baek et al. 2021), I-Tasser (Yang et al. 2015), Modeller (Webb and Sali 2021), and others have consistently delivered the reliable structure prediction from the protein sequence within certain limits while implementing various flavors of protein structure prediction approaches. These limitations can be circumvented by implementing a consensus approach that combines structural predictions from state-of-the-art methods, with at least one method for each type of structural modeling approach. In the described pipeline, AlphaFold2 (Jumper et al. 2021) and RosettaFold (Baek et al. 2021) use the knowledge-based deep learning approaches to predict protein structure and have delivered an exceedingly high accuracy over most of the other counterparts. Similarly, I-Tasser employs an ab initio and homolog-based hybrid approach by implementing various physics and knowledge-based potentials/principles and has consistently improved performance for protein structure prediction over the years, as evidenced by CASP experiments (Yang et al. 2015). Furthermore, when an experimentally solved closely related protein structure to the protein sequence (to be modeled) is available, Modeller, a completely homolog-based protein structure prediction method, produces very reliable results (Webb and Sali 2021). These protein structure prediction tools are available as both servers and standalone applications. Their dual availability enables the user to execute the proposed pipeline with or without sufficient computational resources and maintain optimal accuracy. In general, each of the aforementioned protein structure prediction tools provides five model structures for a given protein sequence, resulting in 20 model structures for an input protein sequence.

13.3.2 Quality Assessment of Predicted Model Structures

A comprehensive protein structure quality assessment is critical in identifying the most reliable predicted structure from a pool of 20 models for the input protein sequence (Singh et al. 2016; Olechnovič and Venclovas 2017; Uziela et al. 2017, 2018; Wang et al. 2021). The current state-of-the-art methods for quantifying the reliability and accuracy of model structures have consistently delivered desirable results (Singh et al. 2016; Olechnovič and Venclovas 2017; Kaushik and Zhang 2021; Mcguffin et al. 2021) and could identify the most suitable model structure from a pool of 20 model structures for the input protein sequence very efficiently (Davis et al. 2007; Laskowski et al. 2012; Cao et al. 2016; Baldassarre et al. 2020; Kaushik and Zhang 2020). A consensus outcome of three state-of-the-art methods, namely ModFold8 (Mcguffin et al. 2021), ProTSAV (Singh et al. 2016), and ProFitFun (Kaushik and Zhang 2020, 2021), was implemented in the proposed pipeline. ModFold variants (the more recent being ModFold8) have been the leading methods for estimating the quality of predicted model structures for years. The ModFold8 (available at https://www.reading.ac.uk/bioinf/ModFOLD) implements advanced deep learning-based contact potentials along with several other scores to quantify the accuracy of predicted model structures. To select the most likely model structure in the proposed pipeline, the single ModFold8 quality assessment scores for the 20 model structures were used (Mcguffin et al. 2021). ProTSAV (available at http://www.scfbio-iitd.res.in/software/proteomics/protsav.jsp), a protein tertiary structure analysis and validation server, is similar to ModFold8 in that it performs a comprehensive quality estimation of protein model structures by integrating ten different parameters derived from different tools to circumvent their limitations and provides an overall structural quality score, named ProTSAV-score (Singh et al. 2016). The ProTSAV-score can be used for comparing and identifying the best model structure from the input protein sequence's 20 model structures. Another method used in this pipeline for protein structure quality assessment is ProFitFun (available at https://github.com/KYZ-LSB/ProTerS-FitFun). It is a machine learning-based approach for quantifying the protein model structure fitness to identify the native-like model structure from multiple model structures. The ProFitFun implements dihedral and solvent accessibility preferences of amino acid residues at tripeptide level that are derived from experimentally solved protein structures. It returns a global quality assessment score for the input protein model structures ranging from 0 to 1, with a higher score indicating a better structure (Kaushik and Zhang 2020, 2021, 2022). The best model structure was determined by averaging the quality assessment scores from ModFold8, ProTSAV, and ProFitFun, with a higher mean value representing a better model structure. The model structure with the highest score was used for further refinement and optimization. It was observed that most of the time the second and third best model structures were very close to the the best structure in terms of the quality scores and structural features. Whenever the top three model structures with similar scores but different structural features were observed, more than a single model structure were considered as the candidate model structures.

13.3.3 Refinement and Optimization of Selected Model Structure

The best model structure chosen based on the consensus quality assessment scores was then subjected to protein structural refinement to improve the quality assessment scores (Heo et al. 2013; Bhattacharya 2019; Chen et al. 2020; Heo et al. 2021). GalaxyRefine, an exhaustive protein structure refinement approach (available at http://galaxy.seoklab.org/refine), was implemented to optimize the best model structure. It performs side chain rebuilding and repacking, followed by overall structure relaxation based on molecular dynamics. The implemented refinement approach in GalaxyRefine improves the structural quality of the predicted structures (Heo et al. 2013). The refined and optimized model structure was then used to predict protein-protein complexes as one of the interacting partners with NAbs. Figure 13.1 depicts the workflow for the proposed consensus approach to predict, identify, refine, and optimize model structures for a target protein sequence.

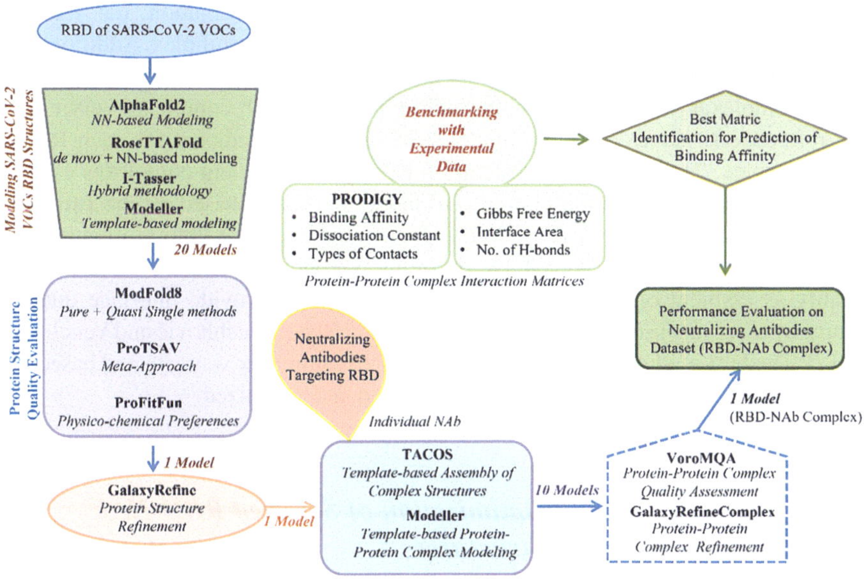

Fig. 13.1 Detailed workflow of the proposed pipeline for fast and reliable screening of neutralizing antibody therapeutics for their neutralization potential against SARS-CoV-2 VOCs

13.4 Protein-Protein Complex Structure Modeling and Refinement

13.4.1 Homolog-Based RBD-NAb Complex Modeling

The increase in experimentally solved protein-protein complex structures has improved the accuracy of various template-based protein-protein complex prediction methods (Maheshwari and Brylinski 2016; Yan et al. 2017; Porter et al. 2019; Park et al. 2021). However, identifying an appropriate template protein-protein complex that can be utilized as a reference for predicting the structure of the target protein-protein complex is paramount for reliable modeling of protein-protein complexes and continues to be a major challenge, particularly in cases of low sequence similarities. Thus, accurate identification of interacting partners in a protein-protein complex via comparative modeling approaches is pivotal for the success of protein-protein complex prediction methods. The structural information of experimentally solved homolog protein-protein complexes can be utilized to create template-based modeling of target protein-protein complexes. In the proposed pipeline, TACOS (Szilagyi and Zhang 2014), a standalone package for template-based protein-protein docking, was implemented to model the RBD-NAbs complexes. TACOS uses a hierarchical approach based on threading, refinement, and simulations to identify the protein complex template structure from PDB based on a target pair of sequences (interacting partners) using multichain threading. Using experimentally characterized wild-type RBD-NAb complex structural information, the TACOS method predicted the five model structures for the RBD-NAb complex. Additionally, a standalone version of Modeller was used for template-based modeling of RBD-NAbs complexes, with corresponding wild-type RBD-NAb complexes serving as the reference structure, and five model structures were predicted for each RBD-NAb complex. The best model structure for the protein-protein complexes can be chosen from a pool of multiple predicted model structures based on the structure quality assessment performed by VoroMQA (available at http://bioinformatics.ibt.lt/wtsam/voromqa). The VoroMQA performs Voronoi tessellation-based model quality assessment of multimeric protein complexes to provide interface quality scores by assessing protein-protein interaction interfaces (Olechnovič and Venclovas 2017). The best model structure for each RBD-NAb complex was selected based on its interface quality score, and it was then refined and optimized.

13.4.2 Refinement and Optimization of Selected RBD-NAb Complex Model Structures

The quality and optimization of predicted model structures determine the extent and reliability of molecular and atomic insights from protein-protein complex models. Implementation of low-resolution energy functions in protein-protein docking and predictions leads to partially optimized structures. The accuracy of such model structure can be improved by using protein complex refinement methods to optimize interface contacts and inter-protein orientations. In the proposed pipeline, GalaxyRefineComplex (available at http://galaxy.seoklab.org/refinecomplex) was implemented to optimize the RBD-NAb complexes. By using iterative side chain repacking and short molecular dynamics-based model protein complex structure relaxation, the GalaxyRefineComplex performs effective sampling (Heo et al. 2016). The sampling is intended to simulate a protein-protein interaction mechanism mediated by side-chain interactions between the partnering proteins (Heo et al. 2016). VoroMQA's reassessment of refined protein-protein complexes may aid in identifying the best model structure for the RBD-NAb protein complex (Olechnovič and Venclovas 2017). The workflow for the proposed consensus approach for modeling RBD-NAb complex structures using template-based protein-protein complex modeling approaches for the interacting protein partners is shown in Fig. 13.1.

13.5 Characterization of Binding Affinities of RBD–NAbs Complexes and Benchmarking

13.5.1 Identification of Protein-Protein Interaction Strength Prediction Metric

A comprehensive understanding of protein-protein interactions in various biological functional complexes can be achieved by a thorough investigation of the structural determinants of partnering proteins' interacting interfaces. Accurate quantification of the extent and strength of interactions among the interacting proteins may provide structural insights into protein-protein interactions. Binding affinity, change in Gibbs free energy (ΔG), dissociation constant (K_d), number and type of different interactions, and so on are examples of quantification metrics. Several tools predict the values of these metrics from the protein complex structure (Krissinel and Henrick 2007; Krissinel 2010; Xue et al. 2016; Vangone and Bonvin 2017; Jiménez-García et al. 2019). However, in the absence of corresponding experimental characterization, the predicted values of these metrics, while serving the rationale of quantifying the strength of interaction, can lead to a quandary, particularly when the two different metrics reflect entirely contradictory interpretations. To avoid this situation, the benchmarking of predicted values of these metrics with experimental values is required to identify the best metrics to reflect the strength of interactions in the protein complex model structures.

PDBbind database suffices as a repository of experimentally characterized binding affinity of protein-protein complexes. It was created to assist computational and statistical investigations in the fields of molecular recognition and drug discovery (Wang et al. 2004, 2005; Liu et al. 2015). To identify an ideal metric that mimics the experimental binding affinities, a dataset of 580 protein complexes, consisting of 400 heterodimeric and 180 heterotrimeric protein-protein complexes, was extracted from the PDBbind.

The most common computational tools that can be used for binding strength prediction of protein-protein complexes are Protein Binding Energy Prediction (PRODIGY) and Proteins, Interfaces, Structures, and Assemblies (PISA). PRODIGY (available at http://www.bonvinlab.org/software) estimates the binding strengths of protein complexes from their structures by implementing its intramolecular properties and contact-based models (Xue et al. 2016; Vangone and Bonvin 2017; Jiménez-García et al. 2019). The output of PRODIGY includes binding free energy prediction (in Kcal mol^{-1}), predicted dissociated constant (K_d), and the number and types of intermolecular interactions. Similarly, PISA analyzes the chemical and structural features of biomolecular interfaces for estimating their free energy of dissociation, H-bonds, disulfide bonds, salt-bridges, interface area (in $Å^2$), and the residues and atoms involved in the interactions (Krissinel and Henrick 2007; Krissinel 2010). In the proposed pipeline, we have used PRODIGY and PISA to calculate the aforementioned parameters for the 580 protein complexes compiled from the PDBbind database. A correlation analysis was performed between experimental binding affinities and the individual binding interaction metric. It is clear from the correlation results that the predicted dissociation constant (K_d) has the highest correlation value with the experimental binding affinities. A Spearman's correlation analysis of predicted dissociation constants, Gibb's free energy, and the number of H-bonds with the experimental binding affinities is summarized in Table 13.1.

13.5.2 Characterizing RBD-NAb Interactions by Predicted Dissociation Constants

The interface of interactions from the refined and optimized RBD-NAb protein complex can be used to predict the dissociation constant by implementing PRODIGY (Xue et al. 2016; Vangone and Bonvin 2017; Jiménez-García et al. 2019), a cutting-edge tool for predicting protein-protein complex binding. Besides the predicted dissociation constant, the PRODIGY predicts several types of contacts among the interacting protein partners, such as contacts involving charged-polar residues, charged-charged residues, apolar-polar residues, polar-polar residues, and apolar-apolar residues intermolecular interactions. To complement PRODIGY in evaluating the protein-protein interactions, PISA predicts the change in Gibbs free energy (ΔG), intermolecular hydrogen bonds, salt bridges, and disulfide bonds (Krissinel and Henrick 2007; Krissinel 2010). As previously stated, the dissociation constant (K_d) was selected as the most relevant predicted metric for simulating experimental binding affinities and was thus used to measure and investigate RBD-NAb binding strengths. A complete workflow of the proposed pipeline is depicted in Fig. 13.1.

Table 13.1 Spearman's correlation analysis of predicted metrics for protein-protein interaction binding strength with the experimental binding affinities

Metric for binding strength	Prediction tool	Correlation coefficient
Predicted dissociation constant	PRODIGY	0.79
Intermolecular contacts	PRODIGY	0.67
Apolar-polar contacts	PRODIGY	0.63
Charged-apolar contacts	PRODIGY	0.42
Charged-polar contacts	PRODIGY	0.38
Number of H-bonds	PDBe-PISA	0.35
Polar-polar contacts	PRODIGY	0.33
Interface area (Å)	PDBe-PISA	0.31
Apolar-apolar contacts	PRODIGY	0.27
Gibb's free energy	PDBe-PISA	0.21
Charged-charged contacts	PRODIGY	0.18

13.6 Application of Proposed Computational Pipeline: A Case Study

To eliminate the possibility of intrinsic bias, the full genomes of SARS-CoV-2 VOCs that were circulating previously (Alpha, Beta, Gamma, and Delta) as well as those that are circulating currently (Omicron) were randomly selected from the EpiCoV™ database along with their GISAID accession identifiers as follows: EPI_ISL_679974 for Alpha, EPI_ISL_2493065 for Beta, EPI_ISL_978506 for Gamma, EPI_ISL_1663516 for Delta, and EPI_ISL_6854348 for Omicron. The genomic sequence for the RBD of Spike protein (corresponding to T333–G526) of each VOC was extracted by using multiple sequence alignment of the complete genomic sequences of all VOCs along with wild-type SARS-CoV-2, using MAFFT, a state-of-the-art tool for performing multiple sequence alignment of genomic or protein sequences that uses fast Fourier transform to deliver alignment of large data in a compute and time-efficient manner (Katoh and Standley 2013).

The consensus approach outlined in Sect. 13.3.1 and 13.3.2 were used to estimate the structure models of Spike's RBD of each VOC. It is worth noting that while modeling the structures using the state-of-the-art prediction methods, corresponding experimental structures were explicitly removed from the reference datasets. This was done to test the accuracy of the proposed pipeline's prediction in the event of future outbreaks. By implementing the consensus approach described in Sect. 3.2, the predicted model structures ($n = 20$) from various state-of-the-art methods were subjected to a comprehensive protein structure quality assessment, and the best model structures for RBD of individual SARS-CoV-2 VOCs were selected for further refinement and optimization using GalaxyRefine (refer to Sect. 13.3.3). A quality assessment overview of the selected models of spike's RBD for each SARS-CoV-2 VOC is provided in Table 13.2.

The refined and optimized predicted structures for each SARS-CoV-2 VOC and the mutations in their RBD over the wild-type are depicted in Fig. 13.2. The predicted model structure of RBD deviated from their experimental counterparts with 1 Å root mean square deviations (RMSD) (as summarized in Table 13.2), indicating that the computational pipeline is highly reliable for generating protein structure models of emerging SARS-CoV-2 variants, where the experimental structures are yet to be solved.

Table 13.2 Summary of quality assessment scores for the refined and optimized model structures of RBD of SARS-CoV-2 VOCs

VOCs	ProTSAV	ModFold8	ProFitFun	Overall quality scores	RMSD (Å)
RBD_{alpha}	0.88	0.64	0.94	0.82	0.36 Å from 7EKF
RBD_{beta}	0.85	0.71	0.92	0.83	0.55 Å from 7EKG
RBD_{gamma}	0.85	0.70	0.91	0.82	0.27 Å from 7EKC
RBD_{delta}	0.83	0.68	0.91	0.81	0.43 Å from 7V8A
$RBD_{omicron}$	0.82	0.65	0.88	0.78	0.85 Å from 7T9L

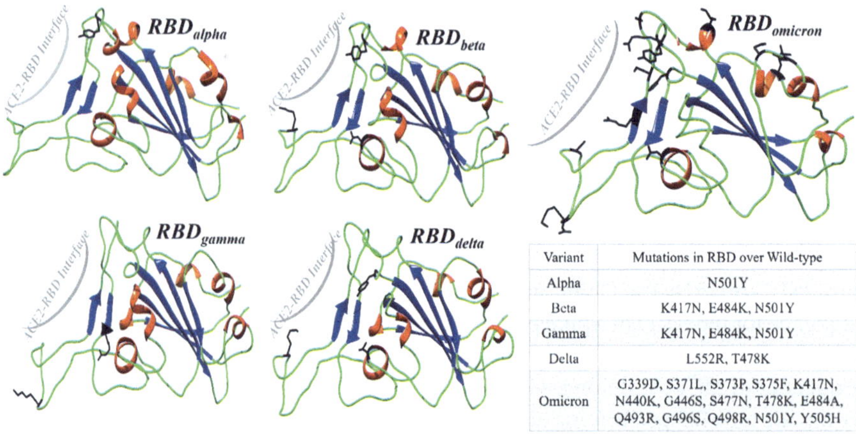

Variant	Mutations in RBD over Wild-type
Alpha	N501Y
Beta	K417N, E484K, N501Y
Gamma	K417N, E484K, N501Y
Delta	L552R, T478K
Omicron	G339D, S371L, S373P, S375F, K417N, N440K, G446S, S477N, T478K, E484A, Q493R, G496S, Q498R, N501Y, Y505H

Fig. 13.2 Depiction of the refined and optimized model structures of spike's RBD using the proposed computational pipeline. The mutations in RBD of different VOCs are depicted in black color

Next, we selected a dataset of 67 NAbs targeting spike's RBD from the CoV-AbDab (coronavirus antibody database) based on two criteria: (i) those antibodies should have their experimentally solved 3D structures, and (ii) they should not have any missing or nonstandard residues at the interfaces. The protein databank identifiers (PDB IDs) of these selected NAbs are provided in Table 13.3. The experimental structures of the selected RBD-NAbs were used as the reference for performing template-based RBD-NAb modeling (prediction for 335 (5 × 67) RBD-NAB complexes).

The refined and optimized model structure for RBD (for each SARS-CoV-2 VOC) was used as an interacting partner along with NAb for modeling the RBD-NAb protein complex by using the template-based protein-protein complex modeling approach, as described in Sect. 13.4.1. The best model structure for RBD-NAb complexes was identified (refer to Sect. 13.4.2). The refined RBD-NAb protein complex structures were reassessed by VoroMQA interface scores and the top-ranked structure from an individual set of RBD-NAb protein complex model structures was chosen to perform further protein-protein structural interaction analysis.

A positive correlation was observed between the binding affinity of the RBD-NAb complex and their neutralization potential (Kaushik et al. 2022; Kumar et al. 2023). Taking this into account, we predicted the binding affinities (via dissociation constants) of RBD of each SARS-CoV-2 VOC with 67 NAbs. The NAbs showing a minimum of twofold increase in binding affinities to RBD as compared to NAB-RBD (wild-type) were categorized as effective NAbs. Based on this criterion, we found that only 7 out of 67 NAbs could be categorized as effective NAbs against SARS-CoV-2 Omicron VOC. Notably, none of these NAbs could be identified as broad-spectrum NAbs (NAbs having the capability to neutralize all the VOCs). The predictions of this approach were found to ideally reflect experimental neutralization data (Cameroni et al. 2022). Among the seven NAbs identified as potent NAbs

Table 13.3 List of selected neutralizing antibodies (NAbs) (retrieved from the CoV-AbDab) targeting RBD in wild-type SARS-CoV-2

NAb	PDB ID	NAb	PDB ID	NAb	PDB ID
10D12	7E3C	COVA1-16	7LQ7	P5A-2F11	7CZY
13G9	7E3K	COVOX-222	7OR9	P5A-2G7	7D03
6D6	7EAN	CV05-163	7LOP	P5A-2G9	7CZT
910-30	7KS9	CV2-75	7M3I	P5A-3A1	7D0C
9G11	7E3B	DH1041	7LAA	P5A-3C12	7D0D
A23-58-1	7LRS	DH1043	7LJR	P5C3	7P40
Ab_58G6	7E3L	Fab_15033	7KLG	PDI-93	7MZJ
ab4	7E39	Fab1-57	7LS9	Regdanvimab	7CM4
Asarnow_3D11	7M7B	Fab2-36	7N5H	REGN10985	7M42
Asarnow_5A6	7KQB	Fab2-4	6XEY	REGN10987	6XDG
AZD-8895	7L7E	Fab2-43	7L56	S2-E12	7K4N
Bamlanivimab	7L3N	Fab2-7	7LSS	S2H14	7JX3
BD23	7BYR	Fab-298	7K9Z	S2H97	7M7W
BD-623	7E7Y	GH-12	7D6I	S2-M11	7K43
BG10-19	7M6E	H4	7L58	S2X35	7R6W
BG1-22	7M6F	HB27	7CYP	SARS2-38	7MKL
BG4-25	7M6D	HbnC3t1p1_C6	7B0B	S-B8	7KN3
BG7-20	7M6H	N-612-056	7S0B	WRAIR-2125	7N4L
C002	7K8S	P2B-1A1	7CZP	WRAIR-2173	7N4J
C022	7RKU	P2B-1A10	7CZQ	P17	7E5S
C118	7RKV	P2B-2F6	7BWJ	CB6	7C01
C121	7K8Y	P5A-1B6	7CZV	P5A-1D2	7CHO
CA521	7E23	P5A-1B9	7CZX		

against Omicron VOC, three have been already experimentally validated as effective NAbs, while four new NAbs (GH-12, P2B-1A1, Asarnow_3D11, and C118) were identified as effective and need experimental validation. In the dataset of 67 NAbs, only 3 were experimentally characterized to have neutralization potential against Omicron VOC. So, the proposed pipeline could efficiently predict (true positives) the experimentally validated NAbs as effective while none as false negative. The predicted effective NAbs could be used as potential therapeutics against the SARS-CoV-2 Omicron variant; however, this is subjected to experimental characterization. Lastly, the computational framework proposed in this chapter can be used as a high-throughput approach for the selection of effective antibody therapeutics against many viruses, including emerging viruses.

13.7 Conclusion and Future Prospective

With a high rate of substitutions among circulating SARS-CoV-2 VOCs and their presence in overlapping geographical areas, new variants with better binding affinity to human ACE2 are likely to emerge in the near future that may compromise the efficacy of existing NAbs against SARS-CoV-2. The identification of the effective NAbs among all existing NAbs necessitates a labor-intensive, time-consuming, and costly extensive experimental screening. Considering the primary use of NAbs in harboring respiratory symptoms, the importance of identifying effective NAbs in a timely and cost-efficient manner is highly necessitated. An approach combining state-of-the-art computational methods into a pipeline is targeted to deliver reliable solutions by narrowing down NAbs candidates, as well as for preemptively exploring the consequences of potential mutations. We show that this strategy allows for fast and targeted experimental confirmation. The consensus-based computational pipeline for reliable binding affinity prediction between the RBD of Spike protein and a panel of Nabs is described. The presented computational pipeline's robustness and reliability are demonstrated by its high correlation with the experimental data of available NAbs for effectiveness against SARS-CoV-2. The developed pipeline has the potential to furnish crucial insights for predicting the neutralization potential of existing NAbs upon the emergence of new variants in the future and could be implemented to forecast the neutralization potential for the most probable mutations. The proposed computational pathway would help the antiviral researchers in the initial screening of multiple NAbs against viruses of interest.

Acknowledgments RK and TL are thankful to Biotech Research Center, Technology Innovation Institute, Abu Dhabi, UAE for its support and resources.

References

Alai S, Gujar N, Joshi M et al (2021) Pan-India novel coronavirus SARS-CoV-2 genomics and global diversity analysis in spike protein. Heliyon 7(3):e06564. https://doi.org/10.1016/j.heliyon.2021.e06564

Augusto G, Mohsen MO, Zinkhan S et al (2022) In vitro data suggest that Indian delta variant B.1.617 of SARS-CoV-2 escapes neutralization by both receptor affinity and immune evasion. Allergy 77(1):111–117. https://doi.org/10.1111/all.15065

Baek M, DiMaio F, Anishchenko I et al (2021) Accurate prediction of protein structures and interactions using a three-track neural network. Science 373(6557):871–876. https://doi.org/10.1126/science.abj8754

Baldassarre F, Hurtado DM, Elofsson A, Azizpour H (2020) GraphQA: protein model quality assessment using graph convolutional networks. Bioinformatics. https://doi.org/10.1093/bioinformatics/btaa714

Bhattacharya D (2019) refineD: improved protein structure refinement using machine learning based restrained relaxation. Bioinformatics 35(18):3320–3328. https://doi.org/10.1093/bioinformatics/btz101

Cameroni E, Bowen JE, Rosen LE, Saliba C, Zepeda SK, Culap K, Pinto D, VanBlargan LA, De Marco A, di Iulio J, Zatta F, Kaiser H, Noack J, Farhat N, Czudnochowski N, Havenar-Daughton C, Sprouse KR, Dillen JR, Powell AE, Chen A, Maher C, Yin L, Sun D, Soriaga L, Bassi J, Silacci-Fregni C, Gustafsson C, Franko NM, Logue J, Iqbal NT, Mazzitelli I, Geffner J, Grifantini R, Chu H, Gori A, Riva A, Giannini O, Ceschi A, Ferrari P, Cippà PE, Franzetti-Pellanda A, Garzoni C, Halfmann PJ, Kawaoka Y, Hebner C, Purcell LA, Piccoli L, Pizzuto MS, Walls AC, Diamond MS, Telenti A, Virgin HW, Lanzavecchia A, Snell G, Veesler D, Corti D (2022) Broadly neutralizing antibodies overcome SARS-CoV-2 Omicron antigenic shift. Nature 602(7898):664–670. https://doi.org/10.1038/s41586-021-04386-2

Cao R, Bhattacharya D, Hou J, Cheng J (2016) DeepQA: improving the estimation of single protein model quality with deep belief networks. BMC Bioinform 17(1):495. https://doi.org/10.1186/s12859-016-1405-y

Cao Y, Wang J, Jian F et al (2022) Omicron escapes the majority of existing SARS-CoV-2 neutralizing antibodies. Nature 602(7898):657–663. https://doi.org/10.1038/s41586-021-04385-3

Cele S, Jackson L, Khoury DS et al (2022) Omicron extensively but incompletely escapes Pfizer BNT162b2 neutralization. Nature 602(7898):654–656. https://doi.org/10.1038/s41586-021-04387-1

Chen M, Chen X, Jin S et al (2020) Protein structure refinement guided by atomic packing frustration analysis. J Phys Chem B 124(48):10889–10898. https://doi.org/10.1021/acs.jpcb.0c06719

Davis IW, Leaver-Fay A, Chen VB et al (2007) MolProbity: all-atom contacts and structure validation for proteins and nucleic acids. Nucleic Acids Res 35(Web Server issue):W375–W383. https://doi.org/10.1093/nar/gkm216

Dejnirattisai W, Shaw RH, Supasa P et al (2022) Reduced neutralisation of SARS-CoV-2 omicron B.1.1.529 variant by post-immunisation serum. Lancet 399(10321):234–236. https://doi.org/10.1016/s0140-6736(21)02844-0

Elbe S, Buckland-Merrett G (2017) Data, disease and diplomacy: GISAID's innovative contribution to global health. Glob Chall 1(1):33–46. https://doi.org/10.1002/gch2.1018

Faria NR, Mellan TA, Whittaker C et al (2021) Genomics and epidemiology of a novel SARS-CoV-2 lineage in Manaus, Brazil. medRxiv. https://doi.org/10.1101/2021.02.26.21252554

GISIAD (2021) Clade and lineage nomenclature aids in genomic epidemiology studies of active hCoV-19 viruses. Retrieved 21/10/2022, 2022, from https://go.nature.com/3pgSIt6

Greaney AJ, Starr TN, Barnes CO et al (2021a) Mapping mutations to the SARS-CoV-2 RBD that escape binding by different classes of antibodies. Nat Commun 12(1):4196. https://doi.org/10.1038/s41467-021-24435-8

Greaney AJ, Starr TN, Gilchuk P et al (2021b) Complete mapping of mutations to the SARS-CoV-2 spike receptor-binding domain that escape antibody recognition. Cell Host Microbe 29(1):44–57.e49. https://doi.org/10.1016/j.chom.2020.11.007

Haas J, Barbato A, Behringer D et al (2018) Continuous Automated Model EvaluatiOn (CAMEO) complementing the critical assessment of structure prediction in CASP12. Proteins 86(Suppl 1):387–398. https://doi.org/10.1002/prot.25431

Han P, Su C, Zhang Y et al (2021) Molecular insights into receptor binding of recent emerging SARS-CoV-2 variants. Nat Commun 12(1):6103. https://doi.org/10.1038/s41467-021-26401-w

Heo L, Park H, Seok C (2013) GalaxyRefine: protein structure refinement driven by side-chain repacking. Nucleic Acids Res 41(Web Server issue):W384–W388. https://doi.org/10.1093/nar/gkt458

Heo L, Lee H, Seok C (2016) GalaxyRefineComplex: refinement of protein-protein complex model structures driven by interface repacking. Sci Rep 6:32153. https://doi.org/10.1038/srep32153

Heo L, Janson G, Feig M (2021) Physics-based protein structure refinement in the era of artificial intelligence. Proteins 89(12):1870–1887. https://doi.org/10.1002/prot.26161

Jiménez-García B, Elez K, Koukos PI et al (2019) PRODIGY-crystal: a web-tool for classification of biological interfaces in protein complexes. Bioinformatics 35(22):4821–4823. https://doi.org/10.1093/bioinformatics/btz437

Jumper J, Evans R, Pritzel A et al (2021) Highly accurate protein structure prediction with AlphaFold. Nature 596(7873):583–589. https://doi.org/10.1038/s41586-021-03819-2

Kannan SR, Spratt AN, Sharma K et al (2022) Omicron SARS-CoV-2 variant: unique features and their impact on pre-existing antibodies. J Autoimmun 126:102779. https://doi.org/10.1016/j.jaut.2021.102779

Katoh K, Standley DM (2013) MAFFT multiple sequence alignment software version 7: improvements in performance and usability. Mol Biol Evol 30(4):772–780. https://doi.org/10.1093/molbev/mst010

Kaushik R, Zhang KYJ (2020) A protein sequence fitness function for identifying natural and non-natural proteins. Proteins 88(10):1271–1284. https://doi.org/10.1002/prot.25900

Kaushik R, Zhang KYJ (2021) ProFitFun: a protein tertiary structure fitness function for quantifying the accuracies of model structures. Bioinformatics. https://doi.org/10.1093/bioinformatics/btab666

Kaushik R, Zhang KYJ (2022) An integrated protein structure fitness scoring approach for identifying native-like model structures. Comput Struct Biotechnol J 20:6467–6472. https://doi.org/10.1016/j.csbj.2022.11.032

Kaushik R, Singh A, Jayaram B (2018) Where informatics lags chemistry leads. Biochemistry 57(5):503–506. https://doi.org/10.1021/acs.biochem.7b01073

Kaushik R, Kumar N, Zhang KYJ et al (2022) A novel structure-based approach for identification of vertebrate susceptibility to SARS-CoV-2: implications for future surveillance programmes. Environ Res 212(Pt C):113303. https://doi.org/10.1016/j.envres.2022.113303

Khoury DS, Steain M, Triccas JA et al (2021) Analysis: a meta-analysis of early results to predict vaccine efficacy against omicron. medRxiv:2021.2012.2013.21267748. https://doi.org/10.1101/2021.12.13.21267748

Krissinel E (2010) Crystal contacts as nature's docking solutions. J Comput Chem 31(1):133–143. https://doi.org/10.1002/jcc.21303

Krissinel E, Henrick K (2007) Inference of macromolecular assemblies from crystalline state. J Mol Biol 372(3):774–797. https://doi.org/10.1016/j.jmb.2007.05.022

Kryshtafovych A, Schwede T, Topf M et al (2021) Critical assessment of methods of protein structure prediction (CASP)-round XIV. Proteins 89(12):1607–1617. https://doi.org/10.1002/prot.26237

Kumar N, Kaushik R, Tennakoon C et al (2021) Evolutionary signatures governing the codon usage bias in coronaviruses and their implications for viruses infecting various bat species. Viruses 13(9). https://doi.org/10.3390/v13091847

Kumar N, Kaushik R, Zhang KYJ, Uversky VN, Sahu U, Sood R, Bhatia S (2023) A novel consensus-based computational pipeline for screening of antibody therapeutics for efficacy against SARS-CoV-2 variants of concern including Omicron variant. Proteins 91(6):798–806. https://doi.org/10.1002/prot.26467

Kwon S, Won J, Kryshtafovych A, Seok C (2021) Assessment of protein model structure accuracy estimation in CASP14: old and new challenges. Proteins 89(12):1940–1948. https://doi.org/10.1002/prot.26192

Laskowski RA, Macarthur MW, Thornton JM (2012) PROCHECK: validation of protein-structure coordinates. International tables for crystallography, pp 684–687

Liu Z, Li Y, Han L et al (2015) PDB-wide collection of binding data: current status of the PDBbind database. Bioinformatics 31(3):405–412. https://doi.org/10.1093/bioinformatics/btu626

Liu L, Iketani S, Guo Y et al (2022) Striking antibody evasion manifested by the Omicron variant of SARS-CoV-2. Nature 602(7898):676–681. https://doi.org/10.1038/s41586-021-04388-0

Maheshwari S, Brylinski M (2016) Template-based identification of protein-protein interfaces using eFindSitePPI. Methods 93:64–71. https://doi.org/10.1016/j.ymeth.2015.07.017

McGuffin LJ, Aldowsari FMF, Alharbi SMA, Adiyaman R (2021) ModFOLD8: accurate global and local quality estimates for 3D protein models. Nucleic Acids Res 49(W1):W425–W430. https://doi.org/10.1093/nar/gkab321

Mlcochova P, Kemp SA, Dhar MS et al (2021) SARS-CoV-2 B.1.617.2 Delta variant replication and immune evasion. Nature 599(7883):114–119. https://doi.org/10.1038/s41586-021-03944-y
Mohapatra RK, Kuppili S, Kumar Suvvari T et al (2022) SARS-CoV-2 and its variants of concern including Omicron: a never ending pandemic. Chem Biol Drug Des 99(5):769–788. https://doi.org/10.1111/cbdd.14035
Olechnovič K, Venclovas Č (2017) VoroMQA: assessment of protein structure quality using interatomic contact areas. Proteins 85(6):1131–1145. https://doi.org/10.1002/prot.25278
Park T, Won J, Baek M, Seok C (2021) GalaxyHeteromer: protein heterodimer structure prediction by template-based and ab initio docking. Nucleic Acids Res 49(W1):W237–w241. https://doi.org/10.1093/nar/gkab422
Planas D, Saunders N, Maes P et al (2022) Considerable escape of SARS-CoV-2 Omicron to antibody neutralization. Nature 602(7898):671–675. https://doi.org/10.1038/s41586-021-04389-z
Porter KA, Desta I, Kozakov D, Vajda S (2019) What method to use for protein-protein docking? Curr Opin Struct Biol 55:1–7. https://doi.org/10.1016/j.sbi.2018.12.010
Rambaut A, Loman N, Pybus O et al (2020) Preliminary genomic characterization of an emergent SARS-CoV-2 lineage in the UK defined by a novel set of spike mutations. Retrieved 25/01/2022, 2022, from https://virological.org/t/preliminary-genomic-characterisation-of-an-emergent-sars-cov-2-lineage-in-the-uk-defined-by-a-novel-set-of-spike-mutations/563
Raybould MIJ, Kovaltsuk A, Marks C, Deane CM (2021) CoV-AbDab: the coronavirus antibody database. Bioinformatics 37(5):734–735. https://doi.org/10.1093/bioinformatics/btaa739
Shu Y, McCauley J (2017) GISAID: global initiative on sharing all influenza data – from vision to reality. Euro Surveill 22(13). https://doi.org/10.2807/1560-7917.Es.2017.22.13.30494
Singh A, Kaushik R, Mishra A et al (2016) ProTSAV: a protein tertiary structure analysis and validation server. Biochim Biophys Acta 1864(1):11–19. https://doi.org/10.1016/j.bbapap.2015.10.004
Szilagyi A, Zhang Y (2014) Template-based structure modeling of protein-protein interactions. Curr Opin Struct Biol 24:10–23. https://doi.org/10.1016/j.sbi.2013.11.005
Tegally H, Wilkinson E, Giovanetti M et al (2021) Detection of a SARS-CoV-2 variant of concern in South Africa. Nature 592(7854):438–443. https://doi.org/10.1038/s41586-021-03402-9
Trevor B, Hodcroft EB, Neher RA (2021) Updated Nextstrain SARS-CoV-2 clade naming strategy. Retrieved 10/21/2022, 2022, from https://nextstrain.org/blog/2021-01-06-updated-SARS-CoV-2-clade-naming
Uziela K, Menéndez Hurtado D, Shu N et al (2017) ProQ3D: improved model quality assessments using deep learning. Bioinformatics 33(10):1578–1580. https://doi.org/10.1093/bioinformatics/btw819
Uziela K, Menéndez Hurtado D, Shu N et al (2018) Improved protein model quality assessments by changing the target function. Proteins 86(6):654–663. https://doi.org/10.1002/prot.25492
Vangone A, Bonvin A (2017) PRODIGY: a contact-based predictor of binding affinity in protein-protein complexes. Bio Protoc 7(3):e2124. https://doi.org/10.21769/BioProtoc.2124
Wang R, Fang X, Lu Y, Wang S (2004) The PDBbind database: collection of binding affinities for protein-ligand complexes with known three-dimensional structures. J Med Chem 47(12):2977–2980. https://doi.org/10.1021/jm030580l
Wang R, Fang X, Lu Y et al (2005) The PDBbind database: methodologies and updates. J Med Chem 48(12):4111–4119. https://doi.org/10.1021/jm048957q
Wang W, Wang J, Li Z et al (2021) MUfoldQA_G: high-accuracy protein model QA via retraining and transformation. Comput Struct Biotechnol J 19:6282–6290. https://doi.org/10.1016/j.csbj.2021.11.021
Webb B, Sali A (2021) Protein structure modeling with MODELLER. Methods Mol Biol 2199:239–255. https://doi.org/10.1007/978-1-0716-0892-0_14
Wei C, Shan KJ, Wang W et al (2021) Evidence for a mouse origin of the SARS-CoV-2 Omicron variant. J Genet Genomics 48(12):1111–1121. https://doi.org/10.1016/j.jgg.2021.12.003
WHO (2022a) Tracking SARS-CoV-2 variants. Retrieved 01/10/2022, 2022, from https://www.who.int/activities/tracking-SARS-CoV-2-variants

WHO (2022b) WHO Tracking SARS-CoV-2 variants. Retrieved 23/11/2022, 2022, from https://www.who.int/en/activities/tracking-SARS-CoV-2-variants
Xue LC, Rodrigues JP, Kastritis PL et al (2016) PRODIGY: a web server for predicting the binding affinity of protein-protein complexes. Bioinformatics 32(23):3676–3678. https://doi.org/10.1093/bioinformatics/btw514
Yan Y, Zhang D, Zhou P et al (2017) HDOCK: a web server for protein-protein and protein-DNA/RNA docking based on a hybrid strategy. Nucleic Acids Res 45(W1):W365–w373. https://doi.org/10.1093/nar/gkx407
Yang J, Yan R, Roy A et al (2015) The I-TASSER Suite: protein structure and function prediction. Nat Methods 12(1):7–8. https://doi.org/10.1038/nmeth.3213
Zahradník J, Tuekprakhon A, Ginn HM et al (2021) Receptor binding and escape from Beta antibody responses drive Omicron-B.1.1.529 evolution. bioRxiv:2021.2012.2003.471045. https://doi.org/10.1101/2021.12.03.471045

Chapter 14
Insights on Various Antiviral Drugs for Treating Patients with COVID-19

Ujjwal Kumar De, Yashpal Singh Malik, Kuldeep Dhama, Balasubramanian Ganesh, Rajkumar Thangarasu, Manikandan Natesan, Rahul Kaushik, Prashant Kumar, Sudesh Gaidhani, Senthilkumar Natesan, and Souvik Ghosh

Abstract The severe acute respiratory syndrome coronavirus 2 (SARS-CoV-2) was the cause of the first pandemic of the twenty-first century, which began in Hubei Province, China, in late December 2019. Researchers across the globe are engaged in developing vaccines and therapeutics to manage COVID-19, but the lack of effective vaccines or drugs with proven therapeutic potential is still a worrying situation, with the total number of cases crossing 34 million and nearly one million deaths as on October 2, 2020. Several existing antiviral drugs, which include nucleoside analogs, protease inhibitors, and viral polymerase inhibitors, have been investigated for potential activity against SARS-CoV-2. Apart from antivirals, several antibiotics and antiparasitic drugs have also shown efficacy in impeding viral propagation and improving related health complications in the host. Virus-induced immunopathology due to the cytokine storm has been tackled using drugs targeting specific cytokines to curtail the symptoms and assist recovery. The repurposed drugs and combined therapies are being evaluated using several drug combinations to lessen the viral load in infected patients. Alongside synthetic drug molecules, natural compounds found to exhibit important immunomodulatory effects in helping the host's immune system to fight SARS-CoV-2 infection are being considered. This chapter provides comprehensive information on the ongoing research and developments focusing on antiviral therapeutics, their mechanisms, and the status of their efficacy in managing COVID-19 and minimizing the rate of viral infection along with their prospects and obstacles.

Keywords Antiviral drugs · COVID-19 · SARS-CoV-2 · Treatment

U. K. De (✉)
Division of Medicine, ICAR-Indian Veterinary Research Institute,
Izatnagar, Bareilly, Uttar Pradesh, India
e-mail: ujjwal.de@icar.gov.in

Y. S. Malik (✉)
College of Animal Biotechnology, Guru Angad Dev Veterinary and Animal Sciences
University, Ludhiana, Punjab, India
e-mail: malikyps@gadvasu.in

N. Kumar et al. (eds.), *Advances in Antiviral Research*, Livestock Diseases and Management, https://doi.org/10.1007/978-981-99-9195-2_14

14.1 Introduction

A new coronavirus infection was found to be the cause of a cluster of pneumonia cases that were identified around the end of 2019 in Wuhan City, Hubei Province. The World Health Organisation (WHO) designated the illness COVID-19 (coronavirus disease 2019) and its causative virus as SARS-CoV-2 (severe acute respiratory syndrome coronavirus 2) in February 2020 as a result of an abrupt increase in cases throughout the world (McIntosh et al. 2020). Currently, 215 countries and territories across the globe have confirmed cases of COVID-19, and the number of infections is steadily escalating in many countries, including India. There are several ways to control CoV infection, including vaccine development, but even with today's sophisticated laboratory technologies, sufficient time for development and translation into the market is required. Before the development of effective vaccines, there was an urgent need for effective drugs to combat COVID-19 and tackle the challenge posed by the pandemic (Dhama et al. 2020a; Malik et al. 2020; Santos et al. 2020; Wu et al. 2020). However, a novel treatment is still an unlikely solution to tackle the pandemic because the process of developing a new drug is laborious and typically takes more than 15 years from the research and development phase to market introduction. Given the rapidly evolving pandemic crisis, and while the candidate vaccines were still under development, many drugs are being repurposed and investigated for their potential use, prophylactically or therapeutically, worldwide.

K. Dhama
Division of Pathology, ICAR-Indian Veterinary Research Institute,
Izatnagar, Bareilly, Uttar Pradesh, India

B. Ganesh · R. Thangarasu · M. Natesan
Laboratory Division, Indian Council of Medical Research -National Institute of Epidemiology, Ministry of Health & Family Welfare, Chennai, Tamil Nadu, India

R. Kaushik
Laboratory for Structural Bioinformatics, Biosystems Dynamics Research Center,
Riken, Japan
e-mail: rahul.kaushik@riken.jp

P. Kumar
Amity Institute of Virology and Immunology, Amity University, Noida, Uttar Pradesh, India
e-mail: pkumar18@amity.edu

S. Gaidhani
Pharmacology, Central Council for Research in Ayurvedic Sciences, Ministry of AYUSH,
Government of India, New Delhi, India

S. Natesan
Indian Institute of Public Health Gandhinagar, Lekawada, Gandhinagar, Gujarat, India
e-mail: snatesan@iiphg.org

S. Ghosh
One Health Center for Zoonoses and Tropical Veterinary Medicine,
Basseterre, St. Kitts, West Indies
e-mail: sghosh@rossvet.edu.kn

Researchers and several agencies are still currently involved in exploring effective COVID-19 drugs and therapeutics and the enrolment of clinical trials (Dhama et al. 2020b; Rabaan et al. 2020).

Recent research outcomes have enriched our understanding of the basic mechanisms of SARS-CoV-2 replication and pathogenesis. This has significantly contributed to the development, repurposing, and interpretation of the mechanisms of action of novel or existing antiviral drugs and biologically active molecules against CoVs (SARS-CoV-1 and Middle East respiratory syndrome coronavirus [MERS-CoV]); many of them have yielded promising outcomes in clinical trials (Dhama et al. 2020a). Keeping in view the sharing of the common genomic structure of COVID-19 and associated Corona viruses, here, we comprehensively summarize the antiviral therapeutics to counter COVID-19. Additionally, we highlight an insight into the potential of homeopathy and the Indian system of medicines (Ayurveda, Yoga & Naturopathy, Unani and Siddha) as possible antiviral approaches for the effective treatment of COVID-19.

14.2 SARS-CoV-2 Antiviral Drugs as Potential Therapeutics

Prophylactics or vaccines represent the main arsenal in preventing any viral infection. The development of new drug(s) targeting the virus or host cell is an alternative option; however, the average time for a novel drug to proceed from the research and development phase to market introduction is at least 15 years; this cannot solve the immediate situation. Therefore, treating SARS-CoV-2 infections using already available drugs that target related viruses sharing similar replication mechanisms is the preferred method for controlling the pandemic (Fig. 14.1).

Here, we discuss the different antiviral drugs that are being evaluated in different clinical centers worldwide to treat COVID-19.

14.2.1 Lopinavir/Ritonavir

The antiviral activity of lopinavir (an antiretroviral drug) is attributed the inhibition of the viral protease enzyme (Fig. 14.2). The antiviral activity of lopinavir against SARS and MERS epidemics through blocking a postentry step of viral replication suggests a promising reference for Covid 19 treatment (Yao et al. 2020). The activity of lopinavir is potentiated when used in combination with another protease inhibitor, ritonavir, which attenuates the activity of cytochrome (CYT) P450 3A and enhances the half-life of lopinavir. Earlier, the lopinavir/ritonavir combined formulation was approved for the treatment of human immunodeficiency virus (HIV). Furthermore, anti-SARS-CoV efficacy was also demonstrated in vitro (Chu et al. 2004). Meanwhile, it was also opined that lopinavir/ritonavir combined therapy can be useful in the early stages of SARS-CoV-2 infection with mild symptoms, but a

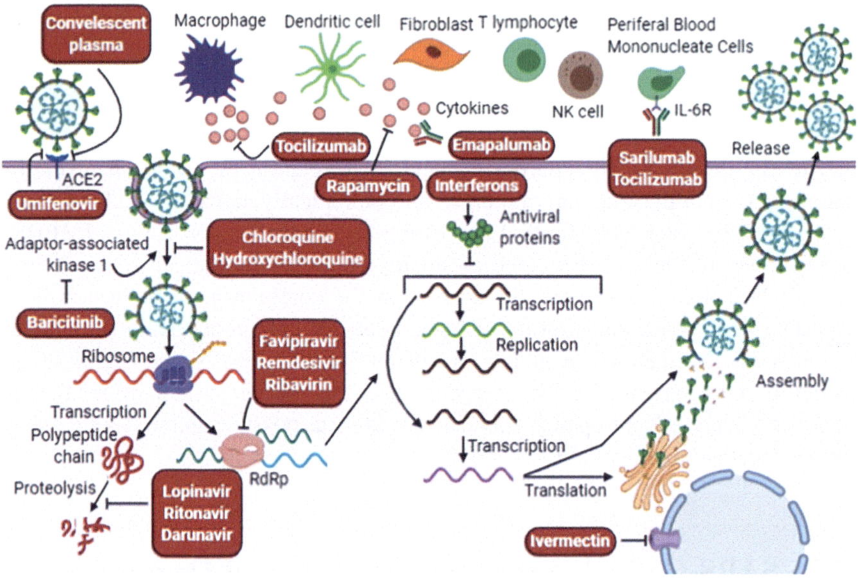

Fig. 14.1 Summary of the possible repurposed drugs under clinical trials against COVID-19 in the context of host pathways and virus replication mechanisms

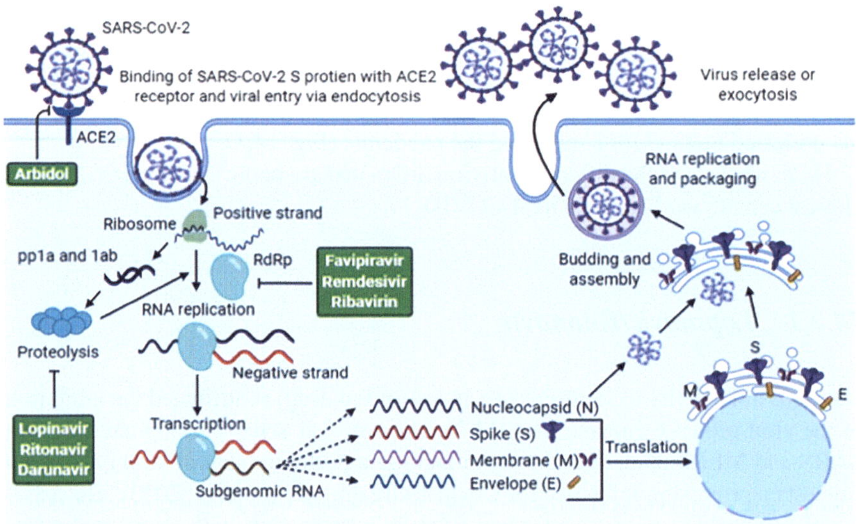

Fig. 14.2 Representing the various therapeutic approaches against SARS-CoV-2 and the mechanism of action of antivirals

larger clinical trial is warranted for further confirmation (Ahsan et al. 2020). In an open-label, prospective, multicenter, randomized, phase 2 clinical trial, the triple antiviral therapy comprising of lopinavir, ritonavir, and interferon beta-1b in adults COVID-19 patients with mild to moderate symptoms was found effective in reducing the duration of viral shedding and hospital stay and ameliorating symptoms in mild to moderate COVID-19 patients (Hung et al. 2020).

14.2.2 *Favipiravir*

Favipiravir (Avigan), an approved drug for the treatment of influenza virus, inhibits the RNA-dependent RNA polymerase (RdRp) in RNA viruses such as SARS-CoV-2 (Fig. 14.2) (Delang et al. 2018). Recently, a preliminary clinical trial indicated better results for favipiravir treatment than those for lopinavir/ritonavir treatment, with minimum adverse effects (Cai et al. 2020; Sandoiu 2020). The results of another multicenter, open-labeled, randomized trial revealed a higher recovery rate and better clinical outcomes in patients who received favipiravir (Chen et al. 2020a, b). Phase II and III trials are being conducted in the USA and Japan, respectively, before its clinical use. However, favipiravir is effective before reaching peak viremia in the body. Since it was previously approved for the treatment of influenza, favipiravir would require government permission for use against COVID-19.

14.2.3 *Remdesivir*

Remdesivir is a promising anti-SARS-CoV-2 medication that was earlier used to treat Ebola virus illness (EVD) (Fig. 14.2) (Frediansyah et al. 2020; Sheahan et al. 2020). During the initial phases of infection, when viral multiplication starts in the upper respiratory tract, it prevents viral replication and is found very effective. Remdesivir, an adenosine analog, is a robust inhibitor of the viral RNA polymerase; it resembles the RNA nucleoside and acts as a chain terminator. The results of a recent compassionate-use of remdesivir trial by Gilead Sciences were very positive: more than 65% of severe COVID-19 patients showed marked improvement of clinical signs (Grein et al. 2020).

14.2.4 *Umifenovir*

Umifenovir (Arbidol), a patented drug for the treatment of influenza and SARS infection, has documented antiviral activity against SARS-CoV-2 in vitro at very low concentrations (−30 μM). It is believed to block viral entry into host cells. A randomized, open-labeled, multicenter trial in humans revealed satisfactory clinical recovery rates in COVID-19 patients. Particularly, COVID-19 patients with hypertension or diabetes showed better improvement when treated with favipiravir than with umifenovir (Chen et al. 2020a, b). However, there are some more ongoing phase IV clinical trials studying umifenovir treatment in COVID-19 patients. Notably, umifenovir is not associated with any significant side effects.

14.2.5 Darunavir

Darunavir (Prezista), another HIV-1 protease inhibitor, has promising anti-SARS-CoV-2 activity. An in vitro study reported that the compound successfully inhibited viral replication at a concentration of 300 μM. In a recent study, darunavir did not show any anti-SARS-CoV-2 activity on human colon carcinoma cell line (Caco-2) cells at clinically relevant concentrations (EC_{50} > 100 μM) (De Meyer et al. 2020). However, Johnson and Johnson reported no evidence to support the activity of darunavir against SARS-CoV-2. Darunavir, an approved drug for use against HIV with a boosting agent, was developed and marketed by Janssen as Prezista such as ritonavir or cobicistat (Johnson and Johnson 2020). However, the results of a single-center, open-labeled, randomized, controlled phase III trial documented that the darunavir/cobicistat combination did not produce any satisfactory results in reducing the symptoms or duration of treatment in COVID-19 patients (ClinicalTrials.gov, Bethesda (MD): National Library of Medicine (USA). 2020).

14.2.6 Cyclophilin Inhibitors/Alisporivir

Cyclophilins, the cellular peptidyl-prolyl cis-trans isomerize enzymes, play a key role in various coronavirus replication, including feline, human coronaviruses SARS-CoV, and MERS-CoV (Liu and Zhu 2020). Keeping in view the role of Cyclophilin A in replication of various coronaviruses, cyclosporine A inhibitors could have a promising role in limiting various coronaviruses (Softic et al. 2020; Liu et al. 2020). Alisporivir, a potent cyclophilin inhibitor and a nonimmunosuppressive analog of Cyclosporine A (CsA), exhibits antiviral activity by inhibiting the postentry step of the SARS-CoV-2 in a dose-dependent manner with complete inhibition at 10 μM as evident from the in vitro study on Vero E6 cells (Softic et al. 2020). The preclinical data showed the strong antiviral and cytoprotective properties of alisporivir in various coronavirus infection models, including SARS-CoV-2. Alisporivir might be a promising antiviral drug candidate for patients with severe forms of SARS-CoV-2 infection (Pawlotsky 2020).

14.3 Antiparasitic Drugs

Considering the present scenario, drug repurposing seems to be a plausible alternative for treating COVID-19 patients. The possibility of utilizing a known antimalarial, chloroquine (CQ), and one of its metabolites, hydroxychloroquine (HCQ), is being explored in different clinical investigations to reduce mortality and morbidity in SARS-CoV-2 infections. One WHO-led solidarity study has examined the use of some of the most promising antiparasitic drugs with potential antiviral effects against the COVID-19 virus (WHO 2020).

14.3.1 Chloroquine and Hydroxychloroquine (HCQ)

Chloroquine interferes with the glycosylation of the SARS-CoV-2 cell receptor by increasing the endosomal pH resulting in the prevention of viral infection. Chloroquine hinders the pre-entry and postentry phases of SARS-CoV-2 and represses cathepsins to cleave the SARS-CoV-2 spike protein through the development of the autophagosome. Furthermore, chloroquine interferes with SARS-CoV-2 molecular crosstalk, virion assembly, budding, and the proteolytic processing of the M protein through the hindrance of MAP-kinase. An in vitro examination of the Vero E6 cell line showed that chloroquine (standard dose) successfully decreases viral replication because of its good penetration of tissues, including the lung (Gao et al. 2020). A clinical trial of >100 Chinese patients infected with COVID-19 showed that chloroquine therapy shortened the duration of symptoms, reduced the exacerbation of pneumonia, and promoted virus-negative seroconversion with no significant side effects compared to the control group (Salvi and Patankar 2020).

In addition, hydroxychloroquine (HCQ) also prevents viral fusion with the host cell by inhibiting protease activity and cleaving spike proteins on the coronavirus surface. Most significantly, HCQ helps to inhibit the "cytokinestorm" responsible for disease progression to ARDS by reducing the activation and differentiation of T-cells, as well as the production of cytokines by these immune cells (Patrì and Fabbrocini 2020). Earlier *in vitro* studies showed that HCQ ($EC_{50} = 0.72$ μM) had a higher potential for antiviral effects than chloroquine ($EC_{50} = 5.47$ μM) for treating SARS-CoV-2 infections (Salvi and Patankar 2020; Patrì and Fabbrocini 2020). Although the earlier *in vivo* studies evaluated the potency of HCQ on SARS-CoV-2 infection, preliminary results were promising (Gao et al. 2020). However, subsequent clinical studies, case reports, and *in vitro* studies have reported conflicting results of the variable results and the known adverse effects on cardiac functions and hypoglycemia and lack of screening of glucose-6-phosphate dehydrogenase (G-6PD) deficient individuals in the population before initiation of prophylactic administration (Kaur et al. 2020). Hence, the FDA cautions against the use of hydroxychloroquine or chloroquine for COVID-19 outside of the hospital setting.

14.3.2 Ivermectin

Ivermectin, an FDA-approved antiparasitic drug and a drug of choice for both medical and veterinary dermatologists for the treatment of cutaneous larva migrans and mange infection, was examined for its antiviral activity against COVID-19 (Sharun et al. 2020; Caly et al. 2020). Ivermectin acts as a specific inhibitor of importin-α/β-mediated nuclear import of viral proteins and thus hinders the replication of several RNA viruses, including yellow fever virus, chikungunya virus, and HIV.[50-51] Recently, an in vitro study on Vero-hSLAM cells showed that a single dose of the drug at 2 h postinfection with SARS-CoV-2 could reduce the viral RNA level markedly (~5000-fold) at 48 h when compared with control within 24–48 h (Caly et al. 2020). A significant synergism between ivermectin and HCQ has been reported for both chemoprophylaxis and the treatment of COVID-19 patients (Patrì and Fabbrocini 2020). Simultaneous administration of HCQ and ivermectin is reported to be effective in a consequential and synergistic way. HCQ inhibits virus entry into the host cell by acting as a first-level barrier, while ivermectin suppresses viral replication if a virus enters the host, thus boosting the antiviral effects of HCQ. In a recent randomized trial, the administration of ivermectin did not reduce the incidence rate of hospital admission or prolonged emergency department observation for Covid-19 among outpatients at high risk for serious illness (Reis et al. 2022).

14.4 Use of Angiotensin-Converting Enzyme 2 (ACE2) Inhibitors

The ACE2 receptor is required for virus entry into the host cell. Inhibition of this is believed to restrict virus replication in the body. Hence, the selective ACE2 inhibitor DX600 may offer some advantageous outcomes in COVID-19 cases; however, the clinical significance of DX600 is yet to be studied (Huang et al. 2003). In a study performed with the recombinant soluble ACE2 protein, it was observed that the intravenous administration of the recombinant protein could block the entry of SARS-CoV-2 by inhibiting the binding of the viral spike protein onto the ACE2 cell surface receptor (Alhenc-Gelas and Drueke 2020; Pang et al. 2020). Angiotensin receptor blockers, such as losartan, have shown their activity against SARS-CoV in a mouse model, and as SARS-CoV is related to SARS-CoV-2 in terms of receptor usage, it is speculated that the use of losartan may reduce the severity of COVID-19 (Kai and Kai 2020). Recently, the possible therapeutic potential of bioengineered probiotics expressing cell membrane-bound or secreted human ACE2 has been proposed to control SARS-CoV-2 infection. It is reported that bioengineered Lactobacillus paracasei could prevent the spread of the virus and prevent viral entry at multiple organs, including the lungs, through systemic effects by sequestering the COVID-19 virus or preventing the spike protein from interacting with host cell-associated receptors (Senapati et al. 2020).

14.5 Use of Antibacterial Drugs

Azithromycin is a macrolide antibiotic drug that acts by binding to the 50S ribosomal subunit to inhibit the mechanism of protein synthesis. It has also been shown to have antiviral activity in rhinovirus infection by stimulating the expression of interferon-stimulated genes; and against Zika virus infection and preventing virus-induced alterations in the fetal brain. Recently, a combination of azithromycin and HQC was shown to be efficient in decreasing the viral load in COVID-19 cases (Gautret et al. 2020), and a trial of azithromycin was later rolled out. The drug combination has an effect on the ACE2–spike complex. A clinical trial has also been initiated by Pfizer using azithromycin against SARS-CoV-2 infection (NCT04321278). Other antibiotics that have been shown to be effective against other viruses are listed in Table 14.1 and could be tested against COVID-19.

14.6 Interferons

IFN-α, an approved broad-spectrum antiviral drug for viral hepatitis, was found to exhibit promising results in COVID-19 patients who received at a dose of five million units through vapor inhalation twice daily for 10 days, either alone or in combination with ribavirin and antiviral drugs (lopinavir/ritonavir) (Lu 2020; Sallard et al. 2020). Previously, IFN-α 2a was found to be effective against MERS-CoV when used as a triple therapy with ribavirin and lopinavir/ritonavir (Sheahan et al.

Table 14.1 Antibiotics that are tested effective against other viruses and could be tested against COVID-19 (Leung et al. 2005; Langeland et al. 1987; Yamagishi et al. 2009; Broyles et al. 2004; Andronova et al. 2010; Baron et al. 2020; Garrigaa et al. 2018; Rothan et al. 2015)

Name of antibiotic	Mechanism of action	Target virus
Ciprofloxacin	Reduces replication of virus	Polyoma BK virus
Neomycin/ streptomycin	Upregulates ISGs	Herpes simplex, influenza, zika
Polyether antibiotic-CP44161	Inhibits proliferation of virus	Varicella-zoster, HSV-1
Distamycin A	Reduction of DNA dependent RNA synthesis	Vaccinia, adenovirus, herpes zoster, herpes simplex
Netropsin	Reduction of DNA dependent DNA/ RNA synthesis	Vaccinia
Teicoplanin	Inhibits Cathepsin L in lysosome and suppresses viral entry	Ebola, MERS-CoV, SARS-CoV
Rifampicin	Inhibits the synthesis of viral polymerase, suppresses late-stage viral protein synthesis, and thus inhibits virus assembly	Poxvirus, Vaccinia virus
Doxycycline	Inhibits virus replication	Chikungunya

2020). Recently, it has been demonstrated that SARS-CoV-2 is more susceptible to IFNs as compared to SARS-CoV-1; the inhalation of IFN-α 2b markedly reduced the infection rate (Shen and Yang 2020). Thus, IFN-α 2a can be considered as a prophylactic agent against SARS-CoV-2 infection (Lokugamage et al. 2020). Previously, IFN-α was reported to have promising antiviral effects on SARS-CoV-1 in resolving lung pathology (Loutfy et al. 2003). A cohort study recently examined the antiviral activity against COVID-19. 77 hospitalized COVID-19 adult patients were treated with nebulized IFN-a2b (5 mU b.i.d.) and arbidol (200 mg t.i.d.), either individually or together, in an uncontrolled exploratory trial. IFN-a2b with or without arbidol significantly reduced the persistence of detectable virus in the upper respiratory tract, and it also reduced the time that inflammatory markers like IL-6 and C-reactive protein (CRP) were elevated in the blood (Zhou et al. 2020).

Decreased IFN-β production is directly linked to increased susceptibility to viral pneumonia (Sallard et al. 2020). Recently, the suppression of IFN-β production in the body during SARS-CoV-2 infection, as a result of escape from the immune system, was reported (Lu 2020). Considering the significance of IFN-, Synairgen, a UK-based biotechnology company, has got approval to undertake a trial with IFN- on COVID-19 patients (Global Data Healthcare 2020). The benefit of IFN- is that it may be administered by the patients themselves and is similar to IFN- in that it can be inhaled. Ruxolitinib, a JAK-STAT pathway inhibitor, is said to have anti-COVID-19 action as well (Stebbing et al. 2020). The treatment of SARS-CoV-2 with recombinant IFNs (IFN-α, IFN-β, and exogenous IFNs) demonstrated the inhibition of protein synthesis and viral replication (Falzarano et al. 2013; Li et al. 2019a, b; Zumla et al. 2016). Among these, IFN-β is reported to exhibit the highest potency, reflecting prophylactic protection and antiviral potential postinfection. Considering these findings, investigating the safety and effectiveness of human and recombinant IFNs, either alone or in combination with other antiviral medications, in SARS-CoV-2-infected patients seems worthwhile (Nile et al. 2020).

14.7 Ayurveda and Other Traditional Systems of Medicine of India (AYUSH)

The core Ayurveda approach toward the management of infectious diseases comprises a two-flanged approach: (1) improving the immune status of an individual and strengthening the host defense mechanism (*Vyadhi*, *Kshamak*, *Kshamtwa*, and *Prakriti Vighata*) and (2) fighting against pathogens and pathogenesis (*Apakrashana*), an antiviral approach (Fig. 14.3). Several medicinal flora, as well as some multi-ingredient classical formulations, have been extensively explored, and principles for their immune-boosting efficacy and antiviral potential have been studied for their prophylactic and therapeutic potential in epidemics of infectious diseases such as COVID-19 (Fig. 14.3) (Rastogi et al. 2020; Jahan and Onay 2020; Panyod et al. 2020; Vellingiri et al. 2020). The following are examples of medicinal plants and

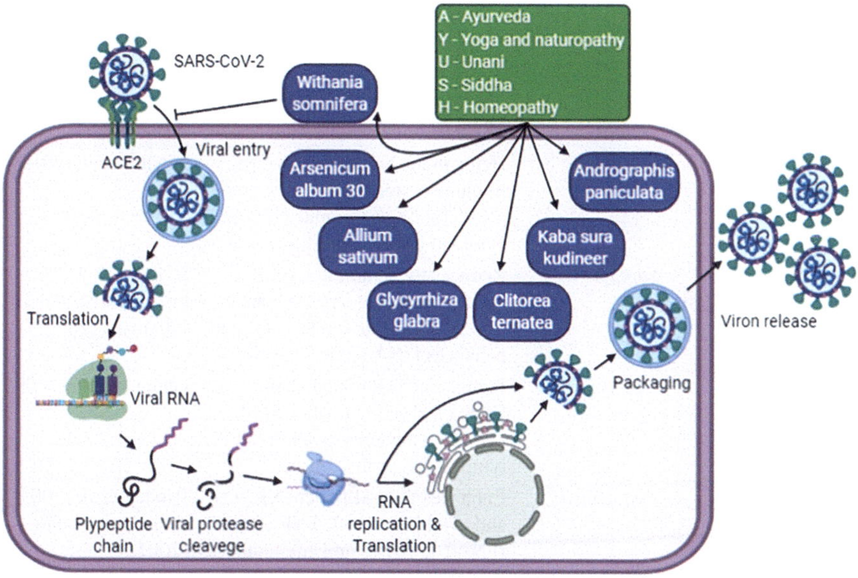

Fig. 14.3 Representing the AYUSH-related medicines against SARS-CoV-2 and the mechanism of action

Ayurveda formulations that have been reported to have potential immune modulatory and antiviral activities (summaries are shown in Tables 14.2, 14.3, and 14.4).

These medicinal plants and Ayurveda formulations could be used for further drug development through a translational approach for generating evidence of Indian traditional medicines as prophylaxis and therapeutic treatments against COVID-19.

14.7.1 Immunomodulation Using Natural Products

The pharmacological properties, specifically their antiviral, antifungal, and anti-inflammatory activities of natural products, have been documented in treating numerous diseases since ancient times. The active component of turmeric, curcumin, has been proven to have antiviral properties and to reduce levels of pro-inflammatory cytokines like IL-1, IL-6, and TNF. It also successfully suppressed the cytokine storm in experimental models afflicted with the Ebola virus (Sordillo and Helson 2015). It has also been demonstrated that curcumin acts as a protease inhibitor and inhibits SARS-CoV-1 replication at an effective concentration of 10 μM (Barnard and Kumaki 2011); hence, it can be speculated that curcumin may be effective in the treatment of SARS-CoV-2 infection, provided its bioavailability is enhanced using a suitable formulation (Manoharan et al. 2020; Zahedipour et al. 2020). Another medicinal plant, neem (*Azadirachta indica*), has been used

Table 14.2 Single plant ingredient-based ayurveda formulations reported for immune modulation activities

Sl. No.	Plant	Name in Indian Ayurvedic system	Mode of immune-boosting action	References
1	*Allium sativum* L.	*Lasun*	Helps in the homeostasis of the immune system and regulates the proliferation of immune cells and cytokine gene expression.	Moutia et al. (2018)
2	*Alstonia scholaris* R.Br.	*Saptaparna*	Reduces the generation of free radicals, reactive oxygen species (ROS), and nitric oxide (NO) and enhances DNA repair capacity	Baliga (2011), Sies (1996), Mistry and Pithawala (2017)
3	*Asparagus racemosus* wild	*Shatavari*	Increases CD3(+) and CD4/CD8(+), T cell activation, Th1/Th2 adjuvant activity and; lymphocyte proliferation	Gautam et al. (2009)
4	*Boerhavia diffusa* L.	*Punarnava*	Enhances natural Killer (NK) cell and cytokines (IL-2, TNF-α, and IFN-γ) activities and attenuates the level of IL-1β and IL-6 production.	Pandey et al. (2005), Manu and Kuttan (2007)
5	*Caesalpinia crista* L.	*Kuberaksha*	Protects red blood cells (RBC) from hemolysis, DNA damage, and delayed-type hypersensitivity	Arif et al. (2009), Shukla (2010), Kumar et al. (2017)
6	*Curcuma longa* L.	*Haridra*	Increases the phagocytic activity of macrophages and stimulates non-specific immunity of the host	Antony et al. (1999), Sengupta et al. (2011)
7	*Ecliptaalba* (L.) Hassk.	*Bringraj*	Increases phagocytic activity and leukocyte counts	Jayathirtha and Mishra (2004)
8	*Glycyrrhiza glabra* L.	*Yashtimadhu*	Stimulates macrophage, up-regulates the expression of CD40, CD86, IL-12, IFN-γ, and IL-10, and decreases IL-4 production.	Nose et al. (1998), Bordbar et al. (2012)
9	*Gymnema sylvestre* (Retz.) R.Br. ex Sm.	*Meshashringi*	Restores innate immune function	Singh et al. (2015)
10	*Moringa oleifera* Lam.	*Shigru*	Stimulates humoral as well as cellular immunity	Nfambi et al. (2015)
11	*Ocimum tenuiflorum* L.	*Tulsi*	Increases TNF-α, IL-2, IFN- γ, IL-4, bone marrow cellularity, total leukocyte count, RBSs T-helper cells, NK-cells, and stimulatory effect on T & B lymphocytes, antibody production in dose-dependent manner	Goel et al. (2010), Hemalatha et al. (2011), Mondal et al. (2011), Jamshidi and Cohen (2017)

Table 14.2 (continued)

Sl. No.	Plant	Name in Indian Ayurvedic system	Mode of immune-boosting action	References
12	*Phyllanthus emblica* L.	*Amla*	Enhances NK cell activity, activates polymorphonuclear and monocyte-macrophase system, lymphocyte proliferation, restores the IL 2 and γ-IFN production, and attenuates the oxidative stress by lowering the lipid peroxidation, reactive oxygen species production, activity of caspase-3, apoptosis, and cytoprotective effect	Suresh and Vasudevan (1994), Sai Ram et al. (2002), Singh et al. (2013), Yamamoto et al. (2016)
13	*Phyllanthus niruri Roxb.* ex Wall.	*Bhumiamalaki*	Activates neutrophils and antibody response, Switching Th2- to a Th1-type immune response, inhibits COX-2 and downstream signaling products of PGE2, TNF-α, and IL-1β. Suppresses the IκB, Ikkα/β inhibitors, NF-κB phosphorylation, IκB degradation, JNK, and ERK, and inhibits the ROS release.	Venkateswaran et al. (1987), Chowdhury et al. (2012), El-Aziz et al. (2012), Muthulakshmi et al. (2016), Harikrishnan et al. (2018)
14	*Picrorhiza kurroa* Royle ex Benth.)	*Katuki*	Enhances phagocytosis, stimulates the cell-mediated and humoral immune responses, and boosts lymphocytosis and cytokine levels (IL-4 and IFN-γ and CD4/ CD8 population)	Sharma et al. (1994), Gupta et al. (2006), Hussain et al. (2013)
15	*Piper longum* L.	*Pippali*	Reduces passive cutaneous anaphylaxis and activates cellular as well as humoral immune responses.	Kulshresta et al. (1969), Sunila and Kuttan (2004)
16	*Swertia chirata*Buch.-Ham. ex Wall.	*Kiratatikta*	Reduces the levels of inflammatory mediators by blocking the expression of COX-2 and phosphorylation of Akt, IKK-β, MAPK, and NF-κB, and activation in LPS-stimulated macrophages.	Hu et al. (2019)
17	*Terminalia chebula Retz.*)	*Haritaki*	Raises the expression of Th1 cytokine, INF γ, CD4+ cells, lymphocyte proliferation, macrophage phagocyte response, decreases IL-4, and increases the delayed-type hypersensitivity	Shivaprasad et al. (2006), Rubab and Ali (2016)

(continued)

Table 14.2 (continued)

Sl. No.	Plant	Name in Indian Ayurvedic system	Mode of immune-boosting action	References
18	*Tinospora sinensis* (Lour.) Merr.)	*Guduchi*	Enhances the phagocytic and intercellular bactericidal capacity of polymorphs by increasing the secretion of lysozyme by macrophage and increasing the IFN-γ, IL-2, IL-4, and IL-1 levels in the peripheral blood mononuclear cells and preventing the deterioration of Kupffer cell activity	Thatte et al. (1992), Nagarkatti et al. (1994), More and Pai (2011), Aranha et al. (2012), Sachan et al. (2019)
19	*Withania somnifera* (L.) Dunal)	*Ashwagandha*	Prevents myelo-suppression, upregulates Th1-dominant polarization, and boosts the humoral as well as cell-mediated immune responses increases the level IFN-γ, IL-2, and granulocyte macrophage colony stimulating factor (GM-CSF). Enhances activities of nitric oxide synthetase and lysosomal enzymes, phagocytic activity	Ziauddin et al. (1996), Iuvone et al. (2003), Bani et al. (2006), Rasool and Varalakshmi (2006)
20	*Zingiber officinale* Roscoe	*Shunthi/Ardrak*	Activates TNF-α and interleukin-1β, decreases age-related oxidative stress markers, and inhibits NO production.	Topic et al. (2002), Ippoushi et al. (2003), Park et al. (2012)

Table 14.3 Multi ingredients-based Ayurveda formulations reported for immune modulation activities

Sl. No.	Name of Ayurvedic preparation	Mode of immune-boosting action	References
1	*Amalaki Rasyana*	Immuno-stimulant activity, cytoprotective activity	Rajani et al. (2012)
2	*Brahma Rasayana*	Inhibition of metastasis and increased the lifespan of the animal	Kumar et al. (1999)
3	*Balacaturbhdra Curna*	Significant effect on cell-mediated and humoral-mediated immune response	Joshi et al. (2016)
4	*Cyavanaprasha*	Augmentation of immune markers (TNF-α, IL-1β, and MIP1α), NK cells, and increases phagocytic activity	Sastry and Gupta (2014)
5	*Pippali Rasayana*	Activation of macrophages and phagocytic activity	Agarwal et al. (1994)

Table 14.4 Medicinal plants reported for antiviral activities in Ayurveda practice

Sl. No.	Plant	Name in Indian Ayurvedic system	Mode of immune-boosting action	References
1	*Andrographis paniculata* (Burm.f.) Nees)	*Kalamegha*	Methanolic extract reported for exhibiting the antiviral inhibitory effect of Dengue virus type 1 (DENV1)-infected Vero E6 cells	Tang et al. (2012)
2	*Azadirachta indica* A. Juss)	*Nimba*	Aqueous extract of neem leaves reported for inhibition of the virus replication in Dengue virus type-2 (DENV2) in suckling mice	Parida et al. (2002)
3	*Cinnamomum cassia* Blume	*Tvakbheda*	Hot water extract is reported for inhibition of viral attachment and internalization in Human Respiratory Syncytial Virus (HRSV)	Yeh et al. (2013)
4	*Ficus religiosa* L.	*Ashvattha*	Water and chloroform bark extracts are reported for direct virus-inactivating activity, the chloroform extract inhibits the entry and attachment of the virus and also limits the production of viral progeny in herpes simplex virus type 2 (HSV-2), and shows inhibitory activity on human respiratory syncytial and rhinovirus infections	Cagno et al. (2015)
5	*Glycyrrhiza glabra* L.	*Yashtimadhu*	Hot water extract of Glycyrrhiza uralensis Fisch. inhibits HRSV by preventing viral attachment, internalization, and stimulating IFN secretion. The extract at the dose of 300 μg/ml significantly decreases the viral loads within the cells and in the suspension. Root extract is effective against Herpes Simplex Virus.	Feng et al. (2013), Sabouri Ghannad et al. (2014)
6	*Mentha piperita* L.	Putiha	Ethanolic extract of leaves exhibits antiviral activity against respiratory syncytial virus (RSV) with a high selectivity index and significantly reduces the production of NO, TNF-α, IL-6, and PGE2 in LPS-stimulated RAW 264.7 cells.	Li et al. (2017)

(continued)

Table 14.4 (continued)

Sl. No.	Plant	Name in Indian Ayurvedic system	Mode of immune-boosting action	References
7	*Peganum harmala* L.	*Harmala*	Ethanolic extract seed shows antiviral activities against the influenza virus in Madin-Darby canine kidney (MDCK) cells. The same was also reported to increase the survival rate and decrease the virus load in the lung without affecting body weight in the mice model.	Li et al. (2019a, b)
8	*Portulaca oleracea* L.	Lonika	Water extract is reported for antiviral activity against the pandemic influenza A virus (IAV); it prevents the binding of the virus to cells and exhibits good virucidal activity.	Haidari et al. (2009)
9	*Punica granatum* L.	Dadima	Polyphenol extract is reported for virucidal activity against the influenza A virus probably by blocking the replication of virus RNA and agglutination of chicken red blood cells.	Haidari et al. (2009)
10	*Senna alexandrina* Mill. (Syn. Sennaangustifolia (Vahl) Batka)	Markandika	Ethanolic extract of leaves is reported antiviral activity against the dengue virus.	Rothan et al. (2014)
11	*Terminalia chebula* Retz.	Haritaki	The ethanolic extract is reported anti-viral activity against HSV-2 by inhibiting virus attachment and penetration to the host cells and in Newcastle disease virus (NDV) in baby hamster kidney (BHK) cells by inhibiting syncytia formation.	Chang et al. (2013), Kesharwani et al. (2017)
12	*Zingiber officinale* Roscoe	Ardraka	Hot water extract of fresh ginger antiviral activity against human respiratory syncytial virus (HRSV) by inducing plaque formation in both HEp-2 and A549 cell lines.	Chang et al. (2013)
13	*Piper longum* L.	Pippali	The ethanol extract is reported for antiviral activity against anti-HBV activity in vitro.	Jiang et al. (2013)

(continued)

Table 14.4 (continued)

Sl. No.	Plant	Name in Indian Ayurvedic system	Mode of immune-boosting action	References
14	*Withania somnifera* (L.) Dunal.):	Ashwagandha	Clinical health benefits have been reported in Human Immunodeficiency Virus-infected patients and to treat the cases of genital herpes occurring due to herpes simplex virus type 1 and 2	Kambizi et al. (2007)
15	*Swertia chirata* Buch -Ham. ex Wall	Kiratatikta	The hydro extract inhibits HSV-1, while the chloroform extract shows antiviral activity against Human immunodeficiency virus (HIV) by attenuating the expression of Viral protein R, which is an active target for HIV disease in Hela cells harboring the TREx plasmid encoding full-length Vpr (TREx-HeLa-Vpr cells).	Verma et al. (2008), So-Yeun et al. (2019)

traditionally in India to treat several diseases. Almost all parts of this plant possess medicinal properties. It has been demonstrated that the methanolic extract of the leaves of this plant inhibits glycoprotein-mediated viral fusion and thus protects against HSV1 infection (Tiwari et al. 2009). In the case of COVID-19, it is speculated that TNFα, as a pleiotropic cytokine, might cause respiratory failure during pneumonia. Nimbolide, which is an active component of the neem tree, has been shown to be an inhibitor of TNFα and to suppress the nuclear translocation of NFκB/HDAC3, thus reducing the cytokine storm in an experimental model of ARDS (Pooladanda et al. 2019). Hence, nimbolide, or neem extract, might be beneficial in the treatment of COVID-19 (Roy and Bhattacharyya 2020). Likewise, ashwagandha (*Withania somnifera*) contains an important natural product, withaferin A, which has been reported to have antiviral activity against HSV and is also beneficial for inflammatory disorders (Grover et al. 2011). Molecular docking studies indicated that withanoside X and quercetin glucoside from *W. somnifera* exhibit anti-SARS-CoV-2 activities by binding to selected proteins, such as 6W01 and 6M0J (Chikhale et al. 2020). Furthermore, withanoside V, an important constituent of *W. somnifera,* has a potential antiviral effect on SARS-CoV-2 by binding to polyproteins essential for viral replication (Tripathi et al. 2020).

There are several immunomodulatory compounds isolated from herbs that have anti-inflammatory activities. One such compound, andrographolide, is isolated from *Andrographis paniculata*. It increases the levels of cytotoxic T-cells, NK cells, and phagocytic cells and has been shown to have antiviral activity (Gupta et al. 2017). Computational analysis has shown that andrographolide and its derivative, chrysin-7-O-β-glucuronide, have therapeutic potential against SARS-CoV-2-induced pneumonia (Pooladanda et al. 2020). Similarly, saikosaponins, which are triterpenoids extracted from several herbs, have exhibited antiviral activity against coronavirus 229E by affecting viral attachment and entry (Cheng et al. 2006). Computational analysis of this ingredient has also shown that it could be a spike protein inhibitor and thus could be used to inhibit SARS-CoV-2 infection (Goswami and Bagchi 2020).

A recent review suggested that naringenin (4′,5,7-trihydroxyflavanone), a flavonoid present in citrus fruits, has the potential to exert therapeutic effects against COVID-19 by inhibiting the function of COVID-19 main protease, 3-chymotrypsin-like protease (3CLpro), and reducing the ACE2 receptor activity (Tutunchi et al. 2020). The plant flavonol quercetin, a US FDA-approved drug ingredient, has been shown to be anti SARS-CoV-1 activity by inhibiting entry into host cells (Yi et al. 2004). Since SARS-CoV-1 and SARS-CoV-2 use the same receptor, it can be speculated that quercetin could be repurposed in treating COVID-19. Recent studies suggest that quercetin, alone or in combination with vitamin C, vitamin D, or estradiol, may serve as a potential drug for the prevention and treatment of COVID-19 (Colunga Biancatelli et al. 2020; Glinsky 2020; Mani et al. 2020). A recent report from China revealed the therapeutic potential of multi-herbal traditional Chinese medicine (TCM) in COVID-19 patients, with liquorice as one of the major components (Ren et al. 2020b). It was observed that this composition of TCM reduced the expression of the SARS-CoV receptor, ACE2, on the surface of susceptible host cells. Prior to this, in 2003, glycyrrhizin was shown to have potent antiviral activity against two clinical isolates of SARS-CoV-1, and viral replication was also supposedly targeted (Cinatl et al. 2003). A diammonium salt of glycyrrhizin, in combination with vitamin C, is also being investigated in a clinical trial for the treatment of COVID-19 (Chen et al. 2020a, b). Besides this, other flavonoids such as naringin, naringenin, hesperetin, hesperidin, and neohesperidin, also showed antiviral and anti-inflammatory action, which can help to control COVID-19 (Filippini et al. 2020; Ngwa et al. 2020). In one in silico study, rutin, diosmin, apiin, and diacetyl curcumin were reported to inhibit the action of the protease of SARS-CoV-2. All of these compounds showed good binding affinity with the target protein (MPro) of SARS-CoV-2 (Adem et al. 2020). However, naringin could effectively prevent cytokine storm by inhibiting the expression of the pro-inflammatory cytokines released during COVID-19 infection (Cheng et al. 2020).

Based on the various data from clinical trials, cohort, or other population studies using TCM for combating contagious respiratory virus diseases, researchers hypothesized that Chinese herbal formulas consisting of Rhizoma Atractylodis Macrocephalae (Baizhu), Radix Astragali (Huangqi), Radix Saposhnikoviae (Fangfeng), Radix Glycyrrhizae (Gancao), Lonicerae Japonicae Flos (Jinyinhua), and Fructus Forsythia (Lianqiao) could be an efficient alternative approach for the prevention of COVID-19 in high-risk populations (Luo et al. 2020a, b). In a systematic review and meta-analysis study, Li et al. (2020) reported that two patients were cured following TCM therapy, indicating the possible wider application of traditional Chinese herbal medicine for patients with COVID-19-induced pneumonia.

Recently, Ang et al. (2020b) systematically reviewed, summarized, and analyzed the herbal formulas endorsed by Chinese guidelines in terms of the composition of herbs and disease stages for the treatment of pediatric COVID-19. They found that the herb Armeniacae Semen, one of the most frequently used herbs among those in the herbal formulas, was recommended for the treatment of pediatric COVID-19. In addition, Glycyrrhizae Radix et Rhizoma was the herb with the highest frequency of use for the treatment of adult COVID-19 infection (Ang et al. 2020a, b). Moreover, Armeniacae Semen has also been found to suppress Th2 cells, which are crucial for immune responses and minimizing airway hyperresponsiveness (Do et al. 2006). Although Glycyrrhizae Radix et Rhizoma possesses both antiviral and anti-inflammatory qualities, the major limitation of the study was the lack of conclusive evidence on the efficacy and potency of the herbal formulas for the treatment of pediatric or adult COVID-19.

In a network pharmacology-based analysis, Yu et al. (2020) explored two traditional Chinese herbal formulas for COVID-19 treatment and opined that both could ameliorate the damage caused by coronavirus infection through increasing nuclear receptor activity. One formula contained a mixture of Rhizoma Atractylodis, Flos Lonicerae, Pericarpium Citri Reticulatae, Rhizoma Phragmitis, Folium Mori, and Radix Astragali seu Hedysari; while the other contained Radix Astragali seu Hedysari, Rhizoma Atractylodis Macrocephalae, Radix Saposhnikoviae, *Cyrtomium fortunei* J. Sm., Flos Lonicerae, *Eupatorium fortunei*Turcz., and Pericarpium Citri Reticulatae. However, a thorough validation is necessary because it is unclear if activating coronavirus membrane fusion, which happens as a result of a receptor-driven ratcheting mechanism, can be used to decrease viral replication. Different herbs were reported to have potential as a treatment for SARS-CoV-2. An in vitro study on Vero E6 cells revealed that LH significantly inhibited SARS-CoV-2 replication and greatly minimized the release of pro-inflammatory cytokines at the mRNA level. Additionally, LH treatment resulted in the abnormal particle morphology of virions in cells. The findings of plant-derived active molecules that could potentially inhibit COVID-19 replication show promise in formulating a novel antiviral drug against COVID-19 (Runfeng et al. 2020).

In the race to design and develop a novel drug, it seems plausible that angiotensin-converting enzyme inhibitors could do the same job as antiviral drugs. Recently, it has been hypothesized that a nicotinic compound could be a potential agent for COVID-19 treatment (Tindle et al. 2020). The nicotinic acetylcholine receptors are considered potential therapeutic targets to mitigate COVID-19 (Changeux et al. 2020).

Angiotensin-converting enzyme 2 (ACE2), the putative receptor for SARS-CoV-2, is reduced by nicotine in multiple organs. Additionally, nicotine alters the renin–angiotensin–aldosterone system, the physiological pathway linked to COVID-19 pathogenesis with pulmonary, cardiovascular, renal, and neurological complications (Oakes et al. 2018; Farsalinos et al. 2020). Furthermore, nicotine alters the severity of COVID-19 disease by suppressing platelet reactivity and macrophage-driven hyperinflammation in the cholinergic anti-inflammatory pathway. The plant metabolites are known to play key roles in disrupting the pathogenic process at different stages. Lipophilic terpenoids (essential oils) are known to interfere with the viral lipid envelope, while phenolics (tannins, rosmarinic acids, flavonoids, etc.) are reported to disrupt the membrane and intraparticle viral proteins. Some plant-derived alkaloids (dictamine, quinine, skimmianine, cinchonine, etc.) are known to act as DNA intercalators, stabilizing nucleic acid levels and restricting viral replication in cells (Wink 2020). The importance of plant metabolites may be speculated from the fact that a major portion of the global medicine market (about 35%) consists of products derived from plants and herbs (Ebenezer et al. 2019). In lieu of a comprehensively proven and approved potent drug or vaccine against COVID-19, some plant secondary metabolites, known to have strong antiviral activities, may be considered as a part of alternative therapies for the treatment of COVID-19. Several previous studies have demonstrated the potential of plant secondary metabolites for the treatment of SARS-CoV infections. A comprehensive study of plant secondary metabolites showed the possibility of anti-SARS-CoV-2 activity from a screening of 149 plant species from 71 different families (Bhuiyan et al. 2020). These plant metabolites were reported to interfere with the protease (papain-like protease and 3CL protease) activity, restrict the viral attachment to host cells, and impede signaling pathways (Ryu et al. 2010; Cho et al. 2013); they may be important molecules for the development of antiviral drugs to combat COVID-19.

Table 14.5 Some other natural products and common dietary supplements having antiviral activity

Common medicinal plants and their components	Mode of antiviral action
Ginger (*Zingiber officinale*)	Inhibits viral attachment and entry [Respiratory Syncytial virus (Lin et al. 2016)
Garlic (*Allium sativum*)	Inhibits viral attachment and entry [Respiratory Syncytial virus (Weber et al. 1992)
Cinnamon (*Cinnamomum cassia*)	Inhibits viral replication (Fatima et al. 2016)
Tulsi (*Ocimum sanctum*)	Has virucidal activity (Ghoke et al. 2018)
Glycyrrhizin (*Glycyrrhizia glabra*)	Reduces the formation of intracellular reactive oxygen species (ROS), thrombin, and the hyperproduction of airway exudates, binds angiotensin-converting enzyme II (ACE2), and generates endogenous interferon (Luo et al. 2020a, b)
Vitamin D	Immunomodulator (Yamshchikov et al. 2009)
Zinc	Inhibits RNA polymerase activity (te Velthuis et al. 2010)
Glycyrrhiza glabra, Alliun sativam	Inhibits viral replication (Keyaerts et al. 2004)
Boerhaavia diffusa, Coriandrum sativum, Cynara scolymus, Embeliaribes Cerasus avium (L.), Coscinium, Punicagranatum cassia occidentalis, Moench, Rubia tinctorum, Citrus aurantium, Berberis integerrima, Peganum harmal, Alcea digitata, Allium sativum Aretmisia annus	Inhibitory effect on ACE (Ziai and Heidari 2015; Hussain et al. 2018; Koch and Pong 2020)

Table 14.6 Proposed herbal formula from traditional Chinese medicine for Covid-19 treatment (Mirzaie et al. 2020)

Sl. No.	Herbal formula	References
1	Huo Xiang Zheng Qi capsule	Li et al. (2022)
2	Lian Hua Qing Wen	Qu et al. (2005)
3	Qing Fei Pai Du Tang	Khan et al. (2020)
4	China gancaoganjiang decoction, qingfei paidu decoction, qingfei touxie fuzheng recipe and sheganmahuang decoction	Ren et al. (2020a)
5	Shuang huang lian oral liquid	Mirzaie et al. (2020)
6	Shen Fu injection	Wang et al. (2020b)
7	Respiratory Detox shot	Chen and Nakamura (2004), Xu and Zhang (2020)
8	Tan Re Qing injection	Gautret et al. (2020)
9	Lian Hua Qing Wen capsule	Yang et al. (2020)
10	Shuang Huang Lian liquid	Yang et al. (2020)
11	Xi Yan Ping injection	Lee et al. (2007)
12	Shan Fu injection	Yang et al. (2020)
13	Jing Yin Granule	Yang et al. (2020)
14	Shen Qi Fu Zheng injection	Jiang et al. (2014)
15	Kang Bing Du granule	Korytkowski et al. (2003)
16	Ke Su Ting Syrup	Yang et al. (2020)

Considering the proposed anti-COVID-19 activities shown by some phytochemicals through meta-analysis and historical data records, there is ample scope for the development of novel antiviral drugs against COVID-19. The activity-guided separation of most active compounds from crude extracts will help to develop potent anti-COVID-19 drugs. Hence, more rigorous clinical trials are required on target species (Tables 14.5 and 14.6).

14.7.2 Homeopathic Treatments for COVID-19

The principle of homeopathy is based on the augmentation of the self-healing powers of the body in a manner that helps medicines work in a holistic way. The primary notion for the mechanism of action is based on the law of similarity. From experimentation in laboratory animal models, it is established that if a patient is treated on a similarity basis, meaning that if an individual presents a particular symptom and there is a medicine that could minimize those symptoms, then that medicine may have the potential to mitigate those symptoms in the patient. In the present COVID-19 pandemic, scientists worldwide have been engaged in the search for effective molecules to treat SARS-CoV-2. Researchers are seeking suitable antivirals, natural products, antiparasitic drugs, monoclonal antibodies, and blood components as therapeutics treatments for COVID-19. Additionally, the Central Council for Research in Homeopathy of the Government of India is researching the possibility of therapeutic homeopathic medicines against COVID-19. Recently, the prophylactic effect of Arsenicum Album has been evaluated against coronavirus, and it was found to increase cellular immunity. Therefore, it is speculated that the enhanced cellular capability of immune cells (macrophages, T cells, B cells, and lymphocytes) by Arsenicum Album could be used against viruses such as SARS-CoV-2. Earlier, the antiviral activity of Arsenicum Album against HIV, induced by increasing the $CD4^+$ cell count, has been demonstrated. Another advantage of this medicine is its special affinity for particular organ systems, such as the respiratory system, and has prophylactic potential in homeopathy for treating asthma, chronic obstructive lung disease, acute and chronic bronchitis, and influenza-like illnesses on the Indian subcontinent. Considering the pathophysiology of COVID-19, Arsenicum Album may be an effective drug candidate. The Ministry of AYUSH (GOI) is presently conducting clinical trials on quarantined COVID-19 patients (https://www.boomlive.in/coronavirus-outbreak/homoeopathy-can-be-used-as-adjuvant-to-covid-19-treatment-dr-anil-khurana-7997).

A homeopathic drug called Boiron, a diluted and dynamized extract of duck liver and heart, has recently been shown to reduce the frequency and length of upper respiratory tract infections in patients with chronic obstructive pulmonary disease (COPD) who have been exposed to influenza. It has also been demonstrated to lessen the frequency of COPD exacerbations in people with the exacerbator phenotype. For COVID-19 patients, it is important to investigate the potential therapeutic efficacy of this homeopathy and other homeopathic remedies (Conde Diez et al. 2019; Kalliantas et al. 2020). Clinical studies are being conducted to evaluate the efficacy and security of customized homeopathic remedies in the treatment and prevention of COVID-19 (Teixeira 2020; Rossi 2020).

14.8 Combinatorial Therapy

Aside from the individual role of several synthetic and natural compounds being investigated by several researchers, the combined effect of therapeutic drugs has been proven effective, such as for HIV. The combinatorial effect of therapeutics is also being investigated for SARS-CoV-2 (Table 14.7).

14.8.1 Combination of Azithromycin and Hydroxychloroquine

HCQ was earlier recommended for use in COVID-19 patients, followed by a recommendation for the use of HCQ in combination with azithromycin. On the basis of a study of 22 subjects, it was claimed that this combination was effective in

Table 14.7 Several other combinations of individually evaluated drugs are under clinical trial to prove their efficacy in human patients

S. No	Drug combination	Dose being evaluated	Clinical trials	Sponsor/ collaborator
1	Oseltamivir + Ritonavir	Ritonavir: 100 mg twice a day for 14 days; Oseltamivir: 75 mg once a day	NCT04261270 [Phase III]	Tongji Hospital
2	Oseltamivir + ASC09F	ASC09F: 400 mg twice a day for 14 days; Oseltamivir: 75 mg once a day	NCT04261270 [Phase III]	Tongji Hospital
3	Oseltamivir + Ritonavir + Hydroxychloroquine	Ritonavir: 200 mg per day; Oseltamivir:300 mg per day; Hydroxychloroquine: 400 mg per day	NCT04303299 [Phase III]	Rajavithi Hospital
4	Lopinavir + Ritonavir + Flavipiravir +	Lopinavir: 800 mg per day; Ritonavir: 200 mg per day; Flavipiravir: 1200 mg every 8 h on day 1 followed by a maintenance dose of 1200 mg twice a day	NCT04303299 [Phase III]	Rajavithi Hospital
5	Ganovo (Danoprevir) + Ritonavir and Interferon	Ganovo: 100 mg twice a day for 14 days; Ritonavir: 100 mg twice a day for 14 days; Interferon: 50 μg twice a day for 14 days	NCT04291729 [Phase 4]	Nineth hospital of Nanchang/ Ascletis Pharmaceuticals Co., Ltd
6	Darunavir + Cobicistat	One tablet of each for 5 days in combination with conventional treatment	NCT04252274 [Phase III]	Shanghai Public Health Clinical Centre

treating SARS-CoV-2-induced pneumonia by decreasing the viral load and helping in the elimination of the virus (Gautret et al. 2020). More recently, in a vigorous study on approximately 3,200,000 users, the use of azithromycin/hydroxychloroquinecombination was refuted because of its adverse effects on heart function and related cardiovascular mortality (Lane et al. 2020). Although the short-term use of hydroxychloroquine is safe, the addition of azithromycin to this treatment has an effect on QT length, leading to heart failure.

14.8.2 Combination of Remdesivir and Chloroquine

As described earlier, remdesivir and CQ have been individually tested against SARS-CoV-2 and were shown to have profound antiviral activities. The combination of CQ and remdesivir has also been found to be effective in the inhibition of SARS-CoV-2 under in vitro conditions (Wang et al. 2020a). Although these drugs have been safely used in human patients, further studies are warranted to elucidate the drug interactions and contraindications.

14.8.3 Combination of Chloroquine, Hydroxychloroquine, and Zinc Supplementation

CQ/HCQ is known to direct extracellular zinc to intracellular lysosomes in addition to its antiviral actions. It prevents virus multiplication and RNA-dependent RNA polymerase activity. Additionally, it is well-known that older persons with chronic lung disease, cardiovascular disease, or diabetes frequently have zinc deficiencies. Therefore, it is hypothesized that CQ/HCQ in combination with zinc supplementation may be more effective than CQ/HCQ alone in lowering COVID-19 morbidity and mortality (Derwand and Scholz 2020).

14.8.4 Combination of Hydroxychloroquine and Ivermectin

Based on the antiviral effects of both HCQ and ivermectin without any drug interactions, some researchers have hypothesized that HCQ and ivermectin could fight COVID-19 infection in a consequential and synergistic manner (Patrì and Fabbrocini 2020).

14.8.5 Combination of Lopinavir/Ritonavir, Arbidol, and Shufeng Jiedu Capsule

A combination therapy comprising ritonavir/lopinavir, arbidol, and Shufeng Jiedu Capsule (a TCM) in four mild to severe COVID-19 patients significantly improved pneumonia-associated symptoms: two patients were confirmed COVID-19 negative and discharged, and one tested negative at the first test. The remaining patient, who had severe pneumonia, showed signs of improvement by the cut-off date for data collection (Wang et al. 2020b).

14.8.6 Combination of Lopinavir, Ritonavir, and Ribavirin

The Indian Council of Medical Research has recommended the off-label emergency use of lopinavir/ritonavir combination for symptomatic COVID-19 patients in India in view of the earlier evidence of the effectiveness of lopinavir/ritonavir against SARS and MERS-CoV (Bhatnagar et al. 2020). In a recent multicenter, prospective, open-label, randomized clinical study, it was reported that triple antiviral therapy, which consists of lopinavir (400 mg), ritonavir (100 mg), and ribavirin (400 mg) combined every 12 h—along with three doses of eight million international units of IFN-1b on alternate days—was both safe and more effective than lopinavir/ritonavir alone (lopinavir 400 mg and ritonavir 100 mg every 12 h) in shortening the virus shedding, alleviating the symptoms, and reducing the hospitalization duration of patients with mild to moderate COVID-19 (Hung et al. 2020).

14.8.7 Combination of Lithium Chloride and Rapamycin

The combination of lithium chloride and rapamycin is intended to reduce the possibility of viral replication as much as possible by limiting viral entry into cells and viral RNA synthesis (lithium affects protein synthesis inhibition via glycogen synthase kinase 3 deactivation), regulating the immune response, and encouraging autophagy without inducing apoptosis (Bou Khalil 2020).

14.8.8 Combination of Baricitinib, Lopinavir/Ritonavir, and Remdesivir

The combination of the anti-inflammatory drug baricitinib, a NAK inhibitor with a particularly high affinity for AAK1, and direct-acting antivirals (lopinavir or ritonavir and remdesivir) has been suggested in the treatment of COVID-19 infection. Baricitinib and these direct-acting antivirals were found to reduce viral infectivity, viral replication, and aberrant host inflammatory response (Stebbing et al. 2020).

14.9 Conclusions and Prospects

The world is facing a highly contagious virus with an enhanced infectivity rate, compared to the previous emergences of fatal coronaviruses, such as SARS-CoV-1 and MERS-CoV. Apart from symptomatic treatments and precautionary measures to contain the virus, there is an immediate need for effective therapeutic molecules to treat critically ill COVID-19 patients and prophylactic agents or immunomodulators to protect healthcare workers and the uninfected human population. A more complete knowledge of the pathogenesis of human coronavirus infection, disease prognosis, and clinical trial outcomes is imperative for identifying appropriate evidence-based therapeutic strategies against COVID-19. The development of a successful therapeutic, vaccine, or drug is of the utmost importance to overcome COVID-19 worldwide.

Globally, many research and development studies and clinical trials for the control and prevention of the COVID-19 pandemic are ongoing. Remdesivir is a nucleoside analog, lopinavir and ritonavir are viral protease inhibitors, and favipiravir is an RdRp inhibitor. Likewise, Cyclophilin A is also an important protein required for various coronavirus replication, and its inhibitor cyclosporine has the ability to suppress coronavirus on a broad spectrum. Alisporivir, a potent cyclophilin inhibitor and a non-immunosuppressive analog of Cyclosporine A (CsA), exhibits antiviral activity by inhibiting the postentry step of the SARS-CoV-2 life cycle. These antiviral agents have the potential to restrict viral propagation in infected individuals and thus could be repurposed to relieve patients suffering from symptoms. A variant of an antiparasitic drug, HCQ, possesses immunomodulatory properties and shows antiviral activity at the pre- and postinfection stages by restricting viral entry into host cells. The restriction of viral entry into the host cells is also achieved by monoclonal antibodies, with a specificity for viral surface proteins as well as for host ACE2 receptors. Immunomodulation by HCQ includes the suppression of the cytokine storm by affecting T-cell activation and differentiation and inhibiting the production of cytokines by these cells. Suppression of the adverse effects of excessive cytokine secretion in COVID-19 patients has also been achieved in clinical trials by using antibodies targeting cytokines, such as IL-1, IL-6, and TNF-α. Ivermectin, another antiparasitic drug, also inhibits viral replication inside infected host cells by affecting the nuclear transport of viral proteins. SARS-CoV-2 exhibits the manifestation of intravascular coagulopathy in severely infected patients, which can be managed by the use of anticoagulants (such as heparin and dipyridamole), as shown in clinical applications to a small number of patients in China. Clinical trials are ongoing for other therapies based on blood constituents, including NK cell- and stem cell-based therapies, which help in clearing viral infection from the host, and immunomodulation by regulating immune components. Blood plasma from recovered COVID-19 patients possesses neutralizing antibodies specific for SARS-CoV-2, which could be used to treat critically ill patients and reduce mortality. Most recently, combinatorial therapy involving different combinations of drug molecules has been shown to effectively manage the disease. Natural products derived from several known medicinal plants, such as turmeric, ginger, Tulsi, garlic, and glycyrrhizin, also possess immunomodulatory and antiviral properties, which could help the human body fight against SARS-CoV-2 and related respiratory viruses. In the recent past, the prophylactic and therapeutic effects of small interfering RNA (siRNA) targeting viral diseases have attracted researchers and pharmaceutical industries to design and develop a highly effective therapy against SARS-CoV and MERS-CoV (Zheng et al. 2004; Sohrab et al. 2018). The rapid development of siRNA targeting different regions of the viral genome of SARS-CoV2 would also be a novel approach for combating Covid-19 disease.

This review provides updated information on various types of promising COVID-19 treatments: candidate drugs, antivirals, antibiotics, antiparasitics, or combinations of one or more treatments; allopathic immunotherapeutic molecules; traditional and age-old AYUSH treatments being practiced in India; and the latest stem cell therapy technologies. Further intense trials of these drugs, involving a significant number of volunteers and infected patients, are required to elucidate effective treatments and preventive medicines, which may help to reduce the overall morbidity and mortality due to COVID-19. Likewise, the experiences and lessons learned during the current pandemic will be appropriately utilized in the future to tackle emerging or re-emerging infectious diseases.

In the current circumstances (non-availability of any approved treatment or vaccine for COVID-19), alternative approaches have turned out to be of the foremost necessity to curb the spread of the disease. This review article is an attempt to provide comprehensive information on various alternative approaches, and their highlights and challenges. We have attempted to deliver accurate and well-versed information, but it is important to bear in mind that the information provided here is for informational purposes only and should not be used for personal treatment without the advice of an appropriately qualified and licensed medical practitioner. Based on the currently available information on various therapeutics being adopted for combating COVID-19, efficient global disease coordination is an inevitable requisite and should be realized through effective collaboration among the world's governments, health authorities, virologists, epidemiologists, physicians, and veterinarians, leading to an accelerated and highly potent novel drug/vaccine discovery. An appropriate drug or prevention strategy against COVID-19, an infectious disease that has caused a global public health disaster, is not too far off.

Acknowledgments All authors acknowledge the help of Ramachandra Bhat, The Director, Institute of pharmacology, Madras Medical College, Park Town, Chennai-600 003, Tamil Nadu, and G.Mariappan, Professor, Dept. of Pharm Chemistry, St Mary's College of Pharmacy, St Francis Street Secunderabad, Telangana 500025. India, for their critical input while writing this manuscript.

References

Adem S, Eyupoglu V, Sarfraz I et al (2020) Identification of potent COVID-19 Main Protease (Mpro) inhibitors from natural polyphenols: an in Silico strategy unveils a hope against CORONA. Preprints:2020030333. https://doi.org/10.20944/preprints202003.0333.v1

Agarwal AK, Singh M, Gupta N et al (1994) Management of giardiasis by an immunomodulatory herbal drug Pippalirasayana. J Ethnopharmacol 44:143–146

Ahsan W, Javed S, Al Bratty M et al (2020) Treatment of SARS-CoV-2: how far have we reached? Drug Discov Ther 14(2):67–72

Alhenc-Gelas F, Drueke TB (2020) Blockade of SARS-CoV-2 infection by recombinant soluble ACE2. Kidney Int 97(6):1091–1093

Andronova VL, Grokhovskiĭ SL, Galegov GA et al (2010) Antiviral properties of the derivatives of netropsin and distamycin against herpes simplex viruses type 1 and variolovaccine. Vopr Virusol 55(6):24–27
Ang L, Lee HW, Choi JY et al (2020a) Herbal medicine and pattern identification for treating COVID-19: a rapid review of guidelines. Integr Med Res 9(2):100407
Ang L, Lee HW, Kim A et al (2020b) Herbal medicine for treatment of children diagnosed with COVID-19: a review of guidelines. Complement Ther Clin Pract 39:101174
Antony S, Kuttan R, Kuttan G (1999) Immunomodulatory activity of curcumin. Immunol Investig 28(5–6):291–303
Aranha I, Clement F, Venkatesh YP (2012) Immunostimulatory properties of the major protein from the stem of the Ayurvedic medicinal herb, guduchi (Tinospora cordifolia). J Ethnopharmacol 139:366–372
Arif T, Mandal TK, Kumara N et al (2009) In vitro and in vivo antimicrobial activities of seeds of Caesalpiniabonduc (Lin.) Roxb. J Ethnopharmacol 123(1):177–180
Baliga MS (2011) Review of the phytochemical, pharmacological and toxicological properties of *Alstonia scholaris* Linn. R. Br (Saptaparna). Chin J Integr Med. https://doi.org/10.1007/s11655-011-0947-0
Bani S, Gautam M, Sheikh FA et al (2006) Selective Th1 up-regulating activity of Withaniasomnifera aqueous extract in an experimental system using flow cytometry. J Ethnopharmacol 107:107–115
Barnard DL, Kumaki Y (2011) Recent developments in anti-severe acute respiratory syndrome coronavirus chemotherapy. Future Virol 6(5):615–631
Baron SA, Devaux C, Colson P et al (2020) Teicoplanin: an alternative drug for the treatment of COVID-19? Int J Antimicrob Agents 55(4):105944
Bhatnagar T, Murhekar MV, Soneja M et al (2020) Lopinavir/ritonavir combination therapy amongst symptomatic coronavirus disease 2019 patients in India: protocol for restricted public health emergency use. Indian J Med Res 151(2):184–189
Bhuiyan FR, Howlader S, Raihan T et al (2020) Plants metabolites: possibility of natural therapeutics against the COVID-19 pandemic. Front Med 7:444
Bordbar N, Karimi MH, Amirghofran Z (2012) The effect of glycyrrhizin on maturation and T cell stimulating activity of dendritic cells. Cell Immunol 280(1):44–49
Bou Khalil R (2020) Lithium chloride combination with rapamycin for the treatment of COVID-19 pneumonia. Med Hypotheses 142:109798
Broyles SS, Kremer M, Knutson BA (2004) Antiviral activity of distamycin A against vaccinia virus is the result of inhibition of postreplicative mRNA synthesis. J Virol 78(4):2137–2141
Cagno V, Civra A, Kumar R et al (2015) Ficus religiosa L. bark extracts inhibit human rhinovirus and respiratory syncytial virus infection in vitro. J Ethnopharmacol 176:252–257
Cai Q, Yang M, Liu D et al (2020) Experimental treatment with favipiravir for COVID-19: an open-label control study. Engineering (Bejing) 10:1192–1198
Caly L, Druce JD, Catton MG et al (2020) The FDA-approved drug ivermectin inhibits the replication of SARS-CoV-2 in vitro. Antivir Res 178:104787
Chang JS, Wang KC, Yeh CF et al (2013) Fresh ginger (Zingiberofficinale) has anti-viral activity against human respiratory syncytial virus in human respiratory tract cell lines. J Ethnopharmacol 145(1):146–151
Changeux JP, Amoura Z, Rey FA, Miyara M (2020) A nicotinic hypothesis for Covid-19 with preventive and therapeutic implications. C R Biol 343(1):33–39. https://doi.org/10.5802/crbiol.8
Chen Z, Nakamura T (2004) Statistical evidence for the usefulness of Chinese medicine in the treatment of SARS. Phytother Res 18(7):592–594
Chen C, Zhang Y, Huang J et al (2020a) Favipiravir versus arbidol for COVID-19: a randomized clinical trial. medRxiv. https://doi.org/10.1101/2020.03.17.20037432
Chen L, Hu C, Hood M et al (2020b) A novel combination of vitamin C, curcumin and glycyrrhizic acid potentially regulates immune and inflammatory response associated with coronavirus infections: a perspective from system biology analysis. Nutrients 12(4):1193

Cheng PW, Ng LT, Chiang LC et al (2006) Antiviral effects of saikosaponins on human coronavirus 229E in vitro. Clin Exp Pharmacol Physiol 33:612–616
Cheng J, Tang Y, Bao B et al (2020) Exploring the active compounds of traditional mongolian medicine Agsirga in iIntervention of novel coronavirus (2019-nCoV) based on HPLC-Q-exactive-MS/MS and molecular docking method. ChemRxiv. https://doi.org/10.26434/chemrxiv.11955273.v2
Chikhale RV, Gurav SS, Patil RB et al (2020) Sars-cov-2 host entry and replication inhibitors from Indian ginseng: an in-silico approach. J Biomol Struct Dyn 22:1–12
Cho JK, Curtis-Long MJ, Lee KH et al (2013) Geranylated flavonoids displaying SARS-CoV papain-like protease inhibition from the fruits of Paulownia tomentosa. Bioorg Med Chem 21:3051–3057
Chowdhury S, Mukherjee T, Mukhopadhyay R et al (2012) The lignan niranthin poisons *Leishmania donovani* topoisomerase IB and favours a Th1 immune response in mice. EMBO Mol Med 4(10):1126–1143
Chu CM, Cheng VCC, Hung IFN et al (2004) Role of lopinavir/ritonavir in the treatment of SARS: initial virological and clinical findings. Thorax 59:252–256
Cinatl J, Morgenstern B, Bauer G et al (2003) Glycyrrhizin, an active component of liquorice roots, and replication of SARS-associated coronavirus. Lancet 361(9374):2045–2046
Colunga Biancatelli RML, Berrill M, Catravas JD et al (2020) Quercetin and vitamin C: an experimental, synergistic therapy for the prevention and treatment of SARS-CoV-2 related disease (COVID-19). Front Immunol 11:1451
Conde Diez S, Viejo Casas A, Garcia Rivero JL et al (2019) Impact of a homeopathic medication on upper respiratory tract infections in COPD patients: results of an observational, prospective study (EPOXILO). Respir Med 146:96–105
De Meyer S, Bojkova D, Cinatl J et al (2020) Lack of antiviral activity of darunavir against SARS-CoV-2. Int J Infect Dis 97:7–10
Delang L, Abdelnabi R, Neyts J (2018) Favipiravir as a potential countermeasure against neglected and emerging RNA viruses. Antivir Res 153:85–94
Derwand R, Scholz M (2020) Does zinc supplementation enhance the clinical efficacy of chloroquine/hydroxychloroquine to win today's battle against COVID-19? Med Hypotheses 142:109815. https://doi.org/10.1016/j.mehy.2020.109815
Dhama K, Khan S, Tiwari R et al (2020a) Coronavirus disease 2019-COVID-19. Clin Microbiol Rev 33(4):e00028–e00020
Dhama K, Sharun K, Tiwari R et al (2020b) COVID-19, an emerging coronavirus infection: advances and prospects in designing and developing vaccines, immunotherapeutics, and therapeutics. Hum Vaccin Immunother 16(6):1232–1238
Do JS, Hwang JK, Seo HJ et al (2006) Nam, Antiasthmatic activity and selective inhibition of type 2 helper T cell response by aqueous extract of semen Armeniacae amarum. Immunopharmacol Immunotoxicol 28(2):213–225
Ebenezer KS, Manivannan R, Punniyamoorthy A et al (2019) Plant secondary metabolites of antiviral properties a rich medicinal source for drug discovery: a mini review. J Drug Deliv Ther 9:161–167
El-Aziz TAA, Mohamed RH, Pasha HF et al (2012) Catechin protects against oxidative stress and inflammatory-mediated cardiotoxicity in adriamycin-treated rats. Clin Exp Med 12:233–240
Falzarano D, Wit E, Rasmussen AL et al (2013) Feldmann, treatment with interferon-α2b and ribavirin improves outcome in MERS-CoV-infected rhesus macaques. Nat Med 19(10):1313–1317
Farsalinos K, Barbouni A, Niaura R (2020) Systematic review of the prevalence of current smoking among hospitalized COVID-19 patients in China: could nicotine be a therapeutic option? Intern Emerg Med 15(5):845–852
Fatima M, Zaidi NU, Amraiz D et al (2016) In vitro antiviral activity of Cinnamomum cassia and its nanoparticles against H7N3 influenza a virus. J Microbiol Biotechnol 26(1):151–159

Feng YC, Wang KC, Chiang LC, Shieh DE, Yen MH, San CJ (2013) Water extract of licorice had anti-viral activity against human respiratory syncytial virus in human respiratory tract cell lines. J Ethnopharmacol 148(2):466–473

Filippini A, D'Amore A, Palombi F et al (2020) Could the inhibition of endo-lysosomal two-pore channels (TPCs) by the natural flavonoid Naringenin represent an option to fight SARS-CoV-2 infection? Front Microbiol 11:970

Frediansyah A, Nainu F, Dhama K et al (2020) Remdesivir and its antiviral activity against COVID-19: a systematic review. Clin Epidemiol Glob Health 9:123–127

Gao J, Tian Z, Yang X (2020) Breakthrough: chloroquine phosphate has shown apparent efficacy in treatment of COVID-19 associated pneumonia in clinical studies. Biosci Trend 14(1):72–73

Garrigaa D, Headey S, Accurso C et al (2018) Structural basis for the inhibition of poxvirus assembly by the antibiotic rifampicin. PNAS 115(33):8424–8429

Gautam M, Saha S, Bani S et al (2009) Immunomodulatory activity of Asparagus racemosus on systemic Th1/Th2 immunity: implications for immunoadjuvant potential. J Ethnopharmacol 121(2):241–247

Gautret P, Lagier JC, Parola P, Meddeb L, Sevestre J, Mailhe M, Doudier B, Aubry C, Amrane S, Seng P (2020) Clinical and microbiological effect of a combination of hydroxychloroquine and azithromycin in 80 COVID 19 patients with at least a sixday follow up: a pilot observational study. Travel Med Infect Dis 2020(34):101663

Ghoke SS, Sood R, Kumar N et al (2018) Evaluation of antiviral activity of Ocimum sanctum and Acacia arabica leaves extracts against H9N2 virus using embryonated chicken egg model. BMC Complement Altern Med 18:174

Glinsky GV (2020) Tripartite combination of candidate pandemic mitigation agents: vitamin D, quercetin, and Estradiol manifest properties of medicinal agents for targeted mitigation of the COVID-19 pandemic defined by genomics-guided tracing of SARS-CoV-2 targets in human cells. Biomed 8(5):129

Global Data Healthcare (2020) Inhaled interferon-beta launches into the fight against the Covid-19 pandemic. https://www.pharmaceutical-technology.com/comment/synairgen-sng001-covid-19-trials/. Accessed 18 Apr 2020

Goel A, Singh DK, Kumar S et al (2010) Immunomodulating property of Ocimum sanctum by regulating the IL-2 production and its mRNA expression using rat's splenocytes. Asian Pac J Trop Med 3:8–12

Goswami T, Bagchi B (2020) Molecular docking study of receptor binding domain of SARS-CoV-2 spike glycoprotein with saikosaponin, a triterpenoid natural product. Preprints:10.26434/chemrxiv.12033774.v1

Grein J, Ohmagari N, Shin D et al (2020) Compassionate use of remdesivir for patients with severe Covid-19. N Engl J Med 382(24):2327–2336

Grover A, Agrawal V, Shandilya A et al (2011) Non-nucleosidic inhibition of herpes simplex virus DNA polymerase: mechanistic insights into the anti-herpetic mode of action of herbal drug withaferin a. BMC Bioinformatics 12(Suppl 13):S22

Gupta A, Khajuria A, Singh J et al (2006) Immunomodulatory activity of biopolymeric fraction RLJ-NE-205 from Picrorhiza kurroa. Int Immunopharmacol 6(10):1543–1549

Gupta S, Mishra KP, Ganju L (2017) Broad-spectrum antiviral properties of rographolide. Arch Virol 162(3):611–623

Haidari M, Ali M, Ward CS, Madjid M (2009) Pomegranate (Punicagranatum) purified polyphenol extract inhibits influenza virus and has a synergistic effect with oseltamivir. Phytomedicine 16(12):1127–1136

Harikrishnan H, Jantan I, Haque MA et al (2018) Anti-inflammatory effects of Phyllanthusamarus Schum. & Thonn. through inhibition of NF-κB, MAPK, and PI3K-Akt signalling pathways in LPS-induced human macrophages. BMC Complement Altern Med 18(1):224

Hemalatha R, Babu KN, Karthik M et al (2011) Immunomodulatory activity and Th1/Th2 cytokine response of *Ocimum sanctum* in myelosuppressedswiss albino mice. Trends Med Res 6:23–31

Hu TY, Ju JM, Mo LH et al (2019) Anti-inflammation action of xanthones from Swertiachirayita by regulatingCOX-2/NF-κB/MAPKs/Akt signaling pathways in RAW 264.7 macrophage cells. Phytomedicine 55:214–221

Huang L, Sexton DJ, Skogerson K et al (2003) Novel peptide inhibitors of angiotensin-converting enzyme 2. J Biol Chem 278:15532–15540

Hung IF, Lung KC, Tso EY et al (2020) Triple combination of interferon beta-1b, lopinavir-ritonavir, and ribavirin in the treatment of patients admitted to hospital with COVID-19: an open-label, randomised, phase 2 trial. Lancet 395(10238):1695–1704

Hussain A, Shadma W, Maksood A et al (2013) Protective effects of Picrorhizakurroa on cyclophosphamide-induced immunosuppression in mice. Pharm Res 5(1):30–35

Hussain F, Jahan N, Rahman K-U et al (2018) Identification of hypotensive biofunctional compounds of *Coriandrum sativum* and evaluation of their angiotensin-converting enzyme (ace) inhibition potential. Oxidative Med Cell Longev 2018:4643736

Ippoushi K, Azuma K, Ito H et al (2003) [6]-gingerol inhibits nitric oxide synthesis in activated J774.1 mouse macrophage and prevents peroxynitrite-induced oxidation and nitration reactions. Life Sci 73(26):3427–3437

Iuvone T, Esposito G, Capasso F et al (2003) Induction of nitric oxide synthatase expression by Withaniasomnifera in macrophages. Life Sci 72:1617–1625

Jahan I, Onay A (2020) Potentials of plant-based substance to inhabit and probable cure for the COVID-19. Turk J Biol 44(3):228–241

Jamshidi N, Cohen MM (2017) The clinical efficacy and safety of tulsi in humans: a systematic review of the literature. Evid Based Complement Alternat Med 2017(2017):9217567

Jayathirtha MG, Mishra SH (2004) Preliminary immunomodulatory activities of methanol extracts of Ecliptaalba and Centellaasiatica. Phytomed 11(4):361–365

Jiang ZY, Liu WF, Zhang XM et al (2013) Anti-HBV active constituents from Piper longum. Bioorg Med Chem Lett 23(7):2123–2127

Jiang W, Shen Q, Chen M et al (2014) Levonorgestrel releasing intrauterine system use in premenopausal women with symptomatic uterine leiomyoma: a systematic review. Steroids 86:69–78

Johnson and Johnson (2020). https://www.jnj.com/lack-of-evidence-to-support-darunavir-based-hiv-treatments-for-coronavirus. Accessed 20 Mar 2020

Joshi AJ, Aparna K, Rajagopala S et al (2016) Evaluation of immunomodulatory activity of Balacaturbhadra- Ayurvedic formulation. Indian J Nat Prod Resour 7(4):293–300

Kai H, Kai M (2020) Interactions of coronaviruses with ACE2, angiotensin II, and RAS inhibitors—lessons from available evidence and insights into COVID-19. Hypertens Res 43:648–654

Kalliantas D, Kallianta M, Karagianni CS (2020) Homeopathy combat against coronavirus disease (Covid-19). Z Gesundh Wiss 5:1–4

Kambizi L, Goosen BM, Taylor MB et al (2007) Anti-viral effects of aqueous extracts of Aloe ferox and Withaniasomnifera on herpes simplex virus type 1 in cell culture. S Afr J Sci 103:359–360

Kaur K, Kaushal S, Kaushal IG (2020) Therapeutic status of hydroxychloroquine in COVID-19: a review. J Anaesthesiol Clin Pharmacol 36(Suppl 1):S160–S165. https://doi.org/10.4103/joacp.JOACP_313_20

Kesharwani A, Polachira SK, Nair R et al (2017) Anti-HSV-2 activity of Terminaliachebula Retz extract and its constituents, chebulagic and chebulinic acids. BMC Complement Alter Med 17(1):110

Keyaerts E, Vijgen L, Maes P et al (2004) In vitro inhibition of severe acute respiratory syndrome coronavirus by chloroquine. Biochem Biophys Res Commun 323(1):264–268

Khan S, Kazmi A, Bashir N et al (2020) COVID-19 infection: origin, transmission, and characteristics of human coronaviruses. J Adv Res 16:91–98

Koch S, Pong W (2020) First up for COVID-19: nearly 30 clinical readouts before end of april, vol 20. BioCentury Inc.

Korytkowski M, Bell D, Jacobsen C et al (2003) A multicenter, randomized, open-label, comparative, two period crossover trial of preference, efficacy, and safety profiles of a prefilled, disposable pen and conventional vial/ syringe for insulin injection in patients with type 1 or 2 diabetes mellitus. Clin Ther 25(11):2836–2848

Kulshresta VK, Singh N, Shrivastava RK (1969) A study of central stimulant effect of Piper longum. Indian J Pharmacol 1(2):8–10

Kumar VP, Kuttan R, Kuttan G (1999) Effect of "rasayanas", a herbal drug preparation on immune responses and its significance in cancer treatment. Indian J Exp Biol 37:27–31

Kumar RS, Narasingappa RB, Joshi CG et al (2017) Caesalpinia Crista Linn. Induces protection against DNA and membrane damage. Phcog Mag 13:S250–S257

Lane JCE, Weaver J, Kosta K et al (2020) Safety of hydroxychloroquine, alone and in combination with azithromycin, in light of rapid wide-spread use for COVID-19: a multinational, network cohort and self-controlled case series study. medRxiv. https://doi.org/10.1101/2020.04.08.20054551

Langeland N, Holmsen H, Lillehaug JR et al (1987) Evidence that neomycin inhibits binding of herpes simplex virus type 1 to the cellular receptor. J Virol 61(11):3388–3393

Lee SH, Chung N, Kwan J et al (2007) Comparison of the efficacy and tolerability of pitavastatin and atorvastatin: an 8-week, multicenter, randomized, open-label, dose-titration study in Korean patients with hypercholesterolemia. Clin Ther 29(11):2365–2373

Leung AYH, Chan MTL, Yuen K-Y et al (2005) Ciprofloxacin decreased Polyoma BK virus load in patients who underwent allogeneic hematopoietic stem cell transplantation. Clin Infect Dis 40(4):528–537

Li Y, Liu Y, Ma A et al (2017) In vitro antiviral, anti-inflammatory, and antioxidant activities of the ethanol extract of Menthapiperita L. Food Sci Biotechnol 26(6):1675–1683

Li CC, Wang XJ, Wang HR (2019a) Repurposing host-based therapeutics to control coronavirus and influenza virus. Drug Discov Today 24:726–736

Li YH, Lai CY, Su MC et al (2019b) Antiviral activity of Portulaca oleracea L. against influenza A viruses. J Ethnopharmacol 2241:112013

Li Y, Liu X, Guo L et al (2020) Traditional Chinese herbal medicine for treating novel coronavirus (COVID 19) pneumonia: protocol for a systematic review and meta-analysis. Syst Rev 9:1–6

Li D, Wu L, Chen F, Liu X, Mei J (2022) Mechanism of Huo-Xiang-Zheng-Qi in preventing and treating COVID-19: a study based on network pharmacology and molecular docking techniques. Nat Prod Commun 17(7). https://doi.org/10.1177/1934578X221102028

Lin LL, Shan JJ, Xie T et al (2016) Application of traditional Chinese medical herbs in prevention and treatment of respiratory syncytial virus. Evid Based Complement Alternat Med 2016:6082729

Liu C, Zhu D (2020) Cyclophilin a inhibitor: potential therapeutics for the treatment of COVID-19. Ann Pharmacol Pharm 5(4):1189

Liu C, von Brunn A, Zhu D (2020) Cyclophilin A and CD147: novel therapeutic targets for the treatment of COVID-19. Med Drug Discov 7:100056. Softic Liu

Lokugamage KG, Hage A, Schindewolf C et al (2020) SARS-CoV-2 is sensitive to type I interferon pretreatment. Version 3. bioRxiv 9. https://doi.org/10.1101/2020.03.07.982264

Loutfy MR, Blatt LM, Siminovitch KA et al (2003) Interferon alfacon-1 plus corticosteroids in severe acute respiratory syndrome: a preliminary study. JAMA 290(24):3222

Lu H (2020) Drug treatment options for the 2019-new coronavirus (2019-nCoV). Biosci Trends 14:69–71

Luo H, Tang QL, Shang YX et al (2020a) Can Chinese medicine be used for prevention of Corona virus disease 2019 (COVID-19)? A review of historical classics, research evidence and current prevention programs. Chin J Integr Med 26(4):243–250

Luo P, Liu D, Li J (2020b) Pharmacological perspective: glycyrrhizin may be an efficacious therapeutic agent for COVID-19. Int J Antimicrob Agents 55(6):105995

Malik YS, Kumar N, Sircar S et al (2020) Coronavirus disease pandemic (COVID-19): challenges and a global perspective. Pathogens 9(7):519

Mani JS, Johnson JB, Steel JC et al (2020) Natural product-derived phytochemicals as potential agents against coronaviruses: a review. Virus Res 284:197989

Manoharan Y, Haridas V, Vasanthakumar KC et al (2020) Curcumin: a wonder drug as a preventive measure for COVID19 management. Indian J Clin Biochem 35:373–375
Manu KA, Kuttan G (2007) Effect of Punarnavine, an alkaloid from Boerhaaviadiffusa, on cell-mediated immune responses and TIMP-1 in B16F-10 metastatic melanoma-bearing mice. Immunopharmacol Immunotoxicol 29(3–4):569–586
McIntosh K, Hirsch MS, Bloom A (2020) Coronavirus disease 2019 (COVID-19). Up to date. https://www.uptodate.com/contents/coronavirus-disease-2019-covid-19. Accessed 5 Mar 2020
Mirzaie A, Halaji M, Dehkordi FS et al (2020) A narrative literature review on traditional medicine options for treatment of coronavirus disease 2019 (COVID-19). Complement Ther Clin Pract 40:101214
Mistry D, Pithawala M (2017) Protective effect of *Alstonia scholaris* Linn. R. Br. against bleomycin induced chromosomal damage in cultured human lymphocytes, in vitro. J Drug Chem Toxicol 41(2):162–168
Mondal S, Varma S, Bamola VD et al (2011) Double-blinded randomized controlled trial for immunomodulatory effects of Tulsi (Ocimum sanctum Linn.) leaf extract on healthy volunteers. J Ethnopharmacol 136(3):452–456
More P, Pai K (2011) Immunomodulatory effects of Tinosporacordifolia (Guduchi) on macrophage activation. Biol Med 3:134–140
Moutia M, Habti N, Badou A (2018) In vitro and in vivo immunomodulator activities of *Allium sativum* L. Evid Based Complement Alternat Med 2018(2018):4984659
Muthulakshmi M, Subramani PA, Michael RD (2016) Immunostimulatory effect of the aqueous leaf extract of Phyllanthusniruri on the specific and nonspecific immune responses of *Oreochromis mossambicus Peters*. Iran J Vet Res 17(3):200–202
Nagarkatti DS, Rege NN, Desai NK et al (1994) Modulation of Kupffer cell activity by Tinosporacordifolia in liver damage. J Postgrad Med 40(2):65–67
Nfambi J, Bbosa GS, Sembajwe LF et al (2015) Immunomodulatory activity of methanolic leaf extract of Moringaoleifera in Wistar albino rats. J Basic Clin Physiol Pharmacol 26:603–611
Ngwa W, Kumar R, Thompson D et al (2020) Potential of flavonoid-inspired phytomedicines against COVID-19. Molecules 25:2707
Nile SH, Nile A, Qiu J et al (2020) COVID-19: pathogenesis, cytokine storm and therapeutic potential of interferons. Cytokine Growth Factor Rev 53:66–70
Nose M, Terawaki K, Oguri K et al (1998) Activation of macrophages by crude polysaccharide fractions obtained from shoots of Glycyrrhizaglabra and hairy roots of Glycyrrhizauralensis in vitro. Biol Pharm Bull 21(10):1110–1112
Oakes JM, Fuchs RM, Gardner JD et al (2018) Nicotine and the renin-angiotensin system. Am J Physiol Regul Integr Comp Physiol 315(5):R895–R906
Pandey R, Maurya R, Singh G et al (2005) Immunosuppressive properties of flavonoids isolated from Boerhaaviadiffusa Linn. Int Immunopharmacol 5(3):541–553
Pang X, Cui Y, Zhu Y (2020) Recombinant human ACE2: potential therapeutics of SARS-CoV-2 infection and its complication. Acta Pharmacol Sin 41:1255–1257
Panyod S, Ho CT, Sheen LY (2020) Dietary therapy and herbal medicine for COVID-19 prevention: a review and perspective. J Tradit Complement Med 10(4):420–427
Parida MM, Upadhyay C, Pandya G et al (2002) Inhibitory potential of neem (AzidarachtaindicaJuss) leaves on dengue virus type-2 replication. J Ethnopharmacol 79:273–278
Park SH, Kyeong MS, Hwang Y et al (2012) Inhibition of LPS binding to MD-2 co-receptor for suppressing TLR4-mediated expression of inflammatory cytokine by 1-dehydro-10-gingerdione from dietary ginger. Biochem Biophys Res Commun 419(4):735–740
Patrì A, Fabbrocini G (2020) Hydroxychloroquine and ivermectin: a synergistic combination for COVID-19 chemoprophylaxis and treatment? J Am Acad Dermatol 82(6):e221
Pawlotsky JM (2020) SARS-CoV-2 pandemic: time to revive the cyclophilin inhibitor alisporivir. Clin Infect Dis 71(16):2191–2194

Pooladanda V, Thatikonda S, Bale S et al (2019) Nimbolide protects against endotoxin-induced acute respiratory distress syndrome by inhibiting TNF-α mediated NF-κB and HDAC-3 nuclear translocation. Cell Death Dis 10:81

Pooladanda V, Thatikonda S, Godugu C (2020) The current understanding and potential therapeutic options to combat COVID-19. Life Sci 254:117765

Qu D, Zheng B, Yao X et al (2005) Intranasal immunization with inactivated SARS-CoV (SARS-associated coronavirus) induced local and serum antibodies in mice. Vaccine 23(7):924–931

Rabaan AA, Al-Ahmed SH, Sah R et al (2020) SARS-CoV-2/COVID-19 and advances in developing potential therapeutics and vaccines to counter this emerging pandemic. Ann Clin Microbiol Antimicrob 19:40

Rajani J, Ashok BK, Galib P et al (2012) Immunomodulatory activity of ĀmalakiRasāyana: an experimental evaluation. Anc Sci Life 32(2):93–98

Rasool M, Varalakshmi P (2006) Immunomodulatory role of *Withania somnifera* root powder on experimental induced inflammation: an in vivo and in vitro study. Vasc Pharmacol 44:406–410

Rastogi S, Pandey DN, Singh RH (2020) COVID-19 pandemic: a pragmatic plan for Ayurveda intervention. J Ayurveda Integr Med 23. S0975-9476(20)30019-X

Reis G, Silva EASM, Silva DCM, Thabane L, Milagres AC, Ferreira TS, Dos Santos CVQ, Campos VHS, Nogueira AMR, de Almeida APFG, Callegari ED, Neto ADF, Savassi LCM, Simplicio MIC, Ribeiro LB, Oliveira R, Harari O, Forrest JI, Ruton H, Sprague S et al (2022) Effect of early treatment with Ivermectin among patients with Covid-19. N Engl J Med 386(18):1721–1731. https://doi.org/10.1056/NEJMoa2115869

Ren JL, Zhang AH, Wang XJ (2020a) Traditional Chinese medicine for COVID-19 treatment. Pharmacol Res 155:104743

Ren X, Shao XX, Li XX et al (2020b) Identifying potential treatments of COVID-19 from traditional Chinese medicine (TCM) by using a data-driven approach. J Ethnopharmacol 258:112932

Rossi EG (2020) The experience of an Italian public homeopathy clinic during the COVID-19 epidemic, March–May 2020. Homeopathy 109(3):167–168

Rothan HA, Zulqarnain M, Ammar YA et al (2014) Screening of antiviral activities in medicinal plants extracts against dengue virus using dengue NS2B-NS3 protease assay. Trop Biomed 31(2):286–296

Rothan HA, Bahrani H, Mohamed Z et al (2015) A combination of doxycycline and ribavirin alleviated chikungunya infection. PLoS One 10(5):e0126360

Roy S, Bhattacharyya P (2020) Possible role of traditional medicinal plant neem (Azadirachta indica) for the management of COVID-19 infection. Int J Res Pharm Sci 11(SPL 2):122–112

Rubab I, Ali S (2016) Dried fruit extract of Terminaliachebula modulates the immune response in mice. Food Agric Immunol 27(1):1–22

Runfeng L, Yunlonge H, Jichengd H et al (2020) Lianhuaqingwen exerts anti-viral and anti-inflammatory activity against novel coronavirus (SARS-CoV-2). Pharmacol Res 156:104761

Ryu YB, Jeong HJ, Kim JH et al (2010) Biflavonoids from Torreya nucifera displaying SARS-CoV 3CLpro inhibition. Bioorganic Med Chem 18:7940–7947

Sabouri Ghannad M, Mohammadi ASS, Faradmal J et al (2014) The effect of aqueous extract of Glycyrrhizaglabra on herpes simplex virus 1. Jundishapur J Microbiol 7(7):e11616

Sachan S, Dhama K, Latheef SK (2019) Immunomodulatory potential of *Tinospora cordifolia* and CPG ODN (tlr21 agonist) against the very virulent, infectious bursal disease virus in spf chicks. Vaccine 7:106

Sai Ram M, Neetu D, Yogesh B et al (2002) Cyto-protective and immunomodulating properties of Amla (Emblicaofficinalis) on lymphocytes: an in-vitro study. J Ethnopharmacol 81(1):5–10

Sallard E, Lescure FX, Yazdanpanah Y et al (2020) Type 1 interferons as a potential treatment against COVID-19. Antivir Res 178:104791

Salvi R, Patankar P (2020) Emerging pharmacotherapies for COVID-19. Biomed Pharmacother 128:110267

Sandoiu A (2020) Is the anti-flu drug Avigan effective in treating COVID-19? https://www.medicalnewstoday.com/articles/anti-flu-drug-effective-in-treating-covid-19. Accessed 18 Apr 2020

Santos J, Brierley S, Gandhi MJ et al (2020) Repurposing therapeutics for potential treatment of SARS-CoV-2: a review. Viruses 12(7):705

Sastry JLN, Gupta A (2014) Quantification of immunity status of dabur chyawanprash – a review part–I (experimental studies). Indian J Appl Res 4:20–24

Senapati S, Dash J, Sethi M et al (2020) Bioengineered probiotics to control SARS-CoV-2 infection. Res Ideas Outcome 6:e54802

Sengupta M, Sharma GD, Chakraborty B (2011) Hepatoprotective and immunomodulatory properties of aqueous extract of Curcuma longa in carbon tetra chloride intoxicated Swiss albino mice. Asian Pac J Trop Biomed 1(3):193–199

Sharma ML, Rao CS, Duda PL (1994) Immunostimulatory activity of Picrorhizakurroa leaf extract. J Ethnopharmacol 41(3):185–192

Sharun K, Dhama K, Patel SK et al (2020) Ivermectin, a new candidate therapeutic against SARS-CoV-2/COVID-19. Version 2. Ann Clin Microbiol Antimicrob 19(1):–23

Sheahan TP, Sims AC, Leist SR et al (2020) Comparative therapeutic efficacy of remdesivir and combination lopinavir, ritonavir, and interferon beta against MERS-CoV. Nat Commun 11:222

Shen K, Yang Y (2020) Diagnosis and treatment of 2019 novel coronavirus infection in children: a pressing issue. World J Pediatr 16(3):219–221

Shivaprasad HN, Kharya MD, Rana AC et al (2006) Preliminary immunomodulatory activities of the aqueous extract of *Terminalia chebula*. Pharm Biol 44(1):32–34

Shukla S (2010) In vivo immunomodulatory activities of aqueous extract of CaesalpiniaBonducella seed. Pharm Biol 48(2):227–230

Sies H (1996) Antioxidants in disease, mechanisms and therapy. Academic Press, New York

Singh MK, Yadav SS, Gupta V et al (2013) Immunomodulatory role of Emblicaofficinalis in arsenic induced oxidative damage and apoptosis in thymocytes of mice. BMC Complement Altern Med 13:193

Singh VK, Dwivedi P, Chaudhary BR et al (2015) Immunomodulatory effect of *Gymnem asylvestre* (r.br.) leaf extract: an in vitro study in rat model. PLoS One 10(10):e0139631

Softic L, Brillet R, Berry F et al (2020) Inhibition of SARS-CoV-2 infection by the cyclophilin inhibitor alisporivir (Debio 025). Antimicrob Agents Chemother 64:e00876–e00820

Sohrab SS, El-Kafrawy SA, Mirza Z et al (2018) Design and delivery of therapeutic sirnas: application to MERS-coronavirus. Curr Pharm Des 24(1):62–77

Sordillo PP, Helson L (2015) Curcumin suppression of cytokine release and cytokine storm. A potential therapy for patients with ebola and other severe viral infections. In Vivo 29(1):1–4

So-Yeun W, Win N, Oo WMN et al (2019) Viral protein R inhibitors from Swertiachirata of Myanmar. J Biosci Bioeng 128(4):445–449

Stebbing J, Phelan A, Griffin I et al (2020) COVID-19: combining antiviral and anti-inflammatory treatments. Lancet Infect Dis 20(4):400–402

Sunila ES, Kuttan G (2004) Immunomodulatory and antitumor activity of Piper longum Linn. andpiperine. J Ethnopharmacol 90:339–346

Suresh K, Vasudevan DM (1994) Augmentation of murine natural killer cell and antibody dependent cellular cytotoxicity activated by Phyllanthusemblica, a new immunomodulator. J Ethnopharmacol 44(1):55–60

Tang L, Ling AP, Koh RY et al (2012) Screening of anti-dengue activity in methanolic extracts of medicinal plants. BMC Complement Altern Med 12:3

te Velthuis AJW, van den Worm SHE, Sims AC et al (2010) Zn^{2+} inhibits coronavirus and Arterivirus RNA polymerase activity *in vitro* and zinc ionophores block the replication of these viruses in cell culture. PLoS Pathog 6(11):e1001176

Teixeira MZ (2020) Clinical research protocol to evaluate the effectiveness and safety of individualized homeopathic medicine in the treatment and prevention of the COVID-19 epidemic. Rev Assoc Med Bras 66(4):405–406

Thatte UM, Kulkarni MR, Dahanukar SA (1992) Immunotherapeutic modification of Escherichia coli peritonitis and bacteremia by Tinosporacordifolia. J Postgrad Med 38(1):13–15

Tindle HA, Newhouse PA, Freiberg MS (2020) Beyond smoking cessation: investigating medicinal nicotine to prevent and treat COVID-19. Nicotine Tob Res 22(9):1669–1670

Tiwari V, Shukla SY, Shukla D (2009) A sugar binding protein cyanovirin-N blocks herpes simplex virus type-1 entry and cell fusion. Antivir Res 84(1):67–75

Topic B, Tani E, Tsiakitzis K et al (2002) Enhanced maze performance and reduced oxidative stress by combined extracts of zingiberofficinale and ginkgo biloba in the aged rat. Neurobiol Aging 23(1):135–143

Tripathi MK, Singh P, Sharma S et al (2020) Identification of bioactive molecule from Withania somnifera (Ashwagandha) as SARS-CoV-2 main protease inhibitor. J Biomol Struct Dyn 8:1–14

Tutunchi H, Naeini F, Ostadrahimi A et al (2020) Naringenin, a flavanone with antiviral and anti-inflammatory effects: a promising treatment strategy against COVID-19. Phytother Res 34(12):3137–3147

Vellingiri B, Jayaramayya K, Iyer M et al (2020) COVID-19: a promising cure for the global panic. Sci Total Environ 725:138277

Venkateswaran PS, Millman I, Blumberg BS (1987) Effects of an extract from Phyllanthus niruri on hepatitis B and woodchuck hepatitis viruses: in vitro and in vivo studies. Proc Natl Acad Sci U S A 84(1):274–278

Verma H, Patil PR, Kolhapure RM et al (2008) Antiviral activity of the Indian medicinal plant extract, Swertiachirata against herpes simplex viruses: a study by in-vitro and molecular approach. Indian J Med Microbiol 26:322–326

Wang M, Cao R, Zhang L et al (2020a) Remdesivir and chloroquine effectively inhibit the recently emerged novel coronavirus (2019-nCoV) in vitro. Cell Res 30:269–271

Wang Z, Chen X, Lu Y, Chen F, Zhang W (2020b) Clinical characteristics and therapeutic procedure for four cases with 2019 novel coronavirus pneumonia receiving combined Chinese and Western medicine treatment. Biosci Trends 14(1):64–68

Weber ND, Andersen DO, North JA et al (1992) In vitro virucidal effects of Allium sativum (garlic) extract and compounds. Planta Med 58:417–423

WHO (2020). https://www.who.int/emergencies/diseases/novel-coronavirus-2019/global-research-on-novel-coronavirus-2019-ncov/solidarity-clinical-trial-for-covid-19-treatments

Wink M (2020) Potential of DNA intercalating alkaloids and other plant secondary metabolites against SARS-CoV-2 causing COVID-19. Diversity 12:175

Wu R, Wang L, Kuo HD et al (2020) An update on current therapeutic drugs treating COVID-19. Curr Pharmacol Rep 2020:1–15

Xu J, Zhang Y (2020) Traditional Chinese medicine treatment of COVID-19. Complement Ther Clin Pract 39:101165

Yamagishi Y, Ueno M, Ueno C et al (2009) Anti-herpes virus activity of polyether antibiotic CP-44161 in vivo. J Antibiot (Tokyo) 62(2):95–98

Yamamoto H, Morino K, Mengistu L et al (2016) Amla enhances mitochondrial spare respiratory capacity by increasing mitochondrial biogenesis and antioxidant systems in a murine skeletal muscle cell line. Oxidative Med Cell Longev 2016:1735841

Yamshchikov AV, Desai NS, Blumberg HM et al (2009) Vitamin D for treatment and prevention of infectious diseases: a systematic review of randomized controlled trials. Endocr Pract 15(5):438–449

Yang Y, Islam MS, Wang J et al (2020) Traditional Chinese medicine in the treatment of patients infected with 2019-new coronavirus (SARS-CoV-2): a review and perspective. Int J Biol Sci 16(10):1708

Yao TT, Qian JD, Zhu WY et al (2020) A systematic review of lopinavir therapy for SARS coronavirus and MERS coronavirus—a possible reference for coronavirus disease-19 treatment option. J Med Virol 92:556–563

Yeh CF, Chang JS, Wang KC, Shieh DE, Chiang LC (2013) Water extract of Cinnamomum cassia Blume inhibited human respiratory syncytial virus by preventing viral attachment, internalization, and syncytium formation. J Ethnopharmacol 147(2):321–326. https://doi.org/10.1016/j.jep.2013.03.010

Yi L, Li Z, Yuan K et al (2004) Small molecules blocking the entry of severe acute respiratory syndrome coronavirus into host cells. J Virol 78:11334–11339

Yu S, Wang J, Shen H (2020) Network pharmacology-based analysis of the role of traditional Chinese herbal medicines in the treatment of COVID-19. Ann Palliat Med 9(2):437–446

Zahedipour F, Hosseini SA, Sathyapalan T et al (2020) Potential effects of curcumin in the treatment of COVID -19 infection. Phytother Res 34(11):2911–2920

Zheng B, Guan Y, Tang Q et al (2004) Prophylactic and therapeutic effects of small interfering RNA targeting SARS-coronavirus. Antivir Ther 9(3):365–374

Zhou Q, Chen V, Shannon CP et al (2020) Interferon-a2b treatment for COVID-19. Front Immunol 11:1061

Ziai S, Heidari M (2015) Inhibitory effects of germinal angiotensin converting enzyme by medicinal plants used in Iranian traditional medicine as antihypertensive. J Kerman Univ Med Sci 16(2):134–144

Ziauddin M, Phansalkar N, Patki P et al (1996) Studies on the immunomodulatory effects of Ashwagandha. J Ethnopharmacol 50(2):69–76

Zumla A, Chan JF, Azhar EL et al (2016) Corona viruses-drug discovery and therapeutic options. Nat Rev Drug Discov 15:327–347

Chapter 15
In Ovo-Based Antiviral Assay for Screening of Herbal Formulations Against Influenza Viruses

Suresh Samadhan Ghoke, Richa Sood, Naveen Kumar, and Sandeep Bhatia

Abstract Multiple formats of in vitro tests and in vivo models enable us to test and screen existing antiviral drugs as well as discover new drugs. Unfortunately, these formats necessitate extensive training, are costly, are labor-intensive, and cause discomfort and distress to laboratory animals. As a result, we provide an *in ovo* model, which is a straightforward, cost-effective, and time-saving paradigm for evaluating possible antiviral drugs or herbal formulations using embryonated chicken eggs (ECEs). In this chapter, the authors describe a meticulous protocol starting from the collection of clinical samples and their transport, virus isolation in embryonated chicken eggs, as well as influenza virus infectivity titration. We also present three additional ways for assessing the antiviral activity of herbal formulations against influenza viruses using an *in ovo* model, which may be used in a variety of laboratories for screening novel drugs or antiviral compounds against influenza viruses.

Keywords Influenza · Antiviral drugs · Herbal extracts · Embryonated chicken eggs · *In ovo* model

15.1 Introduction

Many emerging and reemerging viral infections continue to pose a greater public health risk and have the potential to cause pandemics, wreaking havoc (Sood et al. 2021; Kumar et al. 2022). The risk of viral disease spreading across continents and countries has also increased. Viruses are difficult to control due to their metabolic

---

S. S. Ghoke (✉)
Department of Veterinary Epidemiology and Preventive Medicine, COVAS, Udgir, Latur, Maharashtra, India

National Institute of High Security Animal Diseases, Bhopal, Madhya Pradesh, India
e-mail: sureshghoke@mafsu.in

R. Sood · N. Kumar · S. Bhatia
National Institute of High Security Animal Diseases, Bhopal, Madhya Pradesh, India

N. Kumar et al. (eds.), *Advances in Antiviral Research*, Livestock Diseases and Management, https://doi.org/10.1007/978-981-99-9195-2_15

properties, and there are very few drugs available to treat viral diseases. To date, several chemotherapeutic antivirals have been developed for treating many viruses, including influenza virus; however, the emergence of resistance to the approved antiviral drugs has created interest in antiviral researchers to search for alternative strategies. One of the promising alternative approaches is the search for new antiviral compounds based on traditional knowledge and practices of ethnopharmacology. Therefore, natural products continue to be a valuable source of bioactive compounds for the treatment of infectious diseases, particularly those of viral origin.

Vaccination is the most effective way to prevent viral diseases, but it has its own drawbacks, including low vaccine coverage, reduced effectiveness of current vaccines, and the continued emergence and evolution of new viral strains. Several anti-influenza drugs have been developed and tested over a period of time. The first generation of influenza antivirals, such as ion channel blockers (Amantadine and Rimantidine), and the second generation of Neuraminidase Inhibitors (Oseltamivir, and Zanamivir) were successfully developed, and they have very good efficacy (Kiso et al. 2004; Nabeshima et al. 2012). Despite their excellent antiviral properties, the development of resistance to these drugs has hampered their widespread use (Sood et al. 2018). Immunization with broadly neutralizing antibodies (bNAbs) is being developed as a promising and relatively safer alternative in combating influenza viruses. However, the major challenge associated with such immunotherapies is the emergence of escape variants, compromising their efficacy (Shriver et al. 2015; Kumar et al. 2023).

Standard in vitro tests and practicable models for in vivo testing under optimized conditions are required for the testing of existing drugs as well as the discovery of new anti-influenza drugs. Current methods, such as cell-based assays, are not suitable for monitoring these classes of drugs because these assays require considerable experience and are labor-intensive. Despite having a high predictive value for clinical effectiveness, testing new antiviral drugs in animals is not well standardized, causing pain and suffering to the animals. Therefore, in comparison to in vitro tests and in vivo models, *in ovo* experiments, such as using embryonated chicken eggs (ECE), offer simple, economical, and less time-consuming models for the assessment of antiviral agents including herbal compounds against influenza viruses (Heartl et al. 2004). *In ovo* experiments are in the middle of the in vitro and in vivo studies; hence, they do not encounter direct or indirect legal and ethical aspects of animal protection (Balls et al. 1983; Sauerbrei et al. 2006).

In this chapter, the authors provide a detailed protocol for clinical sample collection, their transport, virus isolation in the embryonated chicken eggs, as well as the infectivity titration of influenza viruses. We also offer three alternative approaches for assessing the herbal extracts' antiviral activity against influenza viruses via *in ovo* model, which can be implemented in various laboratories for screening new drugs or antiviral compounds against influenza viruses.

15.2 Virus Isolation and Its Titration in the Embryonated Chicken Eggs

Embryonated chicken eggs (ECEs) are widely preferred hosts for the isolation of influenza viruses and the production of vaccines (Burnet 1940). Although the ECE system is a typical method for the propagation and isolation of influenza viruses, it can be explored to test the effectiveness of new influenza virus inhibitors or antiviral drugs.

15.2.1 Collection, Transport, and Processing of Clinical Samples

Naso-pharyngeal and cloacal swabs that were collected in suitable viral transport media (glycerol-saline transport medium: make 1:1 (v/v) dilution of autoclaved PBS and sterile glycerol and add 10× antibiotic-antimycotic solution) from the infected chickens should always be transported on ice to maintain the viability of the virus. After receiving them in the designated laboratory, they are first vortexed for 1–2 min. for uniform mixing. It is important to note that the collection tubes containing the swab samples are always opened inside the class II bio-safety cabinet because many of the influenza A viruses are of zoonotic importance (as they can easily transmit to humans). The uniformly mixed and properly marked swabs suspension are then centrifuged at 5000 rpm to remove the debris, and clear supernatant is decanted into separate tubes, labeled, and stored at −80 °C as the sample stock.

15.2.2 Inoculation of Clinical Samples Suspension in ECEs

1. Candle 10-day-old ECEs to check their viability (use of 3 ECEs is recommended for each sample). Transfer the viable ECEs into the class II biosafety cabinet and place them with blunt end up into egg trays after decontaminating the surface by wiping them with 70% ethanol-soaked cotton.
2. Mark the air sac and position of the head of the embryo in each selected ECEs with a pencil.
3. Punch a small hole in the shell of each ECE, opposite to the marking indicating the embryo head position.
4. The sample volume (200 μl) is aspirated into a syringe with a 22-gauge, $1^{1/2}$-inch needle. The needle was then inserted into the hole made at 45° angles as a short stabbing motion; the chorio-allantoic membrane (CAM) was pierced and 200 μl of sample was inoculated into the allantoic cavity.
5. The syringe is then discarded into a proper safety container.

6. The punctured holes are sealed with a drop of glue/fevicol, and the eggs are kept in a humidified (65% relative humidity) incubator at 37 °C for up to 5 days or till the death of the embryo whichever is earlier.
7. The ECEs are candled every day at 12-h intervals. All the dead embryos are chilled at 4 °C.
8. After 5 days of incubation, all the live embryos, if any, are taken out from the incubator and kept for overnight chilling at 4 °C.

15.2.3 Harvesting of the Allantoic Fluid

1. The allantoic fluid from the chilled ECEs is harvested at room temperature in a class II biosafety cabinet. The eggs are wiped with the disinfectant (70% ethanol).
2. The area marked on the outer shell above the air sac of ECEs is broken with a sterile scissor.
3. The shell membrane and CAM are pushed aside with sterile forceps, and allantoic fluid is aspirated with sterile droppers and collected into the sterile collection tubes.
4. The harvested allantoic fluid is then centrifuged at 8000 rpm for 5 min. at 4 °C to remove debris.
5. The clear supernatant is aseptically transferred into sterile collection tubes.
6. The titer of the virus present in the allantoic fluid is estimated by a hemagglutination (HA) test.

15.2.4 Titration of Virus Infectivity

1. The infectivity titer, represented as 50% egg infective dose (EID_{50}), is calculated by Reed and Muench method, 1938.
2. Embryonated chicken eggs 9–11 days old with viable embryos are selected after screening by candling.
3. A tenfold serial dilution (10^{-1} to 10^{-10}) of the stock virus is prepared in 1× PBS. Five eggs are inoculated with 100 μl of each dilution through an allantoic cavity.
4. Eggs are incubated at 37 °C under 65% humidity. The mortality in eggs is recorded for the next 5 days.
5. After 5 days of incubation, the allantoic fluid is harvested from live/dead embryos, and an HA test is performed to ascertain the presence of the virus.
6. EID_{50} is calculated by Reed and Muench's method, 1938.

15.3 Determination of Maximum Nontoxic Concentration of Antiviral Drugs of Herbal Extracts

Toxicity assay of the antiviral drugs or herbal extracts formulation (to be tested) in the proposed models (cell culture and laboratory animals) is the preliminary step for the estimation of maximum nontoxic concentration (MNTC), which is generally performed by looking at morphological alterations in cells (Bhanuprakash et al. 2008) and laboratory/experimental animals (Govorkova et al. 2007). However, toxicity studies using *in ovo* assay (using ECEs) are carried out by observing embryo mortality at different concentrations of the antiviral drugs or herbal formulations within 24 h of inoculation or until the death of the chicken embryo. Therefore, the MNTC of the drugs can be estimated using the Organisation for Economic Co-operation and Development (OECD) guidelines (OECD 2001) for safety testing, and furthermore, it can be used with different assays to assess the antiviral efficacy of drugs or herbal formulations.

15.4 Assessing Herbal Extracts Antiviral Activity Against Influenza Virus Using *In Ovo* Model

The antiviral efficacy of herbal formulations or antiviral drugs can be performed in ECEs by three distinct methods; therapeutic (time-dependent), prophylactic (dose-dependent), and virucidal (dose-dependent) assays. Furthermore, these can either be done by using a dose-dependent or time-dependent manner. In dose-dependent format, tapering dosages are injected through an allantoic cavity in ECEs, and efficacy can be measured by estimating the virus growth following 72 h post-inoculation. In contrast, in the time-dependent format, the virus suspension is inoculated through the allantoic cavity, and eggs are incubated for 2 h, which is followed by the inoculation of herbal formulation in the albumen, again starting from the higher dose and checking the growth after every 24, 48, and 72 h to assess the efficacy of the formulation. The efficacy of each method can be assessed by using a hemagglutination (HA) test, egg embryo mortality, as well as quantification of viral using reverse transcription-quantitative polymerase chain reaction (RT-qPCR).

15.4.1 Dose-Dependent Virucidal Assay

The *in ovo* model offers a convenient way to test multiple herbal extract fractions or drugs. For each treatment group, which includes herbal extracts or drugs, a minimum of five 10-day-old specific pathogen-free ECEs are required to determine their effectiveness against influenza virus. Alongside the main herbal treatment groups, the experiment should also include drug control, vehicle control, and virus control.

In this assay, a precalculated MNTC of the herbal extract is diluted two-fold to a volume of 0.1 mL. This diluted extract is then mixed with 500 EID_{50}/0.1 mL of virus and incubated at a temperature of 37 °C for 2 h. The drug control group consists of a chemotherapeutic antiviral agent incubated with 500 EID_{50}/0.1 mL of the virus, while the virus control group includes 500 EID_{50}/0.1 mL of virus incubated with 0.1 mL of PBS at 37 °C for 2 h. The vehicle control group consists of hydro-methanol incubated with 500 EID_{50}/0.1 mL of virus (Fig. 15.1). Subsequently, 0.1 ml of each virus-herbal extract mixture is inoculated separately into the allantoic cavity of 10-day-old SPF ECEs (5 each) and incubated at 37 °C for 72 h. After the 72-h incubation period, allantoic fluid is collected from the various treatment groups. The presence of the virus in the allantoic fluid can be evaluated using the hemagglutination (HA) test, as described in established methods, and quantified using real-time RT-qPCR (Payungporn et al. 2006) (Fig. 15.1).

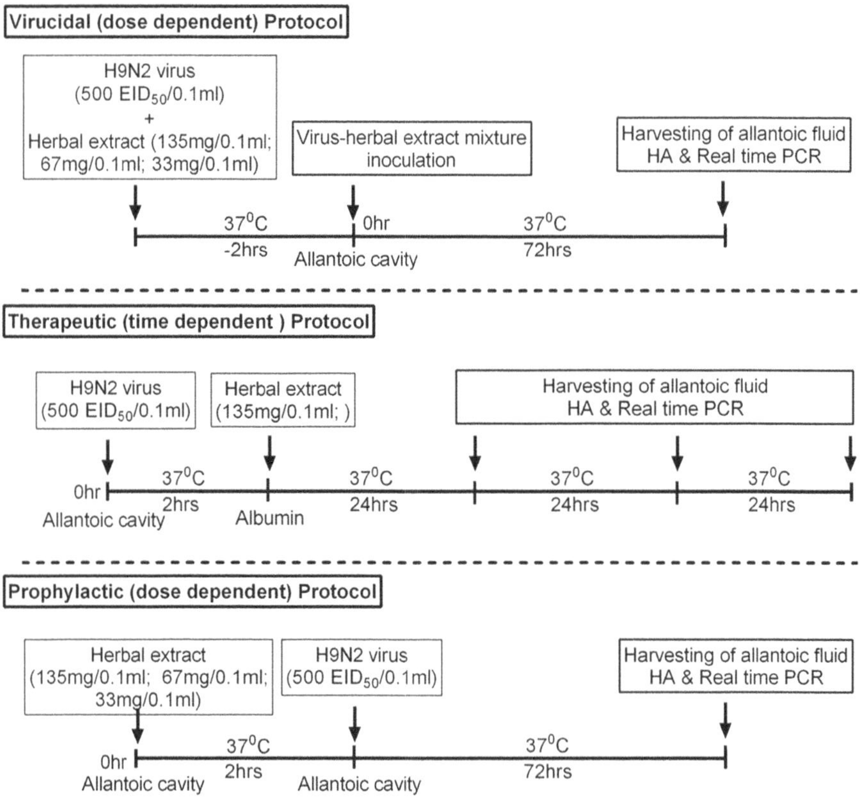

Fig. 15.1 Experimental protocol design for the evaluation of antiviral efficacy of herbal extract against influenza virus using *in ovo* model

15.4.2 Time-Dependant Therapeutic Assay

To evaluate the therapeutic efficacy in a time-dependent manner, it is important to create specific groups. In each treatment group, 500 EID_{50}/0.1 mL of virus is introduced into the allantoic cavity of 10-day-old SPF ECEs and incubated at a temperature of 37 °C for 2 h. Subsequently, the herbal extract's MNTC, which is being tested, is injected separately into the sharp pole of the albumen of these ECEs and incubated at 37 °C. This administration route has been recommended for assessing antiviral drugs, particularly when the drug's bioavailability is low (Wang et al. 2008). In the drug control group, the chemotherapeutic antiviral is administered through the albumen route 2 h after the initial inoculation. Allantoic fluid is collected from each treatment group at 24, 48, and 72 h post-inoculation (Fig. 15.1). The presence of the virus in the allantoic fluid can be determined using the HA test and quantified using real-time RT-qPCR.

15.4.3 Dose-Dependent Prophylactic Assay

To assess the prophylactic efficacy, various groups are established to investigate the impact of herbal extracts at different doses. In each treatment group, the MNTC of different plant extracts is diluted two-fold and administered into the allantoic cavity of 10-day-old SPF ECEs, followed by incubation at 37 °C for 2 h. Subsequently, 500 EID_{50}/0.1 mL of virus is introduced into the allantoic cavity of these eggs and incubated at 37 °C for 72 h (Fig. 15.1). In the drug control group, the specified antiviral drug is administered via the allantoic cavity 2 h after virus inoculation. The presence of the virus in the allantoic fluid can be detected through the HA test and quantified using real-time RT-qPCR.

15.5 Conclusion

Although the embryonated chicken egg model cannot truly replace the cell culture for screening of antiviral compounds in vitro as well as in vivo models, it has multiple advantages over the traditional system, and, therefore, it can prove as a valuable alternative approach to these methods. Some of the key advantages are (1) simple and easy to perform technique, (2) short duration of the experiment, (3) definitive endpoint, (4) low cost allowing for use in screening of a large number of antiviral compounds, and (5) standardized in vivo conditions allowing parallel testing of multiple compounds. Despite several advantages, there are some limitations with the use of *in ovo* model for antiviral drug screening, such as (1) there may be a slight difference between the experimental infection in chick embryos and the natural infection in humans, particularly regarding the mode of infection and disease progression, and may become less acceptable for predicting the efficacy against human influenza, and (2) the course of infection in the ECE route is fulminate; the therapeutic agent can only be administered once. Overall, this model offers an alternative approach for evaluating the efficacy of multiple antiviral drugs or herbal formulations against the influenza virus in the embryonated chicken egg in an economical and effective way.

Acknowledgments We would like to thank and honor the efforts of the Late Dr. Richa Sood, Principal Scientist, ICAR-NIHSAD, Bhopal, without whose supervision, persistence, and encouragement, this innovative research work on model system development would not have happened.

References

Balls M, Ridel RJ, Worden AN (1983) Animals and alternatives in toxicity testing. Academic Press, London. ISBN: 9780120774807

Bhanuprakash, V., Hoshamani, M., Balmurugan, V., Gandhale, P., Naresh, R., Swarup, D. And Singh, R. K. (2008). *In vitro* antiviral activity of plant extracts on goat pox virus replication. Indian J Exp Biol, 46: 120–127. PMID: 18335810

Burnet FM (1940) Influenza virus infections of the chick embryo by amniotic route. Exp Biol 18:353–360. https://doi.org/10.1038/icb.1941.6

Govorkova EA, Ilyushina NA, Boltz DA, Douglas A, Yilmaz N, Webster RG (2007) Efficacy of oseltamivir therapy in ferrets inoculated with different clades of H5N1 influenza virus. Antimicrob Agents Chemother 51:1414–1424. https://doi.org/10.1128/AAC.01312-06

Heartl A, Sauebrei A, Stelzner A, Wutzler P (2004) Influenza infection of the embryonated hen's egg-an alternative model for in vivo evaluation of antiviral compounds. Arzneimittelforschung 54(2):130–134. https://doi.org/10.1055/s-0031-1296948

Kiso M, Mitamura K, Sakai-Tagagawa YS (2004) Resistant influenza A viruses in children treated with oseltamivir: descriptive study. Lancet 364:759–765. https://doi.org/10.1016/s0140-6736(04)16934-1

Kumar N, Uversky VN, Tomar S, Li KSM, Chappell K, Lau SKP (2022) Editorial: emerging and re-emerging viral zoonoses. Front Microbiol 18(13):978259. https://doi.org/10.3389/fmicb.2022.978259

Kumar N, Kaushik R, Zhang KYJ, Uversky VN, Sahu U, Sood R, Bhatia S (2023) A novel consensus-based computational pipeline for screening of antibody therapeutics for efficacy against SARS-CoV-2 variants of concern including Omicron variant. Proteins. https://doi.org/10.1002/prot.26467

Nabeshima S, Kashiwegi K, Ajisaka K, Masui S, Takeoka H (2012) A randomized controlled trial comparing traditional herbal medicine and neuraminidase inhibitors in the treatment of seasonal influenza. J Infect Chemother 18(4):534–543. https://doi.org/10.1007/s10156-012-0378-7

OECD (2001) OECD guidelines for testing of chemicals. Accessed 16 Feb 2023

Payungporn S, Chutinimitkul S, Chaisingh A (2006) Single step multiplex real-time RT-PCR for H5N1 influenza A virus detection. J Virol Methods 131(2):143–147

Reed LJ, Muench H (1938) A simple method of estimating fifty percent endpoints. Hygiene 27:495–497

Sauerbrei A, Heartl A, Brandstaedt A, Schmidtke M, Wutzler P (2006) Utilization of the embryonated egg for *in vivo* evaluation of the anti-influenza virus activity of neuraminidase inhibitors. Med Microbiol Immunol 195:65–71. https://doi.org/10.1007/s00430-005-0002-x

Shriver Z, Trevejo JM, Sasisekharan R (2015) Antibody-based strategies to prevent and treat influenza. Front Immunol 6:315–320. https://doi.org/10.3389/fimmu.2015.00315

Sood R, Kumar N, Bhatia S, Chanu KV, Gupta CL, Pateriya AK, Mishra A, Khandia R, Mawale N, Singh VP (2018) Neuraminidase inhibitors susceptibility profiles of highly pathogenic influenza A (H5N1) viruses isolated from avian species in India (2006–2015). Antivir Res 158:143–146. https://doi.org/10.1016/j.antiviral.2018.08.007

Sood R, Kumar N, Gokhe SS, Pateriya AK, Bhat S, Bhatia S, Panickan S, Mishra A, Murugkar HV, Dixit R, Shrivastava D, Singh P, Tripathi M, Singh VP (2021) Identification and molecular characterization of H9N2 viruses carrying multiple mammalian adaptation markers in resident birds in central-western wetlands in India. Infect Genet Evol 94:105005. https://doi.org/10.1016/j.meegid.2021.105005

Wang JX, Zhou JY, Yang QW, Chen Y, Li X, Piao YA, Li HY (2008) An improved embryonated chicken egg model for the evaluation of antiviral drugs against influenza A virus. J Virol Methods 153:218–222. https://doi.org/10.1016/j.jviromet.2008.06.022

Chapter 16
Path to Cure Hepatitis B: How Far Are We?

Karima Abounouh, Ahd Ouladlahsen, Haya Altawalah, Ali A. Rabaan, Hind Dehbi, Fadila Guessous, and Sayeh Ezzikouri

Abstract Hepatitis B infection is caused by the hepatitis B virus (HBV), which damages the liver and can lead to chronic liver diseases such as cirrhosis and hepatocellular carcinoma (HCC). Worldwide, chronic hepatitis B virus (CHB) infection is considered a serious health problem. The United Nations (UN) has urged all nations to "fight viral hepatitis" by introducing sustainable development goals. The World Health Organization (WHO) responded by introducing an international action plan. In this chapter, although a functional cure for HBV has not yet been achieved, we highlight the progress made by global initiatives and examine the

K. Abounouh
Virology Unit, Viral Hepatitis Laboratory, Institut Pasteur du Maroc, Casablanca, Morocco

Laboratory of Genetics and Molecular Pathology, Medical School, University Hassan II, Casablanca, Morocco

A. Ouladlahsen
Faculté de Médecine et de Pharmacie, Université Hassan II, Casablanca, Morocco

H. Altawalah
Department of Microbiology, Faculty of Medicine, Kuwait University, Kuwait City, Kuwait

Virology Unit, Yacoub Behbehani center, Sabah Hospital, Ministry of Health, Kuwait City, Kuwait

A. A. Rabaan
Molecular Diagnostic Laboratory, Johns Hopkins Aramco Healthcare, Dhahran, Saudi Arabia

H. Dehbi
Laboratory of Genetics and Molecular Pathology, Medical School, University Hassan II, Casablanca, Morocco

F. Guessous
Mohammed VI University of Health Sciences (UM6SS), Casablanca, Morocco

Department of Microbiology, Immunology and Cancer Biology, School of Medicine, University of Virginia, Charlottesville, VA, USA

S. Ezzikouri (✉)
Virology Unit, Viral Hepatitis Laboratory, Institut Pasteur du Maroc, Casablanca, Morocco
e-mail: sayeh.ezzikouri@pasteur.ma

N. Kumar et al. (eds.), *Advances in Antiviral Research*, Livestock Diseases and Management, https://doi.org/10.1007/978-981-99-9195-2_16

obstacles that stand in the way of eliminating the virus while suggesting ways of overcoming them.

Keywords Hepatitis B · HBV · Chronic hepatitis B · Functional cure · Progress · Challenges

Abbreviations

AASLD	American Association for the Study of Liver Diseases
ADV	adefovir
cccDNA	covalently closed circular DNA
CHB	chronic HBV infection
DAAs	direct-acting antivirals
ETV	entecavir
HBV	hepatitis B virus
HBsAg	hepatitis B surface antigen
HBcrAg	hepatitis B core-related antigen
HCC	hepatocellular carcinoma
HTAs	host-targeting agents
IFN	interferon
IFNα	interferon-alfa
LAM	lamivudine
LDT	telbivudine
NAs	nucleos(t)ide analogs
OBI	hepatitis B virus infection
PEG-IFN	pegylated IFN
pgRNA	pregenomic RNA
rcDNA	relaxed circular DNA
RT	the reverse transcriptase domain
SAGE	Strategic Advisory Group of Experts on Immunization
TAF	tenofovir alafenamide fumarate
TDF	tenofovir disoproxil fumarate
UN	United Nations
VLPs	virus-like particles
WHO	World Health Organization

16.1 Introduction

Originally called "Australia Antigen," hepatitis B virus (HBV) was identified in 1965 by Dr. Baruch Blumberg (1977). HBV is a member of the *Orthohepadnaviridae* family, which can cause acute or chronic infections (Seeger and Mason 2000). If untreated, chronic HBV infection (CHB) leads to extensive liver damage and hepatocellular carcinoma (HCC) (Vlachogiannakos and Papatheodoridis 2013). Covalently closed circular DNA (cccDNA), an episomal structure that remains in the cell nucleus as a virus reservoir and template for viral replication, provides the molecular basis for persistent HBV infection (Fanning et al. 2019). HBV DNA can be unintentionally incorporated within the genome of the host cell as a consequence of virus replication (Fanning et al. 2019). Hepatitis B surface antigen (HBsAg) and the transcriptional regulator HBV x protein can be generated by the incorporation of HBV sequences, even though they are unable to support viral replication (Fanning et al. 2019). In 2019, almost 296 million people were chronically infected with HBV, 820,000 died from it, and an estimated 1.5 million new cases were reported each year (WHO 2022). The eradication of HBV transmission is now officially recognized as a global health objective. With the introduction of the sustainable development goals, the United Nations (UN) urged all nations to combat viral hepatitis (UN 2020). The WHO has set an interim target of reducing HBV transmission by 30% and mortality by 10% by 2020, with long-term goals of 90% and 65% by 2030 (WHO 2016). Interferon (IFN) and nucleos(t)ide analogs (NA), two currently available antiviral drugs, effectively reduce HBV replication, but do not eradicate HBV, and even in patients whose virus is completely suppressed, they do not totally eliminate the risk of HCC (Lin et al. 2007; Papatheodoridis et al. 2017; Su et al. 2016).

To increase access to diagnostic and current treatments, a number of unmet clinical needs must be handled. This chapter showcases recent accomplishments and discusses obstacles that need to be addressed to completely eliminate HBV infection.

16.2 Hepatitis B Virus Vaccines

16.2.1 Vaccine History

Since 1982, a safe and effective HBV vaccine has become commercially available. Maurice Hilleman, an American microbiologist at Merck Pharmaceuticals, was responsible for developing the vaccines. Based on the work of Dr. Baruch Blumberg and Dr. Wolf Szmuness, Hilleman created a viral subunit vaccine from the cell membrane surface proteins of the hepatitis B virus instead of using the entire virus (Tulchinsky 2018). From the plasma of chronic HBV-infected volunteers, HBsAg subvirion particles (22-nm particles) were harvested to produce these vaccines (Pattyn et al. 2021). A combination of different treatments (pepsin, urea, and formaldehyde) was used to purify and inactivate any infectious particles in the plasma (Pattyn et al. 2021). First manufactured HBV vaccines were Heptavax B (Merck) and Hevac B (Institut Pasteur) (van Beers et al. 1983), and they were used in immunization programs to target high-risk groups. In a number of randomized, double-blind, placebo-controlled clinical trials, vaccine efficacy was 80–100%, even in patients with anti-HBs concentrations greater than or equal to 10 mIU/mL, and even if anti-HBs concentrations dropped to less than 10 mIU/mL. These studies targeted high-risk individuals (hemodialysis patients, men who have sex with men (MSM), and healthcare workers) (Crosnier et al. 1981; Desmyter et al. 1983; Francis et al. 1982; Hadler et al. 1986; Jack et al. 1999; Stevens et al. 1984; Szmuness et al. 1980, 1982). By 1986, plasma-derived vaccines had been completely replaced by a genetically engineered HBV vaccine containing recombinant HBsAg (McAleer et al. 1984; Nwokediuko 2011; Szmuness et al. 1980). This second-generation HBV vaccine allowed HBsAg to be expressed in other organisms, leading to the production of unlimited vaccine supplies (Emini et al. 1986; Kwon and Lee 2011). Hepatitis B recombinant vaccines revolutionized vaccinology since they were the first vaccines to use virus-like particles (VLPs). The most used recombinant vaccines are yeast-derived, where genetically engineered yeast cells (*Saccharomyces cerevisiae*) containing S gene are manufactured to express HBsAg protein (McAleer et al. 1984). When HBsAg self-assembles into VLPs, the product is used to harvest a protective antiviral vaccine against HBV infection. In healthy newborn infants, children, and young adults, HBV immunization has been shown to insure a seroprotection greater than 95% (Pattyn et al. 2021). Adults older than 40 years failed to achieve the same seroprotection, and the response to HBV vaccination dropped to 60–70% in people over 60 years (Pattyn et al. 2021). A lower response rate might also occur in the presence of HIV infection, chronic disease, genetic factors, smoking, and obesity vaccines (Pattyn et al. 2021). Because of their production in yeast, HBsAg recombinant particles lack HBsAg pre-S domain and glycosylation (Shouval 2003; Zuckerman 2006). Mammalian cell-derived vaccines (e.g., continuous mammalian cell line, like the mouse c127 clonal cell line) are manufactured by transfecting mammalian cells with 2–3 gene coding of HBV envelope proteins, which leads to the production of glycosylated pre-S1 and pre-S2 proteins, besides HBsAg protein (Shouval 2003; Zuckerman 2006). These vaccines (including pre-S1 and pre-S2 proteins) and some new adjuvant vaccines have achieved a greater and more rapid anti-HBs response than second-generation vaccines, particularly in immunocompromised individuals and nonresponders to the main HBsAg protein (Rendi-Wagner et al. 2006; Shouval et al. 2015; Zuckerman et al. 2001). Although they are effective, the high cost of producing them limits their availability and their inclusion in universal immunization programs.

16.2.2 Worldwide HBV Vaccination

16.2.2.1 The Strategies Employed

The aim of HBV immunization programs is to make sure that every person at risk for HBV infection can get access to vaccination. The first adopted strategy to control HBV through vaccination was to target high-risk groups. Unfortunately, this approach was not effective considering that high-risk individuals were the hardest to reach and infected before being vaccinated (Kane 1996). In addition, poor coordination and consent resulted in low hepatitis B vaccine coverage after three doses in most of the high-sensitivity groups (Pattyn et al. 2021). Moreover, due to their indistinguishable risk factors, 30% or more persons with acute HBV infection can be dismissed by vaccination targeting high-risk persons (Pattyn et al. 2021). Since the first strategy hardly had any influence on HBV disease control, a more methodical approach focused on a more compromised age cohort was necessary. When infected at birth or in infancy, children are more likely to progress to chronic HBV infection and other HBV-associated liver diseases when they are older. In fact, 90% of infected newborns (<1 year) and 30% of infected children aged 1 to 4 years contract a chronic HBV infection, as opposed to 5% of individuals infected in adulthood (Michel et al. 2015). Furthermore, another study showed that around 25% of chronically infected infants are susceptible to death from cirrhosis and HBV-related liver cancer, and the risk is greater if the infection occurs at birth (Michielsen et al. 2005). HBV infection during the perinatal phase or early childhood accounts for 21% and 48% of deaths, respectively (Goldstein et al. 2005).

The World Health Organization, jointly with the Strategic Advisory Group of Experts on Immunization (SAGE), proposed global recommendations to prevent and control HBV, and they also urged all countries to include HBV vaccine by 1997 in their national immunization programs (Pattyn et al. 2021). By 2017, the WHO implemented an updated worldwide vaccination strategy insisting on giving all children three doses of HBV vaccine and recommended all countries to make this strategy a standard in their national vaccination programs (Table 16.1) (WHO 2017). To prevent perinatal transmission, all newborns should receive a dose of HBV vaccine within 24 hours of delivery (Schillie et al. 2015). To ensure good execution of birth vaccination, it is essential to provide good maternal and child care with the help of proficient health workers when administering the vaccine (Pattyn et al. 2021). Furthermore, it is important to find safe vaccination approaches for infants delivered at home. The WHO also demands that children and young adults be vaccinated at birth since the threat of developing a chronic infection is higher in these age-groups (Pattyn et al. 2021). In 2018, only 42% of newborns benefited from HBV birth vaccination worldwide; this poor coverage is one of the reasons why chronic HBV infection is persisting and becoming a bigger burden (WHO

Table 16.1 Recommended hepatitis B vaccine schedule for different age-groups

Age-group	Schedule	Dosage	Vaccine
Children (birth-10 years)	Birth, 1–2, and 6–18 months	3 doses (0.5 mL)	Engerix-B (GlaxoSmithKline) or Recombivax HB (Merck)
Adolescents (11–19 years)	0, 1–2, and 4–6 months	3 doses (0.5 mL)	Engerix-B (GlaxoSmithKline) or Recombivax HB (Merck)
Adults (20 years and older)	0, 1, and 4–6 months	3 doses (1.0 mL)	Engerix-B (GlaxoSmithKline) or Recombivax HB (Merck)
18 years and older	0, 1, and 6 months or 0, day 7, day 21–30, and 12 months	3 or 4 doses (1.0 mL)	Twinrix (GlaxoSmithKline)

According to: Schillie et al. (2018)

2019). Where resources are available, adults at high risk of HBV infection and unvaccinated individuals should benefit from catch-up vaccination (WHO 2017).

16.2.2.2 Impact of Global HBV Vaccination

In recent years, many parts of the world have presented irrefutable evidence of the effectiveness of HBV vaccination programs. HBV vaccination was included in 189 (97%) nations' national immunization programs by the end of 2019 (WHO 2019).

The percentage of children receiving the recommended three doses of HepB vaccine during childhood has risen from 1% in 1990 to 85% in 2019, reflecting the considerable progress made worldwide (WHO 2019). Widespread HBV vaccination explains the low frequency of chronic HBV infection in children under five, which has fallen from 4.7% before vaccination to less than 1% in 2019 (Cui et al. 2017; Pattyn et al. 2021). Hepatitis B immunization of neonates and babies between 2000 and 2019 averted 22 million fatalities globally, avoiding one death for every 13 individuals who received the vaccine (Guo et al. 1988). The removal of financial barriers (about $1 per dose) and the logistical support provided by the Global Alliance for Vaccination and Immunization (GAVI), which has allowed the world's low-income countries to access the vaccine since 2001, were two major measures that made decreasing death rates possible (Nayagam et al. 2016; Pattyn et al. 2021). According to mathematical simulations, HBV vaccination programs have prevented 210 million additional HBV infections worldwide since they were launched (Nayagam et al. 2016). The carrier rates and the severity of HBV-related sequelae, including HCC, were significantly reduced in nations that accepted the guidelines (Pattyn et al. 2021). Among medical workers, HBV immunization was quite effective at preventing both chronic infection and HBV infection (Meireles et al. 2015). Nevertheless, there are still a lot of differences between WHO areas. While the African region (75%), the Eastern Mediterranean region (80%), and the European region (81%) have poorer coverage, the Southeast Asian region (87%), the American region (89%), and the Western Pacific region (90%) are already above the global average (Hutin et al. 2018). Improvements in HepB vaccination over the past few decades are expected to avert an additional 16 million deaths from 2020 to 2030 since HBV infection deaths often occur in mid- to late adulthood (Kao 2021).

Despite efforts to implement global vaccination programs, chronic HBV infection continues to pose a serious threat to global public health. Clinical investigations have revealed that none of the potential therapeutic vaccination candidates available today are effective in the management of persistent HBV infection (Lim et al. 2019).

16.3 HBV Therapeutics

16.3.1 Current Therapies

The purpose of CHB therapy is to prolong life, suppress viral transmission, and increase the quality of life by delaying the evolution and progression of severe liver diseases and HCC (Abounouh et al. 2019; European Association For The Study Of The Liver 2012). Over the past two decades, significant progress has been made in the treatment of CHB. To date, nine medications have been recognized for treating CHB, interferon-alfa (IFNα) in its conventional and pegylated IFN (PEG-IFN) form, and monotherapy with seven nucleos(t)ide analogs (NAs): lamivudine (LAM), telbivudine (LDT), adefovir (ADV), entecavir (ETV), tenofovir disoproxil fumarate (TDF), tenofovir alafenamide fumarate (TAF), and besif (only in Korea) (European Association for the Study of the Liver. Electronic address & European Association for the Study of the Liver 2017; Sarin et al. 2016; Terrault et al. 2018; Trepo et al. 2014). One year's treatment with PEG-IFN offers the possibility of immune-mediated regulation of HBV infection, higher rates of HBe seroconversion, and the possibility of antiviral suppression, with elimination of HBsAg in a proportion of patients who retain untraceable HBV DNA (Schinazi and Asselah 2017). Unfortunately, not only should PEG-IFN be administered through subcutaneous injection, it is also associated with common side effects like depression; PEG-IFN is also contraindicated during pregnancy in individuals with cirrhosis or autoimmune diseases (Schinazi and Asselah 2017). Due to its significant drawbacks such as low response rates and high risk of adverse effects, many patients are unwilling or ineligible to take it (European Association for the Study of the Liver. Electronic address & European Association for the Study of the Liver 2017; Lampertico et al. 2015; Papatheodoridis et al. 2008; Terrault et al. 2018). While NAs inhibit HBV by direct antiviral activity, over 95% of patients undergoing treatment with the more recent, very effective NAs, tenofovir disoproxil fumarate (TDF) and entecavir (ETV), attain virological recovery (Schinazi and Asselah 2017). In the past, hepatic regression and HCC incidence have previously been linked to HBV DNA inhibition with long-term lamivudine (LAM) (Lai et al. 1998; Liaw et al. 2004). Possible renal toxicity (particularly with TDF) and drug resistance are two side effects of long-term NA treatment that could be very concerning (Schinazi and Asselah 2017). Whereas resistance used to be a problem with older, less effective NAs such as LAM and adefovir ADV, it is now much less of a problem with the highly potent NAs TDF and ETV (Schinazi and Asselah 2017). NAs are taken orally, and their tolerability is good; it is still uncertain whether these medications are safe when used for lifelong therapy.

Currently, antiviral therapies primarily aim to enhance health outcomes through virologic and biochemical responses (Naggie and Lok 2021). A biochemical response is a significant objective to be attained; it is described as regularization of serum alanine aminotransferase with a reduction in HBV viral load. A virologic response to IFN-based medication is characterized by a serum HBV DNA level of less than 2000 IU/mL when tested six months into therapy and at the completion of treatment (Lee et al. 2020). According to assays with a lower limit of detection of 10–20 IU/mL in the blood, a nonsignificant HBV DNA level throughout NA therapy is considered to be a virologic response (Lee et al. 2020). A definitive recovery is considered to be the loss of long-lasting HBsAg with the elimination of serum HBV DNA and the absence of cccDNA. According to tests with a lower limit of HBsAg detection (0.05 IU/mL), a practical cure is characterized by persistent HBsAg elimination or seroconversion. This is an uncommon phenomenon in the course of Crohn's disease and may be linked to a reduced risk of HCC (Cornberg et al. 2020; Tout et al. 2020). Despite their positive effects, these treatments rarely succeed in completely eliminating the virus, and HBsAg loss is still uncommon (European Association for the Study of the Liver. Electronic address & European Association for the Study of the Liver 2017). In order to sustain viral suppression and avoid the virological resurgence that typically happens following therapy withdrawal, an ongoing treatment is necessary (Anderson et al. 2019). In regions where endemicity is high, the cost of lifelong treatment and monitoring is a major economic issue. A different approach to maximize treatment effectiveness and raise the likelihood of clinical recovery is by combining NAs and Peg-IFN therapy. Knowing that NAs lower viral load and therefore improve the immune response to IFN, it is reasonable to predict that progressive co-administration of NAs and IFN will result in a better therapeutic response than each medication taken separately (Yang and Kao 2015).

In the upcoming years, novel therapy strategies encompassing drugs that target different phases of the HBV life cycle and immunomodulators will be authorized for clinical use. The hope is that the integration of these potential approaches, with the antiviral regiments already in use, will work synergistically to improve host immune responses and eradicate cccDNA, finally curing HBV infection completely.

16.3.2 Future Therapies

As mentioned before, currently available drugs have limited curative ability and fall short to completely eradicate HBV virus by ensuring the absence of HBsAg. Hence, the purpose of novel treatment approaches is to produce a "functional cure" for chronic hepatitis B (CHB) that will eliminate HBV DNA in serum, stabilize liver enzymes, and improve liver histology. In recent years, there has been growing interest in the development of innovative host-targeted agents (HTAs) and direct-acting antivirals (DAAs) for HBV infection (Ezzikouri et al. 2020; Roca Suarez et al. 2021). DAA-based therapy approaches can be categorized according to HBV viral cycle characteristics and the process they target (Table 16.2), in particular, medications that interfere with HBV replicative cycle, especially those that block entrance (Gallay et al. 2019; Watashi et al. 2014; Wedemeyer et al. 2020), HBsAg secretion (Bazinet et al. 2020), capsid assembly/disassembly (Zhao et al. 2019; Zoulim et al. 2020), and reverse transcriptase (Roca Suarez et al. 2021). Conjointly, there are

Table 16.2 Current anti-HBV cures under clinical review

Target	Drug name	Mechanism	Company	Clinical trial number	Clinical phase
Viral entry inhibitors	Myrcludex B (Bulevirtide)	Blocks NTCP	MYR GmbH, Germany	NCT03852433	II
	CRV431	Blocks NTCP and protein folding	Hepion Pharmaceuticals, USA	NCT03596697	I
Viral expression inhibitors	JNJ 3989 (ARO-HBV 1 & 2)	siRNA targets HBV transcripts	J&J with Arrowhead, USA	NCT03365947 NCT04439539 NCT04535544	I/II II II
	VIR-2218	siRNA targets HBV transcripts	Vir Biotech, USA	NCT04507269 NCT04412863	II II
	ARC-520	siRNA targets HBV transcripts	Arrowhead Pharmaceuticals, USA	NCT02577029	II
	RG6346 (DCR-HBVS)	siRNA targets HBV transcripts	Dicerna with Roche, Switzerland	NCT03772249	I
	GSK3228836/ GSK33389404	AO targeting HBV transcripts	Ionis with GlaxoSmithKline, USA	NCT04449029 NCT02981602	II
Capsid assembly inhibitor	GLS4 (ritonavir)	Core protein binding	Sunshine Lake Pharma Co, Ltd., China	NCT04147208	II
	JNJ-6379	Core protein binding	Janssen, Ireland	NCT03361956	II
	ABI-HB0731 (vebicorvir)	Core protein binding	Assembly Biosciences, USA	NCT04454567	II
	ABI-H2158	Core protein binding	Assembly Biosciences, USA	NCT04398134	II
HBsAg secretion inhibitors	REP 2139 and REP 2165	HBsAg binding	Replicor Inc., Canada	NCT02565719	II
Nucleos(t) ide analogs	HS-10234	Polymerase inhibitor	Jiangsu Hansoh Pharmaceutical Co, Ltd	NCT03903796	
Innate immunity activators	GS-9688 (selgantolimod) GS-9620 (vesatolimod)	TLR8 agonist TLR7 agonist	Gilead Sciences, USA	NCT03615066 NCT02166047	II II
	RO7020531	TLR7 agonist	Roche, Switzerland	NCT02956850	I

(continued)

Table 16.2 (continued)

Target	Drug name	Mechanism	Company	Clinical trial number	Clinical phase
Adaptive immunity activators	ASC22 (envafolimab)	Anti-PD-L1 antibody	Ascletis Pharma, PR China	NCT04465890	II
	INO-1800	Therapeutic vaccine: DNA plasmids	Inovio, USA	NCT02431312	I
	TG1050	Therapeutic vaccine: HBV proteins	Transgene, France	NCT02428400	I
	HepTcell	Therapeutic vaccine	Altimmune, USA	NCT02496897	I
	GS-4774	Therapeutic vaccine: DNA vaccines	GlobeImmune with Gilead, USA	NCT02174276	II
	JNJ-64300535	Therapeutic vaccine: DNA vaccines	Janssen, Ireland	NCT03463369	I
cccDNA inhibitors	TALENs	cccDNA disruption	–	–	Preclinical
	CRISPR-Cas9	cccDNA disruption	–	–	Preclinical

Abbreviations: *NTCP* sodium taurocholate co-transporting polypeptide, *siRNA* small-interfering RNA, *HBV* hepatitis B virus, *AO* antisense oligonucleotides, *HBsAg* hepatitis B surface antigen, *TLR8* Toll-like receptor 8, *TLR7* Toll-like receptor 7, *PD-L1*: programmed death-ligand 1, *cccDNA* covalently closed circular DNA

drugs that target HBV gene expression by reducing the levels of viral transcripts and antigens (Gane et al. 2020b, c; Mueller et al. 2018). The intracellular pool of cccDNA is indirectly targeted by both pharmacological groups by focusing on viral expression, HBsAg level reduction, and restoring antiviral immune responses. Medications that target virus particles have been the primary focus of antiviral drug research, and it is supported by the idea that these substances are not as prone to interact negatively with human molecules and therefore stop causing toxicity (Agarwal et al. 2020). Nevertheless, a number of methods aimed at boosting innate and/or adaptive immunity are also being explored, as the immune system plays a crucial role in controlling HBV infection (Ezzikouri et al. 2020; Xia and Liang 2019). This strategy is founded on the fact that off-target and side effects might emerge regardless of how virus-specific the design of DAAs may appear (Janssen et al. 2018). The idea is that medication resistance frequently develops following prolonged use of DAAs (Agarwal et al. 2020). Moreover, the restriction on treatment design is imposed by the tiny HBV genome. Thus, HTAs seem to provide a solution to these dilemmas, and many of them are already being clinically evaluated (Daffis et al. 2021; Gane et al. 2020a; Mueller et al. 2018; Pardoll 2012). The more we understand the mechanisms responsible for HBV resistance, the more obvious it becomes that HBV eradication will certainly require combined therapies.

The combination of treatments must achieve total suppression of HBV replication, *de novo* infection, and circulating HBsAg levels, while also strengthening the immune system to strengthen and sustain HBV-specific adaptive responses. Since this approach showed no improvement above monotherapy when combining NAs and IFN, it is likely that the next strategy is to combine the effects of DAAs and HTAs. For example, combinations of DAAs that target the HBV replicative cycle with those that target HBV expression can increase HBV-specific immune turnover in CHB patients (Janssen et al. 2020). Another combination includes NAs and new immunological therapies that have been proposed as potential options for the treatment of CHB patients (Liu et al. 2014b).

To assess the effect of these therapies, a strong scientific foundation is required to monitor potential drug-drug interactions and begin testing in individuals without severe liver disease (Anderson et al. 2019).

16.4 Ending HBV Infection: Limitations

16.4.1 Occult Hepatitis B Infection (OBI)

Individuals with undetectable HBsAg levels are described as having occult hepatitis B virus infection (OBI), which is characterized by the existence of replicative HBV patterns in the liver, with or without circulating HBV DNA (HBsAg) (Raimondo et al. 2019). Clearance of HBsAg is typically viewed as complete recovery from HBV infection months after acute HBV infection in adults or decades after chronic HBsAg carriage (Lok et al. 2017). Independently of the disappearance of HBsAg, the persistence of cccDNA in long-lived hepatocytes is an obstacle to the eradication of HBV infection and can lead to OBI (Caviglia et al. 2018). It suggests that even with effective host immune control, HBV infection could persist. According to recent reports, the prevalence of OBI in the general population ranges from <1% to <18% (Samadi et al. 2020). Serum HBV DNA levels are typically low and only sometimes found, which prevents liver disease (Cacciola et al. 1999; El Chaar et al. 2010; Huang et al. 2012; Kazemi-Shirazi et al. 2000; Spreafico et al. 2015). Patients could still be at some risk, as OBI may be linked to HCC progression in patients with chronic hepatitis C, whereas other studies have not shown such a link (Shetty et al. 2008; Squadrito et al. 2013; Wang et al. 2018). The question of whether OBI could hasten the development of cirrhosis and HCC in patients with other chronic liver illnesses is still debatable. A possible danger with OBI persons is that HBV can be transmitted through blood transfusion, although by applying nucleic acid testing to filter blood products, the risk could be reduced (Mosley et al. 1995; Raimondo et al. 2013; Satake et al. 2007). When patients are on robust immunosuppressive therapies, such as B-cell-targeting drugs, HBV resurgence in the OBI is increasingly reported (Cheng et al. 2003). Though prophylactic antiviral medication reduces the likelihood of HBV recurrence, the most effective way to stop HBV reactivation in OBI persons is still unknown (Reddy et al. 2015). Therefore, to provide a practical guideline to enhance medical monitoring of OBI, more research is required.

16.4.2 Persistence of the Covalently Closed Circular DNA (cccDNA)

The cccDNA is an essential intermediate for viral replication and HBV genome's persistent form. Following viral entry, partially double-stranded relaxed circular DNA (rcDNA) is carried to the hepatocyte nucleus, where it is transformed into a covalently closed circular minichromosome (cccDNA) (Wei and Ploss 2021). The transcriptional template for HBV mRNAs and pregenomic RNA (pgRNA), which starts viral replication, is the stable episomal cccDNA minichromosome (Tu et al. 2021). The nucleocapsid that encapsulates pgRNA and enables the reverse transcriptase domain (RT) to produce rcDNA is created by newly generated HBc (Tu et al. 2021). The nuclear pool of cccDNA can be restored by either encapsidated rcDNA being recycled back to the nucleus where it is converted into cccDNA, or being encased and discharged from the cells to form infectious Dane particles to infect new cells (Wei and Ploss 2021). Although cccDNA levels in the nucleus are very constant, it is still unclear how its individual molecules are kept alive. Since cccDNA is not attached to chromosomes and does not replicate in a semi-conservative manner, it is still unclear what will occur when cells divide (Dandri and Petersen 2020). According to a static model for cccDNA maintenance, once nuclear rcDNA importation and rcDNA to cccDNA conversion begin to decline, cccDNA maintenance is accomplished by suppressing *de novo* cccDNA production (Dandri et al. 2000; Lutgehetmann et al. 2010; Zhu et al. 2001). In support of this idea, the nuclear pool is only slightly impacted by the use of NAs, which indirectly affects cccDNA levels by preventing the creation of rcDNA from scratch (Lai et al. 2020; Werle-Lapostolle et al. 2004). Furthermore, studies in CHB patients receiving NA-based therapy have suggested that it takes many years for cccDNA to completely leave the body, indicating that nuclear import of newly generated rcDNA plays a limited role in preserving the cccDNA pool and that nuclear cccDNA is extremely stable (Lai et al. 2020; Werle-Lapostolle et al. 2004; Zhu et al. 2001). The latest findings demonstrate that low levels of HBV replication continue following NA therapy, which may indicate that cccDNA pool is maintained despite treatment (Boyd et al. 2016; Gordon et al. 2013). Similarly, in chimeric mice, serum from NA-treated CHB patients with reduced viremia is still contagious (Burdette et al. 2020). Stable nuclear copy numbers are maintained by continuous degradation and *de novo* synthesis of cccDNA, which is consistent with the need for *de novo* infections to maintain the cccDNA pool (Ko et al. 2018). Uncertainty surrounds the underlying process of cccDNA clearance and preservation; cccDNA destruction may take place in conjunction with or separately from infected hepatocyte death (Guidotti et al. 1999; Mason et al. 2005; Murray et al. 2005; Summers et al. 2003; Wieland et al. 2004; Zhou et al. 2000). The pinnacle of HBV research and drug development to eradicate the virus focuses mainly on cccDNA inactivation or even degradation.

It would be necessary to use procedures that are not standardized and liver tissue that is not always accessible to measure the concentration of cccDNA, and that is why serum indicators that could substitute for cccDNA transcriptional activity have been created (Chan et al. 2011). Previous research demonstrated that quantitative HBsAg levels reflect the transcriptional activity of cccDNA and correlate better with cccDNA concentration than serum HBV DNA levels since the correlation in HBeAg-negative patients is weaker than in HBeAg-positive patients, mainly as a result of the possibility that a greater percentage of circulating HBsAg in HBeAg-negative patients comes from integrated HBV DNA rather than cccDNA (Chan et al. 2011; Lin et al. 2010; Thompson et al. 2010). According to new findings, serum HBV RNA levels may serve as a more accurate substitute for cccDNA concentration, and it is thought that the circulating HBV RNA is pgRNA that has been partially reverse-transcribed and encapsidated in virus-like particles (Giersch et al. 2017; Hu and Liu 2017; Huang et al. 2018; Wang et al. 2016). An additional indicator is hepatitis B core-related antigen (HBcrAg), which is made up of hepatitis B core antigen, HBeAg, and p22 core-related antigen, all of which are viral antigens produced from the pre-Core/Core gene (Kimura et al. 2005; Kimura et al. 2002). Numerous studies have demonstrated that serum HBV DNA or HBsAg levels do not relate with cccDNA concentration as HBcrAg levels do (Carey et al. 2020; Chen et al. 2019; Suzuki et al. 2009; Testoni et al. 2019; Wang et al. 2019). Nevertheless, the outcomes from HBeAg-positive individuals are less useful due to the poor sensitivity of the existing HBcrAg assays, which also detect HBeAg. The discovery that cccDNA is not the only source of circulating HBsAg adds another layer of complexity; indeed, in HBeAg-negative patients, integrated HBV DNA may actually be a better significant source of HBeAg (Wooddell et al. 2017). As a result, a therapy could have made the cccDNA transcriptionally inactive and stopped it from making HBsAg, but this does not address the urgent need for a practical cure since HBsAg is still being generated from integrated HBV DNA. Contradictorily, a different treatment could perhaps appear to clear HBsAg when in reality it is still being produced from cccDNA.

16.4.3 Hepatitis B Virus Reactivation

Often, in association with drugs that influence the host's immune system, the previously dormant or inactive hepatitis B virus can wake up (Loomba and Liang 2017). Reactivation can happen in people who have recovered from a previous infection, as well as in asymptomatic persons with CHB infection (Loomba and Liang 2017). Given that a growing number of patients take medicines that could weaken the host immune system enough to trigger HBV reactivation, this HBR issue has gained major attention. Three crucial factors affect the threat of HBR: basic HBV infection level (Borentain et al. 2010; Lau et al. 2002); host characteristics (Yeo et al. 2000, 2003, 2004); and kind of exposure to treatment (Abramson and Chung 2014; Bessone and Dirchwolf 2016; Cheng et al. 2003; Kim et al. 2010; Mendez-Navarro et al. 2011; Yeo et al. 2009). The following criteria describe HBR in each patient category according to the American Association for the Study of Liver Diseases (AASLD) most recent proposals on the prevention, diagnosis, and treatment of chronic HBV (Terrault et al. 2018). HBR is diagnosed in HBsAg-positive patients when there is either: (1) An HBV DNA level $\geq$ 2 log or 100 times higher than the reference point; (2) HBV DNA $\geq$ 3 log, or 1000 IU/mL in patients with previously untraceable HBV DNA; or (3) HBV DNA $\geq$ 4 log, or 10000 IU/mL, if the standard level is unavailable (Terrault et al. 2018). The indications for patients with anti-HBc positivity and HBsAg negativity involve the following: (1) emergence of measurable levels of HBV DNA; (2) return of HBsAg, with the latter perhaps indicating more severe effects of HBR (Terrault et al. 2018).

HBV releases nucleocapsids after entering the hepatocytes, including partly double-stranded viral genome that is repaired into full-length covalently closed circular DNA (cccDNA); even decades after recovery from HBV infection, cccDNA is persistent in the infected hepatocyte and may linger in a dormant condition as a possible repository for HBR (Rehermann et al. 1996).

The prevention of liver-related morbidity and preserving the uninterrupted use of immunosuppressive or chemotherapeutic agents are two major objectives for HBR control. The best way to achieve these goals is to prevent the onset of HIV infection. It is difficult to prevent and treat UC, partly because not all people infected with HBV are diagnosed or aware of their condition and partly because there is no cure for HBV infection (Kao 2021). Although efforts are being made to find a functional cure for HBV infection, HB remains important because its recurrence suggests the persistence of cccDNA.

16.4.4 Evaluation of the Disease's Phase

HBV has an unexpected and complicated natural history, with associated diseases ranging from generally mild to fatal. The illness can be roughly split into different stages: immune reactive, low replicative inactive HBsAg carriers, high replicative, low inflammatory, and reactivation stages (Wu and Chang 2015). Liver damage caused by CHB can vary from minor fibrosis to cirrhosis, which can cause a number of possibly fatal consequences (Desalegn et al. 2017). The prioritization of CHB patients for antiviral medication can be aided by reliable noninvasive testing (Desalegn et al. 2017). Identifying further risks and treatment recommendations requires more detailed molecular and immunological evidence. Despite a lower viral load, Africans carrying the A1 genotype are more likely to develop HCC and are more prone to chronic infection and fibrosis (Kramvis 2018). The basal core promoter and pre-core areas of sub-genotype A1 receive particular mutations that impact mRNA transcription and lower HBeAg expression; the double mutation is a significant risk factor for HCC (Kramvis 2018). These findings indicate that genotype A1 has a significantly effective oncogenic impact despite lower levels of HBV viral load HBeAg negativity, indicating that individuals with ubiquitous HBeAg-negative phenotype with low HBV DNA concentrations should be treated (Kramvis 2018). Clinical variations among genotypes B and C are noticeable although inconclusive (Tseng et al. 2015). Hepatitis B genotype and molecular markers continue to be of great relevance, although clinical, in vitro molecular, and immunological associations are needed (Atsama Amougou et al. 2019; Kuhnhenn et al. 2018; Wong et al. 2018). The percentage of HBeAg-negative patients who have substantial fibrosis despite having significantly lower levels of HBV replication and in whom molecular variations, clonal lineages, and necro-inflammatory changes nevertheless indicate a risk of acute liver failure or oncogenesis; and the eventual progression of HCC is still unknown (Kao 2021). It is important to mention that disease progression can be predicted by HBsAg quantification. A sustained inactive phase is predicted by HBsAg levels below $\leq$1000 IU/mL, HBV DNA concentrations under 2000 IU/mL, and normal serum aminotransferases (Liu et al. 2014a; McMahon 2010). It is possible to distinguish between active HBeAg-negative illnesses and inactive diseases using serum HBV DNA which has been linked to disease progression (Liu et al. 2014a). Throughout the course of treatment, an increase in HBV DNA levels indicates the emergence of drug mistrust (Ghany and Doo 2009).

To assess the impact of absolute suppression of HBV DNA on the course of disease requires comparisons between outcomes associated with identifiable but unmeasurable HBV viral load especially in HBeAg-negative patients.

16.5 Future Perspectives

The search for a functional cure for HBV is now possible, although it will be some time before any of the new combination therapies under development can successfully treat HBV in a significant number of patients after a limited course of treatment. Prospects for an effective cure are reviving with the progress of DAAs that target distinct stages of the HBV life cycle, including viral entry, HBV DNA replication, nucleocapsid assembly, cccDNA transcription, and HBsAg production and secretion. To achieve this goal, collaborative efforts between companies will be needed to enable the most effective drug combinations to be tested, which also means that regulators, scientists, the pharmaceutical industry, and physicians will be invited to work together. Sustained HBsAg clearance must be established, given the unpredictable nature of CHB infection. The vast majority (82% to 92%) of patients with two or more negative HBsAg test results more than six months apart have achieved HBsAg clearance over a follow-up of 2 to 5 years, according to studies of people who cleared HBsAg either spontaneously or following IFN or NA therapy (Lok et al. 2020; Wu et al. 2020; Yip et al. 2017). Whether HBsAg seroclearance occurs naturally or following NA or Peg-IFN therapy, its persistence is equivalent (Kim et al. 2014; Stelma et al. 2017; Yip et al. 2017). The risk of viral recurrence is a crucial factor to consider if treatment is stopped after HBsAg clearance has been proven (2 negative test results spaced 6 months or more apart). Limited data indicate that this is the case, with viral recurrence rates (redetection of HBV DNA by PCR) of less than 2% in patients who stopped NA therapy after HBsAg clearance, as opposed to over 70% in patients who stopped NA therapy after more than two years of HBV DNA inhibition without HBsAg clearance (Kim et al. 2020; Lok et al. 2020; Papatheodoridis et al. 2016, 2018). Whether anti-HBs seroconversion is essential to meet the notion of functional cure is another question that remains unresolved. According to numerous studies, around 8.7% to 44% of people who spontaneously cleared HBsAg with IFN or NA had detectable anti-HBs when HBsAg first became undetectable (Lok et al. 2020; Roushan et al. 2016; Yip et al. 2017). In fact, even after HBsAg elimination, some 56% to 78% of patients still have detectable anti-HBs antibodies (Lok et al. 2020; Yip et al. 2017). Consequently, the requirement for simultaneous anti-HBs seroconversion could lead to lower practical cure rates or require long-term treatment to achieve the criteria for a functional cure (Yip et al. 2017). Whether anti-HBs seroconversion is necessary to maintain sustained HBsAg elimination at the end of treatment is still a matter of debate (Yip et al. 2017). To fully understand the complex mechanisms of HBV replication, despite steady progress in our understanding of the HBV life cycle, a number of gaps need to be filled (Wei and Ploss 2021). To improve innovation in antiviral drugs qualified to eliminate or neutralize the HBV minichromosome and enable HBV cure, it is necessary to better understand the fundamental biology of cccDNA (epigenetic regulation, biogenesis, and turnover) (Dandri and Petersen 2020). The as-yet-unknown function of these particles in the HBV life cycle and pathogenesis may be revealed by a better understanding of how viral RNAs are generated, matured, and released. Globally, all patients living with chronic HBV infection must have access to modern treatments that are both effective and inexpensive. For example, although generic NA is widely available, only a small proportion of eligible patients actually receive it (Nguyen et al. 2020). Therefore, if these treatments are not accessible, and if complementary measures are not put in place to improve diagnosis and access to health care, the creation of new drugs alone will not help eradicate HBV infection. On the other hand, HBV awareness must be promoted in the same way as HIV awareness, which is an essential condition for the development of state-of-the-art, low-cost nucleic acid tests for HBV DNA in combination with HCV RNA and HIV RNA tests. Finally, it will be crucial to identify new targets for antiviral therapy to eradicate HBV, which will require the improvement of model systems that better explain the molecular mechanisms of the HBV replication cycle, in particular the creation and maintenance of cccDNA.

16.6 Conclusion

Chronic HBV infection remains a serious health problem worldwide, despite the deployment of universal immunization programs. The goal of a functional cure for HBV has been made possible by the recent renewed interest in and progress of HBV treatments. Preventing disease progression and improving survival rates are the main goals of CHB therapy. With currently available treatments, IFN or NA, a functional cure of HBV is possible, but the cure rate remains fairly low. Furthermore, the inhibition or destruction of cccDNA represents the culmination of HBV research and drug development for the eradication of HBV. In the interests of improving health, current public knowledge of serious viral infections offers a chance to raise awareness of viral hepatitis to a degree equivalent to that of HIV and COVID-19. We can only hope that in the near future, progress will be made in our ongoing fight to eradicate this virus.

References

Abounouh K, Aitraise I, Benabou A, Boussakri I, Doumir MA, El Boussairi C, El Idrissi S, El Mahdaoui C, Qouar DE, Ennahal A, Fathi S, Hafidi M, Lachker L, Ratib C, Tanouti IA, Maaroufi A, Benjelloun S, Guessous F, Pineau P, Ezzikouri S (2019) Virus-associated human cancers in Moroccan population: from epidemiology to prospective research. Infect Genet Evol 75:103990. https://doi.org/10.1016/j.meegid.2019.103990

Abramson JS, Chung RT (2014) Optimal antiviral prophylaxis against hepatitis B reactivation in patients receiving rituximab-based chemotherapy for lymphoma. JAMA 312(23):2505–2507. https://doi.org/10.1001/jama.2014.16095

Agarwal K, Afdhal N, Coffin C, Fung S, Dusheiko G, Foster G, Elkhashab M, Tam E, Ramji A, Iyer R (2020) Liver toxicity in the Phase 2 Catalyst 206 trial of Inarigivir 400 mg daily added to a nucleoside in HBV EAg negative patients. J Hepatol 73:S125

Anderson RT, Lim SG, Mishra P, Josephson F, Donaldson E, Given B, Miller V (2019) Challenges, considerations, and principles to guide trials of combination therapies for chronic hepatitis B virus. Gastroenterology 156(3):529–533 e524. https://doi.org/10.1053/j.gastro.2018.11.062

Atsama Amougou M, Marchio A, Bivigou-Mboumba B, Noah Noah D, Banai R, Atangana PJA, Fewou Moundipa P, Pineau P, Njouom R (2019) Enrichment in selected genotypes, basal core and precore mutations of hepatitis B virus in patients with hepatocellular carcinoma in Cameroon. J Viral Hepat 26(9):1086–1093. https://doi.org/10.1111/jvh.13131

Bazinet M, Pantea V, Placinta G, Moscalu I, Cebotarescu V, Cojuhari L, Jimbei P, Iarovoi L, Smesnoi V, Musteata T, Jucov A, Dittmer U, Krawczyk A, Vaillant A (2020) Safety and efficacy of 48 weeks REP 2139 or REP 2165, tenofovir disoproxil, and pegylated interferon Alfa-2a in patients with chronic HBV infection naive to nucleos(t)ide therapy. Gastroenterology 158(8):2180–2194. https://doi.org/10.1053/j.gastro.2020.02.058

Bessone F, Dirchwolf M (2016) Management of hepatitis B reactivation in immunosuppressed patients: an update on current recommendations. World J Hepatol 8(8):385–394. https://doi.org/10.4254/wjh.v8.i8.385

Blumberg BS (1977) Australia antigen and the biology of hepatitis B. Science 197(4298):17–25. https://doi.org/10.1126/science.325649

Borentain P, Colson P, Coso D, Bories E, Charbonnier A, Stoppa AM, Auran T, Loundou A, Motte A, Ressiot E, Norguet E, Chabannon C, Bouabdallah R, Tamalet C, Gerolami R (2010) Clinical and virological factors associated with hepatitis B virus reactivation in HBsAg-negative and anti-HBc antibodies-positive patients undergoing chemotherapy and/or autologous stem cell transplantation for cancer. J Viral Hepat 17(11):807–815. https://doi.org/10.1111/j.1365-2893.2009.01239.x

Boyd A, Lacombe K, Lavocat F, Maylin S, Miailhes P, Lascoux-Combe C, Delaugerre C, Girard PM, Zoulim F (2016) Decay of ccc-DNA marks persistence of intrahepatic viral DNA synthesis under tenofovir in HIV-HBV co-infected patients. J Hepatol 65(4):683–691. https://doi.org/10.1016/j.jhep.2016.05.014

Burdette D, Lazerwith S, Yang J, Chan H, Delaney W IV, Feierbach B (2020) Evidence of an infectious virus reservoir in suppressed chronic hepatitis B patients

Cacciola I, Pollicino T, Squadrito G, Cerenzia G, Orlando ME, Raimondo G (1999) Occult hepatitis B virus infection in patients with chronic hepatitis C liver disease. N Engl J Med 341(1):22–26. https://doi.org/10.1056/NEJM199907013410104

Carey I, Gersch J, Wang B, Moigboi C, Kuhns M, Cloherty G, Dusheiko G, Agarwal K (2020) Pregenomic HBV RNA and hepatitis B core-related antigen predict outcomes in hepatitis B e antigen-negative chronic hepatitis B patients suppressed on nucleos(t)ide analogue therapy. Hepatology 72(1):42–57. https://doi.org/10.1002/hep.31026

Caviglia GP, Abate ML, Tandoi F, Ciancio A, Amoroso A, Salizzoni M, Saracco GM, Rizzetto M, Romagnoli R, Smedile A (2018) Quantitation of HBV cccDNA in anti-HBc-positive liver donors by droplet digital PCR: a new tool to detect occult infection. J Hepatol 69(2):301–307. https://doi.org/10.1016/j.jhep.2018.03.021

Chan HL, Thompson A, Martinot-Peignoux M, Piratvisuth T, Cornberg M, Brunetto MR, Tillmann HL, Kao JH, Jia JD, Wedemeyer H, Locarnini S, Janssen HL, Marcellin P (2011) Hepatitis B surface antigen quantification: why and how to use it in 2011 – a core group report. J Hepatol 55(5):1121–1131. https://doi.org/10.1016/j.jhep.2011.06.006

Chen EQ, Wang ML, Tao YC, Wu DB, Liao J, He M, Tang H (2019) Serum HBcrAg is better than HBV RNA and HBsAg in reflecting intrahepatic covalently closed circular DNA. J Viral Hepat 26(5):586–595. https://doi.org/10.1111/jvh.13061

Cheng AL, Hsiung CA, Su IJ, Chen PJ, Chang MC, Tsao CJ, Kao WY, Uen WC, Hsu CH, Tien HF, Chao TY, Chen LT, Whang-Peng J, Lymphoma Committee of Taiwan Cooperative Oncology Group (2003) Steroid-free chemotherapy decreases risk of hepatitis B virus (HBV) reactivation in HBV-carriers with lymphoma. Hepatology 37(6):1320–1328. https://doi.org/10.1053/jhep.2003.50220

Cornberg M, Lok AS, Terrault NA, Zoulim F, EASL-AASLD HBV Treatment Endpoints Conference Faculty (2020) Guidance for design and endpoints of clinical trials in chronic hepatitis B – report from the 2019 EASL-AASLD HBV treatment endpoints conference(double dagger). J Hepatol 72(3):539–557. https://doi.org/10.1016/j.jhep.2019.11.003

Crosnier J, Jungers P, Courouce AM, Laplanche A, Benhamou E, Degos F, Lacour B, Prunet P, Cerisier Y, Guesry P (1981) Randomised placebo-controlled trial of hepatitis B surface antigen vaccine in french haemodialysis units: II, Haemodialysis patients. Lancet 1(8224):797–800. https://doi.org/10.1016/s0140-6736(81)92679-9

Cui F, Shen L, Li L, Wang H, Wang F, Bi S, Liu J, Zhang G, Wang F, Zheng H, Sun X, Miao N, Yin Z, Feng Z, Liang X, Wang Y (2017) Prevention of chronic hepatitis B after 3 decades of escalating vaccination policy. China Emerg Infect Dis 23(5):765–772. https://doi.org/10.3201/eid2305.161477

Daffis S, Balsitis S, Chamberlain J, Zheng J, Santos R, Rowe W, Ramakrishnan D, Pattabiraman D, Spurlock S, Chu R, Kang D, Mish M, Ramirez R, Li L, Li B, Ma S, Hung M, Voitenleitner C, Yon C et al (2021) Toll-like receptor 8 agonist GS-9688 induces sustained efficacy in the woodchuck model of chronic hepatitis B. Hepatology 73(1):53–67. https://doi.org/10.1002/hep.31255

Dandri M, Petersen J (2020) cccDNA maintenance in chronic hepatitis B – targeting the matrix of viral replication. Infect Drug Resist 13:3873–3886. https://doi.org/10.2147/IDR.S240472

Dandri M, Burda MR, Will H, Petersen J (2000) Increased hepatocyte turnover and inhibition of woodchuck hepatitis B virus replication by adefovir in vitro do not lead to reduction of the closed circular DNA. Hepatology 32(1):139–146. https://doi.org/10.1053/jhep.2000.8701

Desalegn H, Aberra H, Berhe N, Gundersen SG, Johannessen A (2017) Are non-invasive fibrosis markers for chronic hepatitis B reliable in sub-Saharan Africa? Liver Int 37(10):1461–1467. https://doi.org/10.1111/liv.13393

Desmyter J, Colaert J, De Groote G, Reynders M, Reerink-Brongers EE, Lelie PN, Dees PJ, Reesink HW (1983) Efficacy of heat-inactivated hepatitis B vaccine in haemodialysis patients and staff. Double-blind placebo-controlled trial. Lancet 2(8363):1323–1328. https://doi.org/10.1016/s0140-6736(83)91089-9

El Chaar M, Candotti D, Crowther RA, Allain JP (2010) Impact of hepatitis B virus surface protein mutations on the diagnosis of occult hepatitis B virus infection. Hepatology 52(5):1600–1610. https://doi.org/10.1002/hep.23886

Emini EA, Ellis RW, Miller WJ, McAleer WJ, Scolnick EM, Gerety RJ (1986) Production and immunological analysis of recombinant hepatitis B vaccine. J Infect 13 Suppl A:3–9. https://doi.org/10.1016/s0163-4453(86)92563-6

European Association For The Study Of The Liver (2012) EASL clinical practice guidelines: management of chronic hepatitis B virus infection. J Hepatol 57(1):167–185. https://doi.org/10.1016/j.jhep.2012.02.010

European Association for the Study of the Liver. Electronic address, e. e. e., & European Association for the Study of the Liver (2017) EASL 2017 Clinical Practice Guidelines on the management of hepatitis B virus infection. J Hepatol 67(2):370–398. https://doi.org/10.1016/j.jhep.2017.03.021

Ezzikouri S, Kayesh MEH, Benjelloun S, Kohara M, Tsukiyama-Kohara K (2020) Targeting host innate and adaptive immunity to achieve the functional cure of chronic hepatitis B. Vaccines (Basel) 8(2):216. https://doi.org/10.3390/vaccines8020216

Fanning GC, Zoulim F, Hou J, Bertoletti A (2019) Therapeutic strategies for hepatitis B virus infection: towards a cure. Nat Rev Drug Discov 18(11):827–844. https://doi.org/10.1038/s41573-019-0037-0

Francis DP, Hadler SC, Thompson SE, Maynard JE, Ostrow DG, Altman N, Braff EH, O'Malley P, Hawkins D, Judson FN, Penley K, Nylund T, Christie G, Meyers F, Moore JN Jr, Gardner A, Doto IL, Miller JH, Reynolds GH et al (1982) The prevention of hepatitis B with vaccine. Report of the centers for disease control multi-center efficacy trial among homosexual men. Ann Intern Med 97(3):362–366. https://doi.org/10.7326/0003-4819-97-3-362

Gallay P, Ure D, Bobardt M, Chatterji U, Ou J, Trepanier D, Foster R (2019) The cyclophilin inhibitor CRV431 inhibits liver HBV DNA and HBsAg in transgenic mice. PLoS One 14(6):e0217433. https://doi.org/10.1371/journal.pone.0217433

Gane E, Dunbar PR, Brooks A, Zhao Y, Tan S, Lau A, Yang J, Gaggar A, Subramanian M, Kottilil S (2020a) Efficacy and safety of 24 weeks treatment with oral TLR8 agonist, selgantolimod, in virally-suppressed adult patients with chronic hepatitis B: a phase 2 study. J Hepatol 73:S52

Gane E, Lim Y-S, Tangkijvanich P, O'Beirne J, Lim TH, Bakardjiev A, Ding X, Connolly L, Huang S, Kim J (2020b) Preliminary safety and antiviral activity of VIR-2218, an X-targeting HBV RNAi therapeutic, in chronic hepatitis B patients. Digital International Liver Congress (Digital ILC 2020)

Gane E, Locarnini S, Lim TH, Strasser S, Sievert W, Cheng W, Thompson A, Given B, Schluep T, Hamilton J (2020c) Short-term treatment with RNA interference therapy, JNJ-3989, results in sustained hepatitis B surface antigen suppression in patients with chronic hepatitis B receiving nucleos(t)ide analogue treatment. Digital International Liver Congress

Ghany MG, Doo EC (2009) Antiviral resistance and hepatitis B therapy. Hepatology 49(S5):S174–S184

Giersch K, Allweiss L, Volz T, Dandri M, Lutgehetmann M (2017) Serum HBV pgRNA as a clinical marker for cccDNA activity. J Hepatol 66(2):460–462. https://doi.org/10.1016/j.jhep.2016.09.028

Goldstein ST, Zhou F, Hadler SC, Bell BP, Mast EE, Margolis HS (2005) A mathematical model to estimate global hepatitis B disease burden and vaccination impact. Int J Epidemiol 34(6):1329–1339. https://doi.org/10.1093/ije/dyi206

Gordon SC, Krastev Z, Horban A, Petersen J, Sperl J, Dinh P, Martins EB, Yee LJ, Flaherty JF, Kitrinos KM, Rustgi VK, Marcellin P (2013) Efficacy of tenofovir disoproxil fumarate at 240 weeks in patients with chronic hepatitis B with high baseline viral load. Hepatology 58(2):505–513. https://doi.org/10.1002/hep.26277

Guidotti LG, Rochford R, Chung J, Shapiro M, Purcell R, Chisari FV (1999) Viral clearance without destruction of infected cells during acute HBV infection. Science 284(5415):825–829. https://doi.org/10.1126/science.284.5415.825

Guo MK, Nopakun J, Messer HH, Ophaug R, Singer L (1988) Retention of skeletal fluoride during bone turnover in rats. J Nutr 118(3):362–366. https://doi.org/10.1093/jn/118.3.362

Hadler SC, Francis DP, Maynard JE, Thompson SE, Judson FN, Echenberg DF, Ostrow DG, O'Malley PM, Penley KA, Altman NL et al (1986) Long-term immunogenicity and efficacy of hepatitis B vaccine in homosexual men. N Engl J Med 315(4):209–214. https://doi.org/10.1056/NEJM198607243150401

Hu J, Liu K (2017) Complete and incomplete hepatitis B virus particles: formation, function, and application. Viruses 9, 56(3). https://doi.org/10.3390/v9030056

Huang CH, Yuan Q, Chen PJ, Zhang YL, Chen CR, Zheng QB, Yeh SH, Yu H, Xue Y, Chen YX, Liu PG, Ge SX, Zhang J, Xia NS (2012) Influence of mutations in hepatitis B virus surface protein on viral antigenicity and phenotype in occult HBV strains from blood donors. J Hepatol 57(4):720–729. https://doi.org/10.1016/j.jhep.2012.05.009

Huang H, Wang J, Li W, Chen R, Chen X, Zhang F, Xu D, Lu F (2018) Serum HBV DNA plus RNA shows superiority in reflecting the activity of intrahepatic cccDNA in treatment-naive HBV-infected individuals. J Clin Virol 99–100:71–78. https://doi.org/10.1016/j.jcv.2017.12.016

Hutin YJ, Bulterys M, Hirnschall GO (2018) How far are we from viral hepatitis elimination service coverage targets? J Int AIDS Soc 21(Suppl 2):e25050. https://doi.org/10.1002/jia2.25050

Jack AD, Hall AJ, Maine N, Mendy M, Whittle HC (1999) What level of hepatitis B antibody is protective? J Infect Dis 179(2):489–492. https://doi.org/10.1086/314578

Janssen HLA, Brunetto MR, Kim YJ, Ferrari C, Massetto B, Nguyen AH, Joshi A, Woo J, Lau AH, Gaggar A, Subramanian GM, Yoshida EM, Ahn SH, Tsai NCS, Fung S, Gane EJ (2018) Safety, efficacy and pharmacodynamics of vesatolimod (GS-9620) in virally suppressed patients with chronic hepatitis B. J Hepatol 68(3):431–440. https://doi.org/10.1016/j.jhep.2017.10.027

Janssen H, Hou J, Asselah T, Chan H, Zoulim F, Tanaka Y, Janczewska E, Nahass R, Bourgeois S, Buti M (2020) Efficacy and safety results of the phase 2 JNJ-56136379 JADE study in patients with chronic hepatitis B: interim week 24 data. J Hepatol 73:S129–S130

Kane MA (1996) Global status of hepatitis B immunisation. Lancet 348(9029):696. https://doi.org/10.1016/S0140-6736(05)65598-5

Kao J-H (2021) Hepatitis B virus and liver disease. Springer Nature. https://doi.org/10.1007/978-981-16-3615-8

Kazemi-Shirazi L, Petermann D, Muller C (2000) Hepatitis B virus DNA in sera and liver tissue of HBsAg negative patients with chronic hepatitis C. J Hepatol 33(5):785–790. https://doi.org/10.1016/s0168-8278(00)80311-6

Kim TW, Kim MN, Kwon JW, Kim KM, Kim SH, Kim W, Park HW, Chang YS, Cho SH, Min KU, Kim YY (2010) Risk of hepatitis B virus reactivation in patients with asthma or chronic obstructive pulmonary disease treated with corticosteroids. Respirology 15(7):1092–1097. https://doi.org/10.1111/j.1440-1843.2010.01798.x

Kim GA, Lim YS, An J, Lee D, Shim JH, Kim KM, Lee HC, Chung YH, Lee YS, Suh DJ (2014) HBsAg seroclearance after nucleoside analogue therapy in patients with chronic hepatitis B: clinical outcomes and durability. Gut 63(8):1325–1332. https://doi.org/10.1136/gutjnl-2013-305517

Kim MA, Kim SU, Sinn DH, Jang JW, Lim YS, Ahn SH, Shim JJ, Seo YS, Baek YH, Kim SG, Kim YS, Kim JH, Choe WH, Yim HJ, Lee HW, Kwon JH, Lee SW, Jang JY, Kim HY et al (2020) Discontinuation of nucleos(t)ide analogues is not associated with a higher risk of HBsAg seroreversion after antiviral-induced HBsAg seroclearance: a nationwide multicentre study. Gut 69(12):2214–2222. https://doi.org/10.1136/gutjnl-2019-320015

Kimura T, Rokuhara A, Sakamoto Y, Yagi S, Tanaka E, Kiyosawa K, Maki N (2002) Sensitive enzyme immunoassay for hepatitis B virus core-related antigens and their correlation to virus load. J Clin Microbiol 40(2):439–445. https://doi.org/10.1128/JCM.40.2.439-445.2002

Kimura T, Ohno N, Terada N, Rokuhara A, Matsumoto A, Yagi S, Tanaka E, Kiyosawa K, Ohno S, Maki N (2005) Hepatitis B virus DNA-negative dane particles lack core protein but contain a 22-kDa precore protein without C-terminal arginine-rich domain. J Biol Chem 280(23):21713–21719. https://doi.org/10.1074/jbc.M501564200

Ko C, Chakraborty A, Chou WM, Hasreiter J, Wettengel JM, Stadler D, Bester R, Asen T, Zhang K, Wisskirchen K, McKeating JA, Ryu WS, Protzer U (2018) Hepatitis B virus genome recycling and de novo secondary infection events maintain stable cccDNA levels. J Hepatol 69(6):1231–1241. https://doi.org/10.1016/j.jhep.2018.08.012

Kramvis A (2018) Molecular characteristics and clinical relevance of African genotypes and subgenotypes of hepatitis B virus. S Afr Med J 108(8b):17–21. https://doi.org/10.7196/SAMJ.2018.v108i8b.13495

Kuhnhenn L, Jiang B, Kubesch A, Vermehren J, Knop V, Susser S, Dietz J, Carra G, Finkelmeier F, Grammatikos G, Zeuzem S, Sarrazin C, Hildt E, Peiffer KH (2018) Impact of HBV genotype and mutations on HBV DNA and qHBsAg levels in patients with HBeAg-negative chronic HBV infection. Aliment Pharmacol Ther 47(11):1523–1535. https://doi.org/10.1111/apt.14636

Kwon SY, Lee CH (2011) Epidemiology and prevention of hepatitis B virus infection. Korean J Hepatol 17(2):87–95. https://doi.org/10.3350/kjhep.2011.17.2.87

Lai CL, Chien RN, Leung NW, Chang TT, Guan R, Tai DI, Ng KY, Wu PC, Dent JC, Barber J, Stephenson SL, Gray DF (1998) A one-year trial of lamivudine for chronic hepatitis B. Asia Hepatitis Lamivudine Study Group. N Engl J Med 339(2):61–68. https://doi.org/10.1056/NEJM199807093390201

Lai CL, Wong DK, Wong GT, Seto WK, Fung J, Yuen MF (2020) Rebound of HBV DNA after cessation of nucleos/tide analogues in chronic hepatitis B patients with undetectable covalently closed. JHEP Rep 2(3):100112. https://doi.org/10.1016/j.jhepr.2020.100112

Lampertico P, Maini M, Papatheodoridis G (2015) Optimal management of hepatitis B virus infection – EASL Special Conference. J Hepatol 63(5):1238–1253. https://doi.org/10.1016/j.jhep.2015.06.026

Lau GK, Leung YH, Fong DY, Au WY, Kwong YL, Lie A, Hou JL, Wen YM, Nanj A, Liang R (2002) High hepatitis B virus (HBV) DNA viral load as the most important risk factor for HBV reactivation in patients positive for HBV surface antigen undergoing autologous hematopoietic cell transplantation. Blood 99(7):2324–2330. https://doi.org/10.1182/blood.v99.7.2324

Lee HW, Lee JS, Ahn SH (2020) Hepatitis B virus cure: targets and future therapies. Int J Mol Sci 22(1):213. https://doi.org/10.3390/ijms22010213

Liaw YF, Sung JJ, Chow WC, Farrell G, Lee CZ, Yuen H, Tanwandee T, Tao QM, Shue K, Keene ON, Dixon JS, Gray DF, Sabbat J, Cirrhosis Asian Lamivudine Multicentre Study Group (2004) Lamivudine for patients with chronic hepatitis B and advanced liver disease. N Engl J Med 351(15):1521–1531. https://doi.org/10.1056/NEJMoa033364

Lim SG, Agcaoili J, De Souza NNA, Chan E (2019) Therapeutic vaccination for chronic hepatitis B: a systematic review and meta-analysis. J Viral Hepat 26(7):803–817. https://doi.org/10.1111/jvh.13085

Lin SM, Yu ML, Lee CM, Chien RN, Sheen IS, Chu CM, Liaw YF (2007) Interferon therapy in HBeAg positive chronic hepatitis reduces progression to cirrhosis and hepatocellular carcinoma. J Hepatol 46(1):45–52. https://doi.org/10.1016/j.jhep.2006.08.021

Lin LY, Wong VW, Zhou HJ, Chan HY, Gui HL, Guo SM, Wang H, Huang L, Bao SS, Xie Q, Chan HL (2010) Relationship between serum hepatitis B virus DNA and surface antigen with covalently closed circular DNA in HBeAg-negative patients. J Med Virol 82(9):1494–1500. https://doi.org/10.1002/jmv.21863

Liu J, Yang HI, Lee MH, Lu SN, Jen CL, Batrla-Utermann R, Wang LY, You SL, Hsiao CK, Chen PJ, Chen CJ, R.E.V.E.A.L.-HBV Study Group (2014a) Spontaneous seroclearance of hepatitis B seromarkers and subsequent risk of hepatocellular carcinoma. Gut 63(10):1648–1657. https://doi.org/10.1136/gutjnl-2013-305785

Liu J, Zhang E, Ma Z, Wu W, Kosinska A, Zhang X, Moller I, Seiz P, Glebe D, Wang B, Yang D, Lu M, Roggendorf M (2014b) Enhancing virus-specific immunity in vivo by combining therapeutic vaccination and PD-L1 blockade in chronic hepadnaviral infection. PLoS Pathog 10(1):e1003856. https://doi.org/10.1371/journal.ppat.1003856

Lok AS, Zoulim F, Dusheiko G, Ghany MG (2017) Hepatitis B cure: from discovery to regulatory approval. J Hepatol 67(4):847–861. https://doi.org/10.1016/j.jhep.2017.05.008

Lok AS, Zoulim F, Dusheiko G, Chan HLY, Buti M, Ghany MG, Gaggar A, Yang JC, Wu G, Flaherty JF, Subramanian GM, Locarnini S, Marcellin P (2020) Durability of hepatitis B surface antigen loss with nucleotide analogue and Peginterferon therapy in patients with chronic hepatitis B. Hepatol Commun 4(1):8–20. https://doi.org/10.1002/hep4.1436

Loomba R, Liang TJ (2017) Hepatitis B reactivation associated with immune suppressive and biological modifier therapies: current concepts, management strategies, and future directions. Gastroenterology 152(6):1297–1309. https://doi.org/10.1053/j.gastro.2017.02.009

Lutgehetmann M, Volz T, Kopke A, Broja T, Tigges E, Lohse AW, Fuchs E, Murray JM, Petersen J, Dandri M (2010) In vivo proliferation of hepadnavirus-infected hepatocytes induces loss of covalently closed circular DNA in mice. Hepatology 52(1):16–24. https://doi.org/10.1002/hep.23611

Mason WS, Jilbert AR, Summers J (2005) Clonal expansion of hepatocytes during chronic woodchuck hepatitis virus infection. Proc Natl Acad Sci U S A 102(4):1139–1144. https://doi.org/10.1073/pnas.0409332102

McAleer WJ, Buynak EB, Maigetter RZ, Wampler DE, Miller WJ, Hilleman MR (1984) Human hepatitis B vaccine from recombinant yeast. Nature 307(5947):178–180. https://doi.org/10.1038/307178a0

McMahon BJ (2010) Hepatitis B surface antigen (HBsAg): a 40-year-old hepatitis B virus seromarker gets new life. Gastroenterology 139(2):380–382. https://doi.org/10.1053/j.gastro.2010.06.026

Meireles LC, Marinho RT, Van Damme P (2015) Three decades of hepatitis B control with vaccination. World J Hepatol 7(18):2127–2132. https://doi.org/10.4254/wjh.v7.i18.2127

Mendez-Navarro J, Corey KE, Zheng H, Barlow LL, Jang JY, Lin W, Zhao H, Shao RX, McAfee SL, Chung RT (2011) Hepatitis B screening, prophylaxis and re-activation in the era of rituximab-based chemotherapy. Liver Int 31(3):330–339. https://doi.org/10.1111/j.1478-3231.2010.02332.x

Michel ML, Bourgine M, Fontaine H, Pol S (2015) Therapeutic vaccines in treating chronic hepatitis B: the end of the beginning or the beginning of the end? Med Microbiol Immunol 204(1):121–129. https://doi.org/10.1007/s00430-014-0381-y

Michielsen PP, Francque SM, van Dongen JL (2005) Viral hepatitis and hepatocellular carcinoma. World J Surg Oncol 3:27. https://doi.org/10.1186/1477-7819-3-27

Mosley JW, Stevens CE, Aach RD, Hollinger FB, Mimms LT, Solomon LR, Barbosa LH, Nemo GJ (1995) Donor screening for antibody to hepatitis B core antigen and hepatitis B virus infection in transfusion recipients. Transfusion 35(1):5–12. https://doi.org/10.1046/j.1537-2995.1995.35195090661.x

Mueller H, Wildum S, Luangsay S, Walther J, Lopez A, Tropberger P, Ottaviani G, Lu W, Parrott NJ, Zhang JD, Schmucki R, Racek T, Hoflack JC, Kueng E, Point F, Zhou X, Steiner G, Lutgehetmann M, Rapp G et al (2018) A novel orally available small molecule that inhibits hepatitis B virus expression. J Hepatol 68(3):412–420. https://doi.org/10.1016/j.jhep.2017.10.014

Murray JM, Wieland SF, Purcell RH, Chisari FV (2005) Dynamics of hepatitis B virus clearance in chimpanzees. Proc Natl Acad Sci U S A 102(49):17780–17785. https://doi.org/10.1073/pnas.0508913102

Naggie S, Lok AS (2021) New therapeutics for hepatitis B: the road to cure. Annu Rev Med 72:93–105. https://doi.org/10.1146/annurev-med-080119-103356

Nayagam S, Thursz M, Sicuri E, Conteh L, Wiktor S, Low-Beer D, Hallett TB (2016) Requirements for global elimination of hepatitis B: a modelling study. Lancet Infect Dis 16(12):1399–1408. https://doi.org/10.1016/S1473-3099(16)30204-3

Nguyen MH, Wong G, Gane E, Kao JH, Dusheiko G (2020) Hepatitis B virus: advances in prevention, diagnosis, and therapy. Clin Microbiol Rev 33(2):10. https://doi.org/10.1128/CMR.00046-19

Nwokediuko SC (2011) Chronic hepatitis B: management challenges in resource-poor countries. Hepat Mon 11(10):786–793. https://doi.org/10.5812/kowsar.1735143X.757

Papatheodoridis GV, Manolakopoulos S, Dusheiko G, Archimandritis AJ (2008) Therapeutic strategies in the management of patients with chronic hepatitis B virus infection. Lancet Infect Dis 8(3):167–178. https://doi.org/10.1016/S1473-3099(07)70264-5

Papatheodoridis G, Vlachogiannakos I, Cholongitas E, Wursthorn K, Thomadakis C, Touloumi G, Petersen J (2016) Discontinuation of oral antivirals in chronic hepatitis B: a systematic review. Hepatology 63(5):1481–1492. https://doi.org/10.1002/hep.28438

Papatheodoridis GV, Idılman R, Dalekos GN, Buti M, Chi H, van Boemmel F, Calleja JL, Sypsa V, Goulis J, Manolakopoulos S, Loglio A, Siakavellas S, Keskin O, Gatselis N, Hansen BE, Lehretz M, de la Revilla J, Savvidou S, Kourikou A et al (2017) The risk of hepatocellular carcinoma decreases after the first 5 years of entecavir or tenofovir in Caucasians with chronic hepatitis B. Hepatology 66(5):1444–1453. https://doi.org/10.1002/hep.29320

Papatheodoridis GV, Rigopoulou EI, Papatheodoridi M, Zachou K, Xourafas V, Gatselis N, Hadziyannis E, Vlachogiannakos J, Manolakopoulos S, Dalekos GN (2018) DARING-B: discontinuation of effective entecavir or tenofovir disoproxil fumarate long-term therapy before HBsAg loss in non-cirrhotic HBeAg-negative chronic hepatitis B. Antivir Ther 23(8):677–685. https://doi.org/10.3851/IMP3256

Pardoll DM (2012) The blockade of immune checkpoints in cancer immunotherapy. Nat Rev Cancer 12(4):252–264. https://doi.org/10.1038/nrc3239

Pattyn J, Hendrickx G, Vorsters A, Van Damme P (2021) Hepatitis B vaccines. J Infect Dis 224(12 Suppl 2):S343–S351. https://doi.org/10.1093/infdis/jiaa668

Raimondo G, Caccamo G, Filomia R, Pollicino T (2013) Occult HBV infection. Semin Immunopathol 35(1):39–52. https://doi.org/10.1007/s00281-012-0327-7

Raimondo G, Locarnini S, Pollicino T, Levrero M, Zoulim F, Lok AS, Taormina Workshop on Occult HBV Infection Faculty Members (2019) Update of the statements on biology and clinical impact of occult hepatitis B virus infection. J Hepatol 71(2):397–408. https://doi.org/10.1016/j.jhep.2019.03.034

Reddy KR, Beavers KL, Hammond SP, Lim JK, Falck-Ytter YT, American Gastroenterological Association Institute (2015) American Gastroenterological Association Institute guideline on the prevention and treatment of hepatitis B virus reactivation during immunosuppressive drug therapy. Gastroenterology 148(1):215–219; quiz e216–217. https://doi.org/10.1053/j.gastro.2014.10.039

Rehermann B, Ferrari C, Pasquinelli C, Chisari FV (1996) The hepatitis B virus persists for decades after patients' recovery from acute viral hepatitis despite active maintenance of a cytotoxic T-lymphocyte response. Nat Med 2(10):1104–1108. https://doi.org/10.1038/nm1096-1104

Rendi-Wagner P, Shouval D, Genton B, Lurie Y, Rumke H, Boland G, Cerny A, Heim M, Bach D, Schroeder M, Kollaritsch H (2006) Comparative immunogenicity of a PreS/S hepatitis B vaccine in non- and low responders to conventional vaccine. Vaccine 24(15):2781–2789. https://doi.org/10.1016/j.vaccine.2006.01.007

Roca Suarez AA, Testoni B, Zoulim F (2021) HBV 2021: new therapeutic strategies against an old foe. Liver Int 41(Suppl 1):15–23. https://doi.org/10.1111/liv.14851

Roushan MR, Mohammadpour M, Baiany M, Soleimani S, Bijani A (2016) Time to seroconversion of HBsAg to anti-HBs in individuals who lost HBsAg during follow-up. Epidemiol Infect 144(12):2648–2653. https://doi.org/10.1017/S0950268816001217

Samadi E, Mirshahabi H, Motamed N, Sadeghi H (2020) Prevalence of occult hepatitis B virus infection in hemodialysis patients using nested PCR. Rep Biochem Mol Biol 9(1):82–88. https://doi.org/10.29252/rbmb.9.1.82

Sarin SK, Kumar M, Lau GK, Abbas Z, Chan HL, Chen CJ, Chen DS, Chen HL, Chen PJ, Chien RN, Dokmeci AK, Gane E, Hou JL, Jafri W, Jia J, Kim JH, Lai CL, Lee HC, Lim SG et al (2016) Asian-Pacific clinical practice guidelines on the management of hepatitis B: a 2015 update. Hepatol Int 10(1):1–98. https://doi.org/10.1007/s12072-015-9675-4

Satake M, Taira R, Yugi H, Hino S, Kanemitsu K, Ikeda H, Tadokoro K (2007) Infectivity of blood components with low hepatitis B virus DNA levels identified in a lookback program. Transfusion 47(7):1197–1205. https://doi.org/10.1111/j.1537-2995.2007.01276.x

Schillie S, Walker T, Veselsky S, Crowley S, Dusek C, Lazaroff J, Morris SA, Onye K, Ko S, Fenlon N, Nelson NP, Murphy TV (2015) Outcomes of infants born to women infected with hepatitis B. Pediatrics 135(5):e1141–e1147. https://doi.org/10.1542/peds.2014-3213

Schillie S, Harris A, Link-Gelles R, Romero Ward J, Nelson N (2018) Recommendations of the Advisory Committee on Immunization Practices for use of a hepatitis B vaccine with a novel adjuvant. MMWR 67(15):455–458

Schinazi RF, Asselah T (2017) From HCV to HBV cure. Liver Int 37(Suppl 1):73–80. https://doi.org/10.1111/liv.13324

Seeger C, Mason WS (2000) Hepatitis B virus biology. Microbiol Mol Biol Rev 64(1):51–68. https://doi.org/10.1128/MMBR.64.1.51-68.2000

Shetty K, Hussain M, Nei L, Reddy KR, Lok AS (2008) Prevalence and significance of occult hepatitis B in a liver transplant population with chronic hepatitis C. Liver Transpl 14(4):534–540. https://doi.org/10.1002/lt.21284

Shouval D (2003) Hepatitis B vaccines. J Hepatol 39(Suppl 1):S70–S76. https://doi.org/10.1016/s0168-8278(03)00152-1

Shouval D, Roggendorf H, Roggendorf M (2015) Enhanced immune response to hepatitis B vaccination through immunization with a pre-S1/pre-S2/S vaccine. Med Microbiol Immunol 204(1):57–68. https://doi.org/10.1007/s00430-014-0374-x

Spreafico M, Berzuini A, Foglieni B, Candotti D, Raffaele L, Guarnori I, Colli A, Maldini FF, Allain JP, Prati D (2015) Poor efficacy of nucleic acid testing in identifying occult HBV infection and consequences for safety of blood supply in Italy. J Hepatol 63(5):1068–1076. https://doi.org/10.1016/j.jhep.2015.06.016

Squadrito G, Cacciola I, Alibrandi A, Pollicino T, Raimondo G (2013) Impact of occult hepatitis B virus infection on the outcome of chronic hepatitis C. J Hepatol 59(4):696–700. https://doi.org/10.1016/j.jhep.2013.05.043

Stelma F, van der Ree MH, Jansen L, Peters MW, Janssen HLA, Zaaijer HL, Takkenberg RB, Reesink HW (2017) HBsAg loss after peginterferon-nucleotide combination treatment in chronic hepatitis B patients: 5 years of follow-up. J Viral Hepat 24(12):1107–1113. https://doi.org/10.1111/jvh.12738

Stevens CE, Alter HJ, Taylor PE, Zang EA, Harley EJ, Szmuness W (1984) Hepatitis B vaccine in patients receiving hemodialysis. Immunogenicity and efficacy. N Engl J Med 311(8):496–501. https://doi.org/10.1056/NEJM198408233110803

Su TH, Hu TH, Chen CY, Huang YH, Chuang WL, Lin CC, Wang CC, Su WW, Chen MY, Peng CY, Chien RN, Huang YW, Wang HY, Lin CL, Yang SS, Chen TM, Mo LR, Hsu SJ, Tseng KC et al (2016) Four-year entecavir therapy reduces hepatocellular carcinoma, cirrhotic events and mortality in chronic hepatitis B patients. Liver Int 36(12):1755–1764. https://doi.org/10.1111/liv.13253

Summers J, Jilbert AR, Yang W, Aldrich CE, Saputelli J, Litwin S, Toll E, Mason WS (2003) Hepatocyte turnover during resolution of a transient hepadnaviral infection. Proc Natl Acad Sci U S A 100(20):11652–11659. https://doi.org/10.1073/pnas.1635109100

Suzuki F, Miyakoshi H, Kobayashi M, Kumada H (2009) Correlation between serum hepatitis B virus core-related antigen and intrahepatic covalently closed circular DNA in chronic hepatitis B patients. J Med Virol 81(1):27–33. https://doi.org/10.1002/jmv.21339

Szmuness W, Stevens CE, Harley EJ, Zang EA, Oleszko WR, William DC, Sadovsky R, Morrison JM, Kellner A (1980) Hepatitis B vaccine: demonstration of efficacy in a controlled clinical trial in a high-risk population in the United States. N Engl J Med 303(15):833–841. https://doi.org/10.1056/NEJM198010093031501

Szmuness W, Stevens CE, Harley EJ, Zang EA, Alter HJ, Taylor PE, DeVera A, Chen GT, Kellner A (1982) Hepatitis B vaccine in medical staff of hemodialysis units: efficacy and subtype cross-protection. N Engl J Med 307(24):1481–1486. https://doi.org/10.1056/NEJM198212093072403

Terrault NA, Lok ASF, McMahon BJ, Chang KM, Hwang JP, Jonas MM, Brown RS Jr, Bzowej NH, Wong JB (2018) Update on prevention, diagnosis, and treatment of chronic hepatitis B: AASLD 2018 hepatitis B guidance. Hepatology 67(4):1560–1599. https://doi.org/10.1002/hep.29800

Testoni B, Lebosse F, Scholtes C, Berby F, Miaglia C, Subic M, Loglio A, Facchetti F, Lampertico P, Levrero M, Zoulim F (2019) Serum hepatitis B core-related antigen (HBcrAg) correlates with covalently closed circular DNA transcriptional activity in chronic hepatitis B patients. J Hepatol 70(4):615–625. https://doi.org/10.1016/j.jhep.2018.11.030

Thompson AJ, Nguyen T, Iser D, Ayres A, Jackson K, Littlejohn M, Slavin J, Bowden S, Gane EJ, Abbott W, Lau GK, Lewin SR, Visvanathan K, Desmond PV, Locarnini SA (2010) Serum hepatitis B surface antigen and hepatitis B e antigen titers: disease phase influences correlation with viral load and intrahepatic hepatitis B virus markers. Hepatology 51(6):1933–1944. https://doi.org/10.1002/hep.23571

Tout I, Loureiro D, Mansouri A, Soumelis V, Boyer N, Asselah T (2020) Hepatitis B surface antigen seroclearance: immune mechanisms, clinical impact, importance for drug development. J Hepatol 73(2):409–422. https://doi.org/10.1016/j.jhep.2020.04.013

Trepo C, Chan HL, Lok A (2014) Hepatitis B virus infection. Lancet 384(9959):2053–2063. https://doi.org/10.1016/S0140-6736(14)60220-8

Tseng TC, Liu CJ, Chen CL, Yang WT, Yang HC, Su TH, Wang CC, Kuo SF, Liu CH, Chen PJ, Chen DS, Kao JH (2015) Higher lifetime chance of spontaneous surface antigen loss in hepatitis B carriers with genotype C infection. Aliment Pharmacol Ther 41(10):949–960. https://doi.org/10.1111/apt.13170

Tu T, Zehnder B, Qu B, Urban S (2021) De novo synthesis of hepatitis B virus nucleocapsids is dispensable for the maintenance and transcriptional regulation of cccDNA. JHEP Rep 3(1):100195

Tulchinsky TH (2018) Maurice Hilleman: creator of vaccines that changed the world. Case Studies Public Health 443

UN (2020) United Nations General Assembly: transforming our world: the 2030 agenda for sustainable development. Retrieved 26 November from https://www.un.org/en/development/desa/population/migration/generalassembly/docs/globalcompact/A_RES_70_1_E.pdf

van Beers D, Liesnard C, Duys M, Etienne J, Cappel R (1983) Preliminary results of hepatitis B vaccination (HEVAC B) in healthy subjects and haemodialysis patients. Dev Biol Stand 54:163–169. https://www.ncbi.nlm.nih.gov/pubmed/6228455

Vlachogiannakos J, Papatheodoridis G (2013) Hepatocellular carcinoma in chronic hepatitis B patients under antiviral therapy. World J Gastroenterol 19(47):8822–8830. https://doi.org/10.3748/wjg.v19.i47.8822

Wang J, Shen T, Huang X, Kumar GR, Chen X, Zeng Z, Zhang R, Chen R, Li T, Zhang T, Yuan Q, Li PC, Huang Q, Colonno R, Jia J, Hou J, McCrae MA, Gao Z, Ren H et al (2016) Serum hepatitis B virus RNA is encapsidated pregenome RNA that may be associated with persistence of viral infection and rebound. J Hepatol 65(4):700–710. https://doi.org/10.1016/j.jhep.2016.05.029

Wang H, Swann R, Thomas E, Innes HA, Valerio H, Hayes PC, Allen S, Barclay ST, Wilks D, Fox R, Bhattacharyya D, Kennedy N, Morris J, Fraser A, Stanley AJ, Gunson R, McLntyre PG, Hunt A, Hutchinson SJ et al (2018) Impact of previous hepatitis B infection on the clinical outcomes from chronic hepatitis C? A population-level analysis. J Viral Hepat 25(8):930–938. https://doi.org/10.1111/jvh.12897

Wang L, Cao X, Wang Z, Gao Y, Deng J, Liu X, Zhuang H (2019) Correlation of HBcrAg with intrahepatic hepatitis B virus Total DNA and covalently closed circular DNA in HBeAg-positive chronic hepatitis B patients. J Clin Microbiol 57(1):1128. https://doi.org/10.1128/JCM.01303-18

Watashi K, Sluder A, Daito T, Matsunaga S, Ryo A, Nagamori S, Iwamoto M, Nakajima S, Tsukuda S, Borroto-Esoda K, Sugiyama M, Tanaka Y, Kanai Y, Kusuhara H, Mizokami M, Wakita T (2014) Cyclosporin a and its analogs inhibit hepatitis B virus entry into cultured hepatocytes through targeting a membrane transporter, sodium taurocholate cotransporting polypeptide (NTCP). Hepatology 59(5):1726–1737. https://doi.org/10.1002/hep.26982

Wedemeyer HSK, Bogomolov PO, Chulanov V, Stepanova T, Viacheslav M, Allweiss L, Dandri M, Ciesek S, Dittmer U, Haefeli W-E, Alexandrov A, Urban S (2020) 48 weeks of high dose (10 mg) bulevirtide as monotherapy or with peginterferon alfa-2a in patients with chronic HBV/HDV coinfection. J Hepatol. https://doi.org/10.1016/S0168-8278(20)30651-6

Wei L, Ploss A (2021) Mechanism of hepatitis B virus cccDNA formation. Viruses 13(8):1463. https://doi.org/10.3390/v13081463

Werle-Lapostolle B, Bowden S, Locarnini S, Wursthorn K, Petersen J, Lau G, Trepo C, Marcellin P, Goodman Z, Delaney WET, Xiong S, Brosgart CL, Chen SS, Gibbs CS, Zoulim F (2004) Persistence of cccDNA during the natural history of chronic hepatitis B and decline during adefovir dipivoxil therapy. Gastroenterology 126(7):1750–1758. https://doi.org/10.1053/j.gastro.2004.03.018

WHO (2016) Global health sector strategy on viral hepatitis, 2016–2021. Retrieved 26 November from https://apps.who.int/iris/bitstream/handle/10665/246177/WHO-HIV-2016.06-eng.pdf

WHO (2017) World hepatitis B vaccines: WHO position paper–July 2017. World Epidemiol Rec. Retrieved 26 November from https://apps.who.int/iris/bitstream/handle/10665/255873/WER9227-369-392.pdf?sequence=1&isAllowed=y

WHO (2019) World Health Organization fact sheet: immunization coverage. WHO, Geneva, Switzerland

WHO (2022, 24 June 2022) Hepatitis B Key facts. Retrieved 26 November from https://www.who.int/news-room/fact-sheets/detail/hepatitis-b#:~:text=In%202021%20WHO%20estimated%20that,on%20setting%20and%20eligibility%20criteria

Wieland SF, Spangenberg HC, Thimme R, Purcell RH, Chisari FV (2004) Expansion and contraction of the hepatitis B virus transcriptional template in infected chimpanzees. Proc Natl Acad Sci U S A 101(7):2129–2134. https://doi.org/10.1073/pnas.0308478100

Wong D, Littlejohn M, Edwards R, Jackson K, Revill P, Gaggar A, Kitrinos K, Subramanian M, Marcellin P, Buti-Ferret M, Janssen H, Gane E, Locarnini S, Thompson A (2018) ALT flares during nucleotide analogue therapy are associated with HBsAg loss in genotype A HBeAg-positive chronic hepatitis B. Liver Int 38(10):1760–1769. https://doi.org/10.1111/liv.13716

Wooddell CI, Yuen MF, Chan HL, Gish RG, Locarnini SA, Chavez D, Ferrari C, Given BD, Hamilton J, Kanner SB, Lai CL, Lau JYN, Schluep T, Xu Z, Lanford RE, Lewis DL (2017) RNAi-based treatment of chronically infected patients and chimpanzees reveals that integrated hepatitis B virus DNA is a source of HBsAg. Sci Transl Med 9(409):eaan0241. https://doi.org/10.1126/scitranslmed.aan0241

Wu JF, Chang MH (2015) Natural history of chronic hepatitis B virus infection from infancy to adult life – the mechanism of inflammation triggering and long-term impacts. J Biomed Sci 22:92. https://doi.org/10.1186/s12929-015-0199-y

Wu Y, Liu Y, Lu J, Cao Z, Jin Y, Ma L, Geng N, Ren S, Zheng Y, Shen C, Chen X (2020) Durability of interferon-induced hepatitis B surface antigen seroclearance. Clin Gastroenterol Hepatol 18(2):514–516 e512. https://doi.org/10.1016/j.cgh.2019.04.020

Xia Y, Liang TJ (2019) Development of direct-acting antiviral and host-targeting agents for treatment of hepatitis B virus infection. Gastroenterology 156(2):311–324. https://doi.org/10.1053/j.gastro.2018.07.057

Yang HC, Kao JH (2015) Viral hepatitis. HBV cure – can we pin our hopes on immunotherapy? Nat Rev Gastroenterol Hepatol 12(3):129–131. https://doi.org/10.1038/nrgastro.2015.8

Yeo W, Chan PK, Zhong S, Ho WM, Steinberg JL, Tam JS, Hui P, Leung NW, Zee B, Johnson PJ (2000) Frequency of hepatitis B virus reactivation in cancer patients undergoing cytotoxic chemotherapy: a prospective study of 626 patients with identification of risk factors. J Med Virol 62(3):299–307. https://doi.org/10.1002/1096-9071(200011)62:3<299::aid-jmv1>3.0.co;2-0

Yeo W, Chan PK, Hui P, Ho WM, Lam KC, Kwan WH, Zhong S, Johnson PJ (2003) Hepatitis B virus reactivation in breast cancer patients receiving cytotoxic chemotherapy: a prospective study. J Med Virol 70(4):553–561. https://doi.org/10.1002/jmv.10430

Yeo W, Zee B, Zhong S, Chan PK, Wong WL, Ho WM, Lam KC, Johnson PJ (2004) Comprehensive analysis of risk factors associating with hepatitis B virus (HBV) reactivation in cancer patients undergoing cytotoxic chemotherapy. Br J Cancer 90(7):1306–1311. https://doi.org/10.1038/sj.bjc.6601699

Yeo W, Chan TC, Leung NW, Lam WY, Mo FK, Chu MT, Chan HL, Hui EP, Lei KI, Mok TS, Chan PK (2009) Hepatitis B virus reactivation in lymphoma patients with prior resolved hepatitis B undergoing anticancer therapy with or without rituximab. J Clin Oncol 27(4):605–611. https://doi.org/10.1200/JCO.2008.18.0182

Yip TC, Wong GL, Wong VW, Tse YK, Lui GC, Lam KL, Chan HL (2017) Durability of hepatitis B surface antigen seroclearance in untreated and nucleos(t)ide analogue-treated patients. J Hepatol. https://doi.org/10.1016/j.jhep.2017.09.018

Zhao N, Jia B, Zhao H, Xu J, Sheng X, Luo L, Huang Z, Wang X, Ren Q, Zhang Y, Zhao X, Cui Y (2019) A first-in-human trial of GLS4, a novel inhibitor of hepatitis B virus capsid assembly, following single- and multiple-ascending-Oral-dose studies with or without ritonavir in healthy adult volunteers. Antimicrob Agents Chemother 64(1):10. https://doi.org/10.1128/AAC.01686-19

Zhou T, Guo JT, Nunes FA, Molnar-Kimber KL, Wilson JM, Aldrich CE, Saputelli J, Litwin S, Condreay LD, Seeger C, Mason WS (2000) Combination therapy with lamivudine and adenovirus causes transient suppression of chronic woodchuck hepatitis virus infections. J Virol 74(24):11754–11763. https://doi.org/10.1128/jvi.74.24.11754-11763.2000

Zhu Y, Yamamoto T, Cullen J, Saputelli J, Aldrich CE, Miller DS, Litwin S, Furman PA, Jilbert AR, Mason WS (2001) Kinetics of hepadnavirus loss from the liver during inhibition of viral DNA synthesis. J Virol 75(1):311–322. https://doi.org/10.1128/JVI.75.1.311-322.2001

Zoulim F, Lenz O, Vandenbossche JJ, Talloen W, Verbinnen T, Moscalu I, Streinu-Cercel A, Bourgeois S, Buti M, Crespo J, Manuel Pascasio J, Sarrazin C, Vanwolleghem T, Shukla U, Fry J, Yogaratnam JZ (2020) JNJ-56136379, an HBV capsid assembly modulator, is well-tolerated and has antiviral activity in a phase 1 study of patients with chronic infection. Gastroenterology 159(2):521–533 e529. https://doi.org/10.1053/j.gastro.2020.04.036

Zuckerman JN (2006) Protective efficacy, immunotherapeutic potential, and safety of hepatitis B vaccines. J Med Virol 78(2):169–177. https://doi.org/10.1002/jmv.20524

Zuckerman JN, Zuckerman AJ, Symington I, Du W, Williams A, Dickson B, Young MD, UK Hepacare Study Group (2001) Evaluation of a new hepatitis B triple-antigen vaccine in inadequate responders to current vaccines. Hepatology 34(4 Pt 1):798–802. https://doi.org/10.1053/jhep.2001.27564

Chapter 17
Unveiling Emerging Avenues in Antiviral Research: Navigating Clinical Trials and Regulatory Landscapes

Sina Salajegheh Tazerji, Pouneh Hajipour, Phelipe Magalhães Duarte, Ahmed Hesham, Heba Alshater, Sarah El-Nakeep, and Mohamed Fawzy

Abstract Combating viral diseases represents a major challenge, especially due to the complexity, whether due to the ability of these pathogens to mutate or resistance to the drugs used as treatment. In general, coping with viruses is classically done preventively through the use of vaccines. However, the use of antiviral drugs such as remdesivir can also be used. The pandemic caused by SASRS-CoV-2 has increased concern about these microorganisms, mainly due to their ability to mutate, climate change, globalization, and anthropic actions. With that, in addition to the surveillance of emerging and reemerging viruses, the pandemic raised other issues, such as the therapeutic and preventive alternatives available for the main viral diseases and also for viruses that are still unknown. Undoubtedly, new prevention and therapeutic tools need to be available due to the complexity and aggressiveness of some diseases. In addition, these tools must have a high degree of safety and significant cost-effectiveness, in addition to being highly effective and practical. For this, conducting clinical trials is essential in view of the maximum benefit and the least harm to the population included in the clinical research. For this, the regulation of the tests becomes unique in order to guarantee the lowest risk and the highest safety. By improving existing tools and new technologies such as artificial intelligence, we will be able to better prepare for the next pandemic, optimizing the time for potential drugs to reach the clinical trial stage, as occurred in the COVID-19 pandemic.

Keywords COVID-19 · SARS-CoV-2 · Outbreak · Pandemic · Aantiviral · Vaccines · Remdesivir · Clinical trials

---

S. Salajegheh Tazerji (✉)
Department of Clinical Sciences, Faculty of Veterinary Medicine, Science and Research Branch, Islamic Azad University, Tehran, Iran

Young Researchers and Elites Club Science and Research Branch, Islamic Azad University, Tehran, Iran
e-mail: sina.salajegheh@srbiau.ac.ir

P. Hajipour
Faculty of Veterinary Medicine, Shahid Bahonar University of Kerman, Kerman, Iran

N. Kumar et al. (eds.), *Advances in Antiviral Research*, Livestock Diseases and Management, https://doi.org/10.1007/978-981-99-9195-2_17

17.1 Antiviral Research: Current Scenario and Future Challenges in the Event of Emergence of New Viruses and Drug Resistance

Viruses are combatted in clinical practice using antivirals and preventive vaccines. However, viral mutations and evolution could help the virus adapt to the surrounding environment despite these eliminating tools. The only successful elimination of a virus by a vaccine in medical history is the eradication of the smallpox virus (Vere Hodge and Field 2011). Many antiviral drugs for SARS-CoV-2 have been introduced since the beginning of the pandemic. The question of viral resistance to drugs like remdesivir was proposed early as one of the causes of the expected clinical failure of the drugs used by clinicians (Vitiello 2022).

17.1.1 Current Situation

Global genetic research on the possible mechanisms for remdesivir resistance is in progress. A recent publication suggests genetic affection as the mechanism involved (Gandhi et al. 2022). Remdesivir is an adenosine analog that blocks the activity of SARS-CoV-2 at nsp12 of the viral polymerase enzyme. It was the first drug to achieve FDA approval for the treatment of COVID-19 infections, given parenterally through IV injection in severely ill hospitalized patients (Kokic et al. 2021). Another antiviral that has recently acquired the approval of FDA is molnupiravir. It is an oral antiviral that decrease COVID-19-related mortality and hospitalization rate in unvaccinated patients if given at an early stage (Jayk Bernal et al. 2022).

P. M. Duarte
Postgraduate Program in Animal Bioscience, Federal Rural University of Pernambuco (UFRPE), Recife, Pernambuco, Brazil

A. Hesham
Department of Chemistry, Suez Canal University, Ismailia, Egypt

Middle East for Vaccines (MEVAC), El-Salihya El-Gededa, El-Sharkia, Egypt

H. Alshater
Forensic Medicine and Clinical Toxicology Department, Menoufia University Hospital, Shbien El-Kom, Egypt

S. El-Nakeep
Hepatogastroentrology Unit, Internal Medicine Department, Faculty of Medicine, Ain Shams University, Cairo, Egypt

M. Fawzy
Department of Virology, Faculty of Veterinary Medicine, Suez Canal University, Ismailia, Egypt

17.1.2 *Future Challenges*

17.1.2.1 Emerging New Viruses

RNA viruses are known for their repeated mutations and resistance to antivirals due to their lack of a proofreading mechanism; this results in epidemic outbreaks, at times turning into pandemics (Smith 2017; Cannalire et al. 2020). In 2020, the WHO prioritized certain viruses as being the most dangerous on the planet with a high risk of mortality and morbidity. In addition, they named "Disease X" as an unknown pathogen that will cause a pandemic. After the 2019 pandemic, COVID-19 became the first on the list (WHO 2022).

The emergence of new variants of SARS-CoV-2 as detected by genetic databases, such as the Global Initiative on Sharing All Influenza Data (GISAID), Nextstrain, and the Phylogenetic Assignment of Named Global Outbreak (PANGO), guide the pharmaceutical research for drug and vaccination modifications to avoid resistance or ineffective end products (Flores-Vega et al. 2022).

17.1.2.2 Antiviral Resistance

The study of the viral genome made clear that viral mutations are insertions, deletions, substitutions, or rearrangements that could help the virus survive the targeting by the host or drugs (Vere Hodge and Field 2011). In COVID-19, spike protein mutations can enhance the spread of the virus, severity of infection, or evasion of host protective immune mechanisms (Flores-Vega et al. 2022).

The general mechanisms of antivirals include inhibition of polymerase or protease enzymes (Vere Hodge and Field 2011). Remdesivir, as a polymerase inhibitor and a monophosphoramidate nucleoside with a wide range of effect on RNA viruses, targets protease and polymerase in SARS-CoV-2 (Cannalire et al. 2020). Only solitary reported cases of clinical antiviral resistance to Remdesivir. A case reported a viral breakthrough with the E802D mutation after initial viral and clinical response. Fortunately, most of the mutations causing resistance are generated in the laboratory (in vitro), without a real clinical impact (Gandhi et al. 2022). In addition, the host response to the infection may depend on the chronicity of the disease, where antiviral resistance may be higher in chronic infections than acute ones (Mason et al. 2018). Elimination of the virus by the host will protect the host from the viral reactivation, as is the case with the influenza virus. Reactivation of a latent or unresolved infection, as in the herpes simplex or varicella-zoster viruses, will result in lifelong viral survival in the host and further spreading of the virus to other individuals, which could be another cause of antiviral failure (Vere Hodge and Field 2011).

Another problem encountered during the pandemic is creating practical therapeutic strategies based on old and new medications. This caused discrepancies in the information presented, as some drugs had > 100 clinical trials done, while others had only one (Robinson et al. 2022).

Clinical implications of antiviral resistance include drug toxicity due to viral resistance leading to the use of a second line of antivirals and progression of infection and increased mortality caused by the use of ineffective drugs (Strasfeld and Chou 2010). The ideal drug should have a high safety profile, high cost-effectiveness, and high efficacy and should be easily and practically applicable (Robinson et al. 2022).

17.2 Applications of Organoids and Spheroids in Antiviral Research

Due to the seriousness and the speed of spread of the pandemic, contamination, severity of symptoms, mortality rate, and the unknown long-term effect of SARS-CoV-2 infection, procedures that contribute to the improvement of vaccines and drugs, as well as repositioning of right now used drugs, are critically required, so the engineered 3D models are utilized as in vitro platforms to study infections, host-virus interactions, and drug (de Melo et al. 2021). Organoid technologies may display a chance for understanding the pathologic mechanisms of SARS-CoV-2 in tissues (İnanç and Erdemli 2021). The advanced expression organoid alludes to cells developing in a defined three-dimensional (3D) environment in vitro to make mini-clusters of cells that self-organize and separate into useful cell sorts. Besides, organoids can be procured from either embryonic stem cells (ESCs), initiated pluripotent stem cells (iPSCs), or neonatal or grown-up stem cells (ASCs) (52, 70) via a handle comparable to the path in which the organ procures its unmistakable organization. ASC-derived organoids are created specifically from postnatal or grown-up tissues either from single ASC or ASC-containing tissue units (Corrò et al. 2020). Moreover, it can also be set up from patient-specific material for disease modeling and precision medicine. Furthermore, ESC/iPSC-derived organoids include stepwise separation conventions utilizing different development variables or inhibitors that take after the formative prompts amid gastrulation and organogenesis that have pluripotent property enables empowering the era of organoids from all three germ layers (Fig. 17.1). Since organoid innovation has ended up an autonomous field of

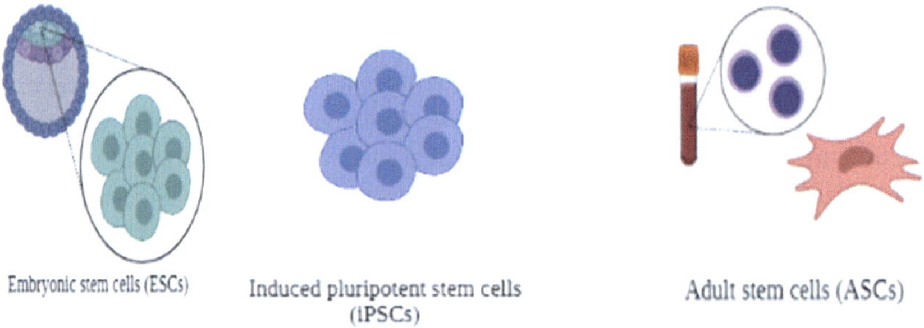

Fig. 17.1 Types of stem cell sources for organoid research stem cells (PSCs), including embryonic stem cells (ESCs), induced PSCs (iPSCs), and human adult stem cells (ASCs) (Lee et al. 2021)

investigation, an exact understanding of organogenesis in a dish will be a noteworthy figure for the inquiry about and utilization of tissue or stem cell-derived organoids (Lee et al. 2021). In the year 1900s, numerous researchers have agonized over the concept of the 3D engineering of organoids to set up the ideal conditions for the cellular composition and connections in tissues watched in vivo. After that organoids have been respected as 3D structures developing from stem cells and comprising of organ-specific cell sorts that self-organize with spatially confined heredity commitment (Kaluthantrige Don and Huch 2021). In 2009, scientists created an advanced interstitial organoid taking after a crypt-villus structure which instruct to felicitous implementation in cultures of other tissue sorts as stomach, pancreas, liver, colon, brain, lung, and heart from human tissue or pluripotent stem cells (Lancaster et al. 2013; Qian et al. 2019). As of now, stem cell-derived organoids are set up from the separation of stem cells refined in a specialized microenvironment that helps the formative preparation of certain tissues. Underneath, we show cases of producing each lineage-derived organoids from stem cells. Scientists detailed the primary self-assembled cerebral organoids and consequent considers established human PSC-derived brain region-specific organoids, including the forebrain, choroid plexus, hippocampus, spiral glial cell, midbrain, hypothalamus, and retina (Lancaster et al. 2013; Qian et al. 2019). Besides, gatherings such as coordinates human forebrain spheroids set up from the combined frame of dorsal and ventral forebrain organoids or thalamic and cortex organoids have too been detailed (Xiang et al. 2019).

Since they may facilitate the modelling of intelligent connections between different brain regions and take into account the assembly of neural circuits, assembleoids represent a kind of advanced innovation in brain organoid research. Other types of organoids, such as skin, inner ear, and salivary gland organoids, have also been recorded (Koehler et al. 2017; Lee et al. 2018; Sui et al. 2020). Takasato et al. produced kidney organoids composed of renal interstitium and endothelium, and a kidney illness shown in PSC-derived kidney organoids by quality altering was too detailed (Takasato et al. 2014, 2016). There are many types of organoids: self-organizing chamber-like cardiac organoids from human PSCs, murine heart organoids with multicellular compositions, cardiac organoids utilized for cardiac disease modeling, and screening for cardiotoxicity and blood vessel organoids for the generation of blood vessel organoids in humans and mice (Nugraha et al. 2019; Filippo Buono et al. 2020; Lee et al. 2020; Hofbauer et al. 2021). Human organoids have differing applications extending from essential inquiries about to medication development such as studying the tissue development as organoids could prolong basic and effortlessly accessible mini-organ systems for studying contributory factors related to complex morphogenesis. Besides, organoids have provided a device for finding basic variables of stem cell separation (Rossi et al. 2018; Hofer and Lutolf 2021). Moreover, organoids too encourage knockout considers of fundamental qualities that would not be doable in vivo due to embryonic lethality, empowering in-depth investigate into the part of qualities involved in human hereditary disarranges. Another application of organoids is the imitation of human pathologies at the organ level and gives a compelling show for human translational ponders to create commonsense applications of logical disclosures in human illnesses. It can reproduce host-pathogen interactions in vitro, and human gastric organoids were utilized to study Helicobacter pylori infection by mimicking the virus infection in the stomach, and also, Zika virus infections were also inspected in brain organoids (Garcez et al. 2016; Pompaiah and Bartfeld 2017; Sutarjono 2019). Norovirus and rotavirus have been matured in human intestinal organoid cultures, and the infections of Salmonella typhi and Clostridium difficile were also repeated in cultured intestinal organoids (Engevik et al. 2015; Ettayebi et al. 2016; Heo et al. 2018). More as of late, organoids have prolonged valuable inquire about apparatuses in response to the SARS-CoV-2 pathogen and many researches exhibited that SARS-CoV-2 could be infected and reproduced in the primary human liver and gut organoids (Fig. 17.2) (Monteil et al. 2020; Lamers et al. 2020). Besides, COVID-19 pathogenesis was confirmed in organoids and researchers have found that the viral receptor, angiotensin-converting enzyme 2, is the component of entrance of SARS-CoV-2 into host cells, which the pathogen could contaminate other tissues past the lung (Hoffmann et al. 2020) (Fig. 17.2). Furthermore, human organoids play a part in improving our understanding of human cancers since organoids reiterate the data of the parental patients especially diverse ASCs which have been cultured as cancer organoids, and organoid biobanks have been gained by cancer patient's tissues, inclusive breast, lung, glioblastoma, neuroendocrine, kidney, gastric, liver, pancreas, bladder, prostate, and colon (Tuveson and Clevers 2019; Botti et al. 2021). Also, human organoids are accessible as a novel stage for investigating genetic diseases. Gene editing is used for studying organogenesis and to probe the role of specific genes during the development (Yi et al. 2021). Besides, scientists can attain organogenesis utilizing isogenic variants with targeted genetic modifications by

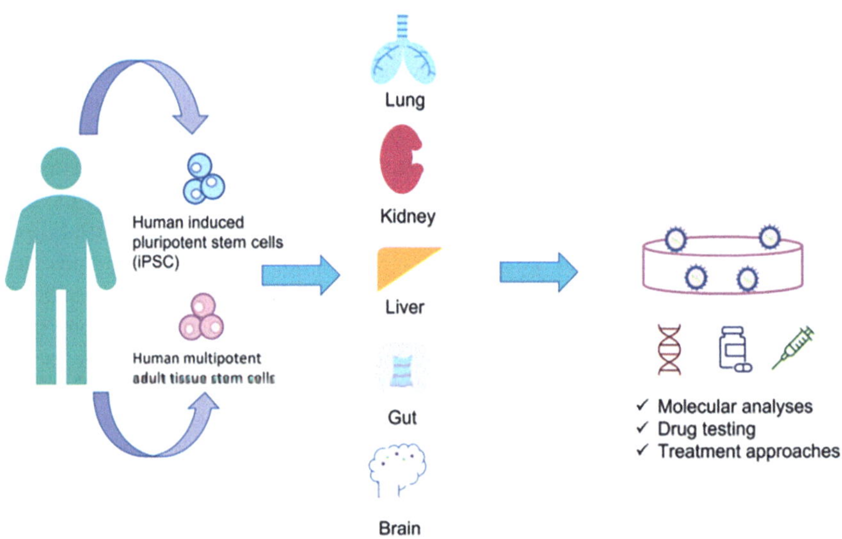

Fig. 17.2 Applications of organoid technology in studies of SARS-CoV-2 (İnanç and Erdemli 2021)

making mutant cell types by CRISPR/Cas9 systems or regulating genes by viral vectors or small interfering RNAs.

Schwank et al. fabricated gene-corrected intestinal organoids derived from biopsies of two diverse cystic fibrosis (CF) patients with a homozygous CFTR F508 deletion using CRISP-Cas9 technology, and then they studied the generated CF patient-derived iPSCs with their homozygous mutations corrected by CRISPR (Yi et al. 2021). So, the genetic disease modeling utilizing organoid technology is a beneficial implement for functional surveys of certain mutations and potential therapeutic strategies for clinical management. Another sort of organoids application is drug screening, especially in the field of cancer organoids. In addition, the United States Food and Drug Administration has as of late embraced an "organ-on-chip" system to screen for organ-specific or multiorgan toxicity of drugs, nourishment added substances, chemicals, makeup, and other compounds (Low et al. 2021). Thus, the utilize of organoid stages for drug toxicity and security testing might too be used since of their adequacy. Besides, organoid-based regenerative medicines have promising potential as an elective approach to cell replacement treatments, and the process of transplantation of kidney organoids in a mouse model fulfilled in vascularization of the mouse kidney with the formation of renal tubules, intestinal organoid, and the transplantation in a murine model resulted in colon regeneration (Fordham et al. 2013; Takasato et al. 2016). With respect to the current drug development impediment of tall costs and time consumption, organoid stages are promising for medicate advancement as a high-throughput approval stage for drugs. The field of organoid therapeutics has been connected for tissue recovery and cell substitution, as well as ready-to-use manufactured organs, the organoid market in many applications is expected to rise, and the demand for commercial human organoids will enhance each year (Takebe et al. 2018). Nowadays, many researchers have tried various diverse approaches for organoid development to study specific neurodevelopmental aims, and they tried to recapitulate the normal brain tissue via incorporating six cell types to improve spheroids including major cell types such as neurons, microglia, astrocytes, and oligodendrocytes, so as to attain better junctions and cell markers of the blood–brain barrier (BBB), which could be employed for drug neurotoxicity assessments (Takebe et al. 2018). Exposure to heavy metals leads to an increase in cell death, so brain organoids were utilized for studying the neural dysfunctions under cadmium exposure. Recently, Gonzalez et al. had modeled Alzheimer's disease (AD) utilizing the organoids manufactured from iPSCs derived from patients affected by familial AD and have explored that organoids exhibited all the pathological hallmarks of AD (Gonzalez et al. 2018). Moreover, da Silva et al. also affirmed hybrid organoids by utilizing human GBM spheroids that could infiltrate inside early-stage cerebral organoids (Gonzalez et al. 2018). Despite their many advantages, organoids have a number of drawbacks, such as the absence of all physiological neural cell types, such as microglia, crosstalk with other tissues during embryonic development, variability between organoids, poor reproducibility, and vascularization as a result of inadequate nutrient supply. Hence, this results in necrosis at the organoid core and limits the viable thickness (Karzbrun and Reiner 2019; Qian et al. 2019). However, a number of groups have recently made an effort to address this by employing various strategies, such as improving nutrient exchange, eliciting the expression of hETV2, a gene linked to the formation of vasculature in developing embryos, and upgrading oxygen supply by forming the air–liquid interface in cerebral organoids, which can lead to improved neuronal survival and the formation of active neuronal networks (Lancaster et al. 2013; Karzbrun and Reiner 2019; Cakir et al. 2019). The forskolin-induced human organoid swelling experiment, which facilitates the examination of individual medication responses in cystic fibrosis, demonstrates how interests and intestinal organoid technologies have currently been used as a viable platform for personalised pharmaceutical approaches (Dekkers et al. 2013; Xinaris et al. 2015). Lately, these organoids were transplanted into immunocompromised mice to evolve mature epithelium and mesenchyme that consist of differentiated intestinal cell lineages, functional brush-border enzymes, and subepithelial and smooth muscle layers. Researchers exhibited the generation of 3D human gastric organoids via directed in vitro differentiation of hESCs and hiPSCs, and they recapitulated in vivo human stomach development by inducing first the PSCs to format definitive endoderm spheroids. Besides, PSCs were utilized to generate retinal organoids by Eiraku et al., who exhibit the generation of retinal epithelium via using 3D floating EB-like aggregates from mESCs cultured (Eiraku et al. 2011). Furthermore, 3D culture systems have been progressed for HiPSC-derived hepatocytes, and the liver organoids can be examined for activity utilizing diverse different detection and imaging systems, these cells have significant value to estimate the potential of drugs for exhibiting liver toxicity, and these liver cells in organoids have advantages as they provide a more relevant cell–cell interaction because of the polarity of the hepatocytes (Eiraku et al. 2011).

Till now, the exposure to organoids to the outside environment is still constrained, and later approaches have proposed the contact of lung organoids to discuss, which empowers the complete differentiation of the grown-up wind current cells and an indeed more realistic examination of the impacts of interaction with pollutants such as poisonous gasses or harmful miniaturized scale and nanoparticles (İnanç and Erdemli 2021), and to date, there are various organoid models to perform the respiratory system and inclusive cultures to mimic proximal and distal airway and alveoli. Besides, researchers working with ECSs exhibited the replication of genetically related lung diseases as fibrosis in organoids. Recently, studies have brought new hope for the organoid technology in immunotherapy and pharmacokinetic testing (Xu et al. 2018).

Thus, the combination of organoid innovation with unused bioengineering methods, such as bioprinting or microfluidics, allows the advancement of unused models of male fruitlessness. Also, 3D societies can be connected to tissue regeneration and building, as of now come to within the urological field for urethral recreation. Within the future, biomaterial researchers and tissue engineers will without a doubt discover novel ways to repair and replace more tissues and organs within the human body, which can certainly have a part of esteem and application in clinics (Patrício et al. 2022).

17.3 Novel Nanocomposites for Drug Delivery Current Advances and Future Potential

The complexity of a few diseases and the characteristic harmfulness of certain drugs has driven to an expanding intrigued within the improvement and optimization of drug delivery frameworks (Begines et al. 2020). A drug delivery framework is a definition or gadget that allows the presentation of dynamic fixings into the body in arrange to move forward not as it was their adequacy but moreover their security by controlling the medicate sum, time and discharge within the location of activity, crossing the organic layers to urge to the restorative target. Thus, in addition to traditional medicinal drug distribution techniques, this also involves the use of vectors to promote the application and dissemination of these drugs into the human body (Bruschi 2015). Besides, the distinctive combinations of vectors and dynamic fixings may permit a wide run of conceivable outcomes for personalization, depending on specific illnesses and patients. The courses utilized to manage and convey dynamic substances to their target tissue are an important factor when treating an infection (Jain 2020a). There are various anatomic routes of administration for drug deliveries that are accessible (Fig. 17.3).

Since old times, people have broadly utilized plant-based natural items as solutions against different illnesses and these natural compounds are considered the base for the discovery of novel drugs. Despite these natural compounds have diverse advantages such as extraordinary chemical diversity, chemical and biological properties with macromolecular specificity and low toxicity and pharmaceutical companies are reluctant to contribute more in common product-based drug revelation and sedate conveyance frameworks and instead explore the accessible chemical compounds libraries to discover novel drugs (Patra et al. 2018). In any case, characteristic compounds are now being screened for treating a few major diseases, including cancer, diabetes, cardiovascular, inflammatory, and microbial infections. This can

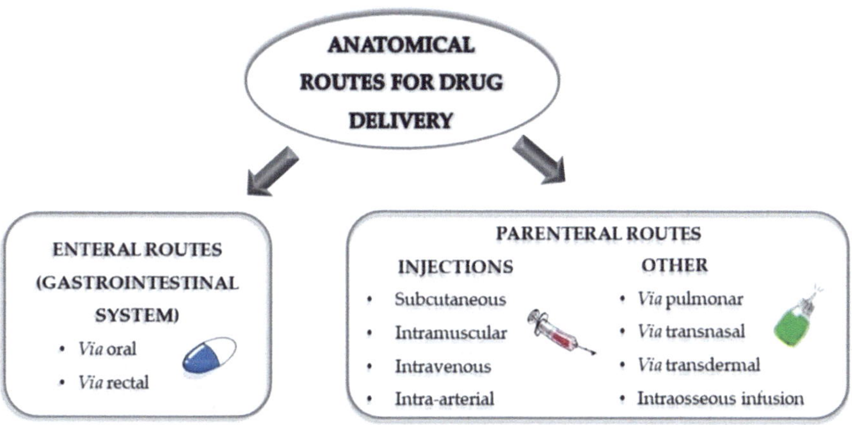

Fig. 17.3 Various anatomic routes for drug delivery (Begines et al. 2020)

be primarily since natural drugs have interesting preferences, such as lower toxicity and side impacts, low-price, and great restorative potential. Hence, many common compounds are not clearing the clinical trial stages since of these issues. The use of expansive measured materials in medicate conveyance postures major challenges, counting in vivo precariousness, destitute bioavailability, and destitute dissolvability, destitute assimilation within the body, issues with target-specific conveyance, and tonic effectiveness, and plausible unfavorable impacts of drugs. Therefore, using unused sedate conveyance frameworks for focusing on drugs to specific body parts may be a choice that might solve these basic issues (Martinho et al. 2011; Thilakarathna and Rupasinghe 2013; Watkins et al. 2015; Jahangirian et al. 2017). To decrease the risks of administration routes, researchers focusing on a better-controlled discharge, keeping up the restorative viability of the dynamic fixing focusing on to the particular location of activity, hence dodging the systemic discharge of the dynamic substance. In this sense, nanotechnology is picking up tall significance, because it may possibly fathom a few of the issues related with the abovementioned conventional organization course. In common, nanoparticles may well be optimized to make strides sedate bioavailability, either by expanding their assimilation through upgraded dissolvability or by encouraging their section through the biologic films (Jain 2020b). Moreover, the advance of the nanoparticles permits the acknowledgment of the particular location of activity, dodging tall systemic concentrations, and decreasing side impacts, and this property has been exceptionally valuable within the conclusion field by combining particular focusing on with the transport and discharge of a differentiate specialist. Polymers are not as it was an extraordinary sort of fabric which will envelop all the abovementioned characteristics, but too, the awesome engineered flexibility they show permits the analyst to customize them agreeing to the prerequisites or last points. The polymeric materials are acquiring great rapport in nanotechnology due to their various range of structures and applications when synthesized from their corresponding monomers, and they can be carried out on biopolymers by chemical derivatization (Wang et al. 2016; Begines et al. 2016, 2019, 2020; Liu et al. 2019). In spite of their preferences and imposing potential, a few bothers in their utilize can be found related to their constrained shape, chemistry, wide measure conveyance agglomeration state, and electromagnetic properties that can lead to destitute verbal bioavailability, flimsiness in circulation, and lacking tissue distribution. Furthermore, in synthesis of polymeric nanoparticles, the utilize of surfactants may be a requirement. The surfactants are amphiphilic natural particles that can self-assemble in arrangement, and most of them are composed by a hydrocarbon chain (hydrophobic part) which is bound to an ionic functional group (forming the cationic part). Besides, there are non-ionic surfactants in which the amphiphilic character is generated by the union of hydrophobic and hydrophilic molecules. Also, low molecular weight polymers are another type of as surfactants, and they are inclusive in the nanocarrier formulation as stabilizer agents during the nanoemulsion procedures. The questions of pharmacokinetics and biodistribution suggested that the use of surfactant surface-modified nanoparticle frameworks would result in increased drug maintenance within the body and accumulation within the target tissue, prolonged time in the blood circulation, and decreased nephrotoxicity and hepatotoxicity as well as lower cardiovascular effects and reduced macrophage uptake (Gao et al. 2016; Joseph and Saha 2017). The multidrug resistance mediated by the human ATP-binding Cassette transporter superfamily as P-glycoprotein (P-gp/ABCB1), multidrug resistance-associated protein 2 (MRP2/ABCC2), and breast cancer-resistant protein (BCRP/ABCG2) have been renowned as the main obstacle toward multiple chemotherapeutic agents (Chen et al. 2016b). Varshochian et al. utilized the PLGA/PVA system to synthesize bevacizumab-loaded micelles for the treatment of ocular neovascularization, and it showed its efficacy in the treatment of retinal and choroidal neovascularization (Varshochian et al. 2015). Ryu et al. synthesized dry tablets which contain dexamethasone-loaded micelles incorporated in an alginate matrix, and they explained that the utilize of this system could enhance 2.6-fold the ocular drug bioavailability (Varshochian et al. 2015). Moreover, Lui et al. explained a similar PLGA/PVA based on micelles to cure choroidal neovascularization, and this system includes polyethyleneimine to incorporate positive charges in the nanoparticles which explores a good anti-angiogenic and a powerful inhibitory efficacy on VEGF secretion from human umbilical vein endothelial cells (HUVEC) (Liu et al. 2019). Chitosan is a natural hydrophilic cationic polysaccharide which showed an excellent mucoadhesion and good properties for drug delivery in mucosa and ophthalmic areas. Ameeduzzafar et al. prepared chitosan-based micelles utilizing sodium tripolyphosphate (TPP) as a crosslinking agent, for the delivery of levofloxacin for ocular infection treatment (Ameeduzzafar et al. 2018). Later on, Silva et al. improved a similar CH/TPP system which includes hyaluronic acid and they encapsulated ceftazidime, a very unstable antibiotic, for utilization in the treatment of eye infections (Silva et al. 2017). Shi et al. prepared a cationic CH grafted methoxy poly(ethylene glycol)-poly("-caprolactone) for enclosing diclofenac (Shi et al. 2015). Also, Huang et al. developed triamcinolone acetonide which used as a drug for eye inflammation, and when combined together with PEG–PCL–PEG block copolymer, they get a nanoformulation that proved higher transcorneal drug permeability in isolated rabbit cornea (Huang et al. 2018). Furthermore, Wen et al. synthesized nanoparticles-loaded in situ gel as dexamethasone delivery system for ocular inflammation, and this drug was contained in lecithin-based micelles and used for an easier administration as ocular eye drop (Wu et al. 2018). Rodriguez-Aller et al. improved a library of various cyclodextrins as nanocarriers of latanoprost, an insoluble prostaglandin utilized for glaucoma treatment, and the results showed that it has a better stability and availability properties, as well as lower eye irritation (Rodriguez-Aller et al. 2015). Lorenzo-Veiga et al. synthesized ocular natamycin nanocarriers which is the only drug approved for fungal keratitis treatment (Rodriguez-Aller et al. 2015). Zorzi et al. improved a PEGylated PV to encapsulate siRNA sequences for the treatment of ocular keratitis caused by Acanthamoeba, and this led to the reduction in corneal damage caused by this disease in a murine model (Zorzi et al. 2019). Gold metallic nanoparticles and their derivatives are the foremost critical examination topic when depicting unused composites which are able to make strides determination and imaging methods. Furthermore, Wang et al. prepared biodegradable polymeric nanoparticle-based on silica-coated gold nanoparticles for photoacoustic imaging, and this technique allows researchers to get images from biologic structures of various shapes and forms, even from organelles (Wang et al. 2016). Furthermore, gadolinium polymeric nanoparticles were prepared with anti-VEGF antibody, which eases delivery to cancer cells in hepatocellular carcinoma (HCC) in order to progress its detection in early phases, and they can be utilized in imaging platform in PAI technique. Besides, Wu et al. explained gadolinium polymeric nanoparticles which are designed for MRI/CT/PAI-guided photothermal therapy, and these nanoparticles absorb NIR light and transform it into heat and produce the in vivo tumor ablation and its eradication (Wu et al. 2018). In addition, Zhou et al. gathered Gd and Au advantages as contrast agents in MR and CT in order to generate new imaging agents for targeted dual mode tumor CT/MR imaging in vivo (Zhou et al. 2018). Later on, McQuade et al. designed a nanoplatform for theranostic purposes on the base of gold and superparamagnetic iron oxide NPs which entrapped within a polymeric micelle (McQuade et al. 2015). Also, Yuan et al. showed the action of pH-sensitive PEG–PLGA–PGlu nanoparticles immersed in curcumin and doxorubicin in drug resistant cancer stem cells and tumor cells of breast tumors which yield perfect results in mice models (Yuan et al. 2018). Also, Zhu et al. prepared a nanosystem on the base of a new galactosamine-conjugated polydopamine-modified copolymer which targets HepG2 cells through ASGP receptor-mediated recognition and inhibit cell proliferation (Chen et al. 2016b). Besides, polymeric nanoparticles have been utilized for the airborne conveyance of chemotherapeutics, qualities, or their combination for lung cancer treatment. Furthermore, nanosized tubular halloysite exhibits various advantages due to its structural characteristics and properties, so it is used in various applications. Hence, attention is focused to the structure and morphology of it and its changes (Fig. 17.4) (Yuan et al. 2015).

Furthermore, Sethi et al. synthesized a crosslinkable lipid shell which contains docetaxel and wortmannin as the prototypical drugs utilized for controlling the drug

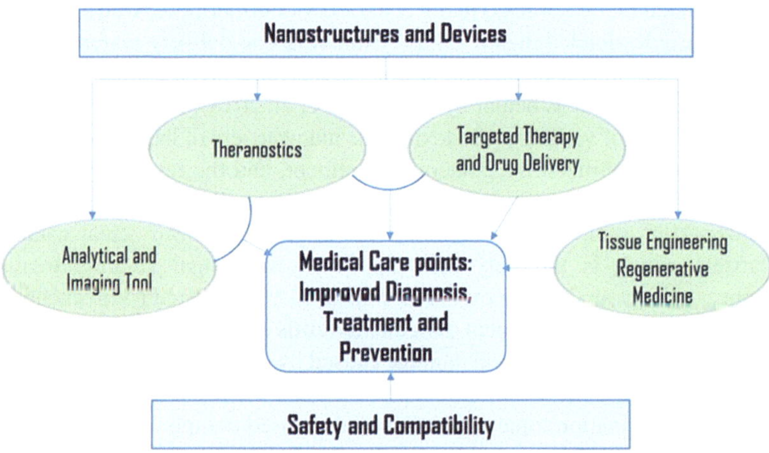

Fig. 17.4 Application of nanomedicine in biomedical research (Yuan et al. 2015)

discharge kinetics, and hence, they exhibited its efficacy in both in vivo and in vitro conditions (Sethi et al. 2014). Besides, Ulbrich et al. designed a drug delivery system on the basis of polymeric and magnetic nanoparticles and showed the effectiveness of attached drugs for cancer cure (Ulbrich et al. 2016). Also, Patil and Devarajan improved insulin-containing alginate nanoparticles with nicotinamide as a permeation agent to reduce serum glucose levels and increase serum insulin levels in diabetic rats (Patil and Devarajan 2016). Moreover, Huang et al. designed injectable hydrogels, which are composed of aldehyde-modified xanthan and carboxymethyl-modified chitosan and contain powerful angiogenic factors, to evolve abdominal wall reconstruction (Huang et al. 2018).

Nanotechnology has been powerfully created in recent years, and all nations, whether created or not, are increasing their ventures in investigate and development in this field. However, analysts dealing with practical applications of nano-drugs face many challenges, including the lack of a precise definition for these substances, the characterization of these nanomaterials in relation to their toxicity and security, and the requirement for strong control. Despite the extensive list of approved nanomedicine, its therapeutic potential is ultimately hampered by the specific administrative regulations' insufficiency for the development and characterization of these nanomaterials. The relationships between the structure and functions of various nanomaterials, as well as their properties, composition, and surface coating, interatomic with organic frameworks. In expansion, it is imperative to assess the plausibility of total and agglomerate arrangement when these nanomedicines are presented into biological systems since they do not reflect the properties of the person molecule; this may produce diverse results and/or unforeseen harmful impacts depending on the nanoformulation (Wacker et al. 2016). Thus, there are many challenges in the application of nanomedicine that speaks to a colossal breakthrough within the abovementioned fields and assures an empowering development within the following decade. Medications will end up more efficient and more secure due to the colossal assortment of nanoparticle plan and functionalization. The records of potential applications grows to the point where the nanocarrier can be customized to best adjust to a certain active fixing, a particular environment and then give fitting medicate area at the location of action, in a controlled way. Be that as it may, it is significant to say that nanoparticle-based medicines are not perfect and have challenges to prevail. To begin with, the number of polymeric materials as of now accessible for their utilization as DDS is still restricted in spite of the fact that the R&D has been moved within the final decade, exceeding expectations, from the small scale to the nanosized scale. Moreover, the specific focus gathered an awesome change in comparison to non-encapsulated drugs, and it could be an exceptionally complex component and speaks to the challenge itself. Overexpression of a specific surface protein is not sufficient to guarantee specific focusing as they are moreover ordinarily communicated in typical cells, and this point is more basic in cancer medicines, where managed drugs usually possess higher harmfulness that may lead to various undesirable auxiliary effects when compared to drugs used in other maladies medications. Besides, when focusing on more particular illnesses and considering the maturing populace, novel details or indications for past blockbuster drugs, counting polymeric nanoparticles, may well be a great recommendation to keep up a productive financial development rate. Accomplishing sensible victory for oral bioavailability of ineffectively ingested lipophilic and hydrophobic drugs, to preserve satisfactory and effective plasma levels over delayed periods of time, still remains a critical challenge. Also, conceiving modern strategies for the fabrication of nanoparticles at sensible costs is a vital portion of this challenge since there is a small number of them that fulfill the appropriated necessities to reach the target and, along these lines, provide the drug in an appropriate way. It is additionally obligatory for these polymeric nanoparticles to be biodegradable or possess a tall capacity to be disposed of on the exterior of the body dodging aggregation, being nontoxic and non-immunogenic. Over the final long time, there has been a worldwide change within the field of nanomedicine, which has driven to a multidisciplinary and collaborative approach with promising comes about and success. The future way of collaborations between hypothetical and exploratory researchers and the pharmaceutical industry, doctors, and administrative organizations will be pivotal and will allow us to actualize the research facility results in the clinic and start the following generation of clinical treatments, attempting to minimize the obliterating results of awful illnesses such as pandemic COVID-19.

17.4 Automated Mass Screening Methods Employed at Pharmaceutical Companies

There are three challenges for mechanized drug design, which are the get-together of synthetically accessible structures, scoring and property expectation, and the systematic optimization of promising particles in adaptive learning cycles, and in the past three decades, various rules, methods, algorithms, and heuristics have been proposed to address each of these issues. Forecasts of the maintainability of customary pharmaceutical revelation and development hones infer the requirement for adjusted techniques for the future (Schneider 2018). Hence, the generation of modern chemical entities with appealing chemical scaffolds has gotten to be attainable, and in spite of the fact that the algorithmic optimization issue can also be considered generally fathomed, the persisting issue of compound scoring, and more exact movement prediction models that expand the capabilities of existing approaches seem to start from advanced machine learning strategies. Therefore, with human organoid measure frameworks, automated synthesis, and brilliantly learning software, we are witnessing an unused wave of fervour regarding the changes in pharmaceutical research and development (Mignani et al. 2016; Gautam and Pan 2016). The automation of science is considered the promise of making way better choices faster, and in drug discovery, robotized frameworks, as of now, have a long and productive history, and robotic screening assays are nowadays the base in the pharmaceutical industry. Moreover, the expansion of various applications of automated frameworks from decision support systems to computational molecular purposes completely fledged automated unions. Various models inclusive traditional rule-based and model-based approaches such as the archetypal DENDRAL system for analyzing mass spectra, LHASA software for manufacture planning, analyzing chemical and biological data similar to Amgen's AADAPT system as well as tools for software for de novo molecular design, and prototypical robotic systems such as ADAM and EVE for automated target (Johnson and Marshall 1992; Lindsay et al. 1993; Cho et al. 2006; Schneider 2013). MacConnell et al. recently discovered a microfluidics-based system for ultra-high-throughput hit deconvolution by series, and it disseminates DNA-encoded compound beads into picolitre-scale droplets, cleaves off the compounds from the beads via ultraviolet (UV) irradiation, and implements a fluorescence-based binding assay (MacConnell et al. 2017). Researchers at AbbVie displayed other techniques that develop an integrated robotic platform for the automated parallel manufacture of small, focused compound libraries and built from commercial components, and they are able to implement liquid handling, evaporation, analytics, purification, and activity testing (Baranczak et al. 2017). Nowadays, artificial intelligence is extending various applications of semi-autonomous or even fully autonomous processes to support project teams to identify and optimize tools in the field of drug discovery. The advantages of automation are diminishing measurement errors and decreasing material consumption, fast feedback loops, and compound optimization and molecular design toward biochemical and biological end points. Moreover, the application of complex cell-based assays to summarise the biology of disease and the process that increases the possibility of identifying substances that seem to be useful in humans (Esch et al. 2015). Another new technique employed in drug design and biological and chemical sciences is high-throughput screening (HTS), which is based on robots, detectors, and software that make an analysis of chemical compounds and give results in a short time, and the resemblance of biological structures is related to toxicity (Szymański et al. 2012). This technique is used in conjunction with analytical techniques such as NMR or coupled methods such as LC-MS/MS, and it is utilized by industrial scientists and academic researchers. It consists of many stages such as target identification, reagent preparation, compound management, assay development and high-throughput library screening, and the efficient nature of its precise focus on a single mechanism to identify the target-specific compounds. Other techniques such as fluorescence resonance energy transfer (FRET) and homogeneous time-resolved fluorescence (HTRF) are accessible for the identification of compounds (Szymański et al. 2012). Furthermore, HTS is utilized by miniaturized cell-based assays which enables chemical libraries to be screened for molecules that have various biological activities. Many molecules can be arrayed such as small molecules, polymers, and antibodies by utilizing robotic spotting technology or soft lithography and then can be assessed for their toxicity (Soen et al. 2003; How et al. 2004; Orner et al. 2004). In addition, advances in systems biology, molecular biology, and bioinformatics have prompted the use of contemporary toxicology techniques, such as computational toxicology, predictive quantitative structure–activity relationship (QSAR) toxicity modelling, and predictive ADME-Tox, which are used in the pharmaceutical industry during the design stages to produce lead compounds with low toxicological potential. Furthermore, these methods authorize improving drug discovery, development, and prediction of toxicity from chemical structure. These days, due to developing societal and moral concerns, researchers (toxicologists, chemists, and modelers) make endeavors to obtain modern bioassay profiling information of potential pertinence to toxicology. Furthermore, other methods such as molecular docking, absorption, distribution, metabolism, and excretion (ADME) are methods that enable the creation of framework to search a database as a virtual library of drugs to extract potential drug candidates (Ojha et al. 2021). All these novel strategies demonstrate to have a critical effect on toxicology; they contribute to significant progress in managing the harmfulness potential of unused compounds. Researchers are presently seeking out new biomarkers for poisonous quality that will contribute to high-throughput preclinical security appraisal. Moreover, HTS techniques have progressed to determine the efficacy of chemicals against modulating the drug-metabolizing enzymes, and it utilizes the activity of cytochrome CYP to enable the establishing of pharmacokinetic drug interaction and is of pivotal importance in the drug development industry. Today, the demand to understand the individual genetic variability in the development of many diseases, drug response, and toxicities exceeds. Thus, the screening of various genetic polymorphisms in large populations is considered a major goal. Laser-desorption ionization time-of-flight mass spectrometry (MALDI-TOF MS) is known as one of these methods which is a vigorous technique for DNA analysis, it has been developed for fast screening of single nucleotide polymorphisms, in vitro studies before preclinical screening low-throughput assays are implemented, and the information was obtained via incubating a tested drug with an suitable system; hence, it can be used to design safer and more metabolically stable drugs. Moreover, Ritter et al. progressed an integrated higher throughput method for the comet assay to determine DNA damage in vivo and in vitro (Ojha et al. 2021). Moreover, high-throughput NMR-based screening is utilized for structural characterization of protein-ligand interactions and identification of compounds that bind to specific protein targets, and there are other techniques such as FIA/DI-MS, MALDI-FT-MS, DI-NMR, HPLC-UV/MS, HPLC-NMR, SFC-MS, and ESI-FT-ICR-MS utilized also in structural characterization of compounds (Tutko et al. 1998; Wang et al. 1998; Hajduk and Burns 2002; Kumar and Clark 2006). Thin-layer chromatography (TLC)-bioautography is considered one of the most important pathogenic strains of fungi and bacteria (Lage et al. 2018). It consists of three techniques such as agar contact method, direct bioautography, and agar medium-overlay assay. Another technique utilized is flow cytometry, which detects live/dead cells, and some researchers employed it to test the effect of compounds/extracts in microorganisms Various researches have exhibited the effectiveness of the flow cytometer for antibacterial testing utilizing combined staining of DNA with propidium iodide (Paparella et al. 2008). Also, colorimetric-based assays utilizing crystal violet or safranin are employed to assess extracts' ability to disrupt biofilm formation on the base of identical procedures, and laser confocal microscopy can also be implemented to detect the biofilm formation (Qian and Xu 2012). Examples of robotics used in high-throughput screening laboratory are illustrated in Fig. 17.5.

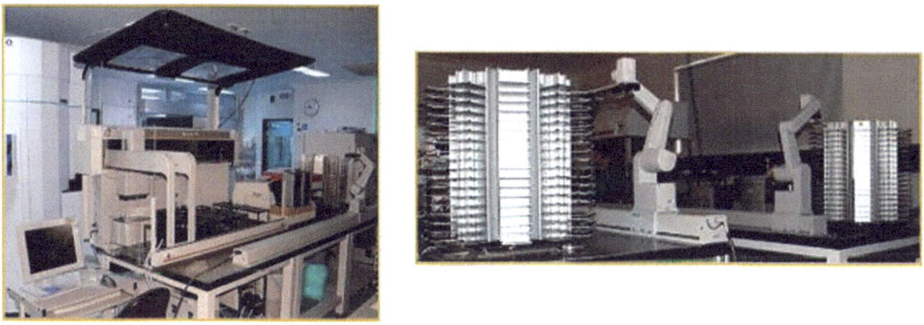

Fig. 17.5 Examples of robotics used in high-throughput screening laboratory

17.5 In Vitro and In Vivo Evaluation Tools of SARS-CoV-2 Antiviral Drugs

The spread of severe acute respiratory syndrome coronavirus SARS-CoV-2 infection has driven to a critical requirement for a remedy that basically underlines sedate repositioning techniques that are able to efficiently recognize unused and dependable medicines by consolidating computational, experimental, and pharmacokinetic ability (Bocci et al. 2020). Furthermore, it resulted in radical security measures like "work from home" and "safe social distancing," which were implemented for weeks or longer in numerous nations and reduced the amount of animal and clinical research conducted globally. The researches on "live" SARS-CoV-2 infections requires Biosafety Level 3 (BSL-3) biocontainment conditions, which place extra burdens on our capacity to distinguish modern effective interventions. Till now, remdesivir drug is only approved by the US FDA (2020). On the other hand, a few studies showed that treatment of mild-to-moderate or mild-to-severe COVID-19 with hydroxychloroquine alone or in combination with azithromycin did not progress clinical status or length of viral shedding in comparison with standard care (Gendrot et al. 2020). Because of disastrous burdens and associated passings, there is a rigid way to quickly find therapeutic intercessions. Thus, this is often a fast-moving widespread, drug repositioning is one of the exceedingly dynamic areas of biomedical research that parallels antibody research and advancement in the fight to halt COVID-19, and many drugs have already been proposed as unused treatments for COVID-19 and are subject to clinical trials (Liu et al. 2020; Lurie et al. 2020; Sanders et al. 2020). Besides, distinguishing viable, low-cost antiviral drugs with limited side impacts, which are adorable instantly, particularly for developing nations, is critically required. A new second-generation tetracycline is doxycycline, which has broad-spectrum antimicrobial, anti-inflammatory, and antiviral activity in vitro, and it was approved as prophylaxis against malaria by the Food and Drug Administration in 1994 (Fig. 17.6) (Michalopoulos 1973; Cazalis et al. 2008; Gaillard et al. 2015). It inhibited the replication of the vesicular stomatitis virus in vitro and the dengue virus via the inhibition of NS2B-NS3 serine protease (Yang et al. 2007; Rothan et al. 2014, 2015; Wu et al. 2015). Besides, it was used as a

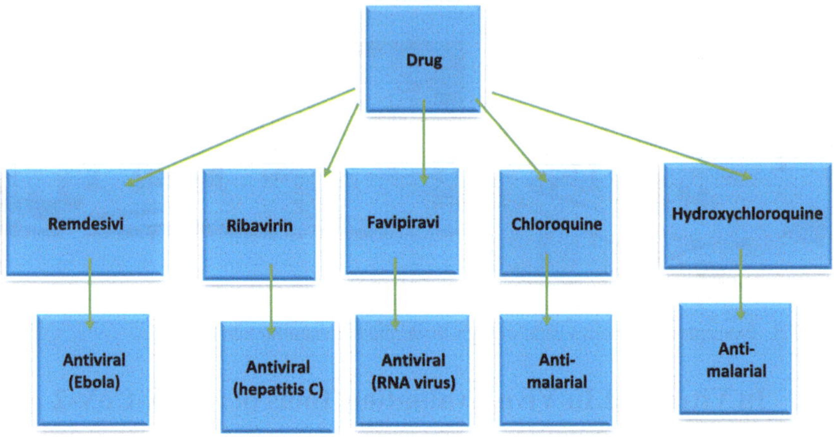

Fig. 17.6 Some of in vitro activity for drug repositioning against COVID-19

potential inhibitor of Crimean-Congo hemorrhagic fever virus nucleoprotein (Rothan et al. 2015). Furthermore, the scientific community proceeds to explore other approved drugs that will exhibit anti-SARS-CoV-2 activity, typically by combining tall throughput screening (HTS) and in silico procedures. Newley, Kuleshov and coworkers have assembled a database for these studies (Kuleshov et al. 2020). It has been found that there is a shortage of consistency between various in vitro studies focusing on the requirement for independent confirmation. Ellinger et al. utilized Caco-2 cells in their assay, while the other studies applied Vero E6 cells when estimating anti-SARS-CoV-2 activity (Ellinger et al. 2020). Moreover, FDA-approved antimicrobial drugs such as Azithromycin, Niclosamide, and Nitazoxanide, which exhibited a promising capability to prohibit the replication of a SARS-CoV-2 isolate, and the evidence showed that various antihistamine and anti-inflammatory drugs could partially decrease SARS-CoV-2 replication in vitro (Mostafa et al. 2020). Prophylactic drugs are other drugs that can be utilized both to piece pathogen's infectious cycle and to boost insusceptibility (Andrade et al. 2020). There are two models of prophylaxis (I) and (II) have been utilized in endemic viral pathologies such as HIV (Krakower et al. 2015). Recently, various preclinical studies, such as in silico, in vitro, and in vivo trials, have been guiding clinical decisions to select the best drug options for the treatment and prophylaxis against SARS-CoV-2 (Fragkou et al. 2020). In this manner, diverse drug classes with prophylactic properties have been repositioned in order to ensure security against viral transmission (Zhou et al. 2020). Thus, this drives to a curious procedure targeting COVID-19 since it can be utilized as an extra boundary to viral spreading, as well as avoiding malady advancement, particularly for patients interior the hazard bunches. Furthermore, the nine drug candidates that presented a selective antiviral index greater than 2 against SARS-CoV-2 Calu-3 cells are salinomycin, Y-320, AZD8055, bemcentinib, dacomitinib, WYE-125132, ebastine, Dp44mT, and cyclosporine

(Limburg et al. 2019). Conversely, clinical tests involving previously endorsed drugs against the novel coronavirus require a variety of sedate choices, encompassing a number of patients across several health conditions and control groups; the evaluation period is justifiable, considering that less than 7 months have passed since the onset of the SARS-CoV-2 episode; and numerous vaccination choices are currently undergoing diverse clinical testing phases in numerous countries, with no guarantee of resolving this global problem. Hence, the utilization of clinical protocol-based scientific evidence information for SARS-CoV-2 prophylaxis, as well as for the daily schedule in healing centers for its treatment, is pivotal for controlling the infection spread, prognostic, and understanding recovery and can in fact offer assistance spare numerous lives. Recently, a high-throughput screening test was implemented to examine the candidates for potential use in COVID-19 treatment from FDA-approved small-molecule drugs. Vero E6 cells and a SARS-CoV-2 strain disconnected in Walk 2020 in Shanghai, and the results of screening assay were subscribed with remdesivir and hydroxychloroquine, which inhibited SARS-CoV-2 infection in vitro (Wang et al. 2020; Peng et al. 2022). Besides, about of 2580 FDA-approved small-molecule drugs were screened for anti-SARS-CoV-2 activity, and 94 compounds from them exhibited vigorous antiviral activity against SARS-CoV-2 that inhibit the infection of SARS-CoV-2. Lately, researchers showed the usage of antiviral derived from algae, and various publications have checked on the bland properties of green growth in these zones (Petit et al. 2021). Talukdar et al. showed the immune response in the cytokines with astaxanthin. Griffithsin, a lectin isolated the from red alga Griffithsia, has shown efficacy on protein S of MERS-CoV (Talukdar et al. 2020b). Astaxanthin and linolenic acid have been showed as potential candidates for the treatment or the prohibition of COVID-19 (Talukdar et al. 2020a). Arthrospira compounds are genus of prokaryotic filamentous cyanobacteria which are mostly utilized as food supplement due to their nutritional properties as they contain high protein content, vitamins, and minerals, and they found to have antiviral activity. Moreover, studies explored that anionic polysaccharides from Arthrospira have antiviral activity against enveloped virus such as VACV. Chen et al. (2016a) showed that Arthrospira extract was capable of inhibiting in vitro influenza viral replication and plaque formation (Chen et al. 2016a). Also, a pre-published work publicly available is focused on the potential of Arthrospira on two viral proteins of SARS-CoV. According to surveys, nanotechnology offers advantages over the traditional framework of conclusion. Examples of these advantages include metallic nanoparticles, micelles, liposomes, and polymeric and lipid nanoparticles. The merits of the use of nanoparticle incorporate the proper embodiment of drugs, moderate discharge of drugs, and assisting within the advancement of properties of sedate particles. Recently, researchers have drawn attention to antiviral synthetic nanoparticles which can mimic the virus and interact strongly with its virulent proteins due to their morphological similarities such as zinc oxide nanoparticle and carbon-based nanomaterials which have an antiviral activity (Tharayil et al. 2021). Moreover, researchers have utilized gold nanoparticles for the evolution of a rapid detection kit for IgG and IgM antibodies against SARS-CoV-2 (Huang et al. 2020). Also, two-dimensional nanoparticles coupled with functional DNA receptor are used to evolve the biosensor for diagnosis of SARS-CoV-2 (Srivastava et al. 2021). Scientists exhibited that chitosan nanoparticle may be used as a transport drug against SARS-CoV-2, and it has very good mucoadhesive properties which enables in the reduction of intestinal reaction caused due to COVID-19 infection (Kolawole et al. 2018). Loczechine et al. exhibited that carbon quantum dots when combined with boric acid have the inhibitory effect on the human coronavirus HCoV-229E (Łoczechin et al. 2019). Additionally, a nano-based biosensor could provide rapid, simple, sensitive SARS-CoV-2 diagnosis and treatment (Morales-Narváez and Dincer 2020). More, surveys showed that gold NPs were utilized for the construction of electrochemical biosensor in order to detect SARS-CoV-2 spike S1 protein antigen. *Vivo* applications showed therapeutic activity of silver nanoparticles and gold nanoparticles. Also, copper ions in the form of sulfates and iodides were utilized as antiviral agents (Horie et al. 2008; Fujimori et al. 2012).

Few considerations were practiced by focusing on the characteristic products and their subsidiaries for their restorative applications and antiviral activities against the SARS-CoV-2, which were demonstrated to be very compelling and zone of research within the coming days (Al-Harrasi et al. 2022). Tannins phenols and polyphenols terpenes have been disclosed to be active participants in the SARS-CoV-2 viruses, and alkaloid-rich plants showed antiviral activity. Furthermore, essential oils and aromatic extracts used as food flavorings and constituents of fragrance compositions showed activity against pathogenic human coronaviruses. So, it gives criteria for in vitro and in vivo testing of the most essential oils, aromatic extracts, and isolated natural compounds (Strub et al. 2022). Thus, there is a critical require for improvement and evaluation of particular antivirals against COVID-19. Investigations into the efficacy of already existing antivirals, potent antivirals, or a combination alongside potent remedies have been conducted. However, the focus should be on planning and adopting secure and effective modalities against SARS-CoV-2. This includes the development of evolving neutralizing antibodies, oligonucleotides targeting SARS-CoV-2 RNA genome, passive antibody transfer, drug repurposing using available antivirals, utilizing antiviral proteases, blocking coronavirus receptors such as ACE2, targeting spike proteins, and exploring combination therapies (Rabaan et al. 2020).

17.6 Clinical Trials and Its Regulation

Medical treatment of diseases in clinical trials is based on the beneficence principle, which stresses "Do no harm," with a maximum benefit to the included population in clinical research (Ogden 2019; Salihu et al. 2015). The regulatory rules in clinical trials aim to yield ethically sound and scientifically rigorous results to guide medical practice. The importance of regulating clinical trials is reflected by the steep increase in the number of registered trials from 5600 in 2000 to 185,000 in 2015, which is before the occurrence of the pandemic (Azuine and Ekejiuba 2015).

17.6.1 Clinical Trial Regulation

Clinical trial development needs 10–15 years on average and a total budget of 1–2 billion dollars in the USA alone in order to yield satisfactory results (Harrer et al. 2019). Phases of clinical trials include preclinical trials (animal or in vitro trials), clinical trials (phase I, II, and III trials), and lastly, post-marketing (phase IV trials) (Umscheid et al. 2011).

17.6.2 Risk of Bias and GRADE Assessments in the Meta-analysis of Randomized Control Trials

Conducting a meta-analysis of multiple RCTs on the same drug will benefit the clinical practice by connecting the outcomes of similar interventions in a larger number of similar populations. This will detect the generalized validity of the drug's effect on the researched population, and it will detect the various important outcomes that may not be examined in some of the trials but examined in others (serious side effects, withdrawal due to side effects, hospitalization days, mortality, survival, etc.).

The tools for assessing these collective clinical trials include the risk of bias and the assessment tools, indicating the level of adherence to the clinical trial regulations (the items are presented in Fig. 17.7) (Viswanathan et al. 2018). Moreover, the GRADE (Grading of Recommendations, Assessment, Development, and Evaluations) is an assessment tool framework for examining the quality of evidence presented in the trial, including risk of bias, imprecision, inconsistency, indirectness, and publication bias (Kirmayr et al. 2021).

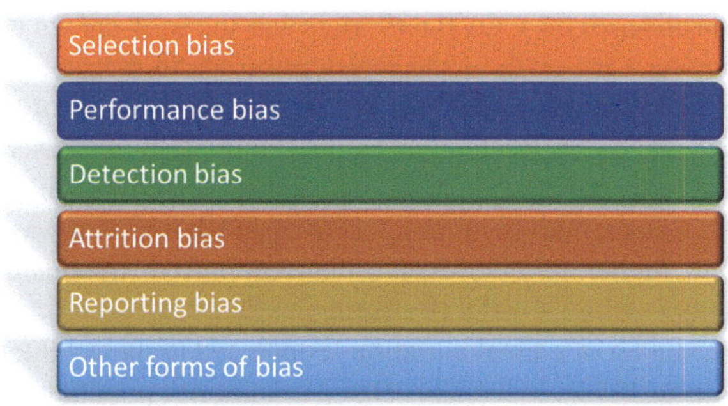

Fig. 17.7 Items used in the ROB assessment tools

17.6.3 Clinical Trials in SARS-CoV-2 or in Other Pandemics

The COVID-evidence database (www.covidevidence.org) was initiated at the beginning of the pandemic to give decision makers insights into how the clinical research strategies were going (Janiaud et al. 2021). However, one of the COVID-19 challenges and future concerns facing the medical field in COVID-19 clinical trials was that the decision makers focused on only the information that would help them make decisions during the pandemic, leaving a plethora of information needed for future dealings with the virus or with other pandemics uncollected (Robinson et al. 2022).

Clinical trials during the pandemic require a faster pace for recruiting and randomizing the study population. This is achieved by new updated methods such as REMAP (Randomized Embedded Multifactorial Adaptive Platform) (Angus et al. 2020). However, REMAP can only be used for critically ill and intensive care patients, while non-hospitalized patients still have no established method for their recruitment and randomization (Robinson et al. 2022).

COVID-19 had a notable impact on the conduction of clinical trials in the USA and Europe, leading to a decrease in the number of clinical trials for non-COVID-19 trials, especially in the USA (Butler et al. 2023).

In addition, during the pandemic, the trial resources, personnel, and sites were used by the public healthcare emergency, and this led to a decline in the feasibility of conducting clinical trials during the pandemic (Flores-Vega et al. 2022). Moreover, the public scare and psychological distress during the pandemic led to a decrease in attendance at the trial sites (Ma et al. 2021).

Another challenge that affected the conduction of COVID-19 clinical trials was the conduction of small-sample trials, which may give results that are specific to certain populations or may mask serious low-incidence adverse events. In COVID-19 trials, only 18% had a sample size of more than 500 patients. In addition, most of the conducted trials were concerned with interventional treatments and only 11% were concerned with preventive methods or medications (Kirmayr et al. 2021). This may increase the load or the time of infection, as COVID-19 is an airborne virus that needs strict evidence-based regulations for its prevention. This will delay the formulation of specific high-evidence preventive regulators for this pandemic.

One of the surprises of the COVID-19 trials was the well-known non-patent drug hydroxychloroquine. One-sixth of the registered trials in the first 100 days examined this drug. Unfortunately, after conducting a meta-analysis of these trials, the drug was found to be ineffective and associated with increased mortality in COVID-19 cases (Kirmayr et al. 2021). Another drug that failed in the post-marketing phase was molnupiravir. In a recent open label trial, it was found to be of less efficacy and higher adverse events than supportive treatment alone in un-hospitalized patients with COVID-19 (Butler et al. 2023).

There had been a lot of ethical violations connected to the conduction of clinical trials in the 1940s to 1970s, especially related to minority populations (children, the elderly, and ethnic or racial minority groups). This left the research populations who were in need of clinical trials with strict regulations unable to trust the system and accept to be recruited in trials. In addition, poorer populations are much more vulnerable to these unethical behaviors in both developing and developed countries (Azuine and Ekejiuba 2015).

In 1993, the National Institute of Health (NIH) created a Revitalization Act, which led to an increase in the recruitment of diverse populations in clinical trials. This was reflected in the constant elevation of their numbers (fourfold increase) across the years. In 2018, the USA discovered a recurring imbalance in non-white Hispanic and black populations when compared to white populations across their clinical trials (Ma et al. 2021).

The informed consent is essential to protecting the research populations from involuntary recruitment or their retention in clinical trials. Low retention rates can lead to high attrition bias in the intended outcome, thus weakening the validity of the study (Salihu et al. 2015; Nunan et al. 2018).

The global clinical trials portal Clinical Trials Government displayed for the treatment of COVID-19 these drugs chloroquine and/or hydroxychloroquine, ivermectin, favipiravir, colchicine, remdesivir, dexamethasone, methylprednisolone, nitazoxanide, azithromycin, camostat, and baricitinib.

Chloroquine was found to be the potent drug that can prevent infection by the virus with EC50: 1.13 μM and EC90:6.90 μM (Vaz et al. 2023).

Moreover, trials involved over 7000 patients in different countries and exhibited that chloroquine and hydroxychloroquine have no influence on reducing mortality, initiation of mechanical ventilation, and the duration of hospitalization in patients with COVID-19 (WHO 2021). After 3 months from starting larger clinical studies, the FDA revoked the emergency authorization of these two drugs used for the treatment of COVID-19 on the basis of the non-reproducibility of clinical results.

Twelve clinical studies showed the best results for the usage of ivermectin, and Babalola et al. showed that the inhibitory effect of ivermectin on SARS-CoV-2 is dose-dependent (Babalola et al. 2022).

Later on, three clinical studies on colchicine displayed that it decreased the risk of the primary endpoint and complications in non-hospitalized patients with COVID-19. However, more studies are needed for the prolonged treatment of patients in critical conditions (Tardif et al. 2021).

In October 2020, three clinical studies approved remdesivir for the treatment of COVID-19 for reducing mortality, hospitalization time, or reining the usage of mechanical ventilation for any of the participants (WHO 2021).

Later on, trials by Wang et al. showed that nitazoxanide can inhibit SARS-CoV-2 in vitro assay displaying an EC_{50} of 2.12 μM, and it was recommended for in vivo studies as a performer drug (Wang et al. 2020).

Further, 1525 participants included in the trials displayed approval for the usage of baricitinib drug in the treatment of COVID-19 by decreasing the mortality rate or progression to invasive mechanical ventilation (Marconi et al. 2021).

Some nanoparticles, lipid-based types in particular, played a significant part in the promotion and manufacturing of the two COVID-19 vaccines—Pfizer and Moderna that are broadly utilized. A survey of 486 clinical trials utilizing nanoparticles between 2002 and 2021, as well as the recent FDA-approved drugs since 2016, was identified (Namiot et al. 2023).

It was noticed that liposomal drugs are the most often merged with other drugs compared with other types of nanoparticles. Furthermore, the most common type of paclitaxel (57 trials out of 62 trials total) is a mixture of numerous nanoparticles loaded with both paclitaxel and NAB-paclitaxel. We anticipate more increments within the number of trials for polymeric and lipid-based nanoparticles. Both progressing clinical trials and FDA-approved drugs recommend the rise of modern types of nanoparticles that ought to be extensively explored. The differences in the influence of adenovirus vector and mRNA vaccines on the mortality will have a major impact on global health, and the additional surveys will exhibit the protective non-specific impact of adenovirus-based vaccines on non-COVID-19 mortality.

Rich countries in Europe and the USA have observed that the more costly mRNA vaccines have been proven to be more effective against COVID-19 in the short term than the less expensive adenovirus vector vaccines. Additionally, the Adenovirus vector vaccines have been linked to a rare blood clotting disorder. These findings primarily concern AstraZeneca (Benn et al. 2023).

Later on, imatinib's clinical trials were primarily observed on patients who had been admitted to the intensive care unit. The findings showed that the medication shortened the time spent in the intensive care unit and on invasive ventilation, decreased mortality after a 90-day follow-up, and decreased pulmonary edoema (Atmowihardjo et al. 2023).

On February 8, 2021, and November 16, 2021, toward seven sites in Saudi Arabia and in Kuwait, the Helmet-COVID trial was used as a trial for an investigator-initiated, pragmatic, and multicenter randomized which could decrease the mortality in patients with acute hypoxemic respiratory failure due to COVID-19 (Arabi et al. 2023).

Clinical trials on Bacillus Calmette-Guérin (BCG) vaccine have attracted more interest because it enhances responsiveness and protection against a range of infections, and it can lower hospitalization, severe disease, and mortality of COVID-19 (Gong et al. 2023).

17.6.4 Future Unresolved Dilemmas During and After Conducting the Clinical Trials

Phase III clinical trials carry the highest cost burden of the RCT (60%). The recruitment takes about one-third of the trial's duration. Thus, any delay in enrollment will result in financial loss and time delay. AI could help in selecting the patients for trial enrollment according to the trial inclusion criteria in large cohort datasets (Harrer et al. 2019). Another problem facing innovative medical therapies after passing the clinical trials is patent pricing. This may lead to unfair supply to poorer geographical areas or to those with lower socioeconomic status in the same country. We have observed this in the unbalanced supply of COVID-19 vaccines to different areas of the world (Robinson et al. 2022).

Moreover, the clinical trial guidelines, SPIRIT-AI, and reporting CONSORT-AI have been modified concerning the use of AI in pathological assessment in order to constitute a framework for using deep neural learning in clinical trials. Issues such as how to use AI, what data sets are preferred, and what errors and failures are encountered in the AI system were addressed in this proposal (McGenity and Treanor 2021).

We can face the next pandemic with new technologies such as biocomputing, nanotechnology, and phytopharming. These innovative technologies can help accelerate the speed of drugs reaching the clinical trial phase in times of urgent need, similar to the COVID-19 pandemic (Ogbadoyi and Umar 2022).

References

Al-Harrasi A, Behl T, Upadhyay T et al (2022) Targeting natural products against SARS-CoV-2. Environ Sci Pollut Res Int 29:42404–42432. https://doi.org/10.1007/s11356-022-19770-2

Ameeduzzafar ISS, Abbas Bukhari SN et al (2018) Formulation and optimization of levofloxacin loaded chitosan nanoparticle for ocular delivery: in-vitro characterization, ocular tolerance and antibacterial activity. Int J Biol Macromol 108:650–659. https://doi.org/10.1016/j.ijbiomac.2017.11.170

Andrade BS, Rangel F de S, Santos NO et al (2020) Repurposing approved drugs for guiding COVID-19 prophylaxis: a systematic review. Front Pharmacol 11:590598. https://doi.org/10.3389/fphar.2020.590598

Angus DC, Berry S, Lewis RJ et al (2020) The REMAP-CAP (randomized embedded multifactorial adaptive platform for community-acquired pneumonia) study. Rationale and design. Ann Am Thorac Soc 17:879–891

Arabi YM, Al Dorzi HM, Aldekhyl S et al (2023) Long term outcomes of patients with COVID 19 treated with helmet noninvasive ventilation or usual respiratory support: follow up study of the Helmet COVID randomized clinical trial. Intensive Care Med 49:302–312

Atmowihardjo LN, Schippers JR, Duijvelaar E et al (2023) Efficacy and safety of intravenous imatinib in COVID 19 ARDS: a randomized, double blind, placebo controlled clinical trial. Crit Care 27:226

Azuine RE, Ekejiuba SE (2015) Overcoming challenges in conducting clinical trials in minority populations: identifying and testing what works. Int J MCH AIDS 3:81–84

Babalola OE, Bode CO, Ajayi AA et al (2022) Ivermectin shows clinical benefits in mild to moderate COVID19: a randomized controlled double-blind, dose-response study in Lagos. QJM Int J Med 114:780–788

Baranczak A, Tu NP, Marjanovic J et al (2017) Integrated platform for expedited synthesis-purification-testing of small molecule libraries. ACS Med Chem Lett 8:461–465. https://doi.org/10.1021/acsmedchemlett.7b00054

Begines B, de-Paz M-V, Alcudia A, Galbis JA (2016) Synthesis of reduction sensitive comb-like polyurethanes using click chemistry. J Polym Sci Part A Polym Chem 54:3888–3900. https://doi.org/10.1002/pola.28367

Begines B, Alcudia A, Aguilera-Velazquez R et al (2019) Design of highly stabilized nanocomposite inks based on biodegradable polymer-matrix and gold nanoparticles for Inkjet Printing. Sci Rep 9:16097. https://doi.org/10.1038/s41598-019-52314-2

Begines B, Ortiz T, Pérez-Aranda M et al (2020) Polymeric nanoparticles for drug delivery: recent developments and future prospects. Nanomater (Basel, Switzerland) 10. https://doi.org/10.3390/nano10071403
Benn CS, Schaltz-Buchholzer F, Nielsen S et al (2023) Randomized clinical trials of COVID-19 vaccines: do adenovirus-vector vaccines have beneficial non-specific effects? I Science 26:106733
Bocci G, Bradfute SB, Ye C et al (2020) Virtual and in vitro antiviral screening revive therapeutic drugs for COVID-19. ACS Pharmacol Transl Sci 3:1278–1292. https://doi.org/10.1021/acsptsci.0c00131
Botti G, Di Bonito M, Cantile M (2021) Organoid biobanks as a new tool for pre-clinical validation of candidate drug efficacy and safety. Int J Physiol Pathophysiol Pharmacol 13:17–21
Bruschi ML (2015) Strategies to modify the drug release from pharmaceutical systems. Woodhead Publishing, Cambridge
Butler CC, Hobbs FDR, Gbinigie OA et al (2023) Molnupiravir plus usual care versus usual care alone as early treatment for adults with COVID-19 at increased risk of adverse outcomes (PANORAMIC): an open-label, platform-adaptive randomised controlled trial. Lancet 401:281–293
Cakir B, Xiang Y, Tanaka Y et al (2019) Engineering of human brain organoids with a functional vascular-like system. Nat Methods 16:1169–1175. https://doi.org/10.1038/s41592-019-0586-5
Cannalire R, Cerchia C, Beccari AR et al (2020) Targeting SARS-CoV-2 proteases and polymerase for COVID-19 treatment: state of the art and future opportunities. J Med Chem 65:2716–2746
Cazalis J, Bodet C, Gagnon G, Grenier D (2008) Doxycycline reduces lipopolysaccharide-induced inflammatory mediator secretion in macrophage and ex vivo human whole blood models. J Periodontol 79:1762–1768. https://doi.org/10.1902/jop.2008.080051
Chen C-C, Hsu L-W, Nakano T et al (2016a) DHL-HisZn, a novel antioxidant, enhances adipogenic differentiation and antioxidative response in adipose-derived stem cells. Biomed Pharmacother 84:1601–1609. https://doi.org/10.1016/j.biopha.2016.10.066
Chen Z, Shi T, Zhang L et al (2016b) Mammalian drug efflux transporters of the ATP binding cassette (ABC) family in multidrug resistance: a review of the past decade. Cancer Lett 370:153–164. https://doi.org/10.1016/j.canlet.2015.10.010
Cho SJ, Sun Y, Harte W (2006) ADAAPT: Amgen's data access, analysis, and prediction tools. J Comput Aided Mol Des 20:249–261. https://doi.org/10.1007/s10822-006-9048-0
Corrò C, Novellasdemunt L, Li VSW (2020) A brief history of organoids. Am J Physiol Cell Physiol 319:C151–C165. https://doi.org/10.1152/ajpcell.00120.2020
de Melo BAG, Benincasa JC, Cruz EM et al (2021) 3D culture models to study SARS-CoV-2 infectivity and antiviral candidates: from spheroids to bioprinting. Biom J 44:31–42. https://doi.org/10.1016/j.bj.2020.11.009
Dekkers JF, Wiegerinck CL, de Jonge HR et al (2013) A functional CFTR assay using primary cystic fibrosis intestinal organoids. Nat Med 19:939–945. https://doi.org/10.1038/nm.3201
Eiraku M, Takata N, Ishibashi H et al (2011) Self-organizing optic-cup morphogenesis in three-dimensional culture. Nature 472:51–56. https://doi.org/10.1038/nature09941
Ellinger B, Bojkova D, Zaliani A, et al (2020) Identification of inhibitors of SARS-CoV-2 in-vitro cellular toxicity in human (Caco-2) cells using a large scale drug repurposing collection
Engevik MA, Engevik KA, Yacyshyn MB et al (2015) Human Clostridium difficile infection: inhibition of NHE3 and microbiota profile. Am J Physiol Gastrointest Liver Physiol 308:G497–G509. https://doi.org/10.1152/ajpgi.00090.2014
Esch EW, Bahinski A, Huh D (2015) Organs-on-chips at the frontiers of drug discovery. Nat Rev Drug Discov 14:248–260. https://doi.org/10.1038/nrd4539
Ettayebi K, Crawford SE, Murakami K et al (2016) Replication of human noroviruses in stem cell-derived human enteroids. Science 353:1387–1393. https://doi.org/10.1126/science.aaf5211
FDA (2020) Coronavirus (COVID-19) update: FDA issues emergency use authorization for potential COVID-19 treatment. FDA. https://www.fda.gov/news-events/press-announcements/

coronavirus-covid-19-update-fda-issues-emergency-use-authorization-potential-covid-19-treatment

Filippo Buono M, von Boehmer L, Strang J et al (2020) Human cardiac organoids for modeling genetic cardiomyopathy. Cells 9. https://doi.org/10.3390/cells9071733

Flores-Vega VR, Monroy-Molina JV, Jiménez-Hernández LE et al (2022) SARS-CoV-2: evolution and emergence of new viral variants. Viruses 14:653

Fordham RP, Yui S, Hannan NRF et al (2013) Transplantation of expanded fetal intestinal progenitors contributes to colon regeneration after injury. Cell Stem Cell 13:734–744. https://doi.org/10.1016/j.stem.2013.09.015

Fragkou PC, Belhadi D, Peiffer-Smadja N et al (2020) Review of trials currently testing treatment and prevention of COVID-19. Clin Microbiol Infect Off Publ Eur Soc Clin Microbiol Infect Dis 26:988–998. https://doi.org/10.1016/j.cmi.2020.05.019

Fujimori Y, Sato T, Hayata T et al (2012) Novel antiviral characteristics of nanosized copper(I) iodide particles showing inactivation activity against 2009 pandemic H1N1 influenza virus. Appl Environ Microbiol 78:951–955. https://doi.org/10.1128/AEM.06284-11

Gaillard T, Madamet M, Pradines B (2015) Tetracyclines in malaria. Malar J 14:445. https://doi.org/10.1186/s12936-015-0980-0

Gandhi S, Klein J, Robertson AJ et al (2022) De novo emergence of a remdesivir resistance mutation during treatment of persistent SARS-CoV-2 infection in an immunocompromised patient: a case report. Nat Commun 13:1547

Gao W, Chen Y, Thompson DH et al (2016) Impact of surfactant treatment of paclitaxel nanocrystals on biodistribution and tumor accumulation in tumor-bearing mice. J Control release Off J Control Release Soc 237:168–176. https://doi.org/10.1016/j.jconrel.2016.07.015

Garcez PP, Loiola EC, Madeiro da Costa R et al (2016) Zika virus impairs growth in human neurospheres and brain organoids. Science 352:816–818. https://doi.org/10.1126/science.aaf6116

Gautam A, Pan X (2016) The changing model of big pharma: impact of key trends. Drug Discov Today 21:379–384. https://doi.org/10.1016/j.drudis.2015.10.002

Gendrot M, Andreani J, Jardot P et al (2020) In vitro antiviral activity of doxycycline against SARS-CoV-2. Molecules 25. https://doi.org/10.3390/molecules25215064

Gong W, Liu Y, Xue Y et al (2023) Two issues should be noted when designing a clinical trial to evaluate BCG effects on COVID-19. Front Immunol 14:1207212

Gonzalez C, Armijo E, Bravo-Alegria J et al (2018) Modeling amyloid beta and tau pathology in human cerebral organoids. Mol Psychiatry 23:2363–2374. https://doi.org/10.1038/s41380-018-0229-8

Hajduk PJ, Burns DJ (2002) Integration of NMR and high-throughput screening. Comb Chem High Throughput Screen 5:613–621. https://doi.org/10.2174/1386207023329996

Harrer S, Shah P, Antony B, Hu J (2019) Artificial intelligence for clinical trial design. Trends Pharmacol Sci 40:577–591

Heo I, Dutta D, Schaefer DA et al (2018) Modelling Cryptosporidium infection in human small intestinal and lung organoids. Nat Microbiol 3:814–823. https://doi.org/10.1038/s41564-018-0177-8

Hofbauer P, Jahnel SM, Papai N et al (2021) Cardioids reveal self-organizing principles of human cardiogenesis. Cell 184:3299–3317.e22. https://doi.org/10.1016/j.cell.2021.04.034

Hofer M, Lutolf MP (2021) Engineering organoids. Nat Rev Mater 6:402–420. https://doi.org/10.1038/s41578-021-00279-y

Hoffmann M, Kleine-Weber H, Schroeder S et al (2020) SARS-CoV-2 cell entry depends on ACE2 and TMPRSS2 and is blocked by a clinically proven protease inhibitor. Cell 181:271–280

Horie M, Ogawa H, Yoshida Y et al (2008) Inactivation and morphological changes of avian influenza virus by copper ions. Arch Virol 153:1467–1472. https://doi.org/10.1007/s00705-008-0154-2

How SE, Yingyongnarongkul B, Fara MA et al (2004) Polyplexes and lipoplexes for mammalian gene delivery: from traditional to microarray screening. Comb Chem High Throughput Screen 7:423–430. https://doi.org/10.2174/1386207043328616

Huang J, Deng Y, Ren J et al (2018) Novel in situ forming hydrogel based on xanthan and chitosan re-gelifying in liquids for local drug delivery. Carbohydr Polym 186:54–63. https://doi.org/10.1016/j.carbpol.2018.01.025

Huang C, Wen T, Shi F-J et al (2020) Rapid detection of IgM antibodies against the SARS-CoV-2 virus via colloidal gold nanoparticle-based lateral-flow assay. ACS Omega 5:12550–12556. https://doi.org/10.1021/acsomega.0c01554

İnanç İ, Erdemli E (2021) Histopathological features of SARS-CoV-2 infection and relationships with organoid technology. J Int Med Res 49:3000605211044382. https://doi.org/10.1177/03000605211044382

Jahangirian H, Lemraski EG, Webster TJ et al (2017) A review of drug delivery systems based on nanotechnology and green chemistry: green nanomedicine. Int J Nanomedicine 12:2957–2978. https://doi.org/10.2147/IJN.S127683

Jain KK (2020a) Role of Nanobiotechnology in drug delivery. Methods Mol Biol 2059:55–73. https://doi.org/10.1007/978-1-4939-9798-5_2

Jain KK (2020b) An overview of drug delivery systems. Methods Mol Biol 2059:1–54. https://doi.org/10.1007/978-1-4939-9798-5_1

Janiaud P, Hemkens LG, Ioannidis JPA (2021) Challenges and lessons learned from COVID-19 trials: should we be doing clinical trials differently? Can J Cardiol 37:1353–1364

Jayk Bernal A, Gomes da Silva MM, Musungaie DB et al (2022) Molnupiravir for oral treatment of Covid-19 in nonhospitalized patients. N Engl J Med 386:509–520

Johnson AP, Marshall C (1992) Starting material oriented retrosynthetic analysis in the LHASA program. 3. Heuristic estimation of synthetic proximity. J Chem Inf Comput Sci 32:426–429. https://doi.org/10.1021/ci00009a005

Joseph E, Saha RN (2017) Investigations on pharmacokinetics and biodistribution of polymeric and solid lipid nanoparticulate systems of atypical antipsychotic drug: effect of material used and surface modification. Drug Dev Ind Pharm 43:678–686. https://doi.org/10.1080/03639045.2016.1278014

Kaluthantrige Don F, Huch M (2021) Organoids, where we stand and where we go. Trends Mol Med 27:416–418. https://doi.org/10.1016/j.molmed.2021.03.001

Karzbrun E, Reiner O (2019) Brain organoids-a bottom-up approach for studying human neurodevelopment. Bioeng (Basel, Switzerland) 6. https://doi.org/10.3390/bioengineering6010009

Kirmayr M, Quilodrán C, Valente B et al (2021) The GRADE approach, part 1: how to assess the certainty of the evidence. Medwave 21:e8109–e8109

Koehler KR, Nie J, Longworth-Mills E et al (2017) Generation of inner ear organoids containing functional hair cells from human pluripotent stem cells. Nat Biotechnol 35:583–589. https://doi.org/10.1038/nbt.3840

Kokic G, Hillen HS, Tegunov D et al (2021) Mechanism of SARS-CoV-2 polymerase stalling by remdesivir. Nat Commun 12:279

Kolawole OM, Lau WM, Khutoryanskiy VV (2018) Methacrylated chitosan as a polymer with enhanced mucoadhesive properties for transmucosal drug delivery. Int J Pharm 550:123–129. https://doi.org/10.1016/j.ijpharm.2018.08.034

Krakower DS, Oldenburg CE, Mitty JA et al (2015) Knowledge, beliefs and practices regarding antiretroviral medications for HIV prevention: results from a survey of healthcare providers in New England. PLoS One 10:e0132398. https://doi.org/10.1371/journal.pone.0132398

Kuleshov MV, Stein DJ, Clarke DJB et al (2020) The COVID-19 drug and gene set library. Patterns (New York, NY) 1:100090. https://doi.org/10.1016/j.patter.2020.100090

Kumar RA, Clark DS (2006) High-throughput screening of biocatalytic activity: applications in drug discovery. Curr Opin Chem Biol 10:162–168. https://doi.org/10.1016/j.cbpa.2006.02.033

Lage OM, Ramos MC, Calisto R et al (2018) Current screening methodologies in drug discovery for selected human diseases. Mar Drugs 16. https://doi.org/10.3390/md16080279

Lamers MM, Beumer J, van der Vaart J et al (2020) SARS-CoV-2 productively infects human gut enterocytes. Science 369:50–54. https://doi.org/10.1126/science.abc1669

Lancaster MA, Renner M, Martin C-A et al (2013) Cerebral organoids model human brain development and microcephaly. Nature 501:373–379. https://doi.org/10.1038/nature12517
Lee J, Böscke R, Tang P-C et al (2018) Hair follicle development in mouse pluripotent stem cell-derived skin organoids. Cell Rep 22:242–254. https://doi.org/10.1016/j.celrep.2017.12.007
Lee J, Sutani A, Kaneko R et al (2020) In vitro generation of functional murine heart organoids via FGF4 and extracellular matrix. Nat Commun 11:4283. https://doi.org/10.1038/s41467-020-18031-5
Lee H, Im JS, Bin CD, Woo D-H (2021) Trends in the global organoid technology and industry: from organogenesis in a dish to the commercialization of organoids. Organoid 1:e11. https://doi.org/10.51335/organoid.2021.1.e11
Limburg H, Harbig A, Bestle D et al (2019) TMPRSS2 is the major activating protease of influenza a virus in primary human airway cells and influenza B virus in human type II pneumocytes. J Virol 93. https://doi.org/10.1128/JVI.00649-19
Lindsay RK, Buchanan BG, Feigenbaum EA, Lederberg J (1993) DENDRAL: a case study of the first expert system for scientific hypothesis formation. Artif Intell 61:209–261. https://doi.org/10.1016/0004-3702(93)90068-M
Liu J, Zhang X, Li G et al (2019) Anti-angiogenic activity of bevacizumab-bearing dexamethasone-loaded PLGA nanoparticles for potential intravitreal applications. Int J Nanomedicine 14:8819–8834. https://doi.org/10.2147/IJN.S217038
Liu C, Zhou Q, Li Y et al (2020) Research and development on therapeutic agents and vaccines for COVID-19 and related human coronavirus diseases. ACS Cent Sci 6:315–331
Łoczechin A, Séron K, Barras A et al (2019) Functional carbon quantum dots as medical countermeasures to human coronavirus. ACS Appl Mater Interfaces 11:42964–42974. https://doi.org/10.1021/acsami.9b15032
Low LA, Mummery C, Berridge BR et al (2021) Organs-on-chips: into the next decade. Nat Rev Drug Discov 20:345–361. https://doi.org/10.1038/s41573-020-0079-3
Lurie N, Saville M, Hatchett R, Halton J (2020) Developing Covid-19 vaccines at pandemic speed. N Engl J Med 382:1969–1973. https://doi.org/10.1056/NEJMp2005630
Ma MA, Gutiérrez DE, Frausto JM, Al-Delaimy WK (2021) Minority representation in clinical trials in the United States: trends over the past 25 years. In: Mayo Clinic proceedings. Elsevier, pp 264–266
MacConnell AB, Price AK, Paegel BM (2017) An integrated microfluidic processor for DNA-encoded combinatorial library functional screening. ACS Comb Sci 19:181–192. https://doi.org/10.1021/acscombsci.6b00192
Marconi VC, Ramanan AV, de Bono S et al (2021) Efficacy and safety of baricitinib for the treatment of hospitalised adults with COVID-19 (COV-BARRIER): a randomised, double-blind, parallel-group, placebo-controlled phase 3 trial. Lancet Respir Med 9:1407–1418
Martinho N, Damgé C, Reis CP (2011) Recent advances in drug delivery systems. J Biomater Nanobiotechnol 2:510. https://doi.org/10.4236/jbnb.2011.225062
Mason S, Devincenzo JP, Toovey S et al (2018) Comparison of antiviral resistance across acute and chronic viral infections. Antiviral Res 158:103–112
McGenity C, Treanor D (2021) Guidelines for clinical trials using artificial intelligence–SPIRIT-AI and CONSORT-AI. J Pathol 253:14–16
McQuade C, Al Zaki A, Desai Y et al (2015) A multifunctional nanoplatform for imaging, radiotherapy, and the prediction of therapeutic response. Small 11:834–843. https://doi.org/10.1002/smll.201401927
Michalopoulos A (1973) A clinical and laboratory study of doxycycline ('Vibramycin'): a broad-spectrum antibiotic. Curr Med Res Opin 1:445–455. https://doi.org/10.1185/03007997309111706
Mignani S, Huber S, Tomás H et al (2016) Why and how have drug discovery strategies in pharma changed? What are the new mindsets? Drug Discov Today 21:239–249. https://doi.org/10.1016/j.drudis.2015.09.007

Monteil V, Kwon H, Prado P et al (2020) Inhibition of SARS-CoV-2 infections in engineered human tissues using clinical-Grade soluble human ACE2. Cell 181:905–913.e7. https://doi.org/10.1016/j.cell.2020.04.004

Morales-Narváez E, Dincer C (2020) The impact of biosensing in a pandemic outbreak: COVID-19. Biosens Bioelectron 163:112274. https://doi.org/10.1016/j.bios.2020.112274

Mostafa A, Kandeil A, Elshaier YAMM et al (2020) FDA-approved drugs with potent in vitro antiviral activity against severe acute respiratory syndrome coronavirus 2. Pharmaceuticals (Basel) 13. https://doi.org/10.3390/ph13120443

Namiot ED, Sokolov AV, Chubarev VN et al (2023) Nanoparticles in clinical trials: analysis of clinical trials, FDA approvals and use for COVID-19 vaccines. Int J Mol Sci 24:787

Nugraha B, Buono MF, von Boehmer L et al (2019) Human cardiac organoids for disease modeling. Clin Pharmacol Ther 105:79–85. https://doi.org/10.1002/cpt.1286

Nunan D, Aronson J, Bankhead C (2018) Catalogue of bias: attrition bias. BMJ Evid Based Med 23:21–22

Ogbadoyi EO, Umar N (2022) The challenges and opportunities for the development of COVID-19 therapeutics and preparing for the next pandemic. Front Drug Discov 17:1–17

Ogden J (2019) Do no harm: balancing the costs and benefits of patient outcomes in health psychology research and practice. J Health Psychol 24:25–37

Ojha H, Sethi M, Kakkar R et al (2021) Chapter 7 – Chem-bioinformatic approach for drug discovery: in silico screening of potential antimalarial compounds. In: Sharma N, Ojha H, Raghav PK, Goyal RK (eds) Chemoinformatics and bioinformatics in the pharmaceutical sciences. Academic, London, pp 207–243

Orner BP, Derda R, Lewis RL et al (2004) Arrays for the combinatorial exploration of cell adhesion. J Am Chem Soc 126:10808–10809. https://doi.org/10.1021/ja0474291

Paparella A, Taccogna L, Aguzzi I et al (2008) Flow cytometric assessment of the antimicrobial activity of essential oils against Listeria monocytogenes. Food Control 19:1174–1182. https://doi.org/10.1016/j.foodcont.2008.01.002

Patil NH, Devarajan PV (2016) Insulin-loaded alginic acid nanoparticles for sublingual delivery. Drug Deliv 23:429–436. https://doi.org/10.3109/10717544.2014.916769

Patra JK, Das G, Fraceto LF et al (2018) Nano based drug delivery systems: recent developments and future prospects. J Nanobiotechnology 16:71. https://doi.org/10.1186/s12951-018-0392-8

Patrício D, Santiago J, Mano JF, Fardilha M (2022) Organoids of the male reproductive system: challenges, opportunities, and their potential use in fertility research. WIREs Mech Dis:e1590. https://doi.org/10.1002/wsbm.1590

Peng H, Ding C, Jiang L et al (2022) Discovery of potential anti-SARS-CoV-2 drugs based on large-scale screening in vitro and effect evaluation in vivo. Sci China Life Sci 65:1181–1197. https://doi.org/10.1007/s11427-021-2031-7

Petit L, Vernès L, Cadoret J-P (2021) Docking and in silico toxicity assessment of Arthrospira compounds as potential antiviral agents against SARS-CoV-2. J Appl Phycol 33:1579–1602. https://doi.org/10.1007/s10811-021-02372-9

Pompaiah M, Bartfeld S (2017) Gastric organoids: an emerging model system to study helicobacter pylori pathogenesis. Curr Top Microbiol Immunol 400:149–168. https://doi.org/10.1007/978-3-319-50520-6_7

Qian P-Y, Xu SY (2012) Antifouling activity of marine natural products BT. In: Fattorusso E, Gerwick WH, Taglialatela-Scafati O (eds) Handbook of marine natural products. Springer Netherlands, Dordrecht, pp 749–821

Qian X, Song H, Ming G-L (2019) Brain organoids: advances, applications and challenges. Development 146. https://doi.org/10.1242/dev.166074

Rabaan AA, Al-Ahmed SH, Sah R et al (2020) SARS-CoV-2/COVID-19 and advances in developing potential therapeutics and vaccines to counter this emerging pandemic. Ann Clin Microbiol Antimicrob 19:1–37

Robinson PC, Liew DFL, Tanner HL et al (2022) COVID-19 therapeutics: challenges and directions for the future. Proc Natl Acad Sci 119:e2119893119

Rodriguez-Aller M, Guinchard S, Guillarme D et al (2015) New prostaglandin analog formulation for glaucoma treatment containing cyclodextrins for improved stability, solubility and ocular tolerance. Eur J Pharm Biopharm Off J Arbeitsgemeinschaft fur Pharm Verfahrenstechnik eV 95:203–214. https://doi.org/10.1016/j.ejpb.2015.04.032
Rossi G, Manfrin A, Lutolf MP (2018) Progress and potential in organoid research. Nat Rev Genet 19:671–687. https://doi.org/10.1038/s41576-018-0051-9
Rothan HA, Mohamed Z, Paydar M et al (2014) Inhibitory effect of doxycycline against dengue virus replication in vitro. Arch Virol 159:711–718. https://doi.org/10.1007/s00705-013-1880-7
Rothan HA, Bahrani H, Mohamed Z et al (2015) A combination of doxycycline and ribavirin alleviated chikungunya infection. PLoS One 10:e0126360. https://doi.org/10.1371/journal.pone.0126360
Salihu HM, Wilson RE, King LM et al (2015) Socio-ecological model as a framework for overcoming barriers and challenges in randomized control trials in minority and underserved communities. Int J MCH AIDS 3:85
Sanders JM, Monogue ML, Jodlowski TZ, Cutrell JB (2020) Pharmacologic treatments for coronavirus disease 2019 (COVID-19): a review. JAMA 323:1824–1836. https://doi.org/10.1001/jama.2020.6019
Schneider G (2013) De novo molecular design. Wiley, Hoboken
Schneider G (2018) Automating drug discovery. Nat Rev Drug Discov 17:97–113. https://doi.org/10.1038/nrd.2017.232
Sethi M, Sukumar R, Karve S et al (2014) Effect of drug release kinetics on nanoparticle therapeutic efficacy and toxicity. Nanoscale 6:2321–2327. https://doi.org/10.1039/c3nr05961h
Shi S, Zhang Z, Luo Z et al (2015) Chitosan grafted methoxy poly(ethylene glycol)-poly(ε-caprolactone) nanosuspension for ocular delivery of hydrophobic diclofenac. Sci Rep 5:11337. https://doi.org/10.1038/srep11337
Silva MM, Calado R, Marto J et al (2017) Chitosan nanoparticles as a mucoadhesive drug delivery system for ocular administration. Mar Drugs 15. https://doi.org/10.3390/md15120370
Smith EC (2017) The not-so-infinite malleability of RNA viruses: viral and cellular determinants of RNA virus mutation rates. PLoS Pathog 13:e1006254
Soen Y, Chen DS, Kraft DL et al (2003) Detection and characterization of cellular immune responses using peptide-MHC microarrays. PLoS Biol 1:E65. https://doi.org/10.1371/journal.pbio.0000065
Srivastava M, Srivastava N, Mishra PK, Malhotra BD (2021) Prospects of nanomaterials-enabled biosensors for COVID-19 detection. Sci Total Environ 754:142363. https://doi.org/10.1016/j.scitotenv.2020.142363
Strasfeld L, Chou S (2010) Antiviral drug resistance: mechanisms and clinical implications. Infect Dis Clin 24:809–833
Strub DJ, Talma M, Strub M et al (2022) Evaluation of the anti-SARS-CoV-2 properties of essential oils and aromatic extracts. Sci Rep 12:14230. https://doi.org/10.1038/s41598-022-18676-w
Sui Y, Zhang S, Li Y et al (2020) Generation of functional salivary gland tissue from human submandibular gland stem/progenitor cells. Stem Cell Res Ther 11:127. https://doi.org/10.1186/s13287-020-01628-4
Sutarjono B (2019) Can we better understand how Zika leads to microcephaly? A systematic review of the effects of the Zika virus on human brain organoids. J Infect Dis 219:734–745. https://doi.org/10.1093/infdis/jiy572
Szymański P, Markowicz M, Mikiciuk-Olasik E (2012) Adaptation of high-throughput screening in drug discovery-toxicological screening tests. Int J Mol Sci 13:427–452. https://doi.org/10.3390/ijms13010427
Takasato M, Er PX, Becroft M et al (2014) Directing human embryonic stem cell differentiation towards a renal lineage generates a self-organizing kidney. Nat Cell Biol 16:118–126. https://doi.org/10.1038/ncb2894

Takasato M, Er PX, Chiu HS et al (2016) Kidney organoids from human iPS cells contain multiple lineages and model human nephrogenesis. Nature 536:238. https://doi.org/10.1038/nature17982

Takebe T, Wells JM, Helmrath MA, Zorn AM (2018) Organoid center strategies for accelerating clinical translation. Cell Stem Cell 22:806–809. https://doi.org/10.1016/j.stem.2018.05.008

Talukdar J, Bhadra B, Dattaroy T et al (2020a) Potential of natural astaxanthin in alleviating the risk of cytokine storm in COVID-19. Biomed Pharmacother 132:110886. https://doi.org/10.1016/j.biopha.2020.110886

Talukdar J, Dasgupta S, Nagle V, Bhadra B (2020b) COVID-19: potential of microalgae derived natural astaxanthin as adjunctive supplement in alleviating cytokine storm. Available SSRN 3579738

Tardif JC, Bouabdallaoui N, L'Allier PL et al (2021) Colchicine for community-treated patients with COVID-19 (COLCORONA): a phase 3, randomised, double-blinded, adaptive, placebo-controlled, multicentre trial. Lancet Respir Med 9:924–932

Tharayil A, Rajakumari R, Kumar A et al (2021) New insights into application of nanoparticles in the diagnosis and screening of novel coronavirus (SARS-CoV-2). Emergent Mater 4:101–117. https://doi.org/10.1007/s42247-021-00182-w

Thilakarathna SH, Rupasinghe HPV (2013) Flavonoid bioavailability and attempts for bioavailability enhancement. Nutrients 5:3367–3387. https://doi.org/10.3390/nu5093367

Tutko DC, Henry KD, Winger BE et al (1998) Sequential mass spectrometry and MSn analyses of combinatorial libraries by using automated matrix-assisted laser desorption/ionization Fourier transform mass spectrometry. Rapid Commun Mass Spectrom 12:335–338. https://doi.org/10.1002/(SICI)1097-0231(19980331)12:6<335::AID-RCM160>3.0.CO;2-T

Tuveson D, Clevers H (2019) Cancer modeling meets human organoid technology. Science 364:952–955. https://doi.org/10.1126/science.aaw6985

Ulbrich K, Holá K, Šubr V et al (2016) Targeted drug delivery with polymers and magnetic nanoparticles: covalent and noncovalent approaches, release control, and clinical studies. Chem Rev 116:5338–5431. https://doi.org/10.1021/acs.chemrev.5b00589

Umscheid CA, Margolis DJ, Grossman CE (2011) Key concepts of clinical trials: a narrative review. Postgrad Med 123:194–204

Varshochian R, Riazi-Esfahani M, Jeddi-Tehrani M et al (2015) Albuminated PLGA nanoparticles containing bevacizumab intended for ocular neovascularization treatment. J Biomed Mater Res A 103:3148–3156. https://doi.org/10.1002/jbm.a.35446

Vaz ES, Vassiliades SV, Giarolla J et al (2023) Review: drug repositioning in the COVID-19 pandemic: fundamentals, synthetic routes, and overview of clinical studies. Eur J Clin Pharmacol 79:723–751

Vere Hodge A, Field HJ (2011) General mechanisms of antiviral resistance (Chapter 13). In: Tibayrenc M (ed) Genetics and evolution of infectious disease. Elsevier, Amsterdam. pp 339–362

Viswanathan M, Patnode CD, Berkman ND et al (2018) Recommendations for assessing the risk of bias in systematic reviews of health-care interventions. J Clin Epidemiol 97:26–34

Vitiello A (2022) Sars-Cov-2 and risk of antiviral drug resistance. Irish J Med Sci 191:2367–2368

Wacker MG, Proykova A, Santos GML (2016) Dealing with nanosafety around the globe-regulation vs. innovation. Int J Pharm 509:95–106. https://doi.org/10.1016/j.ijpharm.2016.05.015

Wang T, Zeng L, Strader T et al (1998) A new ultra-high throughput method for characterizing combinatorial libraries incorporating a multiple probe autosampler coupled with flow injection mass spectrometry analysis. Rapid Commun Mass Spectrom 12:1123–1129. https://doi.org/10.1002/(SICI)1097-0231(19980831)12:16<1123::AID-RCM293>3.0.CO;2-W

Wang Y, Strohm EM, Sun Y et al (2016) Biodegradable polymeric nanoparticles containing gold nanoparticles and Paclitaxel for cancer imaging and drug delivery using photoacoustic methods. Biomed Opt Express 7:4125–4138. https://doi.org/10.1364/BOE.7.004125

Wang M, Cao R, Zhang L et al (2020) Remdesivir and chloroquine effectively inhibit the recently emerged novel coronavirus (2019-nCoV) in vitro. Cell Res 30:269–271

Watkins R, Wu L, Zhang C et al (2015) Natural product-based nanomedicine: recent advances and issues. Int J Nanomedicine 10:6055–6074. https://doi.org/10.2147/IJN.S92162
WHO (2021) Solidarity trial consortium. N Engl J Med 384:497–511
WHO (2022) Prioritizing diseases for research and development in emergency contexts. World Health Organization. https://www.who.int/activities/prioritizing-diseases-for-research-and-development-inemergency-context. Accessed 21 Dec 2022
Wu Z-C, Wang X, Wei J-C et al (2015) Antiviral activity of doxycycline against vesicular stomatitis virus in vitro. FEMS Microbiol Lett 362. https://doi.org/10.1093/femsle/fnv195
Wu B, Lu S-T, Yu H et al (2018) Gadolinium-chelate functionalized bismuth nanotheranostic agent for in vivo MRI/CT/PAI imaging-guided photothermal cancer therapy. Biomaterials 159:37–47. https://doi.org/10.1016/j.biomaterials.2017.12.022
Xiang Y, Tanaka Y, Cakir B et al (2019) hESC-derived thalamic organoids form reciprocal projections when fused with cortical organoids. Cell Stem Cell 24:487–497.e7. https://doi.org/10.1016/j.stem.2018.12.015
Xinaris C, Brizi V, Remuzzi G (2015) Organoid models and applications in biomedical research. Nephron 130:191–199. https://doi.org/10.1159/000433566
Xu H, Lyu X, Yi M et al (2018) Organoid technology and applications in cancer research. J Hematol Oncol 11:116. https://doi.org/10.1186/s13045-018-0662-9
Yang J-M, Chen Y-F, Tu Y-Y et al (2007) Combinatorial computational approaches to identify tetracycline derivatives as flavivirus inhibitors. PLoS One 2:e428. https://doi.org/10.1371/journal.pone.0000428
Yi SA, Zhang Y, Rathnam C et al (2021) Bioengineering approaches for the advanced organoid research. Adv Mater 33:e2007949. https://doi.org/10.1002/adma.202007949
Yuan P, Tan D, Annabi-Bergaya F (2015) Properties and applications of halloysite nanotubes: recent research advances and future prospects. Appl Clay Sci 112–113:75–93. https://doi.org/10.1016/j.clay.2015.05.001
Yuan J-D, ZhuGe D-L, Tong M-Q et al (2018) pH-sensitive polymeric nanoparticles of mPEG-PLGA-PGlu with hybrid core for simultaneous encapsulation of curcumin and doxorubicin to kill the heterogeneous tumour cells in breast cancer. Artif Cells Nanomed Biotechnol 46:302–313. https://doi.org/10.1080/21691401.2017.1423495
Zhou B, Xiong Z, Wang P et al (2018) Targeted tumor dual mode CT/MR imaging using multifunctional polyethylenimine-entrapped gold nanoparticles loaded with gadolinium. Drug Deliv 25:178–186. https://doi.org/10.1080/10717544.2017.1422299
Zhou Y, Hou Y, Shen J et al (2020) Network-based drug repurposing for novel coronavirus 2019-nCoV/SARS-CoV-2. Cell Discov 6:1–18
Zorzi GK, Schuh RS, Maschio VJ et al (2019) Box Behnken design of siRNA-loaded liposomes for the treatment of a murine model of ocular keratitis caused by Acanthamoeba. Colloids Surf B Biointerfaces 173:725–732. https://doi.org/10.1016/j.colsurfb.2018.10.044

The manufacturer's authorised representative in the EU is Springer Nature Customer Service Centre GmbH, Europaplatz 3, 69115 Heidelberg, Germany. If you have any concerns regarding our products, please contact ProductSafety@springernature.com

Printed and bound by CPI Group (UK) Ltd, Croydon, CR0 4YY
01/10/2026
02225899-0015